Integral Geometry
and
Geometric Probability

ENCYCLOPEDIA OF MATHEMATICS and Its Applications

Other volumes in preparation

GIAN-CARLO ROTA, *Editor*
ENCYCLOPEDIA OF MATHEMATICS AND ITS APPLICATIONS
Volume 1

Section: Probability
Mark Kac, *Section Editor*

Integral Geometry and Geometric Probability

Luis A. Santaló
University of Buenos Aires

With a Foreword by Mark Kac
The Rockefeller University

1976

Addison-Wesley Publishing Company
Advanced Book Program
Reading, Massachusetts

London · Amsterdam · Don Mills, Ontario · Sydney · Tokyo

Library of Congress Cataloging in Publication Data

Santaló Sors, Luis Antonio.
Integral geometry and geometric probability.

(Encyclopedia of mathematics and its applications; v. 1 : Section, Probability)
Bibliography: p.
1. Geometric probabilities. 2. Geometry, Integral.
I. Title. II. Series: Encyclopedia of mathematics and its applications ; v. 1.
QA273.5.S26 516'.362 76-41777
ISBN 0-201-13500-0

Printed in the United States of America

ABCDEFGHIJ-HA-79876

Contents

Part II. GENERAL INTEGRAL GEOMETRY

Part III. INTEGRAL GEOMETRY IN E_n

Part IV. INTEGRAL GEOMETRY IN SPACES OF CONSTANT CURVATURE

Editor's Statement

A large body of mathematics consists of facts that can be presented and described much like any other natural phenomenon. These facts, at times explicitly brought out as theorems, at other times concealed within a proof, make up most of the applications of mathematics, and are the most likely to survive changes of style and of interest.

This ENCYCLOPEDIA will attempt to present the factual body of all mathematics. Clarity of exposition, accessibility to the non-specialist, and a thorough bibliography are required of each author. Volumes will appear in no particular order, but will be organized into sections, each one comprising a recognizable branch of present-day mathematics. Numbers of volumes and sections will be reconsidered as times and needs change.

It is hoped that this enterprise will make mathematics more widely used where it is needed, and more accessible in fields in which it can be applied but where it has not yet penetrated because of insufficient information.

GIAN-CARLO ROTA

Foreword

This monograph is the first in a projected series on Probability Theory.

Though its title "Integral Geometry" may appear somewhat unusual in this context it is nevertheless quite appropriate, for Integral Geometry is an outgrowth of what in the olden days was referred to as "geometric probabilities."

Originating, as legend has it, with the Buffon needle problem (which after nearly two centuries has lost little of its elegance and appeal), geometric probabilities have run into difficulties culminating in the paradoxes of Bertrand which threatened the fledgling field with banishment from the home of Mathematics. In rescuing it from this fate, Poincaré made the suggestion that the arbitrariness of definition underlying the paradoxes could be removed by tying closer the definition of probability with a geometric group of which it would have to be an invariant.

Thus a union of concepts was born that was to become Integral Geometry.

It is unfortunate that in the past forty or so years during which Probability Theory experienced its most spectacular rise to mathematical prominence, Integral Geometry has stayed on its fringes. Only quite recently has there been a reawakening of interest among practitioners of Probability Theory in this beautiful and fascinating branch of Mathematics, and thus the book by Professor Santaló, for many years the undisputed leader in the field of Integral Geometry, comes at a most appropriate time.

Complete and scholarly, the book also repeatedly belies the popular belief that applicability and elegance are incompatible.

Above all the book should remind all of us that Probability Theory is measure theory with a "soul" which in this case is provided not by Physics or by games of chance or by Economics but by the most ancient and noble of all of mathematical disciplines, namely Geometry.

MARK KAC
General Editor, Section on Probability

Preface

During the years 1935–1939, W. Blaschke and his school in the Mathematics Seminar of the University of Hamburg initiated a series of papers under the generic title "Integral Geometry." Most of the problems treated had their roots in the classical theory of geometric probability and one of the project's main purposes was to investigate whether these probabilistic ideas could be fruitfully applied to obtain results of geometric interest, particularly in the fields of convex bodies and differential geometry in the large. The contents of this early work were included in Blaschke's book *Vorlesungen über Integralgeometrie* [51].

To apply the idea of probability to random elements that are geometric objects (such as points, lines, geodesics, congruent sets, motions, or affinities), it is necessary, first, to define a measure for such sets of elements. Then, the evaluation of this measure for specific sets sometimes leads to remarkable consequences of a purely geometric character, in which the idea of probability turns out to be accidental. The definition of such a measure depends on the geometry with which we are dealing. According to Klein's famous Erlangen Program (1872), the criterion that distinguishes one geometry from another is the group of transformations under which the propositions remain valid. Thus, for the purposes of integral geometry, it seems natural to choose the measure in such a way that it remains invariant under the corresponding group of transformations. This sequence of underlying mathematical concepts —probability, measure, groups, and geometry—forms the basis of integral geometry.

The original work was limited almost entirely to metric (euclidean and noneuclidean) geometry and the probabilistic ideas were those of the classical geometric probability initiated by Crofton [132, 133] and Czuber [134] in the last century. After 1940 the new methods of differential geometry and group theory made it possible to unify and to generalize several questions in integral geometry, which led to new problems and noteworthy progress in this field. Consideration of a differentiable manifold (instead of euclidean space) and of a transitive transformation group operating on it gave rise to integral geometry in homogeneous spaces, and the whole theory was illuminated by the ideas of the theory of locally compact groups and their invariant measures. The inclusion of the methods of integral geometry within the framework of the theory of homogeneous spaces was the work of A. Weil

[710, 711] and S. S. Chern [105]. However, integral geometry generally has been restricted to Lie's transformation groups—more precisely, to matrix Lie groups—for two reasons. First, because they are the most important from the point of view of their geometric applications, and second, because they lead to more computable results. Further, the resulting simplification of the presentation compensates for the loss of generality.

As main references on integral geometry, after the work of Blaschke [51], we have our early introduction [568] and the books of M. I. Stoka [646, 647]. Also closely related are Hadwiger's books [270, 274]. With regard to the theory of geometric probability there is the book of Deltheil [144] and the nice brochure of M. G. Kendall and P. A. P. Moran [335], where a large number of applications to different fields are brought together. The present book intends to provide a synopsis of the main topics of integral geometry, including their origins and their applications, with the aim of showing how the interplay between geometry, group theory, and probability has become fruitful for all of these fields.

In recent times, mainly due to the work of R. E. Miles [410, 411, 414, 418], the field of integral geometry has been enriched by the introduction of the ideas and tools of stochastic processes. In a symposium on integral geometry and geometric probability held at Oberwolfach (Germany) in June 1969, D. G. Kendall, K. Krickeberg, and R. E. Miles suggested the term "stochastic geometry" to indicate precisely those contents of geometry and group theory that are in a sense related to stochastic processes. This constitutes a promising field, to which the present book may be considered an introduction, at least from its geometric point of view (see [294] and G. Matheron's recent book [401a], where the theory of random sets and applications to practical problems are treated in great detail, using deep topological ideas).

The book presupposes only a basic course in advanced calculus, although some elementary knowledge of differential geometry, group theory, and probability is desirable. Publications in which prerequisite material can be found are always indicated where apt.

Part I is concerned with integral geometry on the euclidean plane. It is treated in an elementary way. Most of the problems are handled by specific techniques and the main results are proved directly and independently in each case. We consider this part fundamental in that it exhibits the power of the methods and their usefulness in various fields. Chapters 1 to 4 are classical in the theory of geometric probability. They are devoted to the current notions on the measure of sets of points and lines in the plane, including some fairly recent results, in order to illustrate the breadth of the field of applications. Chapter 5 deals with sets of strips as an immediate generalization of sets of lines. In Chapters 6 and 7 the kinematic measure in the plane is treated in detail in order to emphasize how the measure on groups can be applied to strictly geometric problems. These chapters prepare the ground for the general approach of Chapters 9 and 10. Chapter 8 deals with some discrete subgroups of the motions group and their interpretation from the integral geometric viewpoint.

Part II presents an account of the necessary elements of the theory of Lie groups and homogeneous spaces in order to obtain the invariant measures in these spaces and their properties. The general theory is exemplified by the groups of affine transformations (Chapter 11) and the group of motions in euclidean space (Chapter 12). Several examples are discussed. For instance, it is shown that the affine invariant measure of sets of planes with reference to a fixed convex body permits a geometric interpretation of some inequalities among various characteristics of the convex body – a typical result of integral geometry (Sections 2 and 3 of Chapter 11).

Part III is concerned with integral geometry in euclidean n-dimensional space. Chapter 13 contains a résumé of the main results on convex sets in n-dimensional space. Chapter 14 is devoted to the measure of linear spaces that intersect a convex set or, more generally, a compact manifold embedded in euclidean space. Several integral formulas are obtained and some applications to the theory of geometric probability are mentioned. Chapter 15 is concerned with the so-called kinematic fundamental formula, which includes most formulas in euclidean integral geometry as special or limiting cases. In Chapter 16 the general theory is applied in detail to three-dimensional euclidean space, especially to the question of the size distribution of particles embedded in a convex body when only two-dimensional sections are available, a problem that has several areas of application and has received considerable attention in recent years, giving rise to so-called stereology [166].

Finally, Part IV deals with integral geometry in spaces of constant curvature (noneuclidean integral geometry), in particular integral geometry on the sphere, and some new trends in integral geometry (integral geometry and foliated spaces, integral geometry in complex spaces, symplectic integral geometry, and integral geometry in the sense of Gelfand and Helgason). A survey of these new trends is given, entirely without proofs, but with detailed references to the literature.

Each chapter ends with a section of notes or notes and exercises, including a number of references and theorems without proof and emphasizing applications. These notes increase the amount of material covered and, with the extensive bibliography, establish the book's encyclopedic character.

LUIS A. SANTALÓ

Integral Geometry and Geometric Probability

PART I

Integral Geometry in the Plane

CHAPTER 1

Convex Sets in the Plane

1. Introduction

Convex sets play an important role in integral geometry. For this reason we will review here their principal properties, especially those which will be needed in the following sections. In this chapter we consider convex sets in the plane. For convex sets in n-dimensional euclidean space, see Chapter 13. For a more complete treatment, refer to the classical books of Blaschke [50] and Bonnesen and Fenchel [63], or to the more modern texts of Benson [27], Eggleston [162], Grünbaum [247], Jaglom and Boltjanski [320], Hadwiger [270], Hadwiger and coauthors [282], and Valentine [683].

A set of points K in the plane is called *convex* if for each pair of points $A \in K$, $B \in K$ it is true that $AB \subset K$, where AB is the line segment joining A and B. For convenience we shall assume throughout that the convex sets are bounded and closed.

A curve with end points P, Q is called convex if its point set, together with the segment PQ, bounds a convex set. If the convex set K is bounded and has interior points, then the boundary of K is called a *closed convex curve.* Throughout, we will denote by ∂K the boundary of the set K. If all the points of K belong to ∂K, then K is a line segment.

We can prove that (a) All convex curves are piecewise differentiable (i.e., they are the union of a countable set of arcs with continuously turning tangent);

ENCYCLOPEDIA OF MATHEMATICS and Its Applications, Gian-Carlo Rota (ed.).
1, Luis A. Santaló, Integral Geometry and Geometric Probability

ISBN 0-201-13500-0

in other words, convex curves have at most a countable set of corners; (b) All bounded convex curves are rectifiable. The length of the boundary ∂K of a convex set K is called the *perimeter* of K.

2. Envelope of a Family of Lines

The envelope of a family of curves $F(x, y, \lambda) = 0$, depending on a parameter λ, is defined as that curve every point of which is a point of contact with a curve of the family. As is well known, the equation of the envelope can be obtained by eliminating the parameter λ from the two equations $F = 0$, $\partial F/\partial\lambda = 0$. We will apply this result to the case of a family of lines.

A line on the plane may be determined by its distance p from the origin and the angle ϕ of the normal with the x axis (Fig. 1.1). The equation of the line is then

$$x \cos \phi + y \sin \phi - p = 0. \tag{1.1}$$

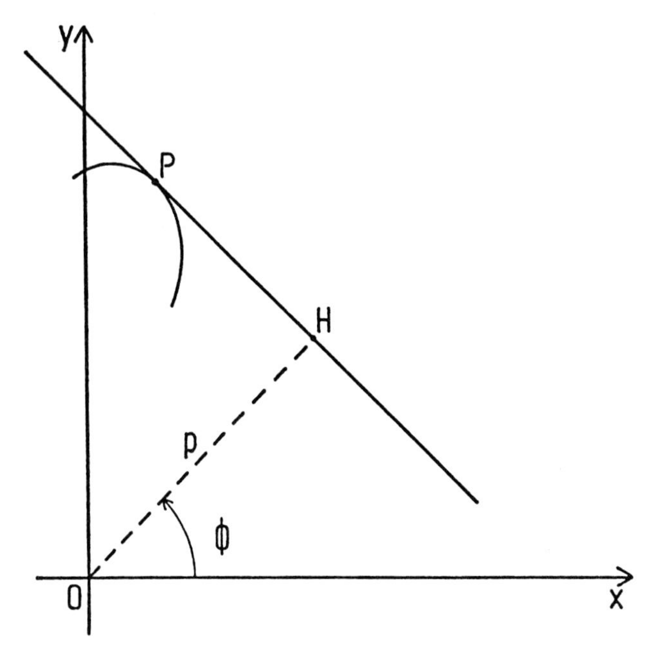

Figure 1.1.

If p is a function $p = p(\phi)$, then (1.1) is the equation of a family of lines and if we assume that $p(\phi)$ is differentiable, the envelope of the family is obtained from (1.1) and the derivative

ISBN 0-201-13500-0

$$-x \sin \phi + y \cos \phi - p' = 0 \qquad (p' = dp/d\phi). \tag{1.2}$$

From (1.1) and (1.2) we get the parametric representation of the envelope of the lines (1.1),

$$x = p \cos \phi - p' \sin \phi, \qquad y = p \sin \phi + p' \cos \phi. \tag{1.3}$$

These formulas give the x, y coordinates of the contact point P of the line with the envelope (Fig. 1.1). Since the coordinates of the point H in which the perpendicular through O intersects the line are $p \cos \phi$, $p \sin \phi$, it follows that

$$HP = p'. \tag{1.4}$$

Assuming that the function p is of class C^2 (recall that class C^n means n times continuously differentiable), from (1.3), it follows that $dx = -(p + p'') \sin \phi \, d\phi$, $dy = (p + p'') \cos \phi \, d\phi$. Hence $ds = |p + p''| \, d\phi$ and the radius of curvature of the envelope becomes $\rho = ds/d\phi = |p + p''|$.

If the envelope is the boundary ∂K of a convex set K and O is an interior point of K, then $p = p(\phi)$ is called the *support function* of K or the support function of the convex curve ∂K with reference to the origin O. The lines (1.1) are then the *support lines* of K. In this case we can prove that $p + p'' > 0$ (see, e.g., [63, p. 18]), and the formulas in the preceding paragraph may be written

$$ds = (p + p'') \, d\phi, \qquad \rho = p + p''. \tag{1.5}$$

It can be proved that a necessary and sufficient condition that the periodic function p be the support function of a convex set K is that $p + p'' > 0$.

From the first equation in (1.5) it follows that the length of a closed convex curve that has support function p of class C^2 is given by

$$L = \int_0^{2\pi} p \, d\phi. \tag{1.6}$$

The assumption that p is of class C^2 can be removed. We can show that formula (1.6) holds for any closed convex curve (see, e.g., [683, p. 161]).

The area of the convex set K can be also evaluated in terms of the support function. Indeed, if we consider K decomposed into elementary triangles of height p and base ds, with the point O as common vertex, we have

$$F = \tfrac{1}{2} \int_{\partial K} p \, ds = \tfrac{1}{2} \int_0^{2\pi} p(p + p'') \, d\phi \tag{1.7}$$

ISBN 0-201-13500-0

and by integration by parts

$$F = \frac{1}{2}\int_0^{2\pi} (p^2 - p'^2)\, d\phi. \tag{1.8}$$

3. Mixed Areas of Minkowski

Let K_1, K_2 be two bounded convex sets on the plane whose support functions with reference to O_1 and O_2 are, respectively, p_1 and p_2, assumed of class C^2. Consider the function $p(\phi) = p_1(\phi) + p_2(\phi)$. The envelope of the lines $x \cos \phi + y \sin \phi - p = 0$ is a closed curve whose radius of curvature is $\rho = p + p'' = (p_1 + p_1'') + (p_2 + p_2'') = \rho_1 + \rho_2$. Since ∂K_1 and ∂K_2 are convex curves, we have $\rho_1 > 0$, $\rho_2 > 0$ and hence $\rho > 0$. Therefore the envelope above is the boundary of a convex set K_{12}. If p_1 and p_2 are not of class C^2 the proof fails, but the result is still true: the function $p = p_1 + p_2$ is always the support function of a convex set K_{12} called the *mixed convex set* of K_1 and K_2 [63, p. 29].

The area of K_{12} has the form

$$F = \frac{1}{2}\int_0^{2\pi} (p^2 - p'^2)\, d\phi = F_1 + F_2 + 2F_{12} \tag{1.9}$$

where F_1, F_2 are the areas of K_1, K_2 and

$$F_{12} = F_{21} = \frac{1}{2}\int_0^{2\pi} (p_1 p_2 - p_1' p_2')\, d\phi \tag{1.10}$$

is the so-called *mixed area of Minkowski* of K_1 and K_2.

Integration by parts, using (1.5), gives

$$F_{12} = \frac{1}{2}\int_0^{2\pi} p_1(p_2 + p_2'')\, d\phi = \frac{1}{2}\int_{\partial K_2} p_1\, ds_2 \tag{1.11}$$

and similarly we have

$$F_{21} = \frac{1}{2}\int_{\partial K_1} p_2\, ds_1 \tag{1.12}$$

where ds_1, ds_2 are the arc elements of ∂K_1, ∂K_2 at the contact points of the support lines normal to the direction ϕ.

ISBN 0-201-13500-0

Note that the mixed area F_{12} does not depend on the origins O_1, O_2. In fact, if O_1 is replaced by $O_1{}^*$ such that $O_1O_1{}^* = a$ and α is the angle of $O_1O_1{}^*$ with the x axis (Fig. 1.2), the support function relative to $O_1{}^*$ is $p_1{}^* = p_1 - a\cos(\phi - \alpha)$, and using (1.10) and integrating by parts we verify that $F_{12}^* = F_{12}$. The same is clearly true if we change O_2.

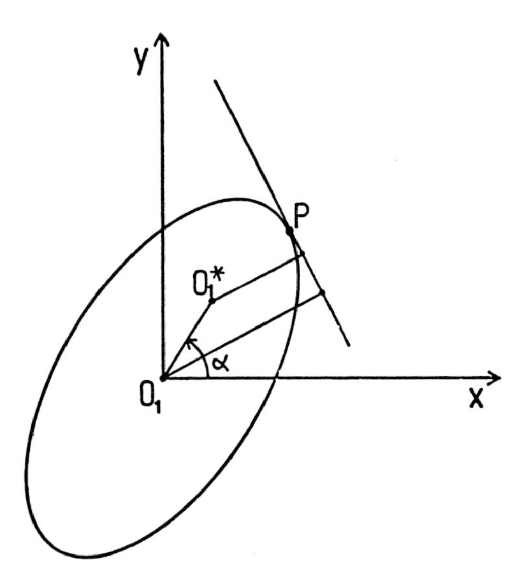

Figure 1.2.

Furthermore, the mixed area F_{12} does not change by translations of K_1 and K_2 since p_1 and p_2 remain unchanged. Assume now that one of the sets, say K_1, is rotated an angle θ about a point O. Let $O_1{}^*$ be the image of O_1. The rotation is equivalent to the rotation of angle θ about O_1 composed by the translation defined by the vector $O_1O_1{}^*$. Thus, because of the invariance of F_{12} by translations, it remains for us to consider rotations about O_1. By the rotation of angle θ about O_1 the new support function of K_1 is $p_1{}^*(\phi) = p_1(\phi - \theta)$ and the mixed area of K_2 and the rotated set $K_1{}^*$ becomes

$$F_{12}(\theta) = \tfrac{1}{2} \int_{\partial K_2} p_1(\phi - \theta)\, ds_2.$$

Integrating with regard to θ and using (1.6) we get

$$\int_0^{2\pi} F_{12}(\theta)\, d\theta = \tfrac{1}{2}L_1L_2. \tag{1.13}$$

This formula will be useful later.

ISBN 0-201-13500-0

Examples 1. If K_1 is a translate of K_2, we may assume $p_1 = p_2$, $ds_1 = ds_2$, and (1.12) gives $F_{12} = F_1 = F_2$. That is, the mixed area of convex sets that are translation congruent is equal to the common area of the sets.

2. If K_2 is a circle of radius r, we have $ds_2 = r\,d\phi$ and (1.11) gives

$$F_{12} = \tfrac{1}{2}r \int_0^{2\pi} p_1\,d\phi = \tfrac{1}{2}rL_1. \tag{1.14}$$

3. Let K_1, K_2 be two line segments of lengths $2a$, $2b$, respectively, and let α be the angle between the lines which contain the segments. Using (1.10) we get $F_{12} = 2ab|\sin\alpha|$. Integrating with respect to α over the range $0, 2\pi$, we get $8ab$, in accordance with (1.13). Note that a line segment is a convex set whose perimeter is twice the length of the segment.

4. Some Special Convex Sets

A line h is said to be a *support line* of the convex set K at a point $P \in \partial K$ if $P \in h$ and K is contained in the closure of one of the two open half planes into which h cuts the plane. If ∂K possesses a tangent at P, then the support line at P coincides with the tangent. Every point on the boundary of a convex set lies on a support line and there are exactly two support lines perpendicular to a given direction.

The *breadth* $\Delta(\phi)$ of K in the direction ϕ is the distance between the two parallel support lines to K that are perpendicular to the direction ϕ and that contain K between them. If $p(\phi)$ is the support function of K, we have $\Delta(\phi) = p(\phi) + p(\phi + \pi)$ and according to (1.6) we have

$$L = \int_0^{\pi} \Delta(\phi)\,d\phi. \tag{1.15}$$

Therefore, the mean value or expected value of Δ is

$$E(\Delta) = L/\pi \tag{1.16}$$

where L is the perimeter of K. Note that the breadth $\Delta(\phi)$ may be defined as the length of the orthogonal projection of K on a line parallel to the direction ϕ. Consequently, (1.16) gives: any closed convex curve ∂K of length L can be projected on a line in such a way that the length of the projection is $\geqslant L/\pi$. It can also be projected so that such a length is $\leqslant L/\pi$.

The least of the breadths of a convex set K is called the *width* of K. We shall represent it by E. The *diameter* of K, represented by D, is the greatest distance between two points of K. It can also be defined as the greatest

ISBN 0-201-13500-0

breadth of K. Since we obviously have $E \leqslant \Delta \leqslant D$, from (1.15) it follows that

$$\pi E \leqslant L \leqslant \pi D. \tag{1.17}$$

We now want to define some special convex sets that have particular interest for our purposes.

Parallel convex sets. The parallel set K_r in the distance r of a convex set K is the union of all closed circular disks of radius r the centers of which are points of K. The boundary ∂K_r is called the outer parallel curve of ∂K in the distance r. Figures 1.3, 1.4, and 1.5 show parallel sets of a segment, a triangle, and an ellipse, respectively.

Figure 1.3.

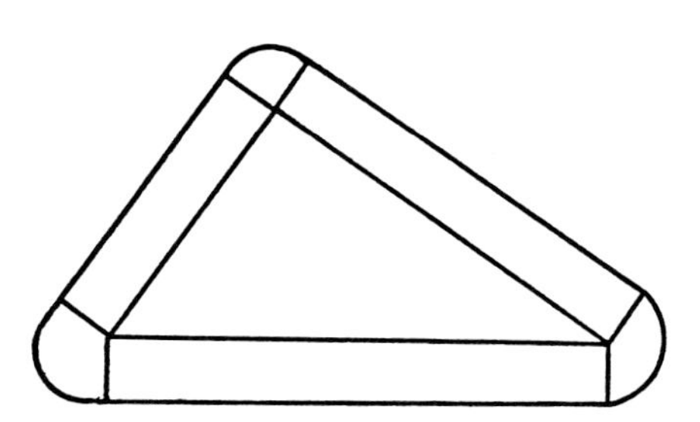

Figure 1.4.

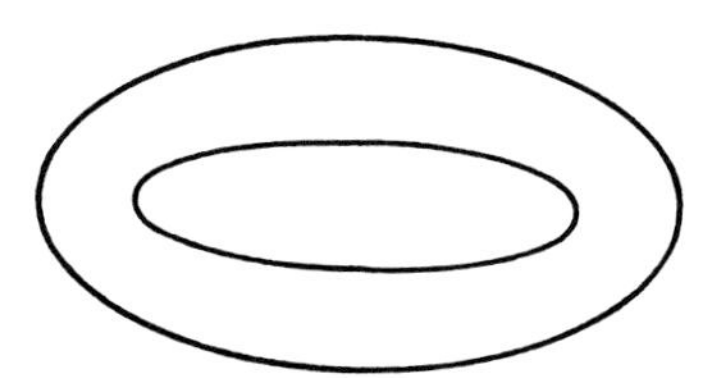

Figure 1.5.

If $p(\phi)$ is the support function of K relative to O, the support function of K_r relative to the same point O is $p(\phi) + r$, and using (1.6) and (1.8) we have that the perimeter and the area of K_r are, respectively,

ISBN 0-201-13500-0

$$L_r = L + 2\pi r, \qquad F_r = F + Lr + \pi r^2. \tag{1.18}$$

If ∂K is of class C^2, the radius of curvature of ∂K_r, by (1.5), is

$$\rho_r = \rho + r. \tag{1.19}$$

For values of r such that $r \leqslant \min \rho$ we can define the *interior parallel set* K_{-r} as the set whose support function is $p(\phi) - r$. Then the length of ∂K_{-r} and the area of K_{-r} are given by the same formulas (1.18) after the substitution $r \to -r$.

Sets of constant breadth. If $\Delta(\phi) = \Delta =$ constant for all ϕ, the convex set K is said to be of constant breadth. In this case we have $E = \Delta = D$ and by (1.15) the perimeter of K is given by the simple formula

$$L = \pi\Delta. \tag{1.20}$$

Furthermore, if ∂K is of class C^2, using (1.5) and the fact that $\Delta = p(\phi) + p(\phi + \pi)$, we have

$$\rho(\phi) + \rho(\phi + \pi) = \Delta. \tag{1.21}$$

Besides the circle, the simpler convex sets of constant breadth are the so-called Reuleaux polygons. Given a linear regular polygon of $2n + 1$ sides ($n = 1, 2, \ldots$), the corresponding Reuleaux polygon is formed by the circular arcs that are subtended by the sides and whose centers are the opposite vertices of the polygon. Figure 1.6 shows the Reuleaux triangle and Reuleaux pentagon. Note that each set parallel to a convex set of constant breadth is also of constant breadth.

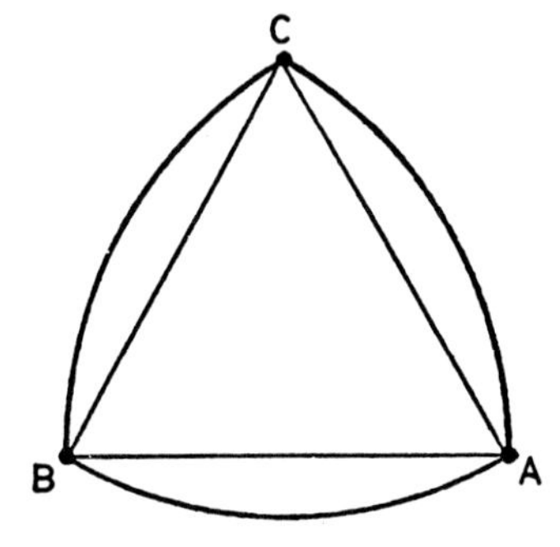

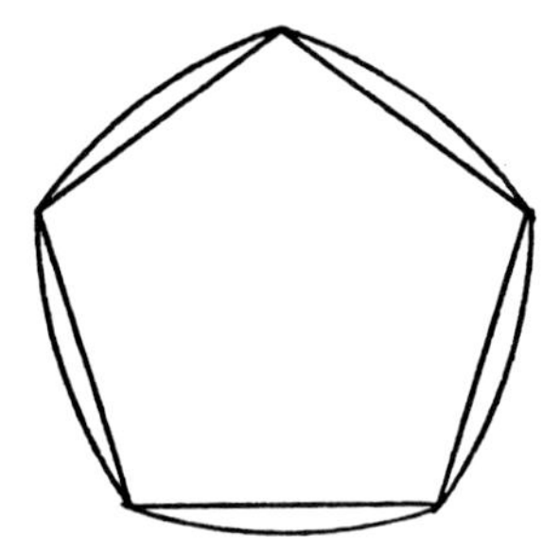

Figure 1.6.

It can be shown that any set of diameter $D \leqslant \Delta$ is a subset of a convex set of constant breadth Δ [63, p. 130].

ISBN 0-201-13500-0

Triangular convex sets. Convex sets of constant breadth can rotate in the interior of a square (i.e., all circumscribed rectangles are congruent squares; Fig. 1.7). There are also convex sets that can rotate inside a fixed equilateral triangle, which is equivalent to saying that all the equilateral triangles that are circumscribed to the set are congruent triangles. These sets are called *triangular sets.* For instance, the shaded spindle in Fig. 1.8 is a triangular set. Each parallel set of a triangular set is also a triangular set.

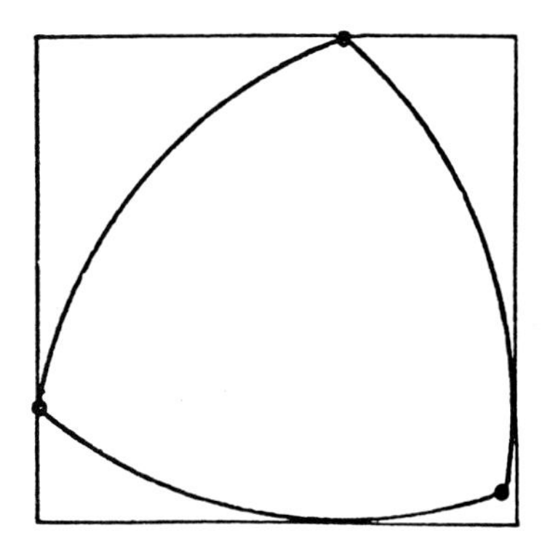

Figure 1.7.

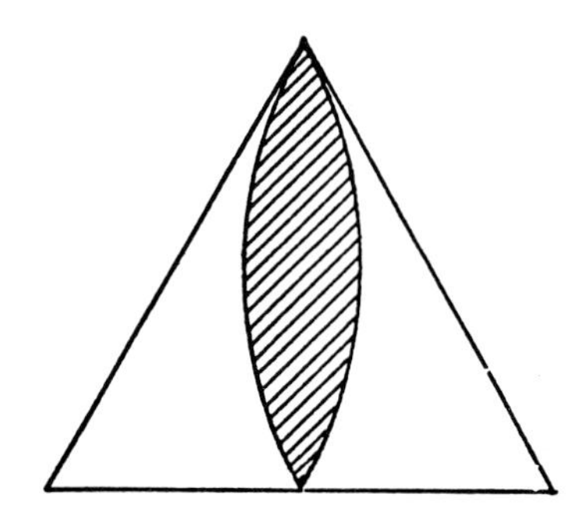

Figure 1.8.

Since the sum of the distances from an interior point of an equilateral triangle to the sides is equal to the height h of the triangle, it follows that the support function of a triangular set satisfies the condition $p(\phi) + p(\phi + 2\pi/3) + p(\phi + 4\pi/3) = h$. From this equality and (1.6) it follows that the perimeter of the triangular sets inscribed in an equilateral triangle of height h is $L = (2\pi/3)h$.

5. Surface Area of the Unit Sphere and Volume of the Unit Ball

Throughout this text we shall denote by O_n the surface area of the n-dimensional unit sphere and by κ_n the volume of the n-dimensional unit ball or solid sphere. Their values are

$$O_n = \frac{2\pi^{(n+1)/2}}{\Gamma((n+1)/2)}, \qquad \kappa_n = \frac{O_{n-1}}{n} = \frac{2\pi^{n/2}}{n\Gamma(n/2)} \tag{1.22}$$

where Γ denotes the gamma function, which satisfies the relations

$$\Gamma(n+1) = n\Gamma(n), \qquad \Gamma(n) = (n-1)! \quad (n \text{ integer}), \quad \Gamma(\tfrac{1}{2}) = \sqrt{\pi}. \tag{1.23}$$

For instance,

$$O_0 = 2, \quad O_1 = 2\pi, \quad O_2 = 4\pi, \quad O_3 = 2\pi^2, \quad \kappa_1 = 2, \quad \kappa_2 = \pi, \quad \kappa_3 = \tfrac{4}{3}\pi.$$

ISBN 0-201-13500-0

6. Notes and Exercises

1. *Minimum problems concerning convex sets.* The area F, perimeter L, diameter D, and width E of a convex set are related by certain inequalities, some of which are the following.

$$L^2 \geqslant 4\pi F, \qquad \sqrt{3}\, F \geqslant E^2, \qquad L \geqslant \pi E, \qquad 2F \geqslant ED,$$

$$L \geqslant 2(D^2 - E^2)^{1/2} + 2E \arcsin(E/D),$$

$$L \leqslant 2(D^2 - E^2)^{1/2} + 2D \arcsin(E/D),$$

$$2F \leqslant E(D^2 - E^2)^{1/2} + D^2 \arcsin(E/D), \qquad 4F \leqslant 2EL - \pi E^2.$$

A complete set of inequalities remains unknown. The determination in each case of the set that satisfies the equality sign is in general a difficult problem. For references and a systematic exposition, see the articles by Kubota and Hemmi [349], Ohmann [462], Santaló [576], and Sholander [606]. Several results on convex sets can be found in the book *Convexity* (V. Klee, ed.; Amer. Math. Soc., Providence, R.I., 1963).

2. *Polar reciprocal convex sets.* Let K be a bounded convex set in the plane. Let O be an interior point of K and let P be the boundary point of K in the direction ϕ from O. The function $h^*(\phi) = |OP|^{-1}$ is the support function of a convex set K^* called the polar reciprocal of K with respect to O. The polar reciprocal convex sets have several applications to the geometry of numbers (see the work of C. G. Lekkerkerker [361] and Cassels [92]). We state the following properties:

(a) The mixed area $F_{12}(K, K^*)$ satisfies $F_{12}(K, K^*) \geqslant \pi$ with equality if and only if K is a circle of center O [192].

(b) If K has O as center of symmetry, then the areas $F(K)$ and $F(K^*)$ satisfy the inequalities

$$\tfrac{1}{2}\pi^2 \leqslant F(K)\cdot F(K^*) \leqslant \pi^2. \tag{1.24}$$

The affine invariant $F(K)\cdot F(K^*)$ has been studied for closed curves not necessarily convex, for which the polar angle is a monotone function of the arc length, and applied to inequalities related to periodic solutions of differential equations by Guggenheimer [255]. For convex curves in the plane, see [296] and for applications to differential equations see [445–448].

The inequalities (1.24) may be extended to centrally symmetric convex sets in n-dimensional euclidean space E_n. Calling $V(K)$ and $V(K^*)$ the respective volumes, we have

$$\kappa_n{}^2 n^{-n/2} \leqslant V(K)\cdot V(K^*) \leqslant \kappa_n{}^2 \tag{1.25}$$

where κ_n is the volume of the unit ball in E_n (1.22). Inequalities (1.25) have their origin in the work of Mahler [387] and were improved by Bambah [19] and Santaló [556]. The right-hand inequality in (1.25) is the best possible for all convex bodies, not necessarily centrally symmetric, and equality holds for the n-dimensional ellipsoid centered at O. For centrally symmetric convex bodies, the left-hand equality in (1.25) holds if and only if K is a parallelotope

ISBN 0-201-13500-0

or a cross polytope. Mahler conjectured that if K ranges over all convex bodies, then $V(K) \cdot V(K^*) \geqslant 4^n n!$ (equality sign if and only if K is a simplex), but this inequality has not been proved. Some improvements can be seen in the work of Dvoretzky and Rogers [157] and Guggenheimer [258]. For applications to the geometry of numbers see [361, pp. 104–110]. The affine invariant $V(K^*)$ has a clear geometrical meaning in affine integral geometry, as we shall see in Chapter 11, Sections 2 and 3.

Exercise 1. Let K be a convex set with boundary ∂K of class C^1. Assuming ∂K oriented, we consider for each point A of ∂K the chord AB such that the angle between AB and the tangent at point A is a constant θ. Assume that K is such that $|AB| = \lambda =$ constant. Show that the length of ∂K is then $L = \pi\lambda/\sin\theta$. If K is of constant breadth and $\theta = \pi/2$, then λ is the breadth and this formula coincides with (1.20).

Exercise 2. Let ∂K be a closed convex curve of class C^1 and assume that through each point $A \in \partial K$ there passes a chord of given length $|AB| = \lambda$. On every such a chord take the point X such that $|AX| = a$, $|XB| = b$ $(a + b = \lambda)$. Show that the area between ∂K and the curve described by X is πab (independent of ∂K). This result is known as the Holditch theorem.

Exercise 3. If the support line perpendicular to the direction ϕ_0 contains a line segment P_1P_2 of ∂K (Fig. 1.9), prove that the support function $p(\phi)$

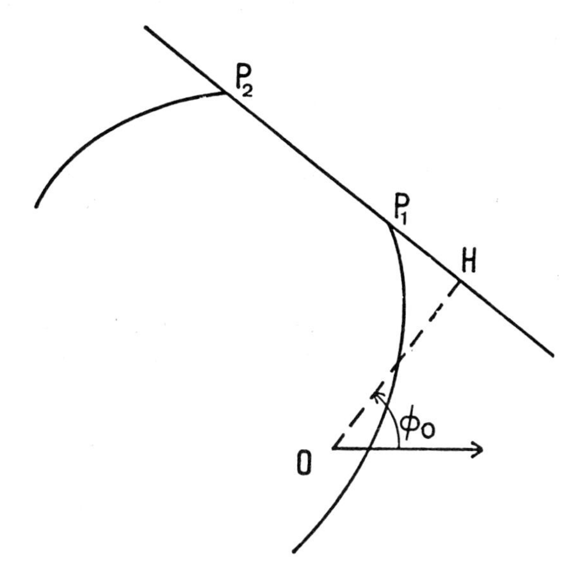

Figure 1.9.

has no derivative at the point ϕ_0, but there exist the right-hand derivative $p_+'(\phi_0)$ and the left-hand derivative $p_-'(\phi_0)$ such that $p_+'(\phi_0) = HP_1$ and $p_-'(\phi_0) = HP_2$ where H is the foot of the perpendicular to the support line from O.

ISBN 0-201-13500-0

CHAPTER 2

Sets of Points and Poisson Processes in the Plane

1. Density for Sets of Points

Let x, y denote rectangular Cartesian coordinates in the plane. In integral geometry and in the theory of geometrical probability the measure of a set of points X is defined as the integral over the set of a differential form $\omega = f(x, y)\, dx \wedge dy$ (provided this integral exists, in the Lebesgue sense), where the function $f(x, y)$ is chosen according to the following criterion: *the measure $m(X)$ is to be invariant under the group of motions in the plane.*

Throughout, we shall use the notation and rules of the exterior calculus, which can be seen in any book of modern calculus or differential geometry (e.g., in those of Fleming [204], Flanders [201, 202], Loomis and Sternberg [368], and H. Cartan [91]).

Let $\mathfrak{M}$ be the group of motions in the plane. With reference to a rectangular Cartesian system of coordinates the equations of a motion $u \in \mathfrak{M}$ are

$$x' = x \cos \alpha - y \sin \alpha + a, \qquad y' = x \sin \alpha + y \cos \alpha + b \tag{2.1}$$

where (a, b) are the components of the translation and α is the rotation of u.

We want to find a function $f(x, y)$ such that the measure $m(X) = \int_X f(x, y)\, dx \wedge dy$ is invariant under $\mathfrak{M}$ for any set X. If $X' = uX$ is the transform of X by the motion u, using that $dx \wedge dy = dx' \wedge dy'$, we have

$$m(X') = \int_{X'} f(x', y')\, dx' \wedge dy' = \int_X f(x', y')\, dx \wedge dy \tag{2.2}$$

ENCYCLOPEDIA OF MATHEMATICS and Its Applications, Gian-Carlo Rota (ed.).
1, Luis A. Santaló, Integral Geometry and Geometric Probability

ISBN 0-201-13500-0

and the condition $m(X') = m(X)$ for any X gives $f(x', y') = f(x, y)$ for all pairs of corresponding points $(x, y) \to (x', y')$. Since any point x, y can be transformed into any other x', y' by a motion (i.e., the group of motions is transitive for points), we have $f(x, y) =$ constant. This justifies the following definition.

The measure of a set of points $P(x, y)$ in the plane is defined by the integral over the set of the differential form

$$dP = dx \wedge dy, \tag{2.3}$$

which is called the density for points.

Up to a constant factor this density is clearly the only one that is invariant under motions. Similarly, for sets of n-tuples of points $P_1, P_2, \ldots, P_n$, assumed independent, we state the following definition.

The density for sets of independent n-tuples of points $P_i(x_i, y_i)$, $i = 1, 2, \ldots, n$ in the plane, is

$$dP_1 \wedge dP_2 \wedge \cdots \wedge dP_n \tag{2.4}$$

where $dP_i = dx_i \wedge dy_i$. This density is unique up to a constant factor.

Densities (2.3) and (2.4) are always taken in absolute value.

If the points P are given in a system of coordinates related to the Cartesian coordinates x, y by the equations $x = h(\xi, \eta)$, $y = g(\xi, \eta)$, the point density takes the form $dP = dx \wedge dy = |J|\, d\xi \wedge d\eta$ where J is the Jacobian determinant $J = h_\xi g_\eta - h_\eta g_\xi$.

Once we have defined the measure of sets of points and sets of n-tuples of points, the probability that a "random" element (point or n-tuple of points) is in the set Y when it is known to be in the set X ($Y \subset X$) is defined by the quotient of measures

$$p = \frac{m(Y)}{m(X)}. \tag{2.5}$$

The elementary notions on geometrical probability that we shall need in the sequel can be seen in the books of Deltheil [144], M. G. Kendall and P. A. P. Moran [335], and M. I. Stoka and R. Theodorescu [654]. For an axiomatic construction see the work of Hadwiger [275] or Rényi [501, 503].

2. First Integral Formulas

We will give an example in the style of Crofton [132, 133] and Lebesgue [357] to show that the simple computation of the density $dP = dx \wedge dy$ in

ISBN 0-201-13500-0

different coordinate systems gives some remarkable integral formulas referring to convex sets in the plane.

Let K_1, K_2 be two bounded convex sets whose support functions relative to the points O_1, O_2 are $p_1(\phi_1)$, $p_2(\phi_2)$. Assume that all the support lines have only one point in common with the corresponding convex set. Let $\tau_i = \phi_i + \pi/2$ $(i = 1, 2)$ be the directions of the support lines that are perpendicular to the directions ϕ_i (Fig. 2.1). The equations of these support lines are

$$(x - x_i) \sin \tau_i - (y - y_i) \cos \tau_i - p_i = 0 \qquad (i = 1, 2) \tag{2.6}$$

where x_i, y_i are the coordinates of O_i.

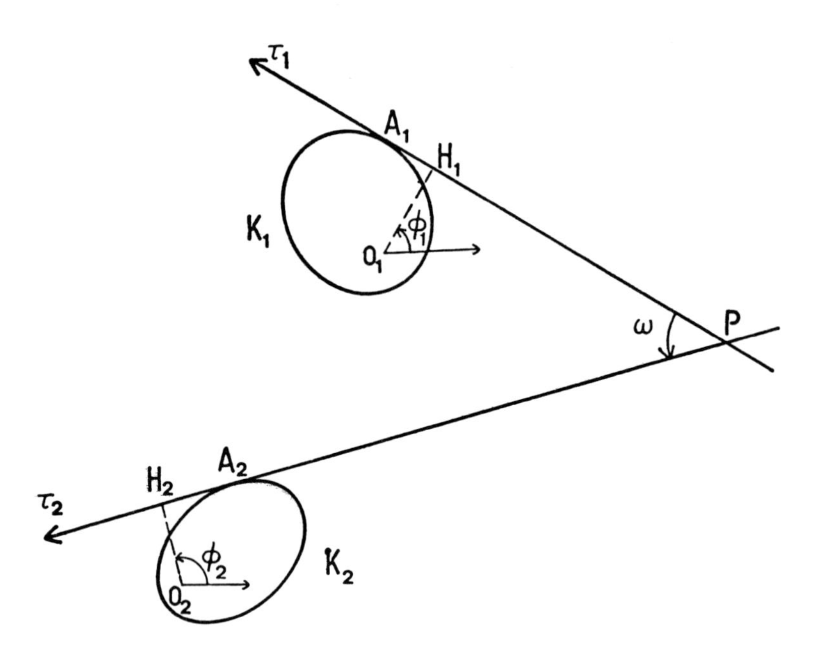

Figure 2.1.

The coordinates of the point P in which the support lines (2.6) intersect are the solutions of system (2.6). We want to express the density dP in terms of τ_1 and τ_2. By differentiation of (2.6) we get

$$\sin \tau_i \, dx - \cos \tau_i \, dy = - t_i \, d\tau_i \tag{2.7}$$

where t_i denotes the length of the tangent PA_i from P to the support point A_i. Indeed, $(x - x_i) \cos \tau_i + (y - y_i) \sin \tau_i = H_iP$ is the projection of O_iP on the support line, and according to (1.4) we have $dp_i/d\tau_i = H_iA_i$.

ISBN 0-201-13500-0

Exterior multiplication of (2.7) for $i = 1, 2$ yields

$$dP = dx \wedge dy = \frac{t_1 t_2}{\sin(\tau_2 - \tau_1)} d\tau_1 \wedge d\tau_2 \quad \text{or} \quad d\tau_1 \wedge d\tau_2 = \frac{\sin(\tau_2 - \tau_1)}{t_1 t_2} dP. \tag{2.8}$$

Let us integrate both terms of this equality over all values of τ_1 and τ_2. The left-hand side gives $4\pi^2$. On the right-hand side we observe that through each point P exterior to K_1 and K_2 there pass two support lines of K_1 and two support lines of K_2. Let t_i, t_1' and t_2, t_2' be the lengths of these lines from P to the corresponding support points and let (t_i, t_j) denote the angle between t_i and t_j. To each point P there corresponds the sum

$$\frac{\sin(t_1, t_2)}{t_1 t_2} + \frac{\sin(t_1, t_2')}{t_1 t_2'} + \frac{\sin(t_1', t_2)}{t_1' t_2} + \frac{\sin(t_1', t_2')}{t_1' t_2'} \tag{2.9}$$

and therefore we have the integral formula

$$\int_{P \notin K_1 \cup K_2} \left[\frac{\sin(t_1, t_2)}{t_1 t_2} + \frac{\sin(t_1, t_2')}{t_1 t_2'} + \frac{\sin(t_1', t_2)}{t_1' t_2} + \frac{\sin(t_1', t_2')}{t_1' t_2'}\right] dP = 4\pi^2 \tag{2.10}$$

where the integral is extended over all the points P exterior to K_1, K_2.

Formula (2.10) contains some particular cases:

(a) If $K_1 \equiv K_2 \equiv K$ and we call t, t' the lengths PA_1, PA_2 (Fig. 2.2), we have $t = t_1 = t_2$, $t' = t_1' = t_2'$ and $(t_1, t_2) = (t_1', t_2') = 0$, $(t_1, t_2') =$

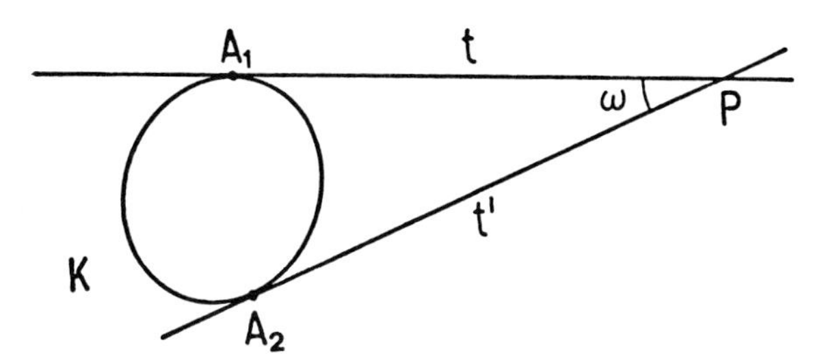

Figure 2.2.

$(t_1', t_2) = \omega =$ angle A_1PA_2. Hence we have

$$\int_{P \notin K} \frac{\sin \omega}{tt'} dP = 2\pi^2. \tag{2.11}$$

(b) If the boundary ∂K of the convex set K has continuous radius of curvature ρ, calling ρ_i $(i = 1, 2)$ the radius of curvature at the support points A_i, we have $\rho_i \, d\tau_i = ds_i$ where ds_i is the arc element of the boundary of K

ISBN 0-201-13500-0

at A_i. Then formula (2.8) can be written

$$\frac{\sin\omega}{tt'}\rho_1\,dP = ds_1 \wedge d\tau_2 \qquad \text{or} \qquad \frac{\sin\omega}{tt'}\rho_1\rho_2\,dP = ds_1 \wedge ds_2. \tag{2.12}$$

(c) Integrating over all points P outside K and taking into account that each point P belongs to two support lines, we get

$$\int_{P\notin K}\frac{\sin\omega}{tt'}(\rho_1+\rho_2)\,dP = 2\pi L, \qquad \int_{P\notin K}\frac{\sin\omega}{tt'}\rho_1\rho_2\,dP = \tfrac{1}{2}L^2. \tag{2.13}$$

The factor $\frac{1}{2}$ in the last term arises from the fact that by interchanging s_1 and s_2 we get the same point P.

EXERCISE. Prove the more general formulas

$$\int_{P\notin K}\frac{\sin^m\omega}{tt'}\,dP = \frac{2\Gamma(m/2)}{\Gamma[(m+1)/2]}\pi^{3/2},$$

$$\int_{P\notin K}\frac{\sin^m\omega}{tt'}(\rho_1+\rho_2)\,dP = \frac{2\Gamma(m/2)}{\Gamma[(m+1)/2]}\sqrt{\pi}\,L. \tag{2.14}$$

Similar formulas have been given by Czuber [135], Masotti-Biggiogero [393–395], and Stoka [647].

3. Sets of Triples of Points

Let $P_i(x_i, y_i)$, $i = 1, 2, 3$, be three independent points in the plane. Assuming that they are not collinear, let $Q(\xi, \eta)$ be the circumcenter of triangle $P_1P_2P_3$ and let (R, α_i) be the polar coordinates of P_i with respect to Q (R being the circumradius of triangle $P_1P_2P_3$). We have

$$x_i = \xi + R\cos\alpha_i, \qquad y_i = \eta + R\sin\alpha_i. \tag{2.15}$$

Putting $dP_i = dx_i \wedge dy_i$, $dQ = d\xi \wedge d\eta$, we obtain by a straightforward calculation the following density for triples of points in the plane.

$$dP_1 \wedge dP_2 \wedge dP_3 = R^3|S|\,dQ \wedge dR \wedge d\alpha_1 \wedge d\alpha_2 \wedge d\alpha_3 \tag{2.16}$$

where

$$S = \sin(\alpha_2-\alpha_1) + \sin(\alpha_3-\alpha_2) + \sin(\alpha_1-\alpha_3)$$

$$= -4\sin\frac{(\alpha_2-\alpha_1)}{2}\sin\frac{(\alpha_3-\alpha_2)}{2}\sin\frac{(\alpha_1-\alpha_3)}{2}. \tag{2.17}$$

ISBN 0-201-13500-0

Noting that $\frac{1}{2}R^2S$ is the area T of triangle $P_1P_2P_3$, we can write

$$dP_1 \wedge dP_2 \wedge dP_3 = 2RT\,dQ \wedge dR \wedge d\alpha_1 \wedge d\alpha_2 \wedge d\alpha_3$$
$$= 2(T/S)\,dQ \wedge dT \wedge d\alpha_1 \wedge d\alpha_2 \wedge d\alpha_3. \qquad (2.18)$$

Let $C \equiv C(P_1, P_2, P_3)$ denote the circumdisk of triangle $P_1P_2P_3$. Consider the set of all ordered triples P_1, P_2, P_3 such that C is contained in a given circle C_ρ of radius ρ. According to (2.16), the measure of this set of triples, calling r, ϕ the polar coordinates of Q with respect to the center of C_ρ, is

$$m(P_1, P_2, P_3; C \subset C_\rho) = \int_0^{2\pi} d\phi \, \frac{1}{4}\int_0^{\rho} (\rho - r)^4 r\,dr \int_0^{2\pi} d\alpha_1 \int_{\alpha_1}^{2\pi} d\alpha_2 \int_{\alpha_2}^{2\pi} |S|\,d\alpha_3 = \frac{\pi^3\rho^6}{5}. \qquad (2.19)$$

The measure of the set of all ordered triples P_1, P_2, P_3 such that $C \subset C_\rho$ and $P_1P_2P_3$ is an acute-angled triangle is

$$\int_0^{2\pi} d\phi \, \frac{1}{4}\int_0^{\rho} (\rho - r)^4 r\,dr \int_0^{2\pi} d\alpha_1 \int_{\alpha_1}^{\alpha_1+\pi} d\alpha_2 \int_{\alpha_1+\pi}^{\alpha_2+\pi} |S|\,d\alpha_3 = \frac{\pi^3\rho^6}{10}.$$

Thus, the probability of an acute-angled triangle in the considered set of triples such that $C \subset C_\rho$ is equal to $\frac{1}{2}$. Taking the limit as $\rho \to \infty$, we get that the probability of a random triangle in the plane (in the sense of density (2.16)) being an acute-angled triangle is $\frac{1}{2}$ (cf. the work of R. E. Miles [411]). Richards [510] has shown that the probability density function for the perimeter L of the triangle $P_1P_2P_3$ whose vertices are three random points in a domain D of area F is $(4\pi^2/21F^2)L^3\,dL$ (asymptotically for large F).

4. Homogeneous Planar Poisson Point Processes

Let D_0, D be two domains of the plane such that $D \subset D_0$. Let F_0, F be the areas of D_0, D, respectively. According to the density $dP = dx \wedge dy$, the probability that a random point of D_0 lies in D is F/F_0. If there are chosen at random n points in D_0, the probability that exactly m of them lie in D is (binomial distribution)

$$p_m = \binom{n}{m}\left(\frac{F}{F_0}\right)^m \left(1 - \frac{F}{F_0}\right)^{n-m}. \qquad (2.20)$$

If D_0 expands to the whole plane and both $n, F_0 \to \infty$ in such a way that $n/F_0 \to \rho$, positive constant, we get

ISBN 0-201-13500-0

$$\lim p_m = \frac{(\rho F)^m}{m!} e^{-\rho F}. \tag{2.21}$$

The right-hand side of (2.21) is the probability function of the *Poisson distribution*; it depends only on the product ρF, which is called the parameter of the distribution. This probability model for points in the plane is said to be a *homogeneous planar Poisson point process of intensity* ρ. It is characterized by the property that the number of points in each domain D is a random variable that depends only on the area F (not on the shape nor the position of D in the plane) and has the distribution (2.21). A systematic study of stochastic point processes, in particular of Poisson processes, in n-dimensional space can be seen in J. R. Goldman's article [232]. From the point of view of integral geometry the most important applications are due to R. E. Miles. We present some of his results given in [411].

1. Let $\Gamma_\theta(\nu, \lambda)$ ($\theta > 0, \lambda > 0, \nu > 0$) denote the so-called gamma distribution, which has the probability density function

$$f(x) = \theta \lambda^{\nu/\theta} x^{\nu-1} \frac{\exp(-\lambda x^\theta)}{\Gamma(\nu/\theta)}, \qquad x \geqslant 0, \tag{2.22}$$

where Γ is the gamma function which satisfies relations (1.23). This means that the probability that the random variable x lies between x and $x + dx$ is equal to $f(x)\,dx$ and thus we have

$$\int_0^\infty f(x)\,dx = 1. \tag{2.23}$$

The moments are

$$\mu_k = \left\{\frac{\Gamma[(\nu + k)/\theta]}{\Gamma(\nu/\theta)}\right\} \lambda^{-k/\theta}, \qquad k = 1, 2, \ldots. \tag{2.24}$$

In particular, the expected value of x is

$$E(x) = \mu_1 = \left\{\frac{\Gamma[(\nu + 1)/\theta]}{\Gamma(\nu/\theta)}\right\} \lambda^{-1/\theta}. \tag{2.25}$$

Let the points of a Poisson process be called particles and consider the set $C(m)$ of triangles whose vertices are particles and that contain exactly m particles in its interior. Let $P_1P_2P_3$ be one such triangle and let T be its area. According to the probability identity prob(area T and m particles in its interior) = prob(area T) $\times$ prob(m particles in its interior, assuming area T) and taking (2.18) and (2.21) into account, we get that the probability density function for the area T of the triangles $P_1P_2P_3$ whose vertices are particles of a Poisson process and that contain exactly m particles has the form

ISBN 0-201-13500-0

$$f_m(T) = K_m \frac{(\rho T)^m}{m!} e^{-\rho T} T, \qquad m = 0, 1, 2, \ldots, \tag{2.26}$$

where K_m is a constant that we determine by condition (2.23). The result is $K_m = \rho^2/(m + 1)$ and hence we have

The distribution of the areas of triangles that contain exactly m particles in a homogeneous planar Poisson point process of intensity ρ is $\Gamma_1(m + 2, \rho)$.

In particular, the mean area of the empty triangles ($m = 0$) is $E(T_0) = 2/\rho$.

2. Consider now the radius R of the circumdisk of triangle $P_1P_2P_3$. According to (2.16) the probability density function of R is proportional to R^3 and hence the probability density function of radii R of circumdisks that contain exactly m particles (ignoring the particles P_1, P_2, P_3 of its boundary) has the form $K_m^*(\rho\pi R^2)^m[\exp(-\rho\pi R^2)/m!]R^3$. Condition (2.23) gives $K_m^* = 2(\pi\rho^2)^2/(m + 1)$ and we get

The distribution of radii R_m of circumdisks to triangles $P_1P_2P_3$ of a homogeneous planar Poisson point process of intensity ρ that contain exactly m particles is $\Gamma_2(2m + 4, \pi\rho)$.

The mean value of R_0 is thus $E(R_0) = 3/2\sqrt{\rho}$. For details and more complete results, see Miles's article [411].

Other point processes have been studied by Krickeberg, for instance, the so-called Cox process or doubly stochastic Poisson process, which plays an important role in the study of point processes that have invariance properties under transformation groups arising in geometry (see the articles by Krickeberg [345, 346]). Related questions are to be found in Matheron's book [401a].

3. *Random tessellations.* A tessellation is an aggregate of cells (bounded convex polygons) that fit together to cover E_2 without overlapping. Regular tessellations in euclidean and hyperbolic planes have been well studied, for instance by Coxeter [128, 130]. The problem arises of defining *random tessellations.* We will consider some that are generated by homogeneous planar Poisson point processes.

Write $P_i(x)$ for the ith nearest particle to the point x and define the following equivalence relation: x is equivalent to y if and only if $P_i(x) = P_i(y)$ for $i = 1, 2, \ldots, n$. Apart from points for which the ith nearest particle is not unique, this equivalence relation divides the plane into disjoint exhaustive sets of equivalent points, which are almost surely bounded convex polygons. These polygons constitute a random tessellation of the plane, which Miles [411] represents by V_n^*.

For those points x for which it is unique, define the neighboring n-set of x as the set of the n particles nearest x. The equivalence relation x is equivalent

ISBN 0-201-13500-0

to y if and only if they have the same neighboring n-set generates the random tessellation represented by V_n. Clearly V_n^* is a refinement of V_n in the sense that each cell V_n is the disjoint union of cells of V_n^*. If $n = 1$, then $V_1^* = V_1$ and we have the so-called Voronoi tessellation, which generalizes to E_n (see [704]).

For each cell of a tessellation we may define the area A, number of vertices N, and perimeter L. Miles [411] has given the following mean values of these scalars for V_n and V_n^*.

(a) For V_n

$$E(A) = [(2n-1)\rho]^{-1}, \qquad E(N) = 6,$$

$$E(L) = \frac{(2n)!}{n!(n-1)!(2n-1)2^{2n-3}\rho^{1/2}} \sim \frac{4}{(\pi n\rho)^{1/2}}.$$

(b) For V_n^*

$$E(A) = \{n[2n - 1 + (64/9\pi^2)(n-1)(n-2)]\rho\}^{-1} \sim \frac{9\pi^2}{64n^3\rho},$$

$$E(N) = \frac{54\pi^2 n + 256(n-1)(n-2)}{9\pi^2(2n-1) + 64(n-1)(n-2)} \sim 4,$$

$$E(L) = \frac{2\sum_1^n[(2m)!/m!(m-1)!2^{2m-2}]}{n[2n - 1 + (64/9\pi^2)(n-1)(n-2)]\rho^{1/2}} \sim \frac{3}{4}\left(\frac{\pi^3}{n^3\rho}\right)^{1/2}$$

where $\sim$ refers to $n \to \infty$.

The higher-order moments are not known, except that $E(A^2) = 1.280\rho^{-2}$ for V_n^*. If v denotes the number of members of V_n^* that contain a member of V_n, the foregoing results give

$$E(v) = n\left(1 + \frac{64(n-1)(n-2)}{9\pi^2(2n-1)}\right) \sim \frac{32n^2}{9\pi^2}.$$

For Voronoi tessellations, Crain [130], by a Monte Carlo technique, has estimated the expected values of N^2, N^3, N^4, L^2, L^3, A^2, A^3. Interesting theoretical and experimental results on random tessellations have been reported by Matschinski ([403] and references therein); see also Ambarcumjan's articles [8, 9]. For random tessellations in E_3 see Chapter 16, Section 4, Note 9.

5. Notes

1. *Some classical integral formulas.* (a) Let K be a bounded convex set in the plane and consider two fixed support lines that form an agnle α ($0 \leqslant \alpha \leqslant \pi$) (Fig. 2.3). These support lines divide the plane into five regions

ISBN 0-201-13500-0

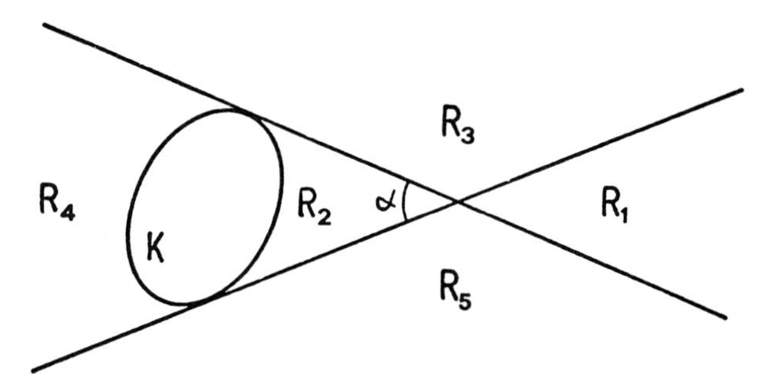

Figure 2.3.

exterior to K, say $R_1, R_2, \ldots, R_5$, as indicated in Fig. 2.3. If I_i represents the integral of $(\sin \omega/tt')\,dP$ extended over R_i, where t, t' are the lengths of the support lines from P to the support point, the following formulas can be proved.

$$I_1 = \tfrac{1}{2}\alpha^2, \qquad I_2 = \tfrac{1}{2}(\pi - \alpha)^2,$$

$$I_3 = I_5 = \tfrac{1}{2}(\pi^2 - \alpha^2), \qquad I_4 = \tfrac{1}{2}\pi^2 + \pi\alpha.$$

These formulas are the work of Crofton [132, 133] and Czuber [134]; they hold good for any $\alpha\,(0 \leqslant \alpha \leqslant \pi)$. Note that $I_1 + I_2 + \cdots + I_5 = 2\pi^2$, as expected according to (2.11).

(b) Let K be a bounded convex set. The points of the plane such that the angle ω between the support lines through it is a constant, say $\omega = \omega_0$, form a closed convex curve C. Show that the integral of the differential form $(\sin \omega/tt')\,dP$ over the ring between C and K has the value $2\pi(\omega - \omega_0)$ [135].

(c) Assuming that t, t', and ω have the same meaning as in cases (a) and (b), we can prove the following formulas [394]:

$$\int_{P \notin K} \omega^2 \left(\frac{1}{t} + \frac{1}{t'}\right) dP = 8\pi L \log 2, \qquad \int_{P \notin K} \sin^2 \omega \left(\frac{1}{t} + \frac{1}{t'}\right) dP = 2\pi L.$$

2. *The probability that a random triangle is obtuse.* Let $p(L)$ be the probability that three points chosen at random in a rectangle with dimensions $1 \times L$ form an obtuse triangle. Then according to Langford [355],

$$p(L) = \frac{1}{3} + \frac{47}{300}\left(L^2 + \frac{1}{L^2}\right) + \frac{\pi}{80}\left(L^3 + \frac{1}{L^3}\right) - \frac{\log L}{5}\left(L^2 - \frac{1}{L^2}\right)$$

if $1 \leqslant L \leqslant 2$, and

$$p(L) = \frac{1}{3} + \frac{1}{L^2}\left(\frac{\pi}{80L} + \frac{47}{300} + \frac{\log L}{5}\right) + \left(\frac{L^2}{10} - \frac{3}{5L^2}\right)\log\frac{L + (L^2 - 4)^{1/2}}{L - (L^2 - 4)^{1/2}}$$

$$+ \frac{L^3}{40}\arcsin\frac{2}{L} - \frac{L^2 \log L}{5} + \frac{47L^2}{300} + \frac{L(L^2 - 4)^{1/2}}{150}\left(-31 + \frac{63}{L^2} + \frac{64}{L^4}\right)$$

ISBN 0-201-13500-0

if $L \geqslant 2$. When the rectangle is a square, $L = 1$, we have $p(1) = (97/150) + (\pi/40) = 0.7252$. Compare this result with those of Section 3. Are they contradictory?

3. *On the convex hull of n randomly given points.* Let K be a bounded convex set with boundary ∂K of class C^2. Let $\kappa = \kappa(s)$ be the curvature of ∂K and assume that it satisfies the inequalities $0 < \kappa < M$ where M is a constant. Let H_n be the convex hull of n randomly given points in K. If F_n is the area of H_n and L_n is the length of its boundary, Rényi and Sulanke [504] have shown the following formulas for their expectations:

$$E(F_n) = F - \frac{\Gamma(8/3)(12F)^{2/3}Q_{1/3}}{10n^{2/3}} + O\left(\frac{1}{n}\right),$$

$$E(L_n) = L - \frac{\Gamma(2/3)(12F)^{2/3}Q_{4/3}}{12n^{2/3}} + O\left(\frac{1}{n}\right)$$

where F and L are the area and perimeter, respectively, of K and

$$Q_m = \int_0^L \kappa^m \, ds.$$

For the number of vertices v of H_n we have

$$E(v) \sim \Gamma(\tfrac{5}{3})(\tfrac{2}{3})^{1/3}(F^{-1/3}Q_{1/3})n^{1/3}.$$

If ∂K is not of class C^2, the computations are more involved. For a square of side a the results are

$$E(F_n) = a^2 - \frac{8a^2 \log n}{3n} + O\left(\frac{1}{n}\right), \qquad E(L_n) = 4a - \frac{a(2-q)(2\pi)^{1/2}}{n^{1/2}} + O\left(\frac{1}{n}\right)$$

where q is the constant

$$q = \int_0^1 \frac{[(1 + t^2)^{1/2} - 1]}{t^{3/2}} \, dt.$$

Consider now the case of a convex polygon K with r vertices. The expected value of the number v of vertices of the convex hull H_n is

$$E(v) = \frac{2r}{3}(\log n + C) + \frac{2}{3}\log\left(F^{-r}\prod_1^r f_i\right) + o(1)$$

where $C = 0.5772\ldots$ is Euler's constant, F is the area of K, and f_i denote the area of the triangle $A_{i-1}A_iA_{i+1}$, where $A_1, A_2, \ldots, A_r$ are the ordered vertices of K. For instance, if K is a triangle, we have $E(v) = 2(\log n + C) + o(1)$, which is independent of the shape of the triangle. The product $F^{-r}\prod f_i$ is an affine invariant of K that is maximal for regular polygons. All these results are the work of Rényi and Sulanke [504].

ISBN 0-201-13500-0

Some extensions to m-dimensional space have been given by H. Raynaud [496]. For instance, if K is the unit m-ball and n points are chosen at random in K, the expected value of the number of faces of its convex hull, as $m \to \infty$, has the infinity order

$$E(v) \sim 2\frac{\kappa_{m-1}^2}{\kappa_m^2}\frac{\Gamma[(m^2+1)/(m+1)]}{(m+1)!}\left[(m+1)\frac{\kappa_m}{\kappa_{m-1}}\right]^{(m^2+1)/(m+1)} n^{(m-1)/(m+1)}$$

where κ_i is the volume of the unit ball of dimension i (1.22).

Other results have been given by Efron [160] and Geffroy [220, 221]. These results refer to the uniform density (2.4) for sets of n-tuples of points. The case of normally distributed points has been considered by Rényi and Sulanke [504]. The more general case of points distributed according only to the assumption of rotational symmetry has been considered by Carnal [84], who gives the asymptotic behavior of the expectations $E(v)$, $E(L_n)$, $E(F_n)$ according to the behavior of the probability that the distance OP of a random point P to the origin O be $> x$ as $x \to 1$ (distribution on the unit disk) or $x \to \infty$ (distribution on the whole plane). See also [736]. The probabilities that the polytopal convex hull of a set of random points in the interior of the unit m-ball possesses a certain number of vertices have been investigated by Miles [420]. For a different approach to the study of random convex hulls see that of L. D. Fisher [196, 197].

4. *Nearest neighbors in a Poisson process.* Let p_n denote the probability that a point of a Poisson process is the nearest neighbor of exactly n other points. The maximum number of points that can have a given point P as nearest neighbor equals 6. This occurs when P is the center of a regular hexagon and each of the six points lies at a vertex. The probability that such a configuration occurs in a Poisson process is zero; hence $p_n = 0$ if $n \geqslant 6$. The exact value of p_n is not known. Roberts [514], by Monte Carlo techniques, has obtained these bounds:

$$p_0 < 0.3785, \qquad p_2 < p_0, \qquad p_n < \frac{p_0}{n-1} < \frac{0.3785}{n-1} \qquad (n = 2, 3, 4, 5).$$

A direct approach to evaluating p_n runs into combinatorial problems that have thus far proved insurmountable.

Consider now a homogeneous Poisson point process of intensity ρ in the plane. Assume that each point is an end point of a linear segment, the orientation of which is uniform over the unit two-dimensional sphere and the length of which has the distribution $dF(s)$ $(0 \leqslant s < \infty)$. Assume that the lengths and orientations are mutually independent. Then the probability that the distance from a point, chosen independently of the process of segments, to the nearest end of a segment exceeds u is $\exp(-\rho B_2(u))$ $(0 \leqslant u < \infty)$, where $B_2(u) = 2u^2\{\pi - \int_0^{2u} [\arccos(s/2u) - (s/2u)(1 - s^2/4u^2)^{1/2}]\, dF(s)\}$. The result is due to Coleman [121]. The same problem in n-dimensional euclidean space E_n has the probability $\exp(-\rho B_n(u))$, where

$$B_n(u) = \left(\frac{2u^n}{n}\right)\left\{O_{n-1} - O_{n-2}\int_0^{2u}\left[\Phi_{n-2}(\alpha)\right.\right.$$

ISBN 0-201-13500-0

$$- \left(\frac{s}{2(n-1)u}\right)\left(1 - \frac{s^2}{4u^2}\right)^{(n-1)/2}\Bigg] dF(s)\Bigg\}$$

where $\Phi_h(\alpha) = \int_0^\alpha \sin^h x\, dx$, $\alpha = \arccos(s/2n)$, and O_h is the surface area of the unit h-dimensional sphere. For related results see those of Holgate [311], Eberhardt [158], and Coleman [120].

5. *Coverage problems on the line.* Suppose that we have $n - 1$ random points $P_1, P_2, \ldots, P_{n-1}$ that are independently and uniformly distributed on the interval 0, L of the real line. The density for $(n-1)$-tuples of points is $dx_1 \wedge dx_2 \wedge \cdots \wedge dx_{n-1}$, where x_i is the coordinate of P_i $(0 \leqslant x_i \leqslant L)$. It can be shown that the probability that exactly m of the n intervals determined by the points $0, x_1, x_2, \ldots, x_{n-1}, L$ have length greater than μ $(m\mu \leqslant L, m \leqslant n)$ is

$$p_{\{m\}} = \binom{n}{m}\left\{\left(1 - \frac{m\mu}{L}\right)^{n-1} - \binom{n-m}{1}\left[1 - \frac{(m+1)\mu}{L}\right]^{n-1} + \binom{n-m}{2}\left[1 - \frac{(m+2)\mu}{L}\right]^{n-1} - \cdots\right\} \tag{2.27}$$

where the sum ends with the last term for which $1 - (m+h)\mu/L$ is positive (see [335]).

In particular, for $m = 0$, we have that the probability that all intervals have length equal to or less than μ $(n\mu \geqslant L)$ is

$$p_{\{0\}} = 1 - \binom{n}{1}\left(1 - \frac{\mu}{L}\right)^{n-1} + \binom{n}{2}\left(1 - \frac{2\mu}{L}\right)^{n-1} - \cdots + (-1)^r\binom{n}{r}\left(1 - \frac{r\mu}{L}\right)^{n-1} \tag{2.28}$$

where r is the greatest integer such that $r\mu \leqslant L$.

The probability that at least m of the intervals will exceed μ in length $(m\mu \leqslant L, m < n)$ is

$$p_m = \binom{n}{m}\left(1 - \frac{m\mu}{L}\right)^{n-1} - \binom{m}{1}\binom{n}{m+1}\left[1 - \frac{(m+1)\mu}{L}\right]^{n-1} + \binom{m+1}{2}\binom{n}{m+2}\left[1 - \frac{(m+2)\mu}{L}\right]^{n-1} - \cdots, \tag{2.29}$$

the sum continuing as long as $(m+h)\mu \leqslant L$ $(h = 0, 1, 2, \ldots)$.

These problems have been considered by a number of authors, among them R. A. Fisher [198]; Garwood [214], who gives an application to vehicular controlled traffic; Baticle [23]; and Levy [365]. See also Kendall and Moran's book [335].

Consider now sets of intervals of length a on the line. Each such interval may be determined by the coordinate of its midpoint (or by the coordinate of any other fixed point in the interval) and the measure of a set of intervals

ISBN 0-201-13500-0

of constant length is defined to be the measure of the set of their midpoints. Then, from (2.28) follows the solution of the following coverage problem on the line:

Let S_0 be a fixed interval of length L on the line. Suppose that we have n random intervals of length a ($na \geqslant L$) that intersect S_0. Then the probability that they completely cover S_0 is

$$p = 1 - \binom{n+1}{1}\left(1 - \frac{a}{L+a}\right)^n + \binom{n+1}{2}\left(1 - \frac{2a}{L+a}\right)^n + \cdots + (-1)^h \binom{n+1}{h}\left(1 - \frac{ha}{L+a}\right)^n \tag{2.30}$$

where h is the greatest integer such that $h \leqslant n+1$, $ha \leqslant L + a$.

The analogous problem on the circle may be stated as follows [630]:

For n arcs of length ϕ chosen independently and at random on the circle of unit radius, the probability that they completely cover the circle is 0 if $n\phi \leqslant 2\pi$ and if $n\phi \geqslant 2\pi$ has the value

$$p_0{}^* = 1 - \binom{n}{1}\left(1 - \frac{\phi}{2\pi}\right)^{n-1} + \binom{n}{2}\left(1 - \frac{2\phi}{2\pi}\right)^{n-1} - \cdots + (-1)^h \binom{n}{h}\left(1 - \frac{h\phi}{2\pi}\right)^{n-1} \tag{2.31}$$

where h is the greatest integer such that $h\phi \leqslant 2\pi$.

Coverage problems in spaces of dimension greater than 1 are much more difficult and in general no exact solution is known (see Chapter 6, Section 7). For other coverage problems see the articles by Robbins [512, 513], Takacs [666], and Votaw [705].

6. *The "parking" problem of Rényi* [502]. Let us consider the interval $(0, x)$ with $x > 1$. An interval of unit length is chosen at random in it. After that, a second interval of unit length is placed at random in the unfilled intervals. This process continues until no additional interval of unit length can be accommodated. Suppose that the number of unit intervals placed in the given interval $(0, x)$ is N. The problem consists in determining the mean value of N as a function of the length x. The problem is clearly that of parking at random cars of unit length on a street of length x. The main result of Rényi is that *the ratio of the mean value of the number N of unit intervals parked at random in the interval of length x to the total length x, as $x \to \infty$, is $C = 0.74759\ldots$.*

The problem has been generalized by several authors. Ney [449] considers the case in which the intervals for parking have their lengths given at random. Dvoretzky and Robbins [156] give certain limit theorems for $N(x)$. Mannion [389] shows that the variance of the occupied portion of the interval as it becomes infinite is proportional to $c_1 x$, where x is the length of the interval and $c_1 = 0.035672$. (See also [440].) Suggestions and bibliographical notes

ISBN 0-201-13500-0

on the parking problem and its extensions to higher dimensions have been given by H. Solomon [616, 618]. For two dimensions see the articles by Palasti [467] and Gani [212]. The three dimensions several problems on random space-filling or random parking have been considered, mainly in connection with models of liquid structure, by Bernal [28], Bernal, Masson, and Knight [29], and D. G. Scott [602].

ISBN 0-201-13500-0

CHAPTER 3

Sets of Lines in the Plane

1. Density for Sets of Lines

A straight line G in the plane is determined by the angle ϕ that the direction perpendicular to G makes with a fixed direction $(0 \leqslant \phi \leqslant 2\pi)$ and by its distance p from the origin O $(0 \leqslant p)$. The coordinates p, ϕ are the polar coordinates of the foot of the perpendicular from the origin onto the line.

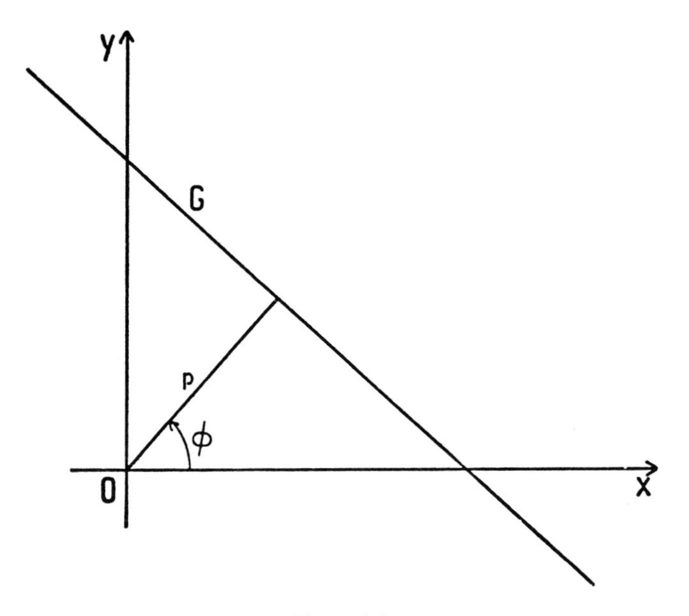

Figure 3.1.

ENCYCLOPEDIA OF MATHEMATICS and Its Applications, Gian-Carlo Rota (ed.).
1, Luis A. Santaló, Integral Geometry and Geometric Probability

ISBN 0-201-13500-0

The equation of G is then (Fig. 3.1)

$$x \cos \phi + y \sin \phi - p = 0. \tag{3.1}$$

The measure of a set X of lines can be defined by any integral of the form

$$m(X) = \int_X f(p, \phi)\, dp \wedge d\phi \tag{3.2}$$

where the function f must be chosen according to certain criteria, depending on the nature of the problem. In integral geometry and in the theory of geometrical probability the chosen criterion establishes that *the measure is to be invariant under the group of motions* $\mathfrak{M}$.

By a motion u (2.1) the line (3.1) transforms into

$$x \cos(\phi - \alpha) + y \sin(\phi - \alpha) - (p - a \cos \phi - b \sin \phi) = 0.$$

Comparing with (3.1), we see that under the motion $u(a, b, \alpha)$ the coordinates p, ϕ of G transform according to

$$p' = p - a \cos \alpha - b \sin \alpha, \qquad \phi' = \phi - \alpha, \tag{3.3}$$

so that $dp \wedge d\phi = dp' \wedge d\phi'$. The measure of the transformed set $X' = uX$ is

$$m(X') = \int_{X'} f(p', \phi')\, dp' \wedge d\phi' = \int_X f(p - a \cos \phi - b \sin \phi, \phi - \alpha)\, dp \wedge d\phi. \tag{3.4}$$

If we want $m(X)$ to equal $m(X')$ for any set X, (3.2) and (3.4) lead to $f(p - a \cos \phi - b \sin \phi, \phi - \alpha) = f(p, \phi)$ and since equality is to hold for all motions u, we must have $f(p, \phi) =$ constant. Choosing this constant equal to unity we have

The measure of a set of lines $G(p, \phi)$ is defined by the integral, over the set, of the differential form

$$dG = dp \wedge d\phi, \tag{3.5}$$

which is called the density for sets of lines.

Up to a constant factor, this density is the only one that is invariant under motions. The density (3.5) will always be taken at absolute value.

We will give an immediate application of density (3.5). Let D be a domain in the plane of area F and let G be a line that intersects D. Multiplying both sides of (3.5) by the length σ of the chord $G \cap D$ and integrating over all the lines G, using the fact that $\sigma\, dp$ is the area element of D, so that the integral with reference to dp for any fixed ϕ ($0 \leqslant \phi \leqslant \pi$) is the area F, we have

ISBN 0-201-13500-0

$$\int_{G \cap D \neq \emptyset} \sigma \, dG = \pi F. \tag{3.6}$$

Other forms of dG. If the line G is determined by other coordinates than p, ϕ, the density dG takes different forms, which can be easily obtained by simple changes of coordinates.

(a) If G is determined by the angle θ that it makes with the x axis and the abscissa x of the point at which it cuts the same axis (Fig. 3.2), we have $p = x \sin \theta$, $\phi = \theta - \pi/2$, and therefore

$$dG = \sin \theta \, dx \wedge d\theta. \tag{3.7}$$

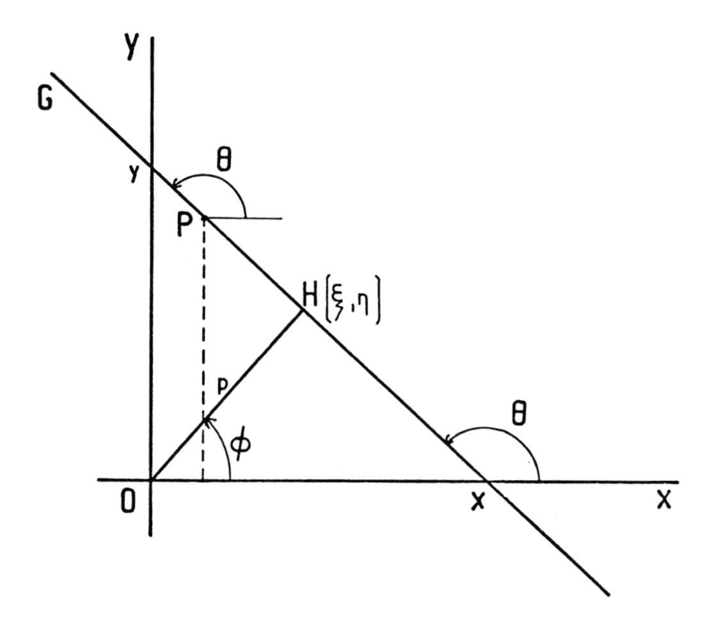

Figure 3.2.

(b) If G is determined by its intercepts x, y on the coordinate axis, we have $p = xy(x^2 + y^2)^{-1/2}$, $\phi = \arctan(x/y)$, and therefore

$$dG = \frac{xy}{(x^2 + y^2)^{3/2}} dx \wedge dy. \tag{3.8}$$

(c) Let G be determined by the coefficients u, v of its equation written in the form $ux + vy + 1 = 0$. Then in (3.8) we have $x = -1/u$, $y = -1/v$, and thus

$$dG = \frac{dx \wedge dv}{(u^2 + v^2)^{3/2}}. \tag{3.9}$$

ISBN 0-201-13500-0

(d) If G is determined by the coordinates ξ, η of the foot of the perpendicular from the origin onto the line, we have $\xi = p \cos \phi$, $\eta = p \sin \phi$, and therefore

$$dG = \frac{d\xi \wedge d\eta}{(\xi^2 + \eta^2)^{1/2}}. \tag{3.10}$$

(e) Let $P(x, y)$ be a point of G and let θ be the angle that G makes with the x axis. We have $p = x \sin \theta - y \cos \theta$, $\phi = \theta - \pi/2$, and thus

$$dG = \sin \theta \, dx \wedge d\theta - \cos \theta \, dy \wedge d\theta. \tag{3.11}$$

All these forms for dG must be taken up to the sign, because we consider all densities to be positive.

2. Lines That Intersect a Convex Set or a Curve

We want to measure the set of lines that meet a fixed bounded convex set K. Take as origin a point $O \in K$ and let $p = p(\phi)$ be the support function of K with reference to O. Applying (3.5) and (1.6), we have

$$m(G; G \cap K \neq 0) = \int_{G \cap K \neq 0} dp \wedge d\phi = \int_0^{2\pi} p \, d\phi = L \tag{3.12}$$

where L is the length of ∂K (perimeter of K). Hence we have

The measure of the set of lines that intersect a bounded convex set K is equal to the length of its boundary.

In terms of geometric probability, we can state

Let K_1 be a convex set contained in the bounded convex set K. The probability that a random line G intersects K_1 if it is known that it intersects K is $p = L_1/L$, where L_1, L are the perimeters of K_1 and K, respectively.

From (3.12) and (3.6) we also deduce that the mean length of the chords that random lines intercept in a convex set K is

$$E(\sigma) = \pi F/L \tag{3.13}$$

where F is the area and L the perimeter of K.

Let C be a piecewise differentiable curve in the plane that has a finite length L. Assume that C is defined by the equations

$$x = x(s), \qquad y = y(s) \tag{3.14}$$

where the parameter s is the arc length. Let G be a line that intersects C at the point x, y and forms with the tangent at this point an angle θ. The co-

ISBN 0-201-13500-0

ordinates s, θ determine the line G and we want to express dG in terms of s, θ. We have

$$\phi = \theta + \tau - (\pi/2) \tag{3.15}$$

where τ is the angle between the tangent to C and the x axis.

Since x, y is a point of G, we have $p = x \cos \phi + y \sin \phi$ and therefore $dp = \cos \phi \, dx + \sin \phi \, dy + (-x \sin \phi + y \cos \phi) \, d\phi$. Since $dx = \cos \tau \, ds$, $dy = \sin \tau \, ds$, we get $dp = \cos(\phi - \tau) \, ds + (-x \sin \phi + y \cos \phi) \, d\phi$. Exterior multiplication by $d\phi$ gives

$$dG = dp \wedge d\phi = \cos(\phi - \tau) \, ds \wedge d\phi$$

and according to (3.15), since $d\phi = d\theta + \tau' \, ds$, we have

$$dG = |\sin \theta| \, ds \wedge d\theta \tag{3.16}$$

where we have written $|\sin \theta|$ because we consider all densities to be positive.

Now integrate both sides of (3.16) over all lines that intersect C. On the right-hand side we have

$$\int_0^L ds \int_0^\pi |\sin \theta| \, d\theta = 2L$$

and on the left-hand side each line G has been counted as many times as it has intersection points with C. Calling this number n, we obtain

$$\int n \, dG = 2L \tag{3.17}$$

where the integral is extended over all lines of the plane, n being 0 for the lines G that do not intersect C. For $n = 2$ we again have the result (3.12).

We have proved (3.17) for piecewise differentiable curves, but it holds good for any rectifiable curve [377, 51]. The left-hand side of (3.17) can be taken as the definition of the length of a continuum of points; it is called the *Favard length* of the continuum [172, 604]. On the importance that the integral in (3.17) be in the Lebesgue sense instead of the Riemann sense, see [171].

Formula (3.17) can be applied for measuring the length of curves. Following Steinhaus [625, 626], we proceed as follows. Take a transparent sheet with a family of equidistant parallels G_i $(i = 1, 2, \ldots)$. Assume that the arc C to be measured cuts G_i in n_i points, so that $s_0 = \sum n_i$ is the total number of intersections. Turning the sheet through an angle $(k/m)\pi$ $(k = 0, 1, \ldots, m - 1)$, we get s_k intersections and the total amount is $N = \sum_0^{m-1} s_k$. Calling a the distance between G_i and G_{i+1}, we get the expression

ISBN 0-201-13500-0

$$\frac{N\pi a}{2m} \tag{3.18}$$

as an approximation of the length of C (since it is an approximation of $\frac{1}{2}\int n\,dG$). The accuracy of (3.18) depends on a and m. For $a \to 0$ and $m \to \infty$, (3.18) tends to the length of C. Estimates of the errors involved in this method are given by Moran [427].

3. Lines That Cut or Separate Two Convex Sets

Let K_1, K_2 be two bounded convex sets in the plane and let L_1, L_2 be the lengths of the boundaries ∂K_1, ∂K_2.

We call *external cover* C_e of K_1 and K_2 the boundary of the convex hull of $K_1 \cup K_2$ (recall that the convex hull of a set of points is the intersection of all the convex sets containing the set). The external cover may be intuitively interpreted as a closed elastic string drawn about K_1 and K_2 (Fig. 3.3). Let L_e denote the length of C_e. If $K_1 \cap K_2 \neq \emptyset$, as in the case of Fig. 3.3, we can also consider the *internal cover* C_i realized by a closed elastic string drawn about K_1 and K_2 and crossing over at a point O placed between K_1 and K_2. Let L_i denote the length of C_i.

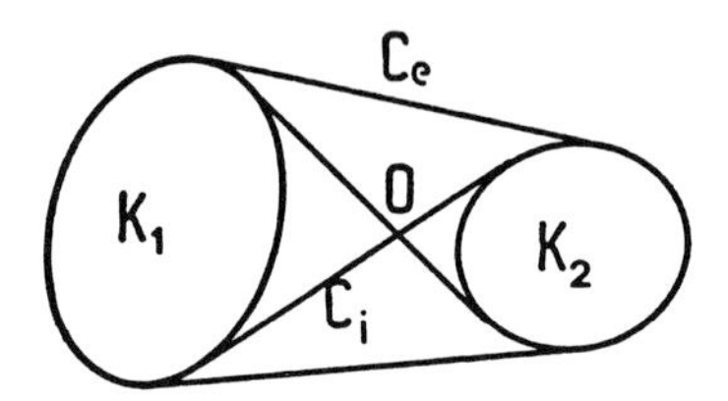

Figure 3.3.

Though they have arcs in common, we will consider ∂K_1, ∂K_2, C_e, C_i as different curves. Except for lines of support that belong to a set of measure zero, every line that meets K_1 and K_2 has two intersection points with each of the curves ∂K_1, ∂K_2, and C_e, and four intersection points with C_i; there are in all ten points of intersection. Let m_{10} be the measure of the set of such lines.

The lines that meet either K_1 or K_2, but not both, have six common points with the curves ∂K_1, ∂K_2, C_i, C_e: let m_6 be the measure of this set of lines. The lines that separate K_1 and K_2 have four intersection points with the curves above (two common points with C_e and two common points with C_i): let m_4 denote the measure of this set of lines.

Since the measure of the set of lines that meet a convex set is equal to its perimeter, the measures m_6', m_6'' of the lines that meet K_1 without meeting

ISBN 0-201-13500-0

K_2, or vice versa, are

$$m_6' = L_1 - m_{10}, \qquad m_6'' = L_2 - m_{10}, \tag{3.19}$$

and therefore we have

$$m_6 = m_6' + m_6'' = L_1 + L_2 - 2m_{10}. \tag{3.20}$$

Since C_e is a closed convex curve, we have

$$m_4 + m_6 + m_{10} = L_e \tag{3.21}$$

and applying (3.17) to the set of the four curves ∂K_1, ∂K_2, C_i, C_e, we obtain

$$4m_4 + 6m_6 + 10m_{10} = 2(L_1 + L_2 + L_e + L_i). \tag{3.22}$$

From (3.19)–(3.22) we get

$$m_4 = L_i - (L_1 + L_2), \qquad m_6' = L_1 - (L_i - L_e),$$

$$m_6'' = L_2 - (L_i - L_e), \qquad m_{10} = L_i - L_e. \tag{3.23}$$

In words,

(a) The measure of all lines that meet K_1 and K_2 is $L_i - L_e$.

(b) The measure of the set of lines that meet K_1 without meeting K_2 is $L_1 - (L_i - L_e)$ and the measure of the set of lines that meet K_2 without meeting K_1 is $L_2 - (L_i - L_e)$.

(c) The measure of the set of lines that separate K_1 and K_2 is $L_i - (L_1 + L_2)$.

We have assumed that $K_1 \cap K_2 = \emptyset$. If they overlap, it is easy to see that results (a)–(c) hold good with the convention of putting $L_i = L_1 + L_2$.

In terms of geometrical probability these results can be stated as follows: Let K_1 and K_2 be two bounded convex sets. If G is a line chosen at random in the plane with the condition that it meets the convex hull of K_1 and K_2, then:

(a) The probability that G meets K_1 and K_2 is

$$p(G \cap K_1 \cap K_2 \neq \emptyset) = (L_i - L_e)/L_e.$$

(b) The probability that G meets K_1 without meeting K_2 is

$$p(G \cap K_1 \neq \emptyset, G \cap K_2 = \emptyset) = 1 - \frac{L_i - L_1}{L_e}$$

and similarly

$$p(G \cap K_1 = \emptyset, G \cap K_2 \neq \emptyset) = 1 - \frac{L_i - L_2}{L_e}.$$

ISBN 0-201-13500-0

(c) The probability that G separates K_1 and K_2 is

$$p(G \cap K_1 = \emptyset, G \cap K_2 = \emptyset, G \cap C_e \neq \emptyset) = \frac{L_i - (L_1 + L_2)}{L_e}.$$

If $K_1 \cap K_2 \neq \emptyset$ in the preceding formulas, we must put $L_i = L_1 + L_2$. If we consider only lines that meet K_1, by applying the same type of argument we obtain the following.

If K_1 and K_2 are two bounded convex sets in the plane (which may or may not overlap), the probability that a random chord of K_1 intersects K_2 is

$$p = (L_i - L_e)/L_1$$

where L_i is the length of the internal cover if $K_1 \cap K_2 = \emptyset$ and $L_i = L_1 + L_2$ if K_1 and K_2 overlap.

The preceding results are the work of Sylvester [665a], who also considered the case of more than two convex sets. About the distribution of the number of intersections of a rectifiable curve C and a random line G (i.e., about the probability p_k that G has k intersection points with C, for every k) very little is known (see the articles by Sulanke [662] and Ambarcumjan [6]).

4. Geometric Applications

1. Let C be a plane closed curve of class C^2, which is oriented, so that there is a prescribed sense of rotation. Let s denote the arc length of C and let $\tau(s)$ denote the angle of the tangent to C with a fixed direction, say the x axis. The curvature of C is defined by $\kappa = d\tau/ds$ and the *total absolute curvature* is defined by the line integral

$$c_a = \int_C |\kappa|\, ds = \int_C |d\tau|. \tag{3.24}$$

Let $\nu(\tau)$ denote the number of unoriented tangents to C that are parallel to the direction τ (e.g., in Fig. 3.4, $\nu = 6$). Since in the last integral (3.24) each direction τ appears $\nu(\tau)$ times, equation (3.24) can be written

$$c_a = \int_0^\pi \nu(\tau)\, |d\tau|. \tag{3.25}$$

On the other hand, if a line G is parallel to the direction τ and meets C in n points P_i, then there are at least n tangents to C that are parallel to G (one for each of the arcs $P_1P_2, P_2P_3, \ldots, P_{n-1}P_n, P_nP_1$) and thus $n(\tau) \leqslant \nu(\tau)$. Using (3.17), we have

ISBN 0-201-13500-0

$$2L = \int_{G \cap C \neq \emptyset} n\, dG \leqslant \int_{G \cap C \neq \emptyset} \nu\, dG = \int_{G \cap C \neq \emptyset} \nu\, dp \wedge d\tau \leqslant \Delta_{\mathrm{m}} \int_0^{\pi} \nu\, |d\tau| = \Delta_{\mathrm{m}} c_{\mathrm{a}} \quad (3.26)$$

where L is the length of C and Δ_{m} its maximal breadth.

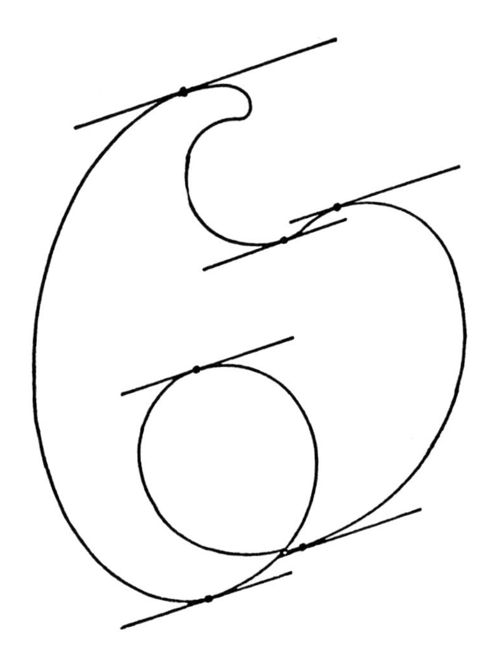

Figure 3.4.

In particular, if C lies inside a circle of radius r, then $\Delta_{\mathrm{m}} \leqslant 2r$ and we have the following theorem of Fáry [169]:

If a closed plane curve of length L and absolute total curvature c_{a} can be enclosed by a circle of radius r, then $L \leqslant rc_{\mathrm{a}}$.

2. Let K be a bounded convex set of area F and perimeter L. Let G, G' be two lines that intersect ∂K at the points corresponding to the arc lengths s, s'. Call θ, θ' the angles of G, G' with the support lines at their intersection points with the boundary ∂K and let σ, σ' be the lengths of the chords intercepted by G, G' in K. We consider the integral

$$I = \int (\sigma \sin \theta' - \sigma' \sin \theta)^2\, ds \wedge d\theta \wedge ds' \wedge d\theta' \qquad (3.27)$$

extended over the ranges $0 \leqslant s \leqslant L$, $0 \leqslant s' \leqslant L$, $0 \leqslant \theta, \theta' \leqslant \pi$. According to (3.16) we have $\sin \theta\, ds \wedge d\theta = dG$, $\sin \theta'\, ds' \wedge d\theta' = dG'$. Now

ISBN 0-201-13500-0

$$\int_0^{\pi} \sigma^2 \, d\theta = 2F, \qquad \int_0^{\pi} \sin^2 \theta \, d\theta = \pi/2 \tag{3.28}$$

where the first integral follows from the formula for area in polar coordinates. An analogous formula holds for G'. Hence

$$I = 2\pi L^2 F - 2 \int_{G \cap K \neq 0} \sigma \, dG \int_{G' \cap K \neq 0} \sigma' \, dG' = 2\pi F(L^2 - 4\pi F)$$

where we have applied (3.6) and the fact that in integral (3.27) each line G, G' is counted twice, once for each intersection with ∂K.

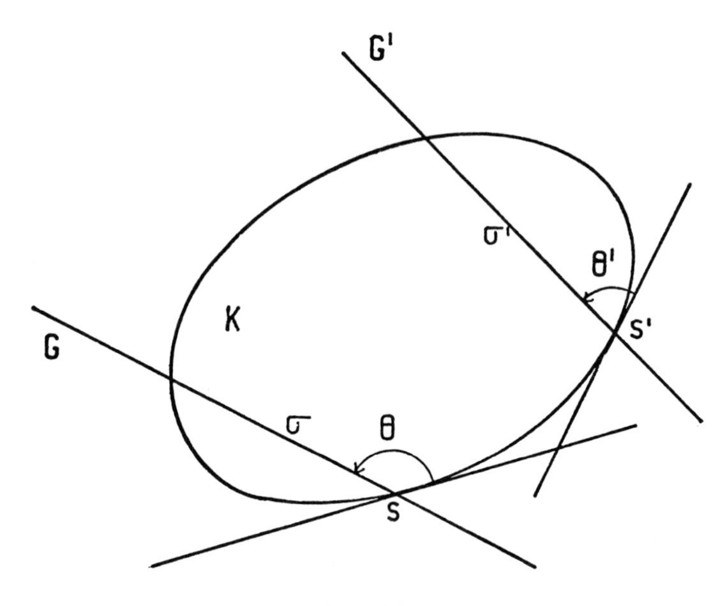

Figure 3.5.

Since clearly $I \geqslant 0$, we have proved the classical *isoperimetric inequality*

$$L^2 - 4\pi F \geqslant 0. \tag{3.29}$$

The equality sign holds if $I = 0$, so that $\sigma/\sin \theta = \sigma'/\sin \theta'$, which means that K is a circle. This proof is due to Blaschke [38].

3. Let C be a curve of length L contained in a closed convex curve C_1 of length L_1. If we consider all lines G that intersect C_1, the average of the number of intersection points of these lines with C will be $E(n) = 2L/L_1$. Since there must be lines G the number of whose intersections with C is not

ISBN 0-201-13500-0

less than $E(n)$, we have: *if a curve of length L lies inside of a closed convex curve of length L_1, there exist lines that cut it in at least $2L/L_1$ points.*

5. Notes and Exercises

1. *Again the isoperimetric inequality for convex curves.* Let C be a closed convex curve of class C^1. Let A_1, A_2 be the points in which C is intersected by the line $G(p, \phi)$ and let ds_1, ds_2 be the arc length elements at A_1, A_2 and θ_1, θ_2 the angles of G with the tangents to C at A_1, A_2. It is easy to see that

$$\sigma\, dG = \sin\theta_1 \sin\theta_2\, ds_1 \wedge ds_2 \tag{3.30}$$

where σ is the length of the chord A_1A_2. Integrating over all oriented lines that meet C, we have

$$\begin{aligned} 2\pi F &= \int_L ds_1 \int_L \sin\theta_1 \sin\theta_2\, ds_2 \\ &= \tfrac{1}{2}\int_L ds_1 \int_L [\cos(\theta_1 - \theta_2) - \cos(\theta_1 + \theta_2)]\, ds_2. \end{aligned}$$

The integral $\int_L \cos(\theta_1 + \theta_2)\, ds_2$ is equal to the projection of C on the tangent line at A_1 and therefore is equal to zero. Thus we have

$$4\pi F = L^2 - 2\int_L ds_1 \int_L \sin^2 \tfrac{1}{2}(\theta_1 - \theta_2)\, ds_2, \tag{3.31}$$

which contains the isoperimetric inequality $L^2 - 4\pi F \geqslant 0$. This proof is due to Pleijel [479].

A more general result can be obtained from the formulas

$$\int_{G \cap C \neq \emptyset} \sigma^n\, dG = \frac{n}{2}\iint_{L\,L} \sigma^{n-1} \cos\theta_1 \cos\theta_2\, ds_1 \wedge ds_2,$$

$$\int_{G \cap C \neq \emptyset} \sigma^n\, dG = \frac{1}{2}\iint_{L\,L} \sigma^{n-1} \sin\theta_1 \sin\theta_2\, ds_1 \wedge ds_2,$$

which yield the following formula of Ambarcumjan [7] (for any integer $n > 0$)

$$2(n+1)\int_{G \cap C \neq \emptyset} \sigma^n\, dG = n\iint_{L\,L} \sigma^{n-1}\, ds_1 \wedge ds_2 - 2n\iint_{L\,L} \sigma^{n-1} \sin^2\left(\frac{\theta_1 - \theta_2}{2}\right) ds_1 \wedge ds_2.$$

As a consequence we have the inequality

$$\int_{G \cap C \neq \emptyset} \sigma^n\, dG \leqslant \frac{n}{2(n+1)}\iint_{L\,L} \sigma^{n-1}\, ds_1 \wedge ds_2 \tag{3.32}$$

which for $n = 1$ gives the isoperimetric inequality.

ISBN 0-201-13500-0

Pleijel's proof has been generalized to nonsimple curves by Banchoff and Pohl [20]. Then F is replaced by the sum of the areas into which the curve divides the plane, each weighted with the square of the winding number, that is,

$$L^2 - 4\pi \int w^2 \, dP \geqslant 0$$

where the integral is over the whole plane. The winding number $w(P)$ of P with respect to the curve can be defined as the number of turns of the vector PX when X goes around the curve. Equality holds if and only if the curve is a circle or a circle traversed several times. Banchoff and Pohl have generalized this isoperimetric inequality to arbitrary dimension and codimension, as we shall see in Chapter 14, Section 7, Note 8.

2. *Distance of two curves.* Let C_1 and C_2 be two curves in the plane and let $N_1(G)$, $N_2(G)$ be the numbers of points at which they are intersected by a variable line G. Define the distance of C_1, C_2 by the integral

$$(C_1, C_2) = \tfrac{1}{2} \int |N_1(G) - N_2(G)| \, dG \tag{3.33}$$

extended over all lines of the plane.

If C_1 and C_2 are rectifiable curves, the distance (C_1, C_2) is finite, since $(C_1, C_2) \leqslant L_1 + L_2$. Show that the distance so defined satisfies the conditions (a) $(C_1, C_1) = 0$; (b) $(C_1, C_2) = (C_2, C_1)$; (c) $(C_1, C_2) + (C_2, C_3) \geqslant (C_1, C_3)$.

If C_1, C_2 are Jordan curves (i.e., homeomorphic images of a line segment or of a circle), then $(C_1, C_2) = 0$ implies $C_1 \equiv C_2$. With the definition above, prove that:

(i) If C_1 and C_2 are closed convex curves such that C_1 lies inside C_2, then $(C_1, C_2) = L_2 - L_1$;

(ii) If C_1, C_2 are closed curves such that C_1 lies outside C_2, then $(C_1, C_2) = L_1 + L_2 - 2(L_i - L_e)$, where L_i, L_e are the lengths of the external and internal covers of C_1, C_2 (Section 3). (See [625].)

3. *Asymmetry of convex curves.* If C is a given curve of length L, its asymmetry relative to a line G_0 may be defined as the distance (C, C^*), where C^* denotes the reflection of C in G_0. The line G_0 that minimizes the asymmetry (C, C^*) is called the *axis of symmetry* of C and the difference $1 - \min(C, C^*)/2L$ is called the *index of symmetry* of C. Compare this definition of Steinhaus [625] with other definitions of symmetry for convex curves (e.g., that of Grünbaum [246]).

4. *Length of order m.* Steinhaus [625] calls the length of order m of a set of points C the integral

$$L_m = \tfrac{1}{2} \int_{G \cap C \neq \emptyset} w_m \, dG \tag{3.34}$$

ISBN 0-201-13500-0

where $w_m = n$ is the number of points of the intersection $G \cap C$ if $n \leqslant m$ and $w_m = m$ if $n > m$. Note that if C is a rectifiable closed curve, then L_2 is the length of the boundary of its convex hull and L_∞ is the ordinary length.

5. *Systems of nonseparable convex sets.* A system of n closed convex sets K_i $(i = 1, 2, \ldots, n)$ in the plane is called *separable* if there is a line G that intersects no K_i and that divides the plane into two half planes, each containing at least one K_i. In the contrary case, the system is said to be *nonseparable.* The following theorem can be proved: Let K_0 denote the convex hull of the nonseparable convex system K_i $(i = 1, 2, \ldots, n)$. Let L_i, D_i, and R_i denote the perimeter, diameter, and radius, respectively, of the circumcircle of the set K_i. Then

$$L_0 \leqslant \sum_1^n L_i, \qquad D_0 \leqslant \sum_1^n D_i, \qquad R_0 \leqslant \sum_1^n R_i.$$

(See the articles by Hadwiger [266] and A. W. Goodman and R. E. Goodman [233].)

6. *Invariance of the line density by reflection and refraction.* Assume that the line G is the trajectory of a ray of light that is incident on a fixed curve C and is reflected according to the law of reflection: the angle of reflection is equal to the angle of incidence. Then if θ is the angle between G and the tangent to C, the angle of the tangent with the reflected line G^* is $\theta^* = -\theta$ (Fig. 3.6) and by making use of expression (3.16) for dG (at absolute value), we have $dG = dG^*$. Hence, *the density dG is invariant under reflections in curves.*

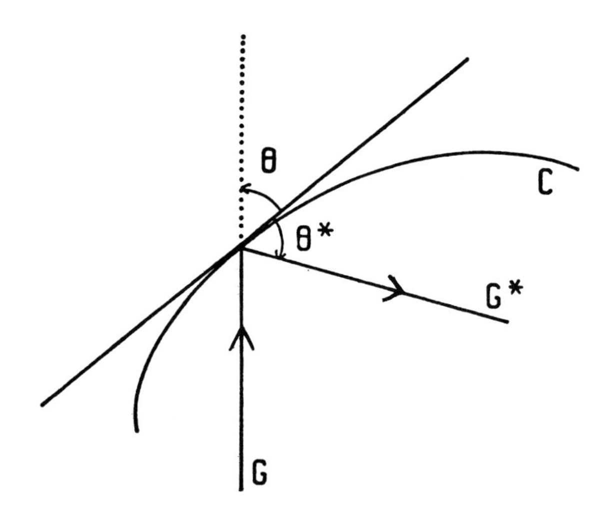

Figure 3.6.

Suppose that the line G_1 is refracted at the curve C that separates two mediums whose indices of refraction are n_1 and n_2, respectively (Fig. 3.7).

ISBN 0-201-13500-0

If i_1, i_2 are, respectively, the angles of incidence and refraction, the classical Snell law of optics says that $\sin i_1/\sin i_2 = n_2/n_1$. On the other hand, the angle of the tangent with G_1 and the angle of the tangent with G_2 are $\theta_1 = \pi/2 - i_1$ and $\theta_2 = \pi/2 - i_2$, so that, using (3.16), we have $dG_1 = -\cos i_1\, ds \wedge di_1 = -ds \wedge d(\sin i_1)$, $dG_2 = -\cos i_2\, ds \wedge di_2 = -ds \wedge d(\sin i_2) = -(n_1/n_2)\, ds \wedge d(\sin i_1) = (n_1/n_2)\, dG_1$. Thus, when a ray is refracted from a medium of index of refraction n_1 to a medium of index of refraction n_2, the density dG_1 is multiplied by the constant factor n_1/n_2.

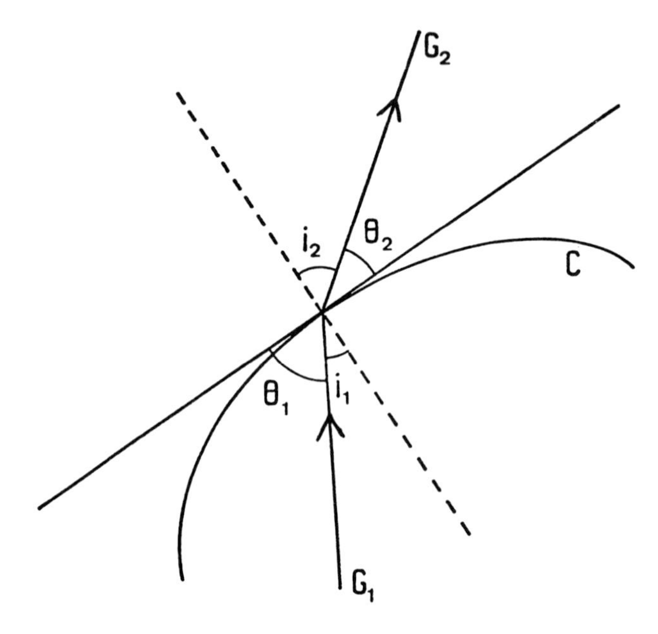

Figure 3.7.

Consider an optical instrument composed of several mediums of indices of refraction $n_1, n_2, \ldots, n_m$ such that the first medium, where the rays enter the instrument, coincides with the last medium, where the rays emerge; that is, $n_1 = n_m$. We have $dG_m = (n_{m-1}/n_m)\, dG_{m-1} = (n_{m-1}n_{m-2})/(n_m n_{m-1})\, dG_{m-2} = \cdots = (n_{m-1} \cdots n_1)/(n_m n_{m-1} \cdots n_2)\, dG_1 = dG_1$. Hence, *line density is preserved during its propagation through optical instruments* [40].

7. *The convex billiard table of G. D. Birkhoff.* Let C be a closed convex curve in the plane. Suppose that a point P moves inside C and collides with C according to the law "the angle of incidence is equal to the angle of reflection." The path of P is determined by the length s of the arc OA (where O is an arbitrary point on C) and the angle α ($0 \leqslant \alpha \leqslant \pi$) of the line $G \equiv AP$ with the tangent to C at A (Fig. 3.8). According to (3.16) the line density can be written $dG = \sin \alpha\, ds \wedge d\alpha$ and since this density is preserved under reflection, we have that the differential form $\sin \alpha\, ds \wedge d\alpha$ is invariant under the motion of point P (see the books by Birkhoff [31] and Arnold and Avez [13]).

ISBN 0-201-13500-0

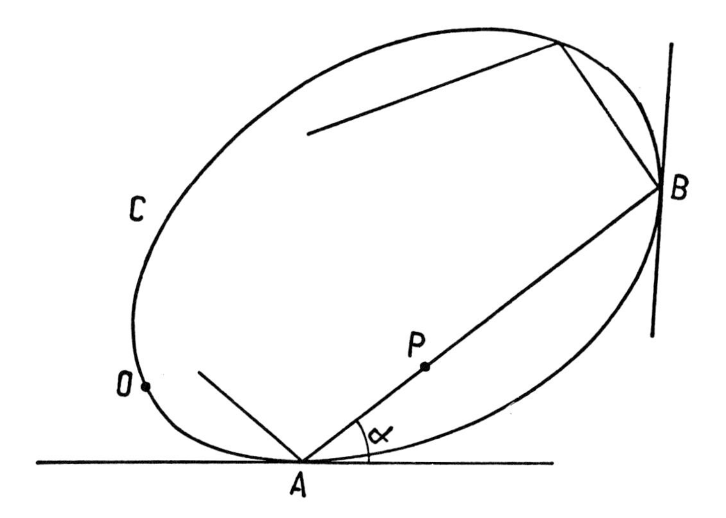

Figure 3.8.

8. *An inequality of Feller.* Let D be a plane domain of area F contained in the unit circle. Suppose that the intersection of D with any line has measure not exceeding a fixed constant $a < 1$. Then necessarily $F < 2a$. This result is the best possible in the sense that there exists a domain such that its intersection with any line has a total length not exceeding a, whereas for its area F we have $F > 2a(1 - \pi^{-2}a^2 - \varepsilon)$ where $\varepsilon > 0$ is arbitrarily small (see [184]). Generalization to the sphere and noneuclidean spaces has been done by Ueno and co-workers [682] and Santalo [563]. If D is convex, we have $F \leqslant \pi a^2/4$ where the equality sign holds if and only if D is a circle [30a].

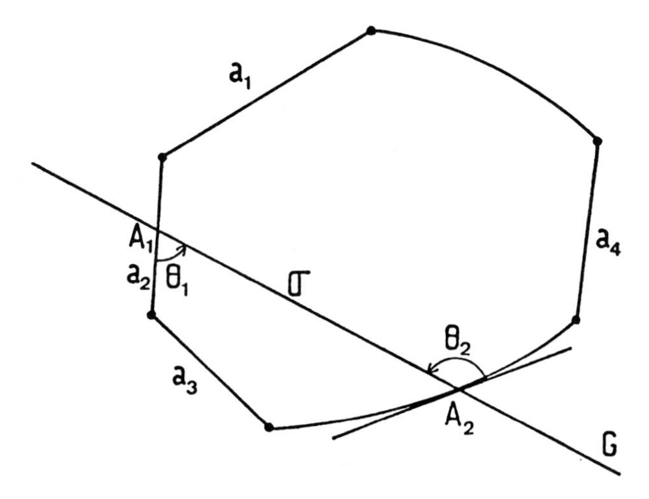

Figure 3.9.

ISBN 0-201-13500-0

9. *An integral formula.* Let C be a closed convex curve of length L that contains m line segments of length $a_1, a_2, \ldots, a_m$, respectively (Fig. 3.9). Let A_1, A_2 be the points at which a line G meets C and call σ the length of the chord A_1A_2 and θ_1, θ_2 the angles that G makes with the support lines of C at A_1 and A_2. Then the following formula holds.

$$\int_{G \cap C \neq \emptyset} \frac{\sigma}{\sin \theta_1 \sin \theta_2} \, dG = \frac{1}{2}\left(L^2 - \sum_1^m a_i^2\right). \tag{3.35}$$

10. *Integral geometry of plane linear graphs.* Sulanke [622] has considered the following kind of problem. Recall that a linear graph is a finite set of points (vertices) together with a finite number of arcs of curve (edges, assumed rectifiable) joining two distinct vertices, such that no two of the edges have an interior point in common and each vertex is the end of at least one edge. To each graph Q is attached the set of real numbers $p_i(Q)$ ($i = 1, 2, \ldots$) that are equal to the probability that a random line intersects Q exactly i times.

THEOREM. *A necessary and sufficient condition for a linear graph to be Eulerian (i.e., a graph such that it is possible to find a path that traverses each arc exactly once, goes through all vertices, and ends at the starting point) is that* $p_i = 0$ *for all i odd.*

Linear graphs with the same set of natural numbers $\{i_1, i_2, \ldots, i_N\}$ such that $p_i > 0$ are said to be of the same type. The problem arises of knowing whether to each set of natural numbers $i_1 < i_2 < \cdots < i_N$ there corresponds a linear graph. Sulanke has solved the problem for $i_N \leqslant 5$ with the exceptions $\{1, 2, 5\}$, $\{1, 4, 5\}$. Recently Hristov has shown that the type $\{1, 2, 5\}$ cannot exist (M R 50, #8295).

Let c be the number of bounded regions into which Q partitions the plane (faces of the graph) and let c_G be the analogous number for the graph $Q + s_G$, where s_G is the line segment of the random line G determined by the first and the last point of the intersection $G \cap Q$. Then, if we assume Q to be connected, the expected value of c_G is $E(c_G) = c - 1 + 2L(Q)/L(Q_*)$ where $L(Q)$ is the total length of Q and $L(Q_*)$ is the length of the convex hull of Q.

If $m(G; N_G(Q) \neq N_G(Q'))$ denotes the measure of the line set G such that the number of intersection points of G with Q is not equal to the number of intersections of G with another linear graph Q', then $\rho(Q, Q') = m(G; N_G(Q) \neq N_G(Q'))$ defines a metric in the space of graphs. Compare with the distance of Note 2. For more details, see [662].

11. *A mean free path.* The mean free path through a plane that contains laminae whose mean area is F_0 and mean perimeter L_0, with a density of D laminae per unit area, is $E(\sigma) = (1 - F_0D)/L_0D$. If the laminae are circular of radius r, we have $E(\sigma) = (1 - \pi r^2 D)/2\pi r D$ and for r small, this value can be approximated by $1/2\pi r D$.

In the space E_3 the analogous mean free path for a moving point through particles of mean volume v and mean area f distributed uniformly at random with a mean density of D particles per unit volume is $E(\sigma) = 4(1 - vD)/fD$. For small particles we can take $E(\sigma) = 4(fD)^{-1}$.

Polya [486] interprets the plane mean free path as the mean visible distance from a point inside a forest of trees of radius r distributed uniformly at

ISBN 0-201-13500-0

random with a density of D trees per unit area and the mean free path in the space as the mean visible distance inside a cloud of snow formed by flakes distributed uniformly at random with mean density D. Santaló [542] has considered the case in which the trajectories are curves of a given shape.

EXERCISE 1. Let AB be a convex arc of length L. Let L_0 be the length of the line segment AB. Show that the probability that a random line that intersects the convex set bounded by the arc AB and the chord AB meets the chord AB is $2L_0/(L + L_0)$.

EXERCISE 2. If it is known that a random line meets a side of a given square, show that the probability that it also meets the opposite side is $p = \sqrt{2} - 1$.

EXERCISE 3. If it is known that a random line meets a diagonal of a given square, the probability that it meets the other diagonal is $p = 2 - \sqrt{2}$.

EXERCISE 4. Let C_i be m closed convex curves of lengths L_i $(i = 1, 2, \ldots, m)$ that are inside a closed convex curve C of length L. Show that there are lines that meet a number of curves C_i that is equal to or greater than the integral part of $(\sum L_i)/L$.

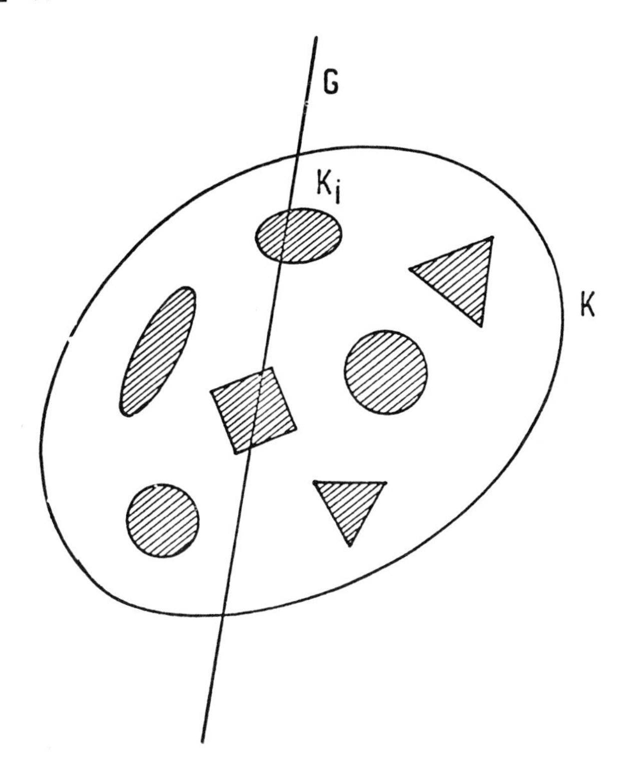

Figure 3.10.

EXERCISE 5. Let K_0, K_1 be two convex sets such that $K_1 \subset K_0$. The probability that n random lines that intersect K_0 do not have common point with K_1 is equal to $(1 - L_1/L_0)^n$, where L_0, L_1 are the perimeters of K_0, K_1.

ISBN 0-201-13500-0

More difficult is the following question: assuming n random lines that intersect K_0, find the probability that a convex set congruent to K_1 may be placed inside K_0 without meeting any one of the given lines. Consider the cases when K_1 is a line segment or a circle.

ISBN 0-201-13500-0

CHAPTER 4

Pairs of Points and Pairs of Lines

1. Density for Pairs of Points

A pair of points $P_1(x_1, y_1)$, $P_2(x_2, y_2)$ may be determined by the four coordinates x_1, y_1, x_2, y_2, as well as by the coordinates p, ϕ of the line G that unites them, together with the distances t_1, t_2 from P_1, P_2 to the foot H of

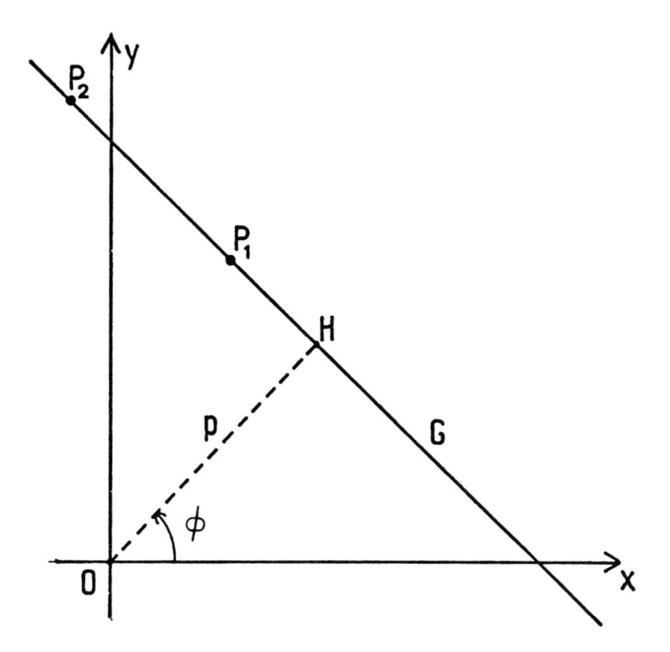

Figure 4.1.

ENCYCLOPEDIA OF MATHEMATICS and Its Applications, Gian-Carlo Rota (ed.).
1, Luis A. Santaló, Integral Geometry and Geometric Probability

ISBN 0-201-13500-0

the perpendicular from the origin O to G (Fig. 4.1). We want to express the product $dP_1 \wedge dP_2 = dx_1 \wedge dy_1 \wedge dx_2 \wedge dy_2$ in terms of the coordinates p, ϕ, t_1, t_2. We have

$$x_i = p\cos\phi - t_i \sin\phi, \qquad y_i = p\sin\phi + t_i\cos\phi \qquad (i = 1, 2). \tag{4.1}$$

Hence

$$dx_i = \cos\phi\, dp - (p\sin\phi + t_i\cos\phi)\, d\phi - \sin\phi\, dt_i,$$

$$dy_i = \sin\phi\, dp + (p\cos\phi - t_i\sin\phi)\, d\phi + \cos\phi\, dt_i,$$

and by exterior multiplication we get

$$dP_i = dx_i \wedge dy_i = p\, dp \wedge d\phi + dp \wedge dt_i - t_i\, d\phi \wedge dt_i \qquad (i = 1, 2)$$

and

$$dP_1 \wedge dP_2 = (t_2 - t_1)\, dp \wedge d\phi \wedge dt_1 \wedge dt_2.$$

Since $dp \wedge d\phi = dG$, we can write

$$dP_1 \wedge dP_2 = |t_2 - t_1|\, dG \wedge dt_1 \wedge dt_2 \tag{4.2}$$

where we have put the absolute value of $t_2 - t_1$ because all densities are assumed to be positive.

2. Integrals for the Power of the Chords of a Convex Set

Let K be a plane bounded convex set and let σ denote the length of the chord determined by the line G on K. Consider the integrals

$$I_n = \int_{G \cap K \neq \emptyset} \sigma^n\, dG, \qquad J_n = \int_{P_1, P_2 \in K} r^n\, dP_1 \wedge dP_2 \tag{4.3}$$

where n is an integer and r means the distance between points P_1 and P_2 of K.

In order to relate the integrals I_n and J_n we apply (4.2) and observe that $|t_1 - t_2| = r$. We have

$$J_n = \int |t_2 - t_1|^{n+1}\, dt_1 \wedge dt_2 \wedge dG$$

$$= \int dG \wedge dt_1 \left[\int_{t_1}^{b} (t_2 - t_1)^{n+1}\, dt_2 + \int_{a}^{t_1} (t_1 - t_2)^{n+1}\, dt_2 \right]$$

ISBN 0-201-13500-0

$$= \frac{1}{n+2} \int dG \int_a^b [(b - t_1)^{n+2} + (t_1 - a)^{n+2}] \, dt_1$$

$$= \frac{2}{(n+2)(n+3)} \int_{G \cap K \neq 0} (b - a)^{n+3} \, dG \tag{4.4}$$

where a, b denote the values of t corresponding to the end points of σ, so that $b - a = \sigma$. We conclude that

$$J_n = \frac{2}{(n+2)(n+3)} I_{n+3}, \tag{4.5}$$

which holds for $n = -1, 0, 1, 2, \ldots$. Formula (4.5) can be written

$$I_n = \frac{n(n-1)}{2} J_{n-3}, \tag{4.6}$$

which holds for $n = 2, 3, 4, \ldots$. For the cases in which $n = 0, 1$, using (3.6) and (3.12), we have

$$I_0 = L, \qquad I_1 = \pi F \tag{4.7}$$

where L is the perimeter and F the area of K.

For $n = 2$, (4.3) and (4.6) become

$$I_2 = \int_{G \cap K \neq 0} \sigma^2 \, dG = \int_{P_1, P_2 \in K} \frac{dP_1 \wedge dP_2}{r}. \tag{4.8}$$

For $n = 3$, because $J_0 = F^2$, (4.6) becomes

$$I_3 = \int_{G \cap K \neq 0} \sigma^3 \, dG = 3F^2, \tag{4.9}$$

which is a remarkable integral formula due to Crofton [133].

For $n = 4$, denoting by $E(r)$ the mean distance between two points of K, that is,

$$E(r) = \frac{1}{F^2} \int_{P_1, P_2 \in K} r \, dP_1 \, dP_2, \tag{4.10}$$

we have, using (4.6),

$$I_4 = \int_{G \cap K \neq 0} \sigma^4 \, dG = 6F^2 E(r). \tag{4.11}$$

ISBN 0-201-13500-0

In the case of a circle of radius R, by direct computation we get

$$I_n = \frac{2 \cdot 4 \cdot \cdots \cdot n}{3 \cdot 5 \cdot \cdots \cdot (n+1)} 2^{n+1} \pi R^{n+1} \qquad \text{for } n \text{ even};$$

$$I_n = \frac{1 \cdot 3 \cdot \cdots \cdot n}{2 \cdot 4 \cdot \cdots \cdot (n+1)} 2^{n} \pi^2 R^{n+1} \qquad \text{for } n \text{ odd}. \tag{4.12}$$

The integrals I_n are not independent. They satisfy certain inequalities. For instance, the inequality $I_0^2 - 4I_1 \geqslant 0$ is the isoperimetric inequality (3.29). In connection with this point, Blaschke [51] proposed the problem of finding necessary and sufficient conditions for a sequence of real numbers to be the sequence I_n associated with a convex set and of determining, when those conditions are satisfied, whether or not the corresponding convex set is unique. In addition to the isoperimetric inequality, the following necessary conditions are known:

$$I_n \geqslant \frac{2 \cdot 4 \cdot \cdots \cdot n}{3 \cdot 5 \cdot \cdots \cdot (n+1)} 2^{n+1} \pi^{-n} I_1^{(n+1)/2} \qquad \text{for } n = 2, 4, 6, \ldots;$$

$$I_n \geqslant \frac{1 \cdot 3 \cdot \cdots \cdot n}{2 \cdot 4 \cdot \cdots \cdot (n+1)} 2^{n} \pi^{-(n+1)} I_1^{(n+1)/2} \qquad \text{for } n = 3, 5, 7, \ldots; \tag{4.13}$$

$$2^8 I_1^3 \geqslant 3^2 \pi^4 I_2^2, \tag{4.14}$$

$$\frac{2^n I_n}{I_0^n} - \binom{k}{1} \frac{2^{n+1} I_{n+1}}{I_0^{n+1}} + \binom{k}{2} \frac{2^{n+2} I_{n+2}}{I_0^{n+2}} + \cdots + (-1)^k \frac{2^{n+k} I_{n+k}}{I_0^{n+k}} > 0, \tag{4.15}$$

for any $n, k = 0, 1, 2, \ldots$;

$$I_{2m} I_{2n} \geqslant I_{m+n}^2, \qquad I_m^{p-n} I_p^{n-m} \geqslant I_n^{p-m}, \tag{4.16}$$

for any integers m, n, p such that $0 \leqslant m \leqslant n \leqslant p$.

Inequalities (4.13) are due to Blaschke [35] and (4.14) is due to Carleman [81]; in both cases the equality sign holds only for circles. Inequalities (4.15) are related to the moment problem of Hausdorff (see the book by Shohat and Tamarkin [605]) and were pointed out by Sulanke ([659]; see also [339]). Inequalities (4.16) are easily deduced from classical inequalities of Schwarz and Holder. About the distribution function of the lengths of chords $G \cap K$, that is, the probability that the chord that a random line G determines on a fixed convex set K has length $\leqslant x$, see the articles by Sulanke [659] and Geciauskas [217, 218]. Mallows and Clark [388] have shown that this distribution function does not characterize the convex set K. They give an example of two noncongruent convex sets with the same distribution function of the chord length. The problem of Blaschke is thus partially solved. The

ISBN 0-201-13500-0

question is related to some methods of pattern recognition (see [443a, 457, 627]).

Examples of mean distances. By direct computation, from definition (4.10), using (4.11), we obtain the following mean distances between two points of a convex set.

For a circle of radius R: $E(r) = (128/45\pi)R$.

For an equilateral triangle of side a:

$$E(r) = \frac{3a}{5}\left(\frac{1}{3} + \frac{1}{4}\log 3\right). \tag{4.17}$$

For a rectangle of sides a, b $(a \geqslant b)$:

$$E(r) = \frac{1}{15}\left[\frac{a^3}{b^2} + \frac{b^3}{a^2} + d\left(3 - \frac{a^2}{b^2} - \frac{b^2}{a^2}\right)\right.$$
$$\left. + \frac{5}{2}\left(\frac{b^2}{a}\log\frac{a+d}{b} + \frac{a^2}{b}\log\frac{b+d}{a}\right)\right] \tag{4.18}$$

where d is the diagonal of the rectangle. In particular, for a square of side a we have $E(r) = (a/15)[\sqrt{2} + 2 + 5\log(1 + \sqrt{2})] = 0.521$. The distribution of random distances within a rectangle and between two rectangles has been investigated by Ghosh [225]. Other results have been obtained by Fairthone [167].

3. Density for Pairs of Lines

Let $G_i(p_i, \phi_i)$, $i = 1, 2$, be two lines that intersect at the point $P(x, y)$ and make angles α_1, α_2 with the x axis. As a kind of "dual" problem of that considered in Section 1, we are going to express the product $dG_1 \wedge dG_2$ in terms of the coordinates x, y, α_1, α_2.

We have (Fig. 4.2) $\phi_i = \alpha_i - (\pi/2)$ and therefore $p_i = x\cos\phi_i + y\sin\phi_i = x\sin\alpha_i - y\cos\alpha_i$ $(i = 1, 2)$. Hence, $d\phi_i = d\alpha_i$ and $dp_i = \sin\alpha_i\, dx - \cos\alpha_i\, dy + (x\cos\alpha_i + y\sin\alpha_i)\, d\alpha_i$, and by exterior multiplication we have $dG_i = dp_i \wedge d\phi_i = \sin\alpha_i\, dx \wedge d\alpha_i - \cos\alpha_i\, dy \wedge d\alpha_i$. From this it follows that

$$dG_1 \wedge dG_2 = |\sin(\alpha_2 - \alpha_1)|\, dP \wedge d\alpha_1 \wedge d\alpha_2 \tag{4.19}$$

which is the formula "dual" to (4.2).

As a first application we are going to evaluate the integral of both sides of (4.19) over all pairs of lines that meet a bounded convex set K. The left-hand side, according to (3.12), is equal to L^2 where L is the perimeter of K. The

ISBN 0-201-13500-0

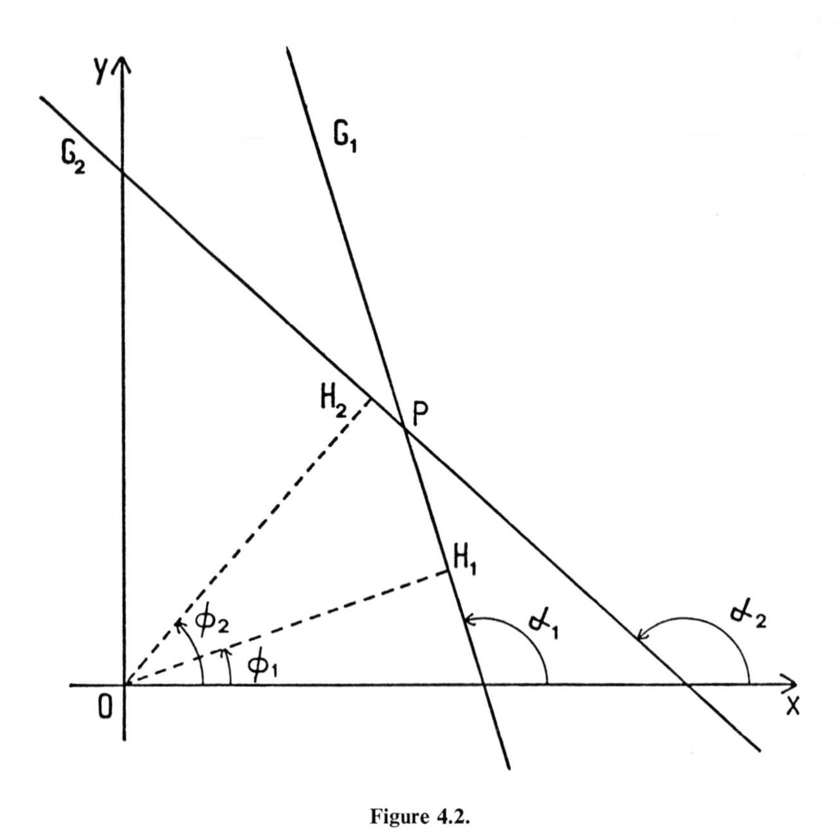

Figure 4.2.

right-hand side may be integrated first over the points $P \in K$ and then over the points outside K. For the points $P \in K$ we have

$$\int_{P \in K} dP \int_0^\pi \int_0^\pi |\sin(\alpha_2 - \alpha_1)| \, d\alpha_1 \wedge d\alpha_2 = 2\pi F \tag{4.20}$$

and for the points outside K, if α and β are the angles formed with the x axis by the support lines of K drawn from P, we have

$$\int_\alpha^\beta \int_\alpha^\beta |\sin(\alpha_2 - \alpha_1)| \, d\alpha_2 \wedge d\alpha_1$$

$$= \int_\alpha^\beta d\alpha_1 \left(\int_\alpha^{\alpha_1} \sin(\alpha_1 - \alpha_2) \, d\alpha_2 + \int_{\alpha_1}^\beta \sin(\alpha_2 - \alpha_1) \, d\alpha_2 \right)$$

ISBN 0-201-13500-0

$$= \int_{\alpha}^{\beta} [2 - \cos(\alpha_1 - \alpha) - \cos(\beta - \alpha_1)]\, d\alpha_1 = 2(\beta - \alpha) - 2\sin(\beta - \alpha). \tag{4.21}$$

If we call $\omega = \beta - \alpha$ the angle between the support lines of K through P, we have

$$\int_{P \notin K} dP \int_{\alpha}^{\beta}\int_{\alpha}^{\beta} |\sin(\alpha_2 - \alpha_1)|\, d\alpha_1 \wedge d\alpha_2 = 2\int_{P \notin K} (\omega - \sin\omega)\, dP \tag{4.22}$$

and since the sum of (4.20) and (4.22) must be equal to L^2, we get the following formula of Crofton

$$\int_{P \notin K} (\omega - \sin\omega)\, dP = \tfrac{1}{2}L^2 - \pi F, \tag{4.23}$$

which holds for any bounded convex set K (Fig. 4.3). See [133, 144, 335].

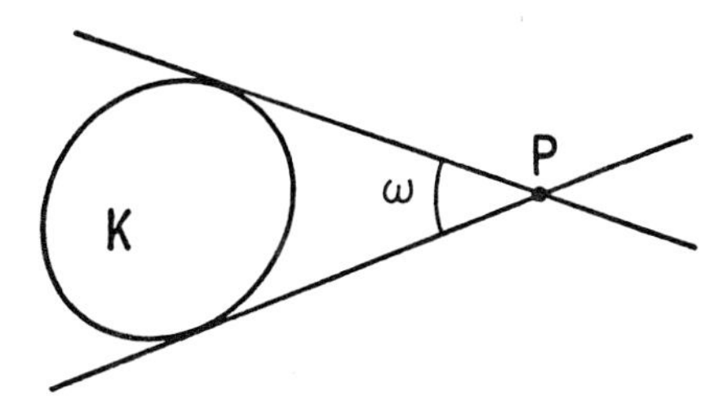

Figure 4.3.

Note that (4.9) and (4.23) result when we try to compute, respectively, the measure of the sets of pairs of points inside K and that of the sets of pairs of lines that meet K. Therefore, these formulas may be considered as corresponding formulas under a certain kind of "duality" in the plane. This fact appears more clearly when we treat the same problem on the sphere [539].

4. Division of the Plane by Random Lines

Let K be a bounded convex set of the plane with interior points. Let F be its area and L its perimeter. Assume n random lines $G_i(p_i, \phi_i)$ that intersect K (Fig. 4.4). These lines will divide K into a number c of regions and will determine a number v of interior vertices (intersection points of pairs of

ISBN 0-201-13500-0

lines that are interior to K) and a number e of edges. In Fig. 4.4, $n = 5$, $c = 11$, $v = 5$, $e = 25$. We want to find the mean values of c, v, e.

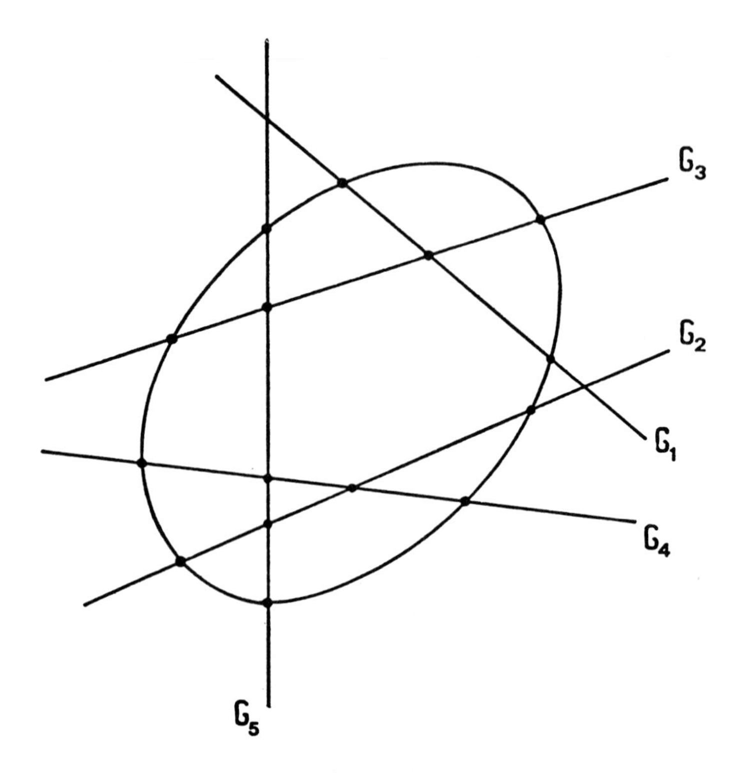

Figure 4.4.

Let us begin with the mean number of vertices v. By definition we have

$$E(v) = \frac{1}{L^n} \int_{G_i \cap K \neq \emptyset} v \, dG_1 \wedge dG_2 \wedge \cdots \wedge dG_n. \tag{4.24}$$

Let v_{hk} ($h \neq k$) denote the function of G_h, G_k that is equal to 1 if $G_h \cap G_k \in K$ and to 0 otherwise (set $v_{ii} = 0$ for completeness). We have $v = \sum v_{hk}$ where the number of summands is $n(n-1)/2$. Calling σ_h the length of the chord $G_h \cap K$, we have

$$\int v_{hk} \, dG_h \wedge dG_k = 2 \int \sigma_h \, dG_h = 2\pi F \tag{4.25}$$

where the integrals are extended over all lines G_h, G_k that meet K. Therefore we have

ISBN 0-201-13500-0

$$\int_{G_i \cap K \neq 0} v \, dG_1 \wedge dG_2 \wedge \cdots \wedge dG_n = \frac{n(n-1)}{2} \int_{G_i \cap K \neq 0} v_{hk} \, dG_1 \wedge \cdots \wedge dG_n$$

$$= n(n-1)\pi F L^{n-2}. \tag{4.26}$$

Thus

For n lines chosen at random so as to meet a bounded convex set K, the mean number of intersection points that are inside K is

$$E(v) = \frac{n(n-1)\pi F}{L^2}. \tag{4.27}$$

We can also find $E(v^2)$. Indeed, we have

$$L^n E(v^2) = \int_{G_i \cap K \neq 0} (v_{12} + v_{13} + \cdots + v_{n-1,n})^2 \, dG_1 \wedge dG_2 \wedge \cdots \wedge dG_n$$

$$= \int_{G_i \cap K \neq 0} \left(\sum_{i<j} v_{ij}^2 \right) dG_1 \wedge \cdots \wedge dG_n$$

$$+ 2 \int_{G_i \cap K \neq 0} \sum_{(i,j) \neq (k,h)} v_{ij} v_{kh} \, dG_1 \wedge \cdots \wedge dG_n$$

$$= 2\pi \binom{n}{2} F L^{n-2} + 2L^{n-4} \binom{n}{4} 3 \int_{\mathrm{I}} v_{ij} v_{kh} \, dG_i \wedge dG_j \wedge dG_k \wedge dG_h$$

$$+ 2L^{n-3} \binom{n}{3} 3 \int_{\mathrm{II}} v_{ij} v_{ih} \, dG_i \wedge dG_j \wedge dG_h \tag{4.28}$$

where in integral I we assume $i < j < k < h$ and the factor 3 arises from the possibilities $v_{ij}v_{kh}$, $v_{ik}v_{jh}$, $v_{ih}v_{jk}$, and in integral II we assume $i < j < h$ and the factor 3 arises from the possibilities $v_{ij}v_{ih}$, $v_{ij}v_{jh}$, $v_{ih}v_{jh}$. Therefore we have

$$L^n E(v^2) = 2\pi \binom{n}{2} F L^{n-2} + 24 L^{n-4} \binom{n}{4} \pi^2 F^2 + 24 L^{n-3} \binom{n}{3} \int_{G \cap K \neq 0} \sigma^2 \, dG. \tag{4.29}$$

Putting $I_2 = \int_{G \cap K \neq 0} \sigma^2 \, dG$, as in (4.8), we get the following result.

$$E(v^2) = 2\pi \binom{n}{2} \frac{F}{L^2} + 24\pi^2 \binom{n}{4} \frac{F^2}{L^4} + 24 \binom{n}{3} \frac{I_2}{L^3}. \tag{4.30}$$

ISBN 0-201-13500-0

For instance, if K is a circle of radius R, we have $I_2 = (16/3)\pi R^3$ (4.12) and we get

$$E(v^2) = \frac{1}{2}\binom{n}{2} + \frac{3}{2}\binom{n}{4} + 16\pi^{-2}\binom{n}{3}. \tag{4.31}$$

Consider the number c of regions into which the n random lines divide K. The chords $G_i \cap K$ and the boundary ∂K form a graph with c regions and $v + 2n$ vertices (v interior vertices and $2n$ vertices on the boundary). Through each interior vertex pass four edges and through each vertex of the boundary pass three edges. Thus, since each edge unites two vertices, the number of edges will be $e = \frac{1}{2}(4v + 6n) = 2v + 3n$ and, applying the Euler formula $v - e + c = 1$, we get

$$c = v + n + 1. \tag{4.32}$$

Note that we have not considered the cases in which more than two lines pass through a point or two of them intersect on the boundary, because these cases have measure zero and do not affect the result. Taking expected values, equality (4.32) yields

$$E(c) = n(n - 1)(\pi F/L^2) + n + 1. \tag{4.33}$$

That is,

For n lines chosen at random so as to meet a bounded convex set K, the mean value of the number of regions into which they divide K is given by (4.33).

We now consider the number of sides of each region. Let e_i be the number of sides of the region c_i ($i = 1, 2, \ldots, c$). The total number of edges is $e = 2v + 3n$ and since $2n$ of them belong to the boundary of K, we have that the number of interior edges is $2v + n$. Thus, since each interior edge is a side of two regions and each edge of the boundary is a side of only one region, we have

$$e^* = \sum_1^c e_i = 2(2v + n) + 2n = 4v + 4n \tag{4.34}$$

and thus

$$E(e^*) = E(\textstyle\sum e_i) = 4E(v) + 4n = \dfrac{4n(n - 1)\pi F}{L^2} + 4n. \tag{4.35}$$

Similarly, the sum of the perimeters of all regions is

$$s = \sum_1^c s_i = 2\sum_1^n \sigma_i + L \tag{4.36}$$

ISBN 0-201-13500-0

and then, since $E(\sigma) = \pi F/L$ (3.13), we have

$$E(s) = 2nE(\sigma) + L = 2n\pi F/L + L. \tag{4.37}$$

For any set of lines $G_1, G_2, \ldots, G_n$ that intersect K, the mean value of the number of sides of the regions into which K is partitioned is $N = e^*/c$. This quotient N is a random variable the mean value of which probably has a very involved expression. Instead of it we consider the following quotient of mean values.

$$E^*(N) = \frac{E(e^*)}{E(c)} = 4 - \frac{4L^2}{n(n-1)\pi F + (n+1)L^2}. \tag{4.38}$$

Similarly, instead of the mean value of the random variable $S = s/c$ (difficult to calculate), we write the following quotient of mean values

$$E^*(S) = \frac{E(s)}{E(c)} = \frac{2\pi nFL + L^3}{n(n-1)\pi F + (n+1)L^2} \tag{4.39}$$

and instead of the mean value of the average $A = F/c$ of the area of the regions, we write the following quotient of mean values

$$E^*(A) = \frac{F}{E(c)} = \frac{FL^2}{n(n-1)\pi F + (n+1)L^2}. \tag{4.40}$$

EXERCISE. Show that for $n = 2$ we have

$$E\left(\frac{F}{c}\right) = \frac{(2L^2 - \pi F)F}{6L^2},$$

so that $E^*(A)/E(A) = 6L^4/(6L^4 + \pi FL^2 - 2\pi^2F^2)$. For instance, if K is a circle, we have $E^*(A)/E(A) = 48/49$. Compute the analogous ratios for N and S.

Limit case: averages for polygons formed by random lines. Let $K(t)$ be a family of convex domains of area $F(t)$ and perimeter $L(t)$ depending upon the parameter t. Assume that for any point P of the plane there is a value t_P of t such that, for all $t > t_P$ we have $P \in K(t)$. That means that $K(t)$ expands over the whole plane as $t \to \infty$.

LEMMA. *Independently of the shape of $K(t)$ we have*

$$\lim_{t\to\infty} \frac{L(t)}{F(t)} = 0. \tag{4.41}$$

Proof. Let $C(t)$ be the greatest circle contained in $K(t)$ and let $R(t)$ be its radius. Let O be the center of $C(t)$ and let $h_t = h_t(\phi)$ denote the support

ISBN 0-201-13500-0

function of $K(t)$ with respect to O. We have $F(t) = \frac{1}{2}\int h_t(\phi)\, ds_t$ where ds_t denotes the arc element of ∂K at the contact point of the support line perpendicular to the direction ϕ and the integral is over ∂K. Since $h_t(\phi) \geqslant R(t)$, we get $F(t) \geqslant \frac{1}{2}R(t)L(t)$ and because $R(t) \to \infty$ as $t \to \infty$, we get (4.41).

Consider a line segment K_0 of length b contained in $K(t)$. If n random lines are assumed to intersect $K(t)$, the probability that exactly m of them meet K is (binomial distribution)

$$p_m = \binom{n}{m}\left(\frac{2b}{L}\right)^m\left(1 - \frac{2b}{L}\right)^{n-m} \tag{4.42}$$

where we have applied that the probability that a random line intersecting $K(t)$ also intersects K_0 is $2b/L$ (Chapter 3, Section 2). Assume that $K(t)$ expands to the whole plane and that $n = n(t) \to \infty$ in such a way that

$$\lim_{t\to\infty} \frac{n(t)}{L(t)} = \frac{\lambda}{2}, \qquad \lambda = \text{positive constant.} \tag{4.43}$$

We say that there is generated in the plane a *Poisson line system* or that we have a homogeneous planar Poisson line process [414]. The limit of p_m is then

$$p_m{}^* = \lim_{t\to\infty} p_m = \frac{(b\lambda)^m}{m!}\, e^{-b\lambda} \tag{4.44}$$

and the mean value of m is $E(m) = \sum_0^\infty m p_m{}^* = b\lambda$; that is, the constant λ is equal to the mean value of the number of lines crossing any line segment of unit length and λ^{-1} is the mean free length of a line.

From (4.27) and (4.30) we have

$$\lim_{t\to\infty} E(v/F) = (\pi\lambda^2)/4, \qquad \lim_{t\to\infty} E(v^2/F^2) = (\pi^2\lambda^4)/16 \tag{4.45}$$

and hence the variance $\text{var}(v/F) = E(v^2/F^2) - (E(v/F))^2 \to 0$.

According to probability theory, from this result we deduce that, with probability 1,

$$\lim_{t\to\infty} \frac{v}{F} = \frac{\pi\lambda^2}{4}. \tag{4.46}$$

This is the mean number of vertices per unit area.

According to (4.43) and (4.41) we have $n/F = (n/L)(L/F) \to 0$ as $t \to \infty$ and using (4.32) we get that the mean area of the regions into which the plane is partitioned by the random lines is

$$E(A) = \lim_{t\to\infty} \frac{F}{c} = \lim_{t\to\infty}(v/F + n/F + 1/F)^{-1} = \frac{4}{\pi\lambda^2}. \tag{4.47}$$

ISBN 0-201-13500-0

For the mean number of sides of each region, from (4.34) and (4.32) we have

$$E(N) = \lim_{t\to\infty} \frac{e^*}{c} = \lim_{t\to\infty} \frac{4v + 4n}{v + n + 1} = 4 \tag{4.48}$$

and for the mean perimeter, using (4.36),

$$E(S) = \lim_{t\to\infty} \frac{s}{c} = \lim_{t\to\infty} \frac{2(\sum \sigma_i/n) + L/n}{v/n + 1 + 1/n} = \frac{4}{\lambda} \tag{4.49}$$

where we have used that $(\sum \sigma_i/n) \to E(\sigma) = \pi F/L$ and $v/n = (v/F)(F/L)(L/n)$. All these equalities hold with probability 1 (almost surely). Hence we can state

If a plane is partitioned into an infinite number of convex polygonal regions by random lines of a homogeneous Poisson line process of parameter λ, then the mean values of the area A, perimeter S, and number of sides N of each region are, respectively,

$$E(A) = 4/\pi\lambda^2, \qquad E(S) = 4/\lambda, \qquad E(N) = 4. \tag{4.50}$$

Note that these mean values coincide with the limit of quotients (4.38)–(4.40) as $t \to \infty$. The parameter λ is the mean number of lines intersecting an arbitrary segment of unit length.

Richards [510] and Miles [407, 418] by probabilistic methods have obtained the mean values of several other measurements of these random polygons. We state without proof the following results of Miles:

$$E(N^2) = (\pi^2 + 24)/2, \qquad E(SN) = (\pi^2 + 8)/\lambda,$$

$$E(AN) = 2\pi/\lambda^2, \qquad E(S^2) = [2(\pi^2 + 4)]/\lambda^2,$$

$$E(AS) = 4\pi/\lambda^3, \qquad E(A^2) = 8/\lambda^4,$$

$$E(NA^2) = \frac{(8\pi^2 - 21)16}{21\lambda^4}, \qquad E(SA^2) = \frac{256\pi^2}{21\lambda^5}, \qquad E(A^3) = \frac{256\pi}{7\lambda^6}.$$

Among the results of Richards [510] are

$$E\left(\int f(r)\, dP_1 \wedge dP_2\right) = (8/\lambda^2) \int_0^\infty r f(r) e^{-\lambda r}\, dr,$$

$$E\left(\int f(p)\, dP_1 \wedge dP_2 \wedge dP_3\right) = (8/21\lambda^2) \int_0^\infty r^3 f(r) \exp(-\lambda r/2)\, dr,$$

where p denotes the perimeter of the triangle $P_1P_2P_3$ and the integral on the

ISBN 0-201-13500-0

left-hand side is extended over a typical polygon (the choice of $f = 1$ yields $E(A^3)$), and

$$E\left(\int f(r)\, ds_1 \wedge ds_2\right) = (2/\lambda)\int_0^\infty (4 + \pi^2\lambda r) f(r) e^{-r}\, dr$$

where ds_1, ds_2 are the arc elements on the boundary of a typical region and r is the distance between them.

The rigorous proofs of these results are not easy; they can be found in a series of papers by Miles [414, 415, 418] and Matheron [399]. The origin of the problem is an earlier result of Goudsmit [234]. The extension to the hyperbolic plane has been carried out by Santaló and Yañez [592]. For the division of E_3 by random planes see Chapter 16, Section 4, Note 8.

The Poisson line process is only one example, perhaps the most interesting due to its simplicity, of different line processes that can be defined in the plane. The study of these line processes and the random divisions of the plane into regions originated by them seems to be a very promising field. We refer the reader to the papers of Davidson [140] and Krickeberg [345, 346] and to *Stochastic Geometry*, edited by E. F. Harding and D. G. Kendall [294] (cf. Note 8, Section 5).

5. Notes

1. *Integral formulas for convex sets.* Crofton's formulas (4.9) and (4.23) are typical examples of a number of integral formulas referring to convex sets in the plane that are obtained by calculating in different ways certain sets of points and lines related to the convex set. We want to give some other examples.

(a) Let K be a bounded convex set with boundary ∂K of class C^1. Let P be a point not in K and call t_1 and t_2 the lengths of the tangent lines through P, from P to the contact point, and ω the angle between these tangents. Then if f is an integrable function over ∂K and f_1, f_2 are its values at the contact points of the tangent lines through P, we have

$$\int_{P \notin K} \left(\frac{f_1}{t_1} + \frac{f_2}{t_2}\right) \sin^2\left(\frac{\omega}{2}\right) dP = \frac{1}{2} L \int_0^{2\pi} f\, d\phi \tag{4.51}$$

where ϕ is the direction of the tangent line. Applying (4.51) for particular values of f ($f =$ constant, $f = \rho^2$, $f = p, \ldots$), we obtain several integral formulas as particular cases (see the results of Lebesgue [357], Masotti-Biggiogero [396], and Stoka [647]).

If A_1, A_2 are the contact points of the tangent lines through P and D denotes the diameter of the circle determined by A_1, A_2, P, Masotti-Biggiogero [392] proved that

ISBN 0-201-13500-0

$$\int \frac{\rho_1 \rho_2}{D^3}\, dP = L, \qquad \int \rho_1 \rho_2 \sin^3 \omega\, dP = 3F^2,$$

$$\int \frac{\rho_1 \rho_2}{\sigma^2} \sin^3 \omega\, dP = \pi F, \qquad \int \frac{\rho_1 + \rho_2}{D^3}\, dP = 4\pi$$

where ω is the angle A_1PA_2, σ is the chord A_1A_2, and ρ_i are the radii of curvature at A_i $(i = 1, 2)$. The integrals are extended over the whole of the plane outside K.

Integral formulas of an analogous kind may be obtained using the Fourier series expansions of the functions $p(\phi)$ and $\rho(\phi)$, as was done by Hurwitz [319]. For instance, if τ denotes the angle of the tangent to ∂K with a fixed direction $(0 \leqslant \tau \leqslant 2\pi)$, putting

$$a_2 = \frac{1}{\pi} \int_0^{2\pi} \rho \cos 2\tau\, d\tau, \qquad a_2' = \frac{1}{\pi} \int_0^{2\pi} \rho \sin 2\tau\, d\tau,$$

we have

$$\int_{P \notin K} \sin^3 \omega\, dP = (3/4)L^2 + (\pi^2/4)(a_2^2 + a_2'^2).$$

For any convex set K we can prove that

$$(16 - \pi^2)F \leqslant \int_{P \notin K} (\omega^2 - \sin^2 \omega)\, dP \leqslant (4/\pi)L^2 - \pi^2 F$$

where the equality sign holds on the left-hand side for circles and on the right-hand side for sets of constant breadth [270, 396].

Let $x = x(s)$, $y = y(s)$ be the parametric equations of a closed convex curve ∂K of length 2π and let a_n, b_n and c_n, d_n be the coefficients of the Fourier expansions for $x(s)$ and $y(s)$, respectively. Then the functionals $\Phi_n(\partial K) = a_n^2 + b_n^2 + c_n^2 + d_n^2$ $(n \geqslant 1)$ satisfy noteworthy inequalities, for instance, $\Phi_1(\partial K) \geqslant (336/27\pi) - 3/2 = 1.282\ldots$ and $\Phi_n(\partial K) \leqslant 8/\pi n^4$ $(n \geqslant 2)$ (see [145a]).

(b) Let G_1, G_2, G_3 be three lines that form the triangle of vertices P_1, P_2, P_3. Let D be the diameter of the circle determined by these points. Then (according to Blaschke [51])

$$dP_1 \wedge dP_2 \wedge dP_3 = D^3\, dG_1 \wedge dG_2 \wedge dG_3. \tag{4.52}$$

More generally, consider the ordered set of n points P_i $(i = 1, 2, \ldots, n)$ and denote by G_i the line P_iP_{i+1} $(i = 1, 2, \ldots, n-1)$, $G_n = P_nP_1$, by L_i the length of P_iP_{i+1}, and by α_i the angle of G_{i-1} and G_i. Then we can prove that [418]

$$\bigwedge L_i\, dp_i \wedge d\phi_i = \bigwedge |\sin \alpha_i|\, dP_i. \tag{4.53}$$

ISBN 0-201-13500-0

(c) Let D be a plane domain with smooth boundary ∂D. Let ds_1 and ds_2 denote the arc elements at points P_1 and P_2 of ∂D and call σ the distance between P_1 and P_2. Then the area of D is given by

$$F^2 = -\tfrac{1}{4} \int_{\partial D \times \partial D} \sigma^2 \cos \psi \, ds_1 \wedge ds_2 \tag{4.54}$$

where ψ is the angle between the outward normals at P_1 and P_2. This formula is the work of Redei and Nagy [497] and has been extended to domains of n-dimensional euclidean space by Hadwiger [267] and Ruben [518].

2. *Functions and their integrals over straight lines.* Note that if $f(x, y) = 1/(1 - x^2 - y^2)^{1/2}$ for $x^2 + y^2 < 1$, then $\int f(x, y)\, ds = \pi$ when the integral is extended over any chord of the unit circle. Rényi raised the question: does there exist a convex curve ∂K other than the circle and an integrable function f defined on the interior of ∂K and not identically zero such that $\int f\, ds =$ constant over all chords of K?

The solution was given by J. W. Green [235], who proved the following theorem.

Let ∂K be a convex curve and f a summable function over its interior, with $\int f\, ds = k$ over all chords of K. Then either $f = 0$ almost everywhere or else K is a circle and $f = k\pi^{-1}(1 - x^2 - y^2)^{-1/2}$.

If $k = 0$, we get the theorem: *If $f(x, y)$ is a continuous and summable function over the plane and the line integral $\int f\, ds$ is zero when extended over any infinite straight line, then $f \equiv 0$.*

The generalization to higher dimensions has been considered by Darling [139].

3. *Density for points and lines that are incident.* Let $P(x, y)$ be a point of a line $G(p, \phi)$. If t denotes the length of the segment HP from the foot of the perpendicular to G from the origin to P and θ is the angle of G with the x axis, we have $t = x \cos \theta + y \sin \theta$. Thus $dt = \cos \theta \, dx + \sin \theta \, dy + (-x \sin \theta + y \cos \theta)\, d\theta$. Exterior multiplication with expression (3.11) for dG gives, up to the sign, $dG \wedge dt = dP \wedge d\theta$, where $dP = dx \wedge dy$. This formula assumes $0 \leqslant \theta \leqslant \pi$. If G is an oriented line, we can set

$$dG^* \wedge dt = dP \wedge d\theta \tag{4.55}$$

and then θ varies in the interval $0 \leqslant \theta \leqslant 2\pi$. Equation (4.55) is the density for sets of pairs (P, G^*) of points P and oriented lines G^* such that P belongs to G^*.

4. *On the mean length of the chords of a convex set.* There are different ways of defining the mean value of the length of random chords of a bounded convex set K. They depend on what is meant by a chord given at random. We give some examples.

(i) If we assume that the line G that contains the chord is chosen at random in the sense of integral geometry, we know that the mean length of the chord has the value $E(\sigma) = \pi F/L$ (3.13).

ISBN 0-201-13500-0

(ii) Suppose that ∂K is a perfectly reflecting wall and that a particle is projected from a point P of K into the direction θ (Fig. 4.5). When the particle arrives at ∂K, say at point A_1, it is reflected and describes the chord A_1A_2; then it is reflected again and successively describes the polygonal line $A_1A_2\ldots$. Our aim is to find the mean value of the length of the chords A_1A_2, $A_2A_3,\ldots$ when P is chosen at random in K and θ is chosen at random

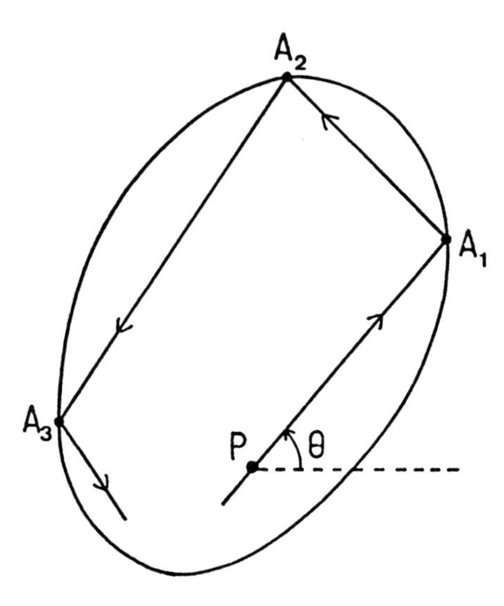

Figure 4.5.

between 0 and 2π. Calling σ the length of the chord that contains PA_1, we define this mean value by

$$E_1(\sigma) = \frac{\int \sigma\, dP \wedge d\theta}{\int dP \wedge d\theta} = \frac{1}{2\pi F}\int \sigma\, dP \wedge d\theta \tag{4.56}$$

where the integrals are extended over all $P \in K$ and to the range $0 \leqslant \theta \leqslant 2\pi$. According to (4.55) we have $\int \sigma\, dP \wedge d\theta = \int \sigma\, dG^* \wedge dt = 2\int \sigma^2\, dG = 2I_2$ where I_2 is defined by (4.8). Consequently we have

$$E_1(\sigma) = I_2/\pi F. \tag{4.57}$$

The invariant I_2 depends on the shape of K and generally lacks a simple expression. In order to compare $E(\sigma)$ with $E_1(\sigma)$ we set $E(\sigma - E(\sigma))^2 = E(\sigma^2) - (E(\sigma))^2$ and therefore, putting $\operatorname{var}(\sigma) = E(\sigma - E(\sigma))^2$, we can write $E(\sigma^2) = (E(\sigma))^2 + \operatorname{var}(\sigma)$ and since $I_2 = E(\sigma^2)L$, we have

$$E_1(\sigma) = E(\sigma) + \frac{L}{\pi F}\operatorname{var}(\sigma) = \frac{\pi F}{L} + \frac{L}{\pi F}\operatorname{var}(\sigma).$$

As a consequence we have $E(\sigma) \leqslant E_1(\sigma)$.

ISBN 0-201-13500-0

For instance, for a circle of radius R we have

$$E(\sigma) = (\pi/2)R, \qquad E_1(\sigma) = (16/3\pi)R. \tag{4.58}$$

For the square of unit side we have

$$E(\sigma) = \pi/4 = 0.785\ldots, \qquad E_1(\sigma) = (4/3\pi)(3\log(1+\sqrt{2}) - \sqrt{2} + 1) = 0.546. \tag{4.59}$$

This problem has applications in architectural acoustics. See [22, 335, 336].

(iii) Another possibility of defining the mean free path in convex domains is the following. Assume a point P chosen at random on the boundary of K and a direction PP_1 chosen at random about P (Fig. 4.6). Let σ be the length of the chord PP_1. The mean value of σ is then

$$E_2(\sigma) = \frac{1}{\pi L}\int_0^L\int_0^\pi \sigma \, ds \wedge d\theta \tag{4.60}$$

where s is the arc of ∂K at P.

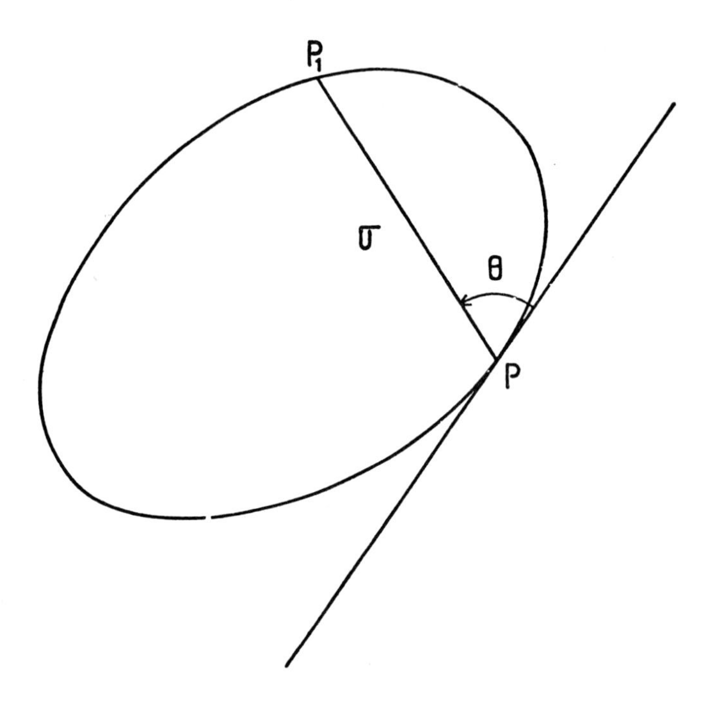

Figure 4.6.

For example, for a circle of radius R the result (given by Horowitz [315]) is $E_2(\sigma) = 4R/\pi$ and for a rectangle of sides a, b

ISBN 0-201-13500-0

$$E_2(\sigma) = \frac{a}{a+b} f(a, b) + \frac{b}{a+b} f(b, a)$$

where

$$f(a, b) = \frac{a}{\pi} \log\left(\frac{a^2 + b^2}{a} + \frac{b}{a}\right) + \frac{2b}{\pi} \log\left[\frac{a + (a^2 + b^2)^{1/2}}{b}\right]$$
$$- \frac{b^2}{\pi a}\left[\frac{(a^2 + b^2)^{1/2}}{b} - 1\right].$$

(iv) Let P, Q be two points chosen at random on ∂K. The mean value of the distances $\sigma = |PQ|$ is

$$E_3(\sigma) = L^{-2} \int_{\partial K \times \partial K} \sigma \, ds_1 \wedge ds_2 \tag{4.61}$$

where ds_1 and ds_2 are the arc elements of ∂K at P and Q. For a given length L, the mean value $E_3(\sigma)$ attains its maximum only for circles. More generally, if the function $g(\sigma)$ is increasing and concave, the mean value

$$E_3(g) = L^{-2} \int_{\partial K \times \partial K} g(\sigma) \, ds_1 \wedge ds_2$$

attains its maximum, for L fixed, only for circles [375]. In particular, choosing $g(\sigma) = \sigma$, we find that the mean length of chords of a closed curve of length L does not exceed $2L/\pi^2$.

(v) Finally, if the random chord is defined by two points P_1, P_2 chosen independently at random inside K, we have

$$E_4(\sigma) = F^{-2} \int_{K \times K} \sigma \, dP_1 \wedge dP_2. \tag{4.62}$$

Using (4.2), (4.4), and (4.11), we have $E_4(\sigma) = 2E(r)$, where $E(r)$ denotes the mean distance between two points of K.

Some details about the preceding results and their generalization to 3-space have been given by Coleman ([119]; see also the articles by Geciauskas [217, 218]).

5. *Sylvester's problem.* A classical problem of geometrical probability, interesting in integral geometry, is that of Sylvester (see [144, 335]):

Find the probability that four points P_0, P_1, P_2, P_3 chosen at random inside a convex set K form a convex quadrilateral; that is, that none of the points is inside the triangle formed by the other three.

Consider the complementary probability that the quadrilateral is not convex. This can occur in four different ways, according to which of the

ISBN 0-201-13500-0

points P_i occurs inside the triangle formed by the other three. The measure of the set of quadrilaterals such that P_3 lies inside the triangle $P_0P_1P_2$ is

$$T_2 = \int_{P_i \in K} T_{012}\, dP_0 \wedge dP_1 \wedge dP_2$$

where T_{012} denotes the area of the triangle with vertices P_0, P_1, P_2. The other three cases have the same measure, so that the total measure of cases in which the quadrilateral is not convex is $4T_2$. The corresponding probability will be $4T_2/F^4$ (where F is the area of K) and the probability of a convex quadrilateral is

$$p = 1 - (4T_2/F^4).$$

Since T_2 contains only areas and area elements, and thus is invariant under unimodular affinities, in order to compute the probability p it suffices to take a simpler figure equivalent to K under affinities. For instance, p has the same value for all ellipses, the same value for all triangles, and the same value for all parallelograms. A direct computation gives these results:

$$\text{ellipse,} \quad T_2 = 35F^4(48\pi^2)^{-1}; \qquad \text{triangle,} \quad T_2 = F^4/12;$$

$$\text{parallelogram,} \quad T_2 = 11F^4/144;$$

$$\text{regular } n\text{-sided polygon,} \quad T_2 = \frac{9\cos^2\alpha + 52\cos\alpha + 44}{36n^2\sin^2\alpha} F^4;$$

where $\alpha = 2\pi/n$. The last result is due to Alikoski [3]. From these results we can write the corresponding probabilities.

Blaschke [33] has shown the inequalities

$$\frac{35}{48\pi^2} \leqslant \frac{T_2}{F^4} \leqslant \frac{1}{12}, \tag{4.63}$$

which say that the probability of Sylvester's problem has the greatest value for ellipses and the smallest value for triangles.

In n-dimensional euclidean space Sylvester's problem leads to the integral $T_n = \int T_{012\ldots n}\, dP_0 \wedge dP_1 \wedge \cdots \wedge dP_n$ where $T_{01\ldots n}$ means the volume of the simplex with vertices $P_0, P_1, \ldots, P_n$ and the integral is extended over all $(n+1)$-tuples of points of a convex body K. For the n-dimensional unit ball, direct computation gives, up to the factor F^{n+2},

$$T_n\,(\text{unit ball}) = \binom{n+1}{\frac{1}{2}(n+1)}^{n+1} \binom{(n+1)^2}{\frac{1}{2}(n+1)^2}^{-1} 2^{-n} \tag{4.64}$$

where if $n+1$ is odd we must put

$$\binom{n+1}{\frac{1}{2}(n+1)} = \frac{2^{2n+4}}{(n+2)\pi} \binom{n+2}{\frac{1}{2}(n+2)}^{-1}.$$

ISBN 0-201-13500-0

For example

$$T_1 = \frac{1}{3}, \qquad T_2 = \frac{35}{48\pi^2}, \qquad T_3 = \frac{9}{715}.$$

The value above for T_n(unit ball) is due to J. F. C. Kingman [337]. The first inequality of Blaschke generalizes to n-dimensional euclidean space; that is, the ellipsoid has the greatest value of the probability of Sylvester's problem. The result is due to Groemer [244, 245]. For historical remarks concerning this problem see [338]; see also [498]. Miles [420] has generalized Sylvester's problem in the sense of finding the probabilities $p(n, m)$ that the polygonal convex hull of $n + 3$ random points of the unit n-ball possesses m vertices. Values of special interest are

$$p(2, 3) = 15/16\pi^2 = 0.0949\ldots, \qquad p(2, 4) = 65/12\pi^2 = 0.5488\ldots,$$

$$p(2, 5) = 1 - 305/48\pi^2 = 0.35619\ldots, \quad p(3, 4) = 9/200\pi^2 = 0.004559\ldots,$$

$$p(3, 5) = \left(\frac{27}{143}\right) - \left(\frac{9}{100\pi^2}\right) = 0.17969\ldots,$$

$$p(3, 6) = \left(\frac{116}{143}\right) + \left(\frac{9}{200\pi^2}\right) = 0.81575\ldots.$$

Other particular cases have been computed by Miles [420] but the general value of $p(n, m)$ is not known.

6. *Lines that intersect a convex set.* Let G_i, $i = 1, 2, \ldots, n$, be n random lines that intersect a convex set K. We know that the mean number of intersection points of the lines G_i that are interior to K is $E(v) = n(n - 1)\pi F/L^2$ (4.27). More difficult is the problem of finding the probability p_h, for $h = 0, 1, 2, \ldots, n(n - 1)/2$, that exactly h intersection points lie inside K. The expression of these probabilities requires new invariants of K besides the area F, the perimeter L, and the invariants I_n of Section 2. For instance, for $n = 3$ there appears the integral

$$U = \int_{G_1 \cap G_2 \in K} u(G_1, G_2)\, dG_1 \wedge dG_2$$

where $u(G_1, G_2)$ denotes the length of the convex quadrilateral whose vertices are the intersections of G_1 and G_2 with ∂K. Applying the results of Chapter 3, Section 3, we can prove that for $n = 3$

$$p_0 = \frac{1}{L^3}(U + I_2 + L^3 - 6\pi FL), \qquad p_1 = \frac{3}{L^3}(2\pi FL - U),$$

$$p_2 = \frac{3}{L^3}(U - I_2), \qquad p_3 = \frac{1}{L^3}(2I_2 - U).$$

ISBN 0-201-13500-0

If ∂K is of class C^2, the probability $p_{n(n-1)/2}$ that all the intersection points lie inside K satisfies the inequality

$$p_{n(n-1)/2} < \frac{n!}{(2n)!}\left(\frac{bL}{2}\right)^n$$

where b is the maximal value of the curvature of ∂K. These results are due to Sulanke [660].

7. *Random polygons in a ring domain.* Let K_0 be a bounded convex domain of the plane containing in its interior another convex domain K_1. Let G_i $(i = 1, 2, \ldots, n)$ be n independent lines, drawn at random, meeting K_0 but not K_1. Let H_i be the closed half plane that is bounded by G_i and contains K_1. Let K_n be the intersection of all the H_i. Then Rényi and Sulanke [505] have proved:

(i) The probability that K_n is an open set is of the order of γ^n as $n \to \infty$, where γ is a constant, $0 < \gamma < 1$, that depends on K_0 and K_1. We will write $p(K_n \text{ open}) = O(\gamma^n)$. In particular, if K_0 and K_1 are parallel convex sets, then $p(K_n \text{ open}) = n/2^{n-1}$.

(ii) If K_1 reduces to a point, the expected value of the number of vertices V_n of K_n is $E(V_n) = \pi^2/2 + O(\gamma^n)$ $(0 < \gamma < 1)$. Thus, $E(V_n) \to \pi^2/2$ as $n \to \infty$.

(iii) If ∂K_1 has bounded continuous curvature $\kappa > 0$, then

$$E(V_n) = \left[\left(\frac{2}{3(L_0 - L_1)}\right)^{1/3} \Gamma\left(\frac{5}{3}\right)\int_0^{L_1} \kappa^{2/3}\, ds\right] n^{1/3} + O(1)$$

where L_0 and L_1 are the perimeters of K_0 and K_1, respectively.

(iv) If K_1 is a convex r-gon, then $E(V_n) = \frac{2}{3} r \log n + O(1)$.

For partial generalization to more dimensions see the work of W. M. Schmidt [598] and Sulanke and Wintgen [665]. The same problem with a general distribution function for lines on the plane has been treated by Ziezold [736], who by "duality" relates this problem to that of the expected number of vertices of the convex hull of n random points in a convex domain (see Chapter 2, Section 5, Note 3).

8. *Stochastic processes of lines.* An oriented line G in the plane E_2 can be regarded as a point (p, ϕ) in the cylinder $C(p, \phi; -\infty < p < \infty, 0 < \phi < 2\pi)$. Let $\mathfrak{M}$ be the group of motions on C induced by the group of motions (translations and rotations) on E_2. Let B_q be the ball $B_q = \{(p, \phi); |p| < q\}$. A line process in E_2 can be considered as a point process on C, that is, a nonnegative integer-valued random Borel measure Z on C such that (a) for all $q \geqslant 0$, $Z(B_q)$ is almost surely finite; (b) Z has almost surely no atoms of mass greater than 1. Davidson [140] considers line processes, denoted by $LP4$, satisfying the further conditions: (c) for all $q \geqslant 0$, $E(Z^2(B_q)) < \infty$ and for all $T \in \mathfrak{M}$, $E(Z(A)) = E(Z(TA))$ (first-order stationarity), $E(Z(TA_1)Z(TA_2)) = E(Z(A_1)Z(A_2))$ (second-order stationarity), provided there is a q such that $A_1 \cup A_2 \subset B_q$; (*d*) Z has almost surely no parallel (or antiparallel) lines. Those members of $LP4$ that also satisfy the condition that (e) the finite-dimensional distributions of Z are stationary under $\mathfrak{M}$, that is,

ISBN 0-201-13500-0

$$\text{Prob}\left(\bigcap_{1}^{k}\left(Z(A_i) = n_i\right)\right) = \text{Prob}\left(\bigcap_{1}^{k}\left(Z(TA_i) = n_i\right)\right)$$

for all $k, n_1, n_2, \ldots, n_k, A_1, \ldots, A_k$, and $T \in \mathfrak{M}$, form the class $LP5$.

Davidson [140] and Krickeberg [345] have shown that every $Z \in LP4$ is second-order stationary under reflections of E_2 (equivalently, under translations of C). For Z satisfying (a), (b), and (c), define the intensity λ of Z as the mean number of points of Z in any set of measure unity. Davidson shows that given $Z \in LP4$, a second process $Z^* \in LP4$ can be constructed such that Z^* is doubly stochastic Poisson (dsP) with the same intensity and covariant measure as Z. Although Davidson obtained an element of $LP4$ that is not dsP, all processes constructed by him to satisfy the restrictions on $LP5$ are dsP and he conjectures that dsP does exhaust $LP5$. If this is so, we can use Miles' results [407, 410, 414] on distributions associated with the Poisson process to get expressions for the same distributions associated with any $Z \in LP5$.

Davidson [140] has discussed the applications of line processes to the structure of paper and to road networks. Paper may be regarded as a (closely packed) random process of long fibers. The strength of paper surely lies in the matting density of its fibers; it is thus of interest to find what line processes have the greatest specific density of intersections per unit area. This specific density is defined as the ratio of the mean number of intersections per unit area to the mean number of pairs of distinct lines per unit area and is invariant under changes in the thickness of the paper. Davidson shows that the strongest paper will be produced by the so-called mixed Poisson process, which is just the ordinary homogeneous Poisson process with a random change of global density. Other results on random fibrous networks have been obtained by Dodson [147, 148] and Corte [124].

With reference to road networks, the problem is to put a network of (conceptually) infinite straight roads in the plane so as to make the average road journey as short as possible. Let A be a random point on the network and consider all the points B on the network at a distance d from A. Let $r(d, A)$ denote the mean distance of the points B from A via the roads of the network. The *coefficient of inefficiency* $m(d)$ of the network is defined as the mean value, over A, of $r(d, A)$ divided by d. For the rectangular grid, it can be shown that $m(d) \geqslant (2 + \pi)/4$ and approaches this limit as $d \to \infty$. For a Poisson line process, $m(d)$ is finite for all d, and $m(d) \to 1$ as $d \to \infty$. In this sense, a Poisson process is the most efficient for long journeys.

The expected number of intersections per unit area between two sets of random parallel lines in the plane has applications in problems of road traffic theory [213]. Other results on points and lines processes are to be found in the work of Papangelou [469] and Daley [136]. For the general theory see the book of G. Matheron [401a].

ISBN 0-201-13500-0

CHAPTER 5

Sets of Strips in the Plane

1. Density for Sets of Strips

By a *strip* of breadth a we mean the closed part of the plane consisting of all points that lie between two parallel lines at a distance a from each other.

We will represent the strips by the letter B. The position of a strip B can be determined by the position of its midparallel. Calling p, ϕ the coordinates of such a line, the density for sets of strips of fixed breadth will be (Fig. 5.1)

$$dB = dp \wedge d\phi. \tag{5.1}$$

Clearly this density is unique, up to a constant factor, if it is to be invariant under the group of motions of the plane.

Let K be a bounded convex set. If $B \cap K \neq \emptyset$, the midparallel of B intersects the parallel set $K_{a/2}$ of K in the distance $a/2$. Conversely, if the midparallel of B intersects $K_{a/2}$, then B intersects K. Therefore, using (1.18) and (3.12), we have that *the measure of the set of strips of breadth a that intersect a convex set K is*

$$m(B; B \cap K \neq \emptyset) = \int_{B \cap K \neq \emptyset} dB = L + \pi a \tag{5.2}$$

where L is the perimeter of K. In particular, we have

(i) The measure of all strips of breadth a that contain a fixed point P is

$$m(B; P \in B) = \pi a. \tag{5.3}$$

(ii) The measure of all strips of breadth a that meet a line segment S of length s is

ENCYCLOPEDIA OF MATHEMATICS and Its Applications, Gian-Carlo Rota (ed.).
1, Luis A. Santaló, Integral Geometry and Geometric Probability

ISBN 0-201-13500-0

$$m(B; B \cap S \neq \theta) = 2s + \pi a. \tag{5.4}$$

(iii) The measure of all strips of breadth a that meet a connected domain, not necessarily convex, is given by the same formula (5.2), where L denotes in this case the perimeter of the convex hull of the domain.

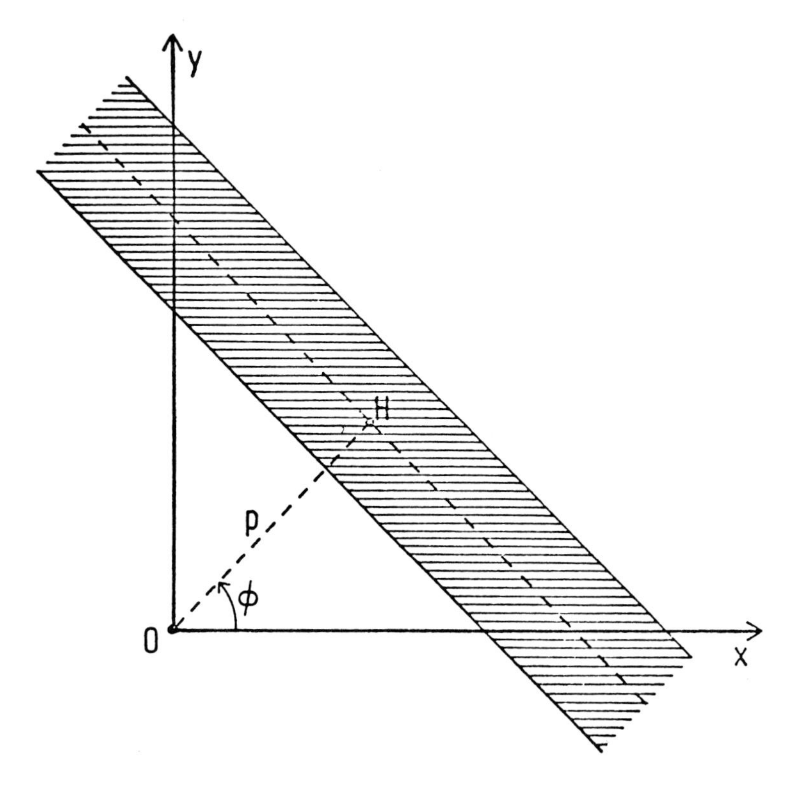

Figure 5.1.

More involved is the measure of the set of strips that contain a given set. However, the result is simple if the set is a convex set of diameter $D \leqslant a$. In this case the measure is the difference between measure (5.2) and the measure of all strips whose boundary meets K, which is equal to $2L$. Thus

$$m(B; K \subset B) = \pi a - L. \tag{5.5}$$

Note that this measure is not negative because $L \leqslant \pi D$ and $D \leqslant a$.

The preceding results may be stated in terms of geometrical probability as follows.

(a) Let K_1 be a convex set contained in a bounded convex set K. The probability that a random strip of breadth a that meets K intersects K_1 is

ISBN 0-201-13500-0

$$p = \frac{L_1 + \pi a}{L + \pi a} \tag{5.6}$$

where L_1 and L are the perimeters of K_1 and K, respectively.

If the diameter of K_1 is less than or equal to a, then the probability that the strip B contains K_1 is

$$p = \frac{\pi a - L_1}{\pi a + L}. \tag{5.7}$$

If K_1 reduces to a point, then (5.7) applies with $L_1 = 0$.

(b) Consider N convex sets K_i ($i = 1, 2, \ldots, N$) contained in a bounded convex set K (Fig. 5.2). Let L be the perimeter of K and L_i the perimeter of K_i.

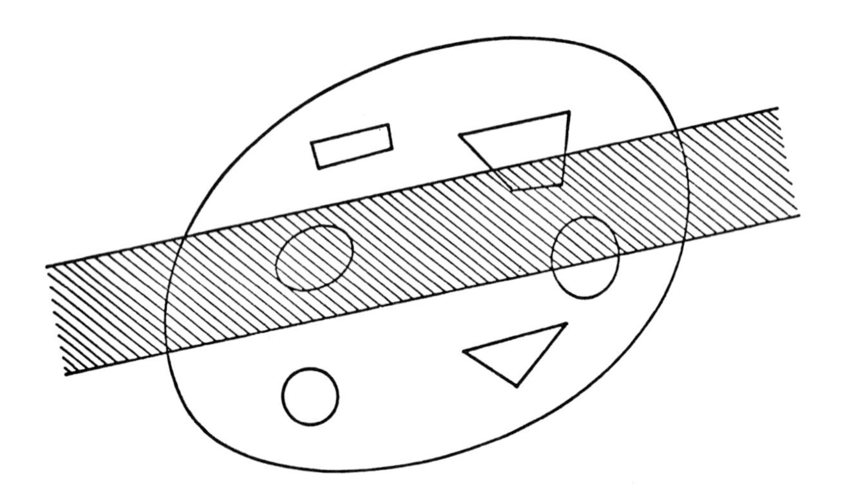

Figure 5.2.

If n denotes the number of sets K_i that are intersected by the strip B (in Fig. 5.2, $n = 3$), we have

$$\int_{B \cap K \neq \emptyset} n\, dB = \sum_1^N m(B; B \cap K_i \neq \emptyset) = \sum_1^N L_i + \pi N a. \tag{5.8}$$

If the diameters of all K_i are equal to or less than a, and n_i denotes the number of sets K_i that are covered by the strip B, from (5.5) we deduce

$$\int_{B \cap K \neq \emptyset} n_i\, dB = \pi N a - \sum_1^N L_i. \tag{5.9}$$

ISBN 0-201-13500-0

From (5.5), (5.8), and (5.9) we have

Let K_i $(i = 1, 2, \ldots, N)$ be N convex sets that are contained in a bounded convex set K. The mean number of sets K_i that are met by a strip of breadth a that intersects K at random is

$$E(n) = \frac{\sum_1^N L_i + \pi N a}{L + \pi a}. \tag{5.10}$$

If the diameters of the sets K_i are all less than or equal to a, then the mean number of sets K_i that are covered by the strip is

$$E(n_i) = \frac{\pi N a - \sum_1^N L_i}{\pi a + L}. \tag{5.11}$$

2. Buffon's Needle Problem

Let us go back to the problem (a) of the preceding section. Assume that K is a convex set of constant breadth D and K_1 is any convex set contained in K. Recall that any convex set of diameter $D_1 \leqslant D$ can be contained in K (Chapter 1, Section 4). The probability that a strip B of breadth a that intersects K will meet K_1 is given by (5.6). Instead of assuming that K is fixed and the strip

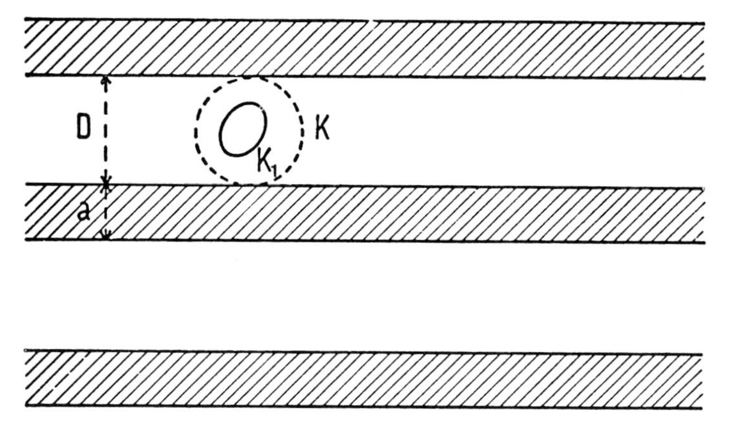

Figure 5.3.

B is given at random, we can consider that the set K, together with the set K_1, is placed at random on a plane on which are ruled parallel strips B at a distance D from each other (Fig. 5.3). Then it is sure that K will meet one and only one strip (except for tangent positions of measure zero) and the probability that K_1 will meet a strip will be given by (5.6); that is, using that $L = \pi D$, we have

ISBN 0-201-13500-0

$$p = \frac{L_1 + \pi a}{\pi(a + D)}. \tag{5.12}$$

Clearly, the existence of the set K is not necessary and we may state that if a convex set K_1 of diameter $D_1 \leqslant D$ and perimeter L_1 is placed at random on the plane, the probability that it intersects a strip is given by (5.12).

If $a = 0$ and K_1 reduces to a line segment of length l, we have $L_1 = 2l$ and (5.12) gives the solution of the classical *Buffon needle problem*:

The probability that a needle of length l placed at random on a plane on which are ruled parallel lines at a distance $D \geqslant l$ apart will intersect one of these lines has the value $p = 2l/\pi D$.

Note. The Buffon needle problem was stated and solved by Buffon in his *Essai d'Arithmétique Morale* (1777) and is considered the first problem in the theory of geometric probability. The formula $p = 2l/\pi D$ gives the possibility of estimating π by the random process of throwing a needle N times on the plane where the parallel lines are drawn. If in n of these throws the needle intersects one parallel line, then $p^* = n/N$ is an estimate of p and $\pi^* = 2l/p^*D$ is the corresponding estimate for π. From the theory of probabilities it is known that the standard error of p in N falls is $[p/(1 - p)/N]^{1/2}$. Since $\delta\pi^* = (2l/p^2D)\,\delta p$, it follows that by N large the standard error of π^* is $\pi[(\pi D - 2l)/2lN]^{1/2}$ and this formula indicates that the more precise estimate of π corresponds to the greatest possible value of l, that is, $l = D$.

The experiment has been carried out several times and the results are quoted in the literature (e.g., by Kendall and Moran [335]). However, all the published results are better than should be expected, as has been pointed out by Gridgeman [238]. This author shows that in order to get the value of π with d exact decimal figures, with a probability of 95%, it is necessary that $N \sim 90.10^{2d}$ (assuming $D = 1$), a value much greater than that announced by experimenters (see also the results of Mantel [390]). For applications of the needle problem and other integral geometric results to the construction of pattern recognition devices, see the work of Novikoff [457]. For applications to estimate the length of curves see [205]. The assumptions that underlie the treatment of Buffon's problem have been analyzed by Kac, Van Kampen, and Wintner [328]; see also [318] and [601].

3. Sets of Points, Lines, and Strips

Let D be a domain in the plane, not necessarily convex, of area F. The density for sets of pairs of points and strips (P, B), assuming the independence of P and B, is $dP \wedge dB$. The measure of the set of pairs (P, B) such that $P \in B \cap D$ will thus be

ISBN 0-201-13500-0

$$m(P, B; P \in B \cap D) = \int_{P \in B \cap D} dP \wedge dB. \tag{5.13}$$

To calculate this integral we fix P and apply (5.3). The result is

$$m(P, B; P \in B \cap D) = \pi a \int_{P \in D} dP = \pi a F \tag{5.14}$$

where a is the breadth of B. On the other hand, if we first fix B and call f

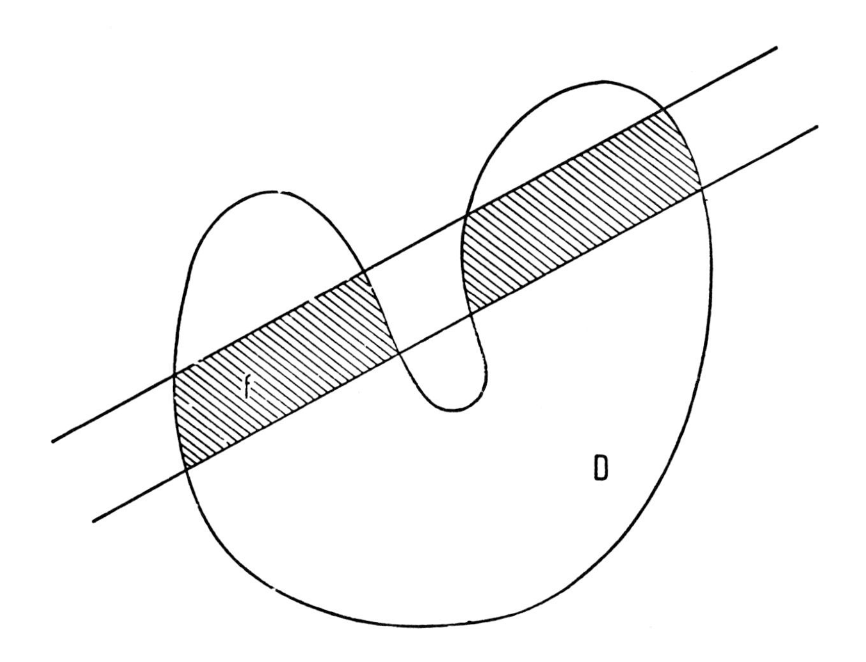

Figure 5.4.

the area of $B \cap D$ (Fig. 5.4), we get

$$m(P, B; P \in B \cap D) = \int_{B \cap D \neq 0} f\, dB \tag{5.15}$$

and thus

$$\int_{B \cap D \neq 0} f\, dB = \pi a F. \tag{5.16}$$

Let L be the perimeter of the convex hull of D. Then $m(P; P \in D) = F$ and $m(B; B \cap D \neq 0) = L + \pi a$ and we can state

ISBN 0-201-13500-0

If P and B are chosen at random such that $P \in D$ and $B \cap D \neq \emptyset$, the probability that P belongs to $B \cap D$ is

$$p = \frac{\pi a}{L + \pi a} \tag{5.17}$$

and the mean value of the area f of the intersection $B \cap D$ is

$$E(f) = \frac{\pi a F}{L + \pi a}. \tag{5.18}$$

Let K be a convex set and consider the sets of pairs G, B (line and strip) such that $G \cap B \cap K \neq \emptyset$. The measure of this set is the integral of the form $dG \wedge dB$ over the set $G \cap B \cap K \neq \emptyset$. The integration can be accomplished either by first fixing G and then integrating over the strips B or by first fixing B and then integrating over the lines G. The first way gives, using (5.4),

$$m(G, B; G \cap B \cap K \neq \emptyset) = \int_{G \cap K \neq \emptyset} (2\sigma + \pi a)\, dG = 2\pi F + \pi a L \tag{5.19}$$

where σ is the length of the chord $G \cap K$. The second way gives

$$m(G, B; G \cap B \cap K \neq \emptyset) = \int_{B \cap K \neq \emptyset} u\, dB \tag{5.20}$$

where u denotes the perimeter of $B \cap K$ (Fig. 5.5). From (5.19) and (5.20)

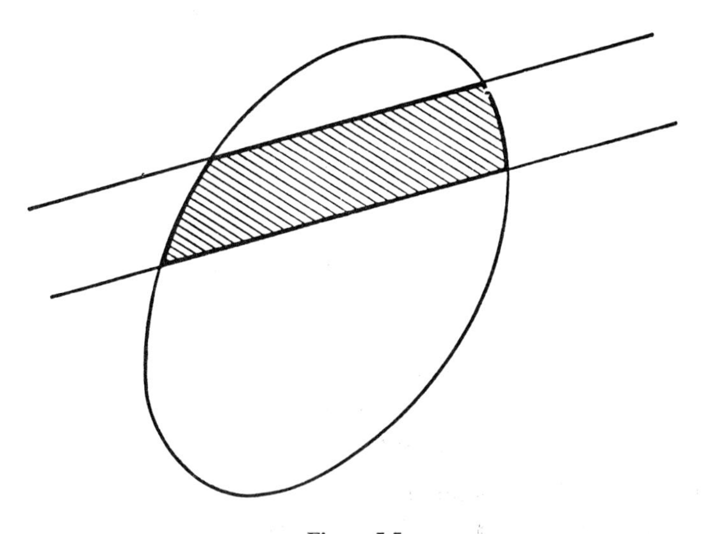

Figure 5.5.

ISBN 0-201-13500-0

we deduce

$$\int_{B \cap K \neq \emptyset} u \, dB = 2\pi F + \pi L. \tag{5.21}$$

These results can be stated as follows.

Let G and B be a line and a strip of breadth a chosen at random such that $G \cap K \neq \emptyset$ and $B \cap K \neq \emptyset$. The probability that $G \cap B \cap K \neq \emptyset$ is

$$p = \frac{2\pi F + \pi a L}{L(L + \pi a)}. \tag{5.22}$$

If $a = 0$, we get the probability that two random chords of K will intersect inside K, that is, $p = 2\pi F/L^2$.

The mean length of the boundary of $B \cap K$ is

$$E(u) = \frac{2\pi F + \pi a L}{L + \pi a}. \tag{5.23}$$

The density for pairs of independent strips B_1, B_2 is $dB_1 \wedge dB_2$ and thus the measure of the set of pairs of strips B_1, B_2 such that $B_1 \cap B_2 \cap K \neq \emptyset$ is

$$m(B_1, B_2; B_1 \cap B_2 \cap K \neq \emptyset) = \int_{B_1 \cap B_2 \cap K \neq \emptyset} dB_1 \wedge dB_2 = \int_{B_1 \cap K \neq \emptyset} (u_1 + \pi a_2) \, dB_1$$

$$= 2\pi F + \pi a_1 L + \pi a_2 (L + \pi a_1) \tag{5.24}$$

where we have applied (5.2) and (5.21) and a_1, a_2 are the breadths of strips B_1 and B_2, respectively.

Therefore we have

If B_1, B_2 are two random strips intersecting the convex set K, the probability that $B_1 \cap B_2 \cap K \neq \emptyset$ is

$$p = \frac{2\pi F + \pi L(a_1 + a_2) + \pi^2 a_1 a_2}{(L + \pi a_1)(L + \pi a_2)}. \tag{5.25}$$

For strips whose breadth is random variable, see [216].

4. Some Mean Values

Let B_i $(i = 1, 2, \ldots, n)$ be n random strips of breadth a that intersect a convex set K. We want to find the mean value of the area f_r of the part of K

ISBN 0-201-13500-0

that is covered by exactly r strips. Consider the integral

$$I_r = \int dP \wedge dB_1 \wedge dB_2 \wedge \cdots \wedge dB_n \tag{5.26}$$

extended over the set of strips B_i that intersect K and over the points P that are covered exactly by r strips. We have

$$I_r = \binom{n}{r} \int_{P \in B_i} dP \wedge dB_1 \wedge \cdots \wedge dB_r \int_{P \notin B_h} dB_{r+1} \wedge \cdots \wedge dB_n \tag{5.27}$$

where $i = 1, 2, \ldots, r$ and $h = r + 1, \ldots, n$. Since the measure of the strips that contain P is equal to πa and the measure of the strips that do not contain P is $(L + \pi a) - \pi a = L$, we have

$$I_r = \binom{n}{r} (\pi a)^r L^{n-r} F. \tag{5.28}$$

On the other hand, if in order to calculate I_r we first keep fixed the strips $B_1, B_2, \ldots, B_n$, we have

$$I_r = \int f_r \, dB_1 \wedge dB_2 \wedge \cdots \wedge dB_n, \tag{5.29}$$

the integral being extended over all strips that meet K. From (5.28) and (5.29) we have

$$\int f_r \, dB_1 \wedge dB_2 \wedge \cdots \wedge dB_n = \binom{n}{r} (\pi a)^r L^{n-r} F. \tag{5.30}$$

Since the measure of the set of n-tuples of strips that intersect K is equal to $(L + \pi a)^n$, we can state

The mean value of the area f_r that is covered exactly r times for n randomly given strips of breadth a that intersect a convex set K is

$$E(f_r) = \frac{\binom{n}{r} (\pi a)^r L^{n-r} F}{(L + \pi a)^n}, \qquad r = 0, 1, 2, \ldots, n. \tag{5.31}$$

If $n \to \infty$ and simultaneously $a \to 0$ according to the law $na = \alpha =$ constant (i.e., the total breadth of the n strips remains constant and equal to α), we get

$$E(f_r) \to \frac{1}{r!} \left(\frac{\pi\alpha}{L}\right)^r F \exp\left(-\frac{\pi\alpha}{L}\right). \tag{5.32}$$

ISBN 0-201-13500-0

For instance, the mean value of the area that is not covered by any strip ($r = 0$), in the limit, becomes $E(f_0) = F \exp(-\pi\alpha/L)$.

In other words these results may be stated as follows.

If it is assumed that n strips of breadth a cross at random a convex set K, the probability that a given point of K belongs exactly to r strips is equal to

$$p_r = \binom{n}{r} \frac{(\pi a)^r L^{n-r}}{(L + \pi a)^n}, \qquad r = 0, 1, \ldots, n. \tag{5.33}$$

If $n \to \infty$, $a \to 0$ in such a manner that $na = \alpha$ (constant), then

$$p_r \to \frac{1}{r!}\left(\frac{\pi\alpha}{L}\right)^r \exp\left(-\frac{\pi\alpha}{L}\right).$$

5. Notes

1. *Generalization of the Buffon needle problem.* Assume that the needle has a length L greater than the distance D between parallels. Then the probability that it will cross at least one parallel line is

$$p = \frac{2}{\pi} \arccos \frac{D}{L} + \frac{2}{\pi D}\left[L - (L^2 - D^2)^{1/2}\right]. \tag{5.34}$$

More precisely, if we assume that $D = 1$ and $L = n + L'$ $(0 \leqslant L' \leqslant 1)$, the probability that the needle will cross exactly h parallel lines $(1 \leqslant h \leqslant n + 1)$ is

$$p_h = \frac{2}{\pi}\left[(h+1)\alpha_{h+1} - 2h\alpha_h + (h-1)\alpha_{h-1}\right]$$

$$+ \frac{2L}{\pi}(\cos\alpha_{h+1} - 2\cos\alpha_h + \cos\alpha_{h-1}) \tag{5.35}$$

where α_i is the angle such that $L \sin\alpha_i = i$ for $i = 1, 2, \ldots, n$, and $\alpha_{n+1} = \pi/2$. For $h = n + 1$, $p_{n+1} = 2L\pi^{-1}\cos\alpha_n + 2n\pi^{-1}(\alpha_n - \pi/2)$. See [335].

2. *Buffon's problem with a broken line.* Assume the plane ruled with equidistant parallel lines a distance D apart. A broken line $\gamma = BAC$ composed of two sides AB, AC of lengths $|AB| = a$ and $|AC| = b$ is dropped at random on the plane. Assuming that the greatest side of the triangle ABC is less than D, prove as an exercise that the probabilities that γ has zero, one, or two intersection points with the parallel lines are, respectively,

$$p_0 = 1 - \frac{a+b+c}{\pi D}, \qquad p_1 = \frac{2c}{\pi D}, \qquad p_2 = \frac{a+b-c}{\pi D}$$

where $c = |BC|$ (Fig. 5.6).

ISBN 0-201-13500-0

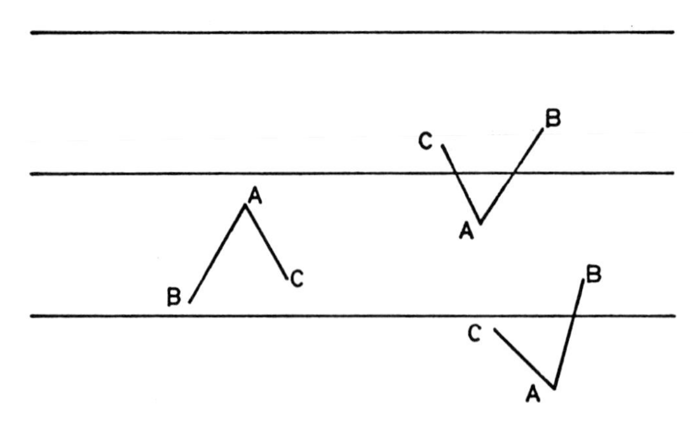

Figure 5.6.

3. *White and black random strips.* Let K be a convex set of area F and perimeter L that we suppose to be white. Assume that K is crossed at random by a black strip B_1 of breadth a. After that, K is crossed at random by a white strip B_2 of the same breadth a, which erases the part of B_1 that it meets. The process continues, drawing at random alternately black and white strips of the same breadth a (Fig. 5.7). After $n + 1$ black strips and n white strips have been drawn, the mean value of the black area of K is

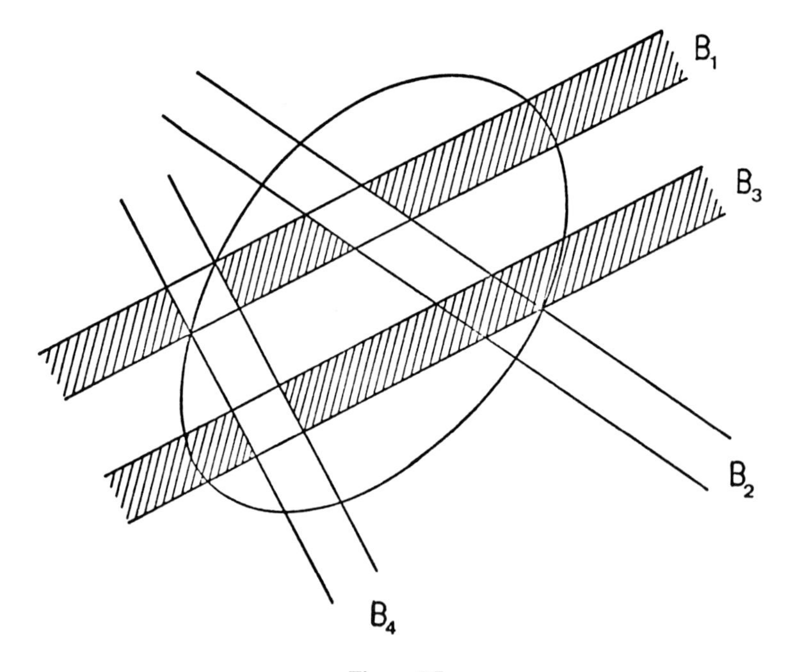

Figure 5.7.

ISBN 0-201-13500-0

$$E(f_{2n+1}) = \frac{\pi a F}{L + \pi a}\left[1 + \left(\frac{L}{L + \pi a}\right)^2 + \left(\frac{L}{L + \pi a}\right)^4 + \cdots + \left(\frac{L}{L + \pi a}\right)^{2n}\right] \tag{5.36}$$

and after $n + 1$ black and $n + 1$ white strips have been drawn, the mean black area becomes

$$E(f_{2n+2}) = \frac{\pi a F L}{(L + \pi a)^2}\left[1 + \left(\frac{L}{L + \pi a}\right)^2 + \left(\frac{L}{L + \pi a}\right)^4 + \cdots + \left(\frac{L}{L + \pi a}\right)^{2n}\right]. \tag{5.37}$$

For $n \to \infty$ we get

$$E(f_{2n+1}) \to \frac{F(L + \pi a)}{2L + \pi a}, \qquad E(f_{2n+2}) \to \frac{FL}{2L + \pi a}. \tag{5.38}$$

(See [534].)

ISBN 0-201-13500-0

CHAPTER 6

The Group of Motions in the Plane: Kinematic Density

1. The Group of Motions in the Plane

We have imposed on the densities for points and lines in the plane the condition of being invariant under the group of motions. We shall now consider in detail this group of motions, which we will represent by $\mathfrak{M}$, as an introduction to the general study of groups of matrices that appears in Chapter 9.

Assume the euclidean plane referred to a rectangular system of Cartesian coordinates. A motion is defined as a transformation $u: P(x, y) \to P'(x', y')$ represented by the equations

$$x' = x \cos \phi - y \sin \phi + a, \qquad y' = x \sin \phi + y \cos \phi + b \tag{6.1}$$

where a, b, ϕ are parameters that have the following respective ranges:

$$-\infty < a < \infty, \qquad -\infty < b < \infty, \qquad 0 \leqslant \phi \leqslant 2\pi. \tag{6.2}$$

If K is a set of points and $K' = uK$ is the image of K under the motion u, we say that K and K' are congruent.

It is easy to give a geometric interpretation of parameters a, b, ϕ. Let $(O; x, y)$ denote the rectangular frame of origin O and x, y axes (Fig. 6.1). Assume that by the motion u the frame $(O; x, y)$ is mapped onto the frame $(O'; x', y')$; then a, b are the coordinates of O' and ϕ is the angle between the x axis and the x' axis.

The unit element of the group is the identity $a = 0$, $b = 0$, $\phi = 0$. Instead of equations (6.1), it is sometimes useful to represent the motion u by the

ENCYCLOPEDIA OF MATHEMATICS and Its Applications, Gian-Carlo Rota (ed.).
1, Luis A. Santaló, Integral Geometry and Geometric Probability

ISBN 0-201-13500-0

matrix

$$u = \begin{pmatrix} \cos\phi & -\sin\phi & a \\ \sin\phi & \cos\phi & b \\ 0 & 0 & 1 \end{pmatrix}. \tag{6.3}$$

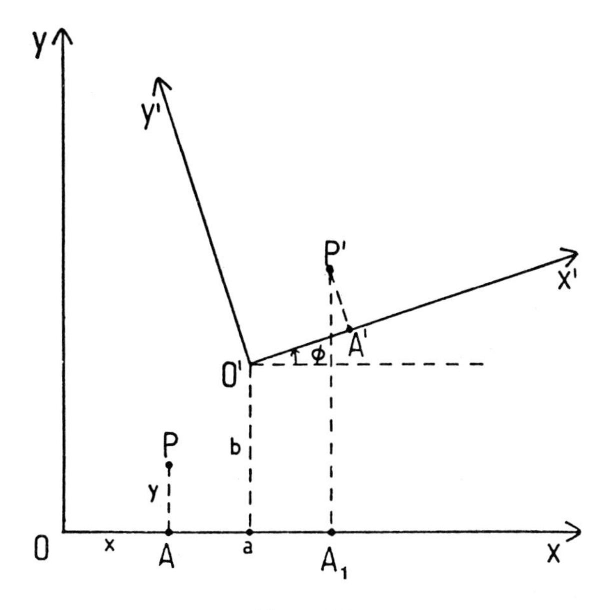

Figure 6.1.

Then the motion $u_2 u_1$ is represented by the product matrix $u_2 u_1$ and the inverse motion u^{-1} is represented by the inverse matrix

$$u^{-1} = \begin{pmatrix} \cos\phi & \sin\phi & -a\cos\phi - b\sin\phi \\ -\sin\phi & \cos\phi & a\sin\phi - b\cos\phi \\ 0 & 0 & 1 \end{pmatrix}. \tag{6.4}$$

Hence the group of motions $\mathfrak{M}$ can be defined as the group of matrices of the form (6.3) with the ordinary matrix product as composition law. We will use the same symbol to represent a motion and the corresponding matrix.

Each motion is determined by a point of the 3-space (a, b, ϕ). This space, with the equivalence relation $(a, b, \phi) \sim (a, b, \phi + 2k\pi)$ (k any integer) is the space of the group $\mathfrak{M}$ and it will be represented by the same letter $\mathfrak{M}$.

Each motion $s \in \mathfrak{M}$ defines two endomorphisms of $\mathfrak{M}$: the *left translation*

$$L_s \colon u \to su; \tag{6.5}$$

and the *right translation*

ISBN 0-201-13500-0

$$R_s: u \to us. \tag{6.6}$$

For instance, if

$$s = \begin{pmatrix} \cos\phi_0 & -\sin\phi_0 & a_0 \\ \sin\phi_0 & \cos\phi_0 & b_0 \\ 0 & 0 & 1 \end{pmatrix}, \tag{6.7}$$

we have

$$L_s: \begin{pmatrix} \cos\phi & -\sin\phi & a \\ \sin\phi & \cos\phi & b \\ 0 & 0 & 1 \end{pmatrix}$$

$$\to \begin{pmatrix} \cos(\phi+\phi_0) & -\sin(\phi+\phi_0) & a\cos\phi_0 - b\sin\phi_0 + a_0 \\ \sin(\phi+\phi_0) & \cos(\phi+\phi_0) & a\sin\phi_0 + b\cos\phi_0 + b_0 \\ 0 & 0 & 1 \end{pmatrix}, \tag{6.8}$$

which can be written

$$L_s \begin{cases} a \to a\cos\phi_0 - b\sin\phi_0 + a_0, \\ b \to a\sin\phi_0 + b\cos\phi_0 + b_0, \\ \phi \to \phi + \phi_0. \end{cases} \tag{6.9}$$

Analogously we have

$$R_s \begin{cases} a \to a_0\cos\phi - b_0\sin\phi + a, \\ b \to a_0\sin\phi + b_0\cos\phi + b, \\ \phi \to \phi_0 + \phi. \end{cases} \tag{6.10}$$

2. The Differential Forms on $\mathfrak{M}$

A differential form of order 1 or a 1-form on $\mathfrak{M}$ (also called a Pfaffian form) is any expression of this type

$$\omega(u) = \alpha(u)\,da + \beta(u)\,db + \gamma(u)\,d\phi \tag{6.11}$$

where $\alpha(u)$, $\beta(u)$, $\gamma(u)$ are functions of class C^∞ defined on the space $\mathfrak{M}$, that is, infinitely differentiable functions of the coordinates a, b, ϕ of the point $u \in \mathfrak{M}$.

All differential 1-forms on $\mathfrak{M}$ at the point u with the addition and product by a scalar defined in the natural way ($\omega_1(u) + \omega_2(u) = (\alpha_1 + \alpha_2)\,da + (\beta_1 + \beta_2)\,db + (\gamma_1 + \gamma_2)\,d\phi$, $\lambda\omega = \lambda\alpha\,da + \lambda\beta\,db + \lambda\gamma\,d\phi$) constitute a vector space of dimension 3, called the vector space of the 1-forms at u (or the cotangent space to $\mathfrak{M}$ at the point u) and represented by T_u^*. The 1-forms

ISBN 0-201-13500-0

da, db, $d\phi$ and any set of three independent linear combinations of them constitute a basis for T_u^*. The left and right translations L_s and R_s induce on T_u^* the maps

$$L_s^*: \omega(u) \to \omega(su), \qquad R_s^*: \omega(u) \to \omega(us) \tag{6.12}$$

where in order to write $\omega(su)$ and $\omega(us)$ in explicit form the following equations, consequences of (6.9) and (6.10), must be used.

$$L_s^* \begin{cases} da \to \cos\phi_0\, da - \sin\phi_0\, db, \\ db \to \sin\phi_0\, da + \cos\phi_0\, db, \\ d\phi \to d\phi; \end{cases} \tag{6.13}$$

$$R_s^* \begin{cases} da \to -(a_0 \sin\phi_0 + b_0 \cos\phi)\, d\phi + da, \\ db \to (a_0 \cos\phi - b_0 \sin\phi)\, d\phi + db, \\ d\phi \to d\phi. \end{cases} \tag{6.14}$$

An important problem is that of finding the 1-forms on $\mathfrak{M}$ that are invariant under L_s^* and those that are invariant under R_s^*. They are called left-invariant and right-invariant 1-forms, respectively. To this end, note that the matrix

$$\Omega_L = u^{-1}\, du \tag{6.15}$$

is invariant under left translations, because

$$L_s^* \Omega_L = (su)^{-1}\, d(su) = u^{-1} s^{-1} s\, du = u^{-1}\, du = \Omega_L \tag{6.16}$$

and therefore the elements of Ω_L are left-invariant 1-forms. From (6.3) and (6.4) we deduce

$$\Omega_L = u^{-1}\, du = \begin{pmatrix} 0 & -d\phi & \cos\phi\, da + \sin\phi\, db \\ d\phi & 0 & -\sin\phi\, da + \cos\phi\, db \\ 0 & 0 & 0 \end{pmatrix}. \tag{6.17}$$

Hence the 1-forms

$$\omega_1 = \cos\phi\, da + \sin\phi\, db, \qquad \omega_2 = -\sin\phi\, da + \cos\phi\, db,$$
$$\omega_3 = d\phi \tag{6.18}$$

are left-invariant 1-forms.

On the other hand, $\omega_1, \omega_2, \omega_3$ are independent forms (since the determinant of the coefficients of da, db, $d\phi$ is not zero) and thus any linear combination of them with constant coefficients is also an invariant 1-form under L_s^*. We will prove that the converse is also true, that is, *any 1-form on* $\mathfrak{M}$ *that is left invariant is a linear combination with constant coefficients of the 1-forms* (6.18).

ISBN 0-201-13500-0

For a proof, we observe that since $\omega_1, \omega_2, \omega_3$ are independent, they form a basis for T_u^* and therefore each 1-form $\omega(u)$ may be written as $\omega(u) = \alpha(u)\omega_1 + \beta(u)\omega_2 + \gamma(u)\omega_3$. If ω is invariant under L_s^*, we have

$$\omega(su) = \alpha(su)\omega_1(su) + \beta(su)\omega_2(su) + \gamma(su)\omega_3(su) = \omega(u)$$

and since $\omega_i(su) = \omega_i(u)$ $(i = 1, 2, 3)$, we have

$$(\alpha(su) - \alpha(u))\omega_1(u) + (\beta(su) - \beta(u))\omega_2(u) + (\gamma(su) - \gamma(u))\omega_3(u) = 0.$$

Because of the independence of $\omega_1, \omega_2, \omega_3$ it follows that $\alpha(su) = \alpha(u)$, $\beta(su) = \beta(u)$, $\gamma(su) = \gamma(su)$, which implies (since s is any point of $\mathfrak{M}$) that α, β, γ are constants. Thus, we have solved the problem of finding all left-invariant 1-forms on $\mathfrak{M}$.

To find the right-invariant 1-forms we consider the matrix

$$\Omega_R = du\, u^{-1} \tag{6.19}$$

which is right invariant, because

$$R_s^*\Omega_R = d(us)(us)^{-1} = du\, ss^{-1}u^{-1} = du\, u^{-1} = \Omega_R.$$

According to (6.3) and (6.4) we have

$$\Omega_R = \begin{pmatrix} 0 & d\phi & b\, d\phi + da \\ d\phi & 0 & -a\, d\phi + db \\ 0 & 0 & 0 \end{pmatrix} \tag{6.20}$$

and therefore we have the following set of right-invariant 1-forms.

$$\omega^1 = b\, d\phi + da, \qquad \omega^2 = -a\, d\phi + db, \qquad \omega^3 = d\phi. \tag{6.21}$$

Any linear combination with constant coefficients of $\omega^1, \omega^2, \omega^3$ will be a right-invariant 1-form and the same proof as above shows that, conversely, *any right-invariant 1-form on $\mathfrak{M}$ is a linear combination with constant coefficients of the forms* (6.21).

Finally, observe that by differentiating the identity $uu^{-1} = e =$ unit matrix, we get $du \cdot u^{-1} + u \cdot du^{-1} = 0$ and therefore

$$du^{-1} = -u^{-1}\, du\, u^{-1}. \tag{6.22}$$

From this equality and (6.15), (6.19) we get

$$\Omega_L(u^{-1}) = -\Omega_R(u), \tag{6.23}$$

which is an important relation that will be useful later.

ISBN 0-201-13500-0

3. The Kinematic Density

Since ω_1, ω_2, ω_3 are left-invariant 1-forms, the exterior product

$$dK = \omega_1 \wedge \omega_2 \wedge \omega_3 = da \wedge db \wedge d\phi \tag{6.24}$$

is a left-invariant 3-form. Furthermore, up to a constant factor, dK is the unique left-invariant 3-form in $\mathfrak{M}$. Indeed, if $\psi = f(a, b, \phi)\, da \wedge db \wedge d\phi = f(u)\omega_1 \wedge \omega_2 \wedge \omega_3$ is a left-invariant 3-form, we have $f(su)\omega_1(su) \wedge \omega_2(su) \wedge \omega_3(su) = f(u)\omega_1(u) \wedge \omega_2(u) \wedge \omega_3(u)$ and since $\omega_i(su) = \omega_i(u)$ $(i = 1, 2, 3)$, it follows that $f(su) = f(u)$. Since any u may be sent into any su by a suitable left translation s, the function f must have the same value for all points of $\mathfrak{M}$; thus, it is a constant.

From (6.21) it follows that

$$\omega^1 \wedge \omega^2 \wedge \omega^3 = da \wedge db \wedge d\phi = dK \tag{6.25}$$

that is, the differential form dK is also right invariant. The same argument as above shows that, up to a constant factor, it is the unique 3-form that is invariant under right translations. Finally, from (6.22) we deduce

$$dK(u^{-1}) = -\ dK(u); \tag{6.26}$$

that is, up to the sign, dK is also invariant under inversion of the motion. Since we will always take the densities in absolute value, the change of sign in (6.26) is immaterial and we can state that *the 3-form* (6.24) *is invariant under left and right translations and under inversion of the motion. It is called the kinematic density for the group of motions in the plane.*

The kinematic density dK is the invariant volume element of the space of the group of motions $\mathfrak{M}$. By integrating dK over a domain on $\mathfrak{M}$, we get the measure of the corresponding set of motions (kinematic measure). We now consider some examples in order to see the geometrical meaning of the kinematic measure and its invariance properties.

Consider a rectangle $K = OABC$ and a fixed domain K_0, as shown in Fig. 6.2. We ask for the measure of the set of motions u such that $uK \cap K_0 \neq \emptyset$, that is, the set of motions that carry K into a position in which it intersects K_0. This measure is the integral of $dK = da \wedge db \wedge d\phi$ over the points $O'(a, b)$ and the angles ϕ such that $uK \cap K_0 \neq \phi$. The left invariance of this measure means that instead of K_0 we could take the domain sK_0, the image of K_0 by the motion s, the measure being the same in both cases. In other words, the measure of motions such that $uK \cap K_0 \neq \emptyset$ is equal to the measure of motions such that $uK \cap sK_0 \neq \emptyset$, for any fixed s.

The right invariance means that instead of K we could take sK and the measure of the motions such that $u(sK) \cap K_0 \neq \emptyset$ is also the same as before.

ISBN 0-201-13500-0

It follows that the kinematic measure does not depend on the initial position of K or K_0, so that instead of the measure of a set of motions we can say that the problem is that of finding "the measure of the set of rectangles congruent to K that have common point with the domain K_0." This terminology, based on "sets of congruent figures," is sometimes more intuitive than that based on "sets of motions," though they are obviously equivalent.

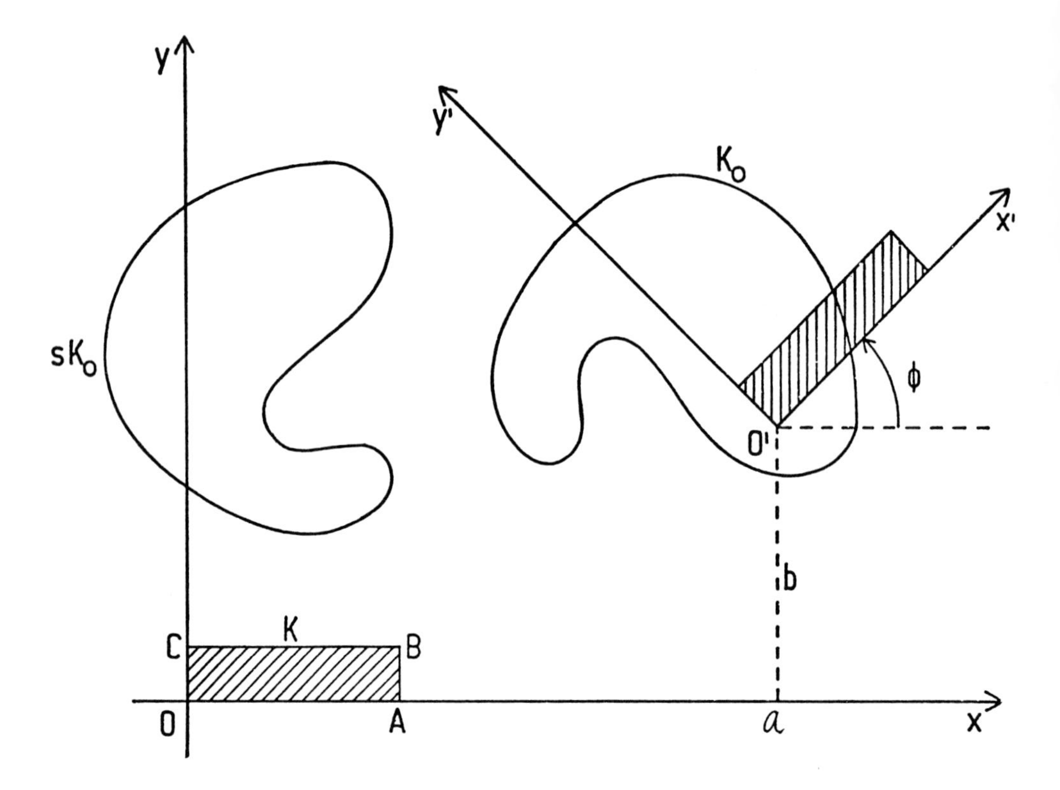

Figure 6.2.

The invariance of dK by inversion of the motion means that the measure of the set of motions such that $uK \cap K_0 \neq \emptyset$ is equal to the measure of the set of motions $u' = u^{-1}$ such that $K \cap u'K_0 \neq \emptyset$. For instance, if K reduces to a single point $P_0(0, 0)$ and we put $uP_0 = P(a, b)$, we have

$$m(u; uP_0 \in K_0) = \int_{uP_0 \in K_0} da \wedge db \wedge d\phi = 2\pi \int_{uP_0 \in K_0} da \wedge db = 2\pi F_0 \quad (6.27)$$

where F_0 is the area of K_0. If we consider the set of inverse motions u' such that $P_0 \in u'K_0$, the result should be the same and we have

ISBN 0-201-13500-0

$$m(u'; P_0 \in u'K_0) = \int_{P_0 \in u'K_0} dK_0 = 2\pi F_0 \tag{6.28}$$

where we have used dK_0 in order to indicate that the moving figure is now the domain K_0. Equation (6.28) is a simple but useful integral formula.

Remark. From the preceding paragraphs it follows that to compute the measure of a set of congruent figures, or a set of positions of a figure K, we must choose a frame $(O'; x', y')$ fixed in K (*moving frame*) and then compute the integral of $dK = da \wedge db \wedge d\phi$ over the set, where a, b are the coordinates of O' with respect to the *fixed frame* $(O; x, y)$ and ϕ is the angle between the x axis and x' axis.

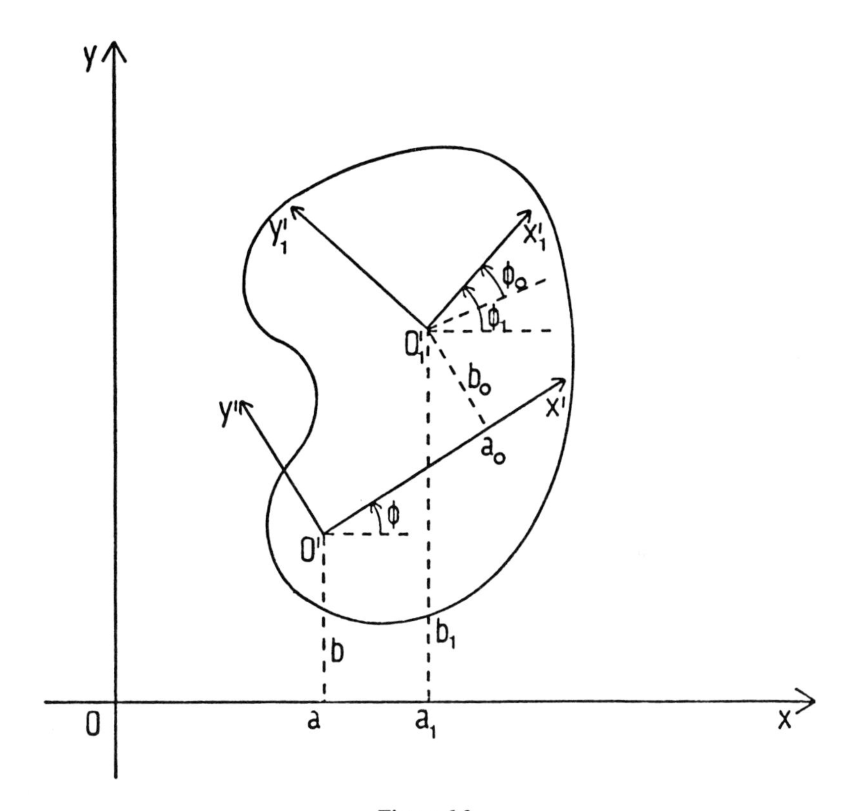

Figure 6.3.

The choice of the moving frame is arbitrary. Indeed, if instead of $(O'; x', y')$ we choose the frame $(O_1'; x_1', y_1')$ such that (Fig. 6.3)

$$a_1 = a + a_0 \cos\phi - b_0 \sin\phi, \qquad b_1 = b + a_0 \sin\phi + b_0 \cos\phi,$$
$$\phi_1 = \phi + \phi_0, \tag{6.29}$$

ISBN 0-201-13500-0

where a_0, b_0 are the coordinates of O_1' relative to the frame $(O'; x', y')$ and ϕ_0 is the angle between $O'x'$ and $O_1'x_1'$, we note that (6.29) are the equations of a right translation (6.10) and therefore the kinematic density remains unchanged. In other words, *the right invariance of dK is equivalent to the invariance by a change of moving frame.* This property allows the choice of the more suitable moving frame for each particular case.

Other expressions for the kinematic density. Let $(P; x', y')$ be a moving frame defined by the point $P(a, b)$ and the angle ϕ that the line Px' makes with the x axis. If we define this moving frame by new coordinates, we will have new expressions for dK. For instance, let $(P; x', y')$ be determined by the

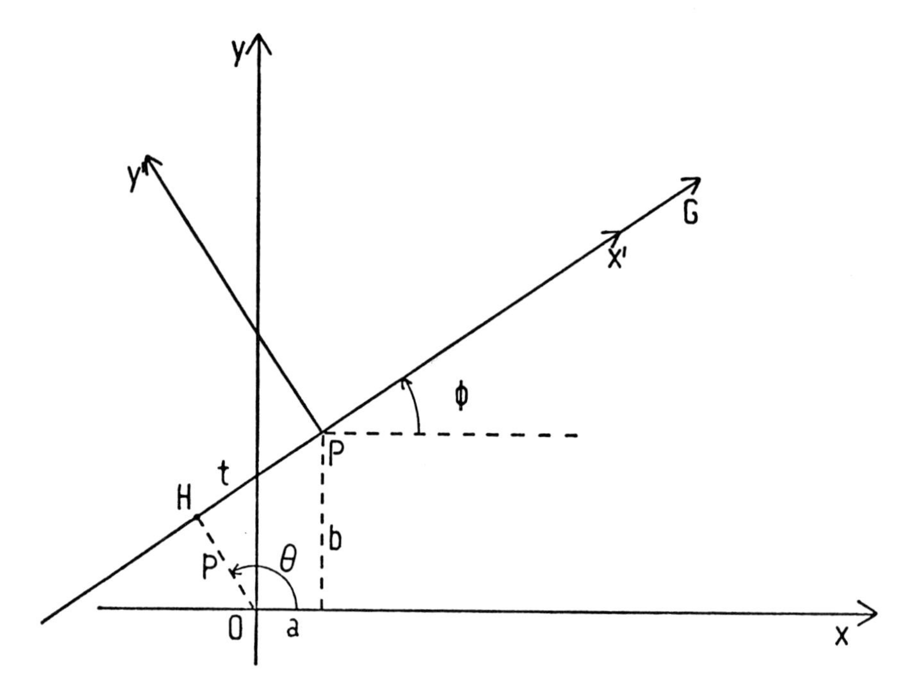

Figure 6.4.

line Px', which we will denote by $G(p, \theta)$ and the distance $t = PH$ from P to the foot of the perpendicular drawn from O to G (Fig. 6.4). The transformation formulas are

$$a = p\cos\theta + t\sin\theta, \qquad b = p\sin\theta - t\cos\theta, \qquad \phi = \theta - \pi/2.$$

Hence $dK = da \wedge db \wedge d\phi = dp \wedge d\theta \wedge dt$, or

$$dK = dG^* \wedge dt \tag{6.30}$$

ISBN 0-201-13500-0

where we write G^* in order to indicate that G must be considered oriented, because a change of orientation does not superpose the frame $(P; x', y')$ onto itself. Each unoriented line is the support of two oriented lines.

If we set $dP = da \wedge db$, (6.30) can be written

$$dP \wedge d\phi = dG^* \wedge dt. \tag{6.31}$$

Another expression for dK is the following. Except for translations, which depend on two parameters (a, b) and therefore belong to a set of measure zero, each motion u is a rotation about an invariant point called the center of rotation of u. Let $Q(\xi, \eta)$ be the center of rotation of the motion that maps $(O; x, y)$ onto $(O'; x', y')$ and let ϕ be the angle of this rotation. The coordinates a, b, ϕ and ξ, η, ϕ are related by the equations

$$a = (1 - \cos\phi)\xi + \sin(\phi)\eta, \qquad b = -\sin(\phi)\xi + (1 - \cos\phi)\eta$$

as follows putting $x = x' = \xi$, $y = y' = \eta$ into equations (6.1). From these transformation formulas it follows easily that

$$dK = 4\sin^2(\phi/2)\, d\xi \wedge d\eta \wedge d\phi. \tag{6.32}$$

This expression is not applicable to translations, since then Q is a point at infinity.

4. Sets of Segments

Let K_0 be a fixed convex set of area F_0 and perimeter L_0. Consider an oriented segment K of length l. We want to calculate the measure of the set of segments congruent to K that meet K_0 (Fig. 6.5). Choosing expression (6.30) for the kinematic density, letting G be the line that contains segment

Figure 6.5.

ISBN 0-201-13500-0

K, and calling σ the length of the chord $G \cap K_0$, we have (using (3.6) and (3.12)),

$$m(K; K \cap K_0 \neq \emptyset) = \int_{K \cap K_0 \neq \emptyset} dG^* \wedge dt = \int_{G \cap K_0 \neq \emptyset} (\sigma + l)\, dG^* = 2\pi F_0 + 2lL_0. \tag{6.33}$$

Thus, *the measure of all oriented segments of length l having a point in common with a convex set of area F_0 and perimeter L_0 is equal to* $2\pi F_0 + 2lL_0$.

The problem of finding the measure of the segments of a constant length that are contained in K_0 has no simple solution and depends largely on the shape of K_0. For a circle C of diameter $D \geqslant l$, a direct computation gives

$$m(K; K \subset C) = (\pi/2)[\pi D^2 - 2D^2 \arcsin(l/D) - 2l(D^2 - l^2)^{1/2}] \tag{6.34}$$

and for a rectangle R of sides a, b ($l \leqslant a$, $l \leqslant b$) we have

$$m(K; K \subset R) = 2(\pi ab - 2(a + b)l + l^2). \tag{6.35}$$

For a convex polygon, under certain conditions for l, this measure will be computed in (6.44) below.

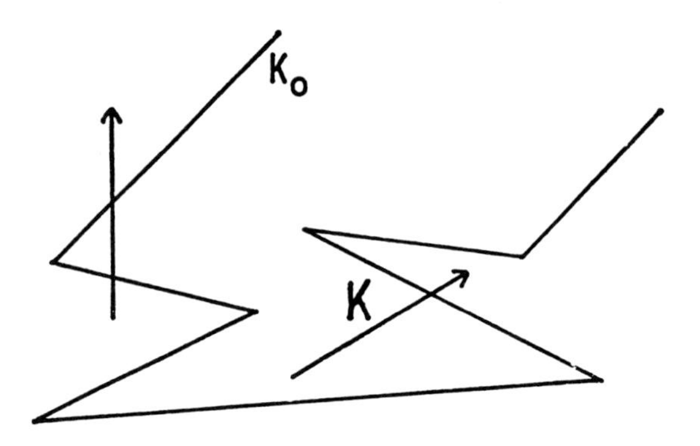

Figure 6.6.

If K_0 reduces to a segment of length l_0 the measure (6.33) reduces to $4ll_0$. If K_0 is a polygonal line of total length L_0, writing this measure for each side of K_0 and summing over all sides, we get

$$\int_{K \cap K_0 \neq \emptyset} n\, dK = 4lL_0 \tag{6.36}$$

ISBN 0-201-13500-0

where n denotes the number of sides of K_0 that are intersected by K for each position of this segment (Fig. 6.6).

We now evaluate the measure of the set of oriented segments K of length l that intersect both sides of a given angle A (Fig. 6.7). We denote by A indis-

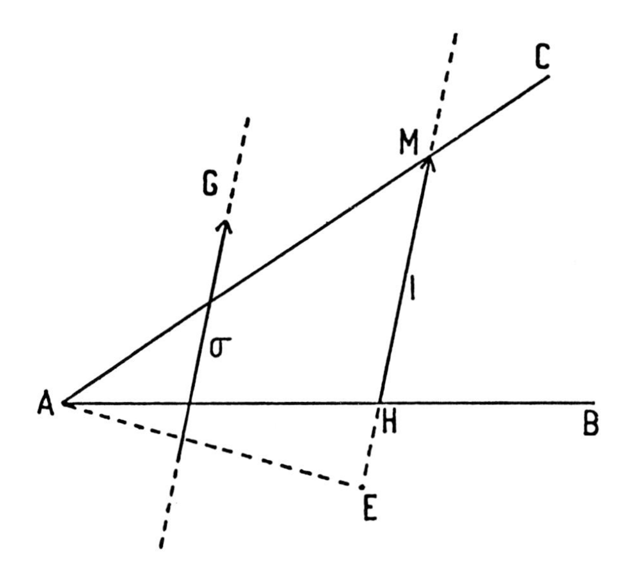

Figure 6.7.

tinctly the vertex of the angle and its measure. Denoting by σ the chord cut by the angle A from the line G that contains K, we have

$$m(K; K \cap AB \neq \emptyset, K \cap AC \neq \emptyset) = \int dG^* \wedge dt = 2 \int_{\sigma \leqslant l} (l - \sigma)\, dG. \quad (6.37)$$

Further

$$\int_{\sigma < l} l\, dG = l \int_{\sigma < l} dp \wedge d\phi = l \int |AE|\, d\phi = 2 \int_0^{\pi - A} T\, d\phi \quad (6.38)$$

where T is the area of the triangle AHM defined by a chord HM of length l perpendicular to the direction ϕ.

On the other hand, we have

$$\int_{\sigma \leqslant l} \sigma\, dG = \int_{\sigma \leqslant l} \sigma\, dp \wedge d\phi = \int_0^{\pi - A} T\, d\phi. \quad (6.39)$$

ISBN 0-201-13500-0

Therefore

$$m(K;\, K \cap AB \neq \theta,\, K \cap AC \neq \theta) = 2 \int_{0}^{\pi - A} T\, d\phi. \tag{6.40}$$

In order to evaluate this integral we observe that

$$2T = (l^2/\sin A) \sin \phi \sin(A + \phi) \tag{6.41}$$

and consequently

$$m = \frac{l^2}{\sin A} \int_{0}^{\pi - A} \sin \phi \sin(A + \phi)\, d\phi = \frac{l^2}{2} [1 + (\pi - A) \cot A]. \tag{6.42}$$

Thus, *the measure of the set of oriented segments of length l that intersect both sides of an angle A is given by* (6.42).

Sets of segments interior to a given convex polygon. Let K_0 be a convex polygon and K an oriented segment of length l such that it cannot intersect two nonconsecutive sides of K_0.

Let m_i ($i = 0, 1, 2$) be the measure of the set of positions of K in which it has exactly i points in common with the boundary of K_0 (m_0 is the measure of the set of segments K that are interior to K_0). Formulas (6.33), (6.36), and (6.42) can be written

$$\begin{aligned} m_0 + m_1 + m_2 &= 2\pi F_0 + 2lL_0, \\ m_1 + 2m_2 &= 4lL_0, \\ m_2 &= (l^2/2) \sum_{A_i} [1 + (\pi - A_i) \cot A_i], \end{aligned} \tag{6.43}$$

respectively, and therefore

$$\begin{aligned} m_0 &= 2\pi F_0 - 2lL_0 + \frac{l^2}{2} \sum_{A_i} [1 + (\pi - A_i) \cot A_i], \\ m_1 &= 4lL_0 - l^2 \sum_{A_i} [1 + (\pi - A_i) \cot A_i], \\ m_2 &= \frac{l^2}{2} \sum_{A_i} [1 + (\pi - A_i) \cot A_i]. \end{aligned} \tag{6.44}$$

For unoriented segments these formulas should be divided by 2.

ISBN 0-201-13500-0

5. Convex Sets That Intersect Another Convex Set

Let K_1 be a convex set of area F_1 and perimeter L_1 and let K_0 be a convex set of area F_0 and perimeter L_0. We want to calculate the measure of the set of convex sets congruent to K_1 that have some common point with K_0. In other words, we wish the measure of the set of positions of K_1 in which it intersects K_0. The position of K_1 is defined by the coordinates x_1, y_1 of a point $P_1 \in K_1$ and the angle ϕ that a direction P_1A fixed in K_1 makes with a direction P_0x fixed in the plane (Fig. 6.8).

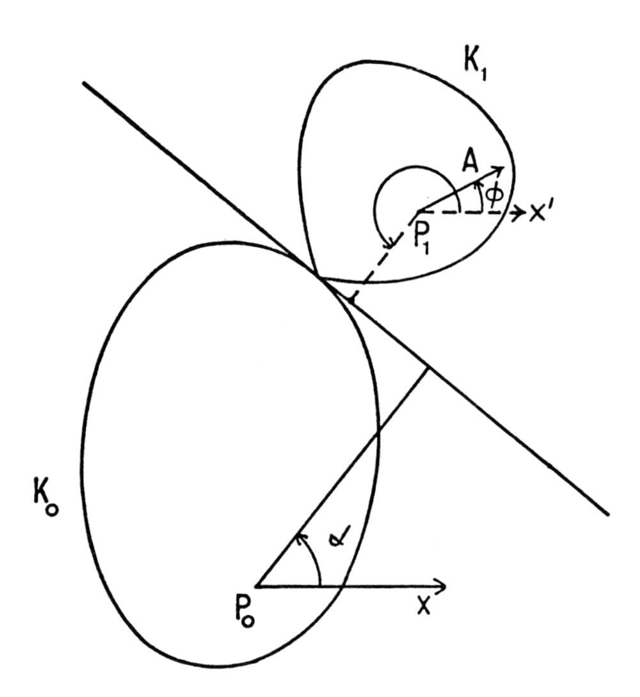

Figure 6.8.

The kinematic density is $dK_1 = dx_1 \wedge dy_1 \wedge d\phi$ where we put dK_1 (instead of dK) to indicate that it refers to the moving set K_1. We want to calculate

$$m(K_1; K_1 \cap K_0 \neq \theta) = \int_{K_1 \cap K_0 \neq \theta} dK_1 = \int_{K_1 \cap K_0 \neq \theta} dP_1 \wedge d\phi. \qquad (6.45)$$

Let $p_0(\alpha)$ and $p_1(\alpha)$ be the support functions of K_0 and K_1 with respect to the origins $P_0(x_0, y_0) \in K_0$, $P_1(x_1, y_1) \in K_1$ and the parallel reference lines P_0x, P_1x', respectively. If we keep ϕ fixed and translate K_1 so that it is in contact with K_0 externally (Fig. 6.8), then the locus of the translated point

ISBN 0-201-13500-0

P_1 traces a new curve, which is the boundary of the convex set defined by the support function

$$p(\alpha) = p_0(\alpha) + p_1(\alpha + \pi). \tag{6.46}$$

Since $p_1(\alpha + \pi)$ is the support function of the set $K_1{}^*$ obtained by reflecting K_1 in P_1, according to Chapter 1, Section 3, the area of the set whose support function is $p(\alpha)$ is $F_0 + F_1 + 2F_{01}^*$ where F_{01}^* is the mixed area of K_0 and $K_1{}^*$. Thus, *the measure of the set of translates of* K_1 *such that* $K_0 \cap K_1 \neq \emptyset$ *is* $F_0 + F_1 + 2F_{01}^*$. This is the result of integrating (6.45) with respect to dP_1.

Further integration with respect to $d\phi$ gives

$$m(K_1; K_1 \cap K_0 \neq \emptyset) = \int_0^{2\pi} (F_0 + F_1 + 2F_{01}^*)\, d\phi = 2\pi(F_0 + F_1) + L_0 L_1 \tag{6.47}$$

where we have applied (1.13). Note that K_1 and $K_1{}^*$ have the same perimeter. We have proved

The measure of all positions of a convex set K_1 *in which it intersects another fixed convex set* K_0 *is*

$$m(K_1; K_1 \cap K_0 \neq \emptyset) = \int_{K_1 \cap K_0 \neq \emptyset} dK_1 = 2\pi(F_0 + F_1) + L_0 L_1. \tag{6.48}$$

Particular cases. 1. If K_1 is a segment of length l, then $F_1 = 0$, $L_1 = 2l$, and (6.48) yields (6.33).

2. If K_1 is a circle of radius R, we can choose for the point P_1 the center of K_1 and then $\int dK_1 = 2\pi \int dP_1$. Formula (6.48) becomes

$$\int_{K_0 \cap K_1 \neq \emptyset} dP_1 = F_0 + L_0 R + \pi R^2, \tag{6.49}$$

which is the second of formulas (1.18), as it should be.

Convex sets contained in another convex set. In general, there is no simple expression for the measure of the positions of a convex set K_1 in which it is contained in a fixed convex set K_0. However, the solution is simple under the hypothesis that ∂K_1 and ∂K_0 have continuous radii of curvature and the greatest radius of curvature of ∂K_1 is less than or equal to the least radius of curvature of ∂K_0. Indeed, with the same notation as above, if K_1 is translated to all positions in which it is interior to K_0, then the locus of the point P_1 is the convex set whose support function is

$$p(\alpha) = p_0(\alpha) + p_1(\alpha). \tag{6.50}$$

ISBN 0-201-13500-0

Note that $p(\alpha)$ is the support function of a convex set because $p + p'' = (p_0 + p_0'') - (p_1 + p_1'') > 0$ (see Chapter 1, Section 2). By (1.8) the area of this convex set is

$$\frac{1}{2}\int_0^{2\pi} (p^2 - p'^2)\, d\phi = F_0 + F_1 - 2F_{01}. \tag{6.51}$$

Integrating over all rotations of K_1 and taking (1.13) into account, we get

$$m(K_1; K_1 \subset K_0) = 2\pi(F_0 + F_1) - L_0 L_1. \tag{6.52}$$

We can state

If K_0 and K_1 are bounded convex sets whose boundaries ∂K_0 and ∂K_1 have continuous radii of curvature and the greatest radius of curvature of ∂K_1 is less than or equal to the least radius of curvature of ∂K_0, then the measure of all convex sets congruent to K_1 that are contained in K_0 is given by (6.52). *If we consider only translations of K_1, then such a measure is given by* (6.51).

If ρ_m is the least radius of curvature of ∂K_0 and K_1 is a circle of radius $R \leqslant \rho_m$, choosing for the point P_1 the center of K_1, from (6.52) it follows that the area filled by the centers of the circles K_1 that are contained in K_0, that is, the area of the convex set interior parallel to K_0 in the distance R, is (see Chapter 1, Section 4)

$$(2\pi)^{-1} m(K_1; K_1 \subset K_0) = F - LR + \pi R^2. \tag{6.53}$$

If ρ_M is the greatest radius of curvature of ∂K_0 and K_1 is a circle of radius $R \geqslant \rho_M$, then (6.53) gives the area filled by the centers of the circles K_1 that completely cover K_0.

6. Some Integral Formulas

1. Let K_0, K_1 be two plane domains, not necessarily convex, of area F_0 and F_1, respectively. Assume that K_0 is fixed and K_1 is moving. Let dK_1 be the kinematic density for K_1. If $P(x, y)$ is a point in the plane and $dP = dx \wedge dy$ its density, we consider the integral

$$I = m(P, K_1; P \in K_0 \cap K_1) = \int_{P \in K_0 \cap K_1} dP \wedge dK_1 \tag{6.54}$$

over all positions of P and K_1 in which P belongs to $K_0 \cap K_1$. If first we leave P fixed, using (6.28), we have

ISBN 0-201-13500-0

$$I = \int_{P \in K_0} dP \int_{P \in K_1} dK_1 = 2\pi F_1 \int_{P \in K_0} dP = 2\pi F_0 F_1; \tag{6.55}$$

leaving K_1 fixed first, we have

$$I = \int_{K_1 \cap K_0 \neq \emptyset} dK_1 \int_{P \in K_1 \cap K_0} dP = \int_{K_1 \cap K_0 \neq \emptyset} f_{01}\, dK_1 \tag{6.56}$$

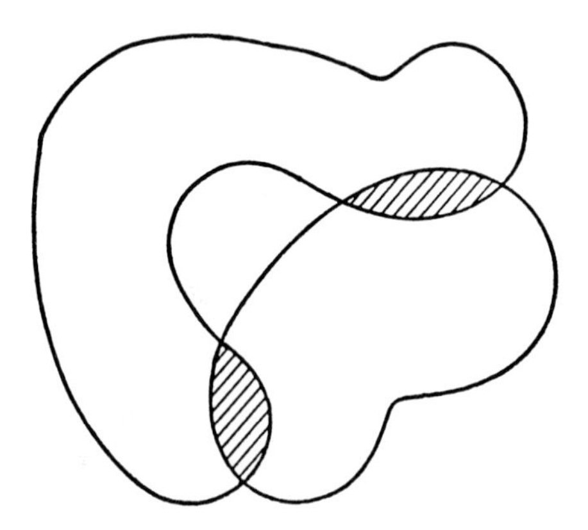

Figure 6.9.

where f_{01} denotes the area of $K_1 \cap K_0$. From (6.55) and (6.56) we get (Fig. 6.9)

$$\int_{K_1 \cap K_0 \neq \emptyset} f_{01}\, dK_1 = 2F_0 F_1. \tag{6.57}$$

2. Let K_0, K_1 be plane domains of area F_0, F_1, respectively, and such that their boundaries are rectifiable curves of lengths L_0 and L_1. Let $A(s_0)$ be a point of ∂K_0 (s_0 being the arc length) and consider the integral

$$J_1 = \int_{A \in K_1} ds_0 \wedge dK_1.$$

If we first leave A fixed, we have

$$J_1 = \int_{\partial K_0} ds_0 \int_{A \in K_1} dK_1 = 2\pi F_1 L_0, \tag{6.58}$$

and if we first leave K_1 fixed and l_{01} denotes the length of the arc of ∂K_0 that is interior to K_1, we get

ISBN 0-201-13500-0

$$J_1 = \int_{K_1 \cap K_0 \neq \emptyset} dK_1 \int_{A \in K_1} ds_0 = \int_{K_1 \cap K_0 \neq \emptyset} l_{01}\, dK_1. \tag{6.59}$$

From (6.58) and (6.59) we deduce (Fig. 6.10),

$$\int_{K_0 \cap K_1 \neq \emptyset} l_{01}\, dK_1 = 2\pi F_1 L_0. \tag{6.60}$$

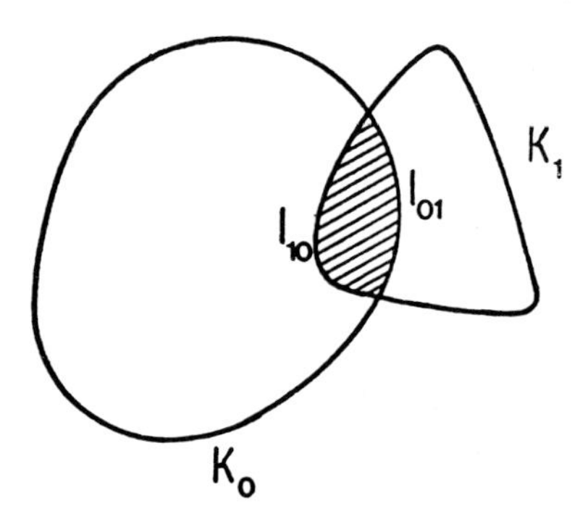

Figure 6.10.

By the invariance property of the kinematic measure under inversion of the motion, if l_{10} denotes the length of the arc of ∂K_1 that is interior to K_0, we have the analogue of (6.60), namely

$$\int_{K_0 \cap K_1 \neq \emptyset} l_{10}\, dK_1 = 2\pi F_0 L_1. \tag{6.61}$$

Adding (6.60) and (6.61), we get

$$\int_{K_0 \cap K_1 \neq \emptyset} L_{01}\, dK_1 = 2\pi(F_1 L_0 + F_0 L_1) \tag{6.62}$$

where L_{01} is the length of the boundary of $K_0 \cap K_1$.

3. Let K_0, K_1, K_2 be three bounded convex sets in the plane. Assume that K_0 is fixed and K_1, K_2 are moving with the kinematic densities dK_1 and dK_2, respectively. By successive application of (6.48) we have

$$m(K_1, K_2; K_0 \cap K_1 \cap K_2 \neq \emptyset) = \int_{K_0 \cap K_1 \cap K_2 \neq \emptyset} dK_1 \wedge dK_2$$

ISBN 0-201-13500-0

$$= \int_{K_0 \cap K_1 \neq \emptyset} [2(F_2 + f_{01}) + L_2 L_{01}]\, dK_1$$

$$= (2\pi)^2(F_1F_2 + F_0F_1 + F_0F_2) + 2\pi(F_0L_1L_2 + F_1L_0L_2 + F_2L_0L_1) \quad (6.63)$$

where we have applied formulas (6.56) and (6.62). The generalization of (6.63) to n moving convex domains is straightforward [523]. Related results have been obtained by Stoka [648, 650, 651] and Filipescu [190]. For the extension to E_n of (6.57) and to Lebesgue measurable sets, see [17, 18].

7. A Mean Value; Coverage Problems

Let K_0 be a fixed convex set of area F_0 and perimeter L_0 and let $K_1, K_2, \ldots, K_n$ be n congruent convex sets of area F and perimeter L. Assume that the sets K_i are dropped at random on the plane in such a way that they intersect K_0.

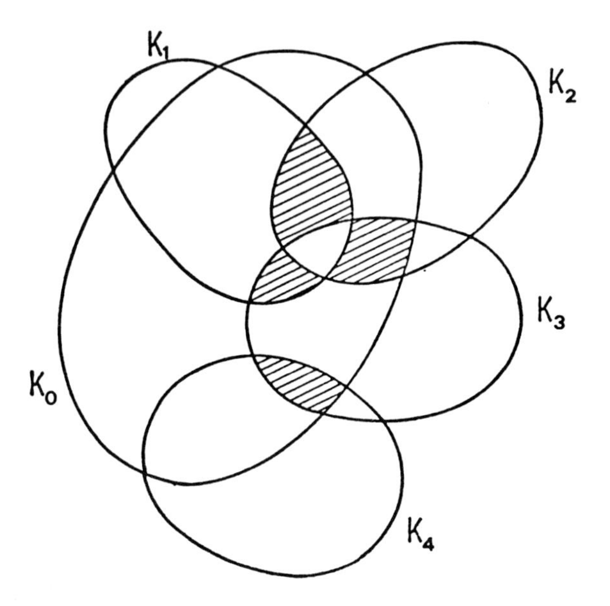

Figure 6.11.

We seek the mean value of the area f_r filled by those points of K_0 that are covered exactly by r sets K_i ($r = 0, 1, 2, \ldots, n$). In Fig. 6.11 the shaded area corresponds to $r = 2$, $n = 4$. To this end let us consider the integral

$$I_r = \int dP \wedge dK_1 \wedge dK_2 \wedge \cdots \wedge dK_n \quad (6.64)$$

over all points P of K_0 that are covered exactly by r sets K_i and to all positions

ISBN 0-201-13500-0

of K_i such that $K_i \cap K_0 \neq \emptyset$. If we leave P fixed, we have

$$I_r = \binom{n}{r} \int_{P \in K_0} (2\pi F)^r (2\pi F_0 + L_0 L)^{n-r} \, dP = \binom{n}{r} (2\pi F)^r (2\pi F_0 + L_0 L)^{n-r} F_0. \tag{6.65}$$

On the other hand, if we first leave $K_1, K_2, \ldots, K_n$ fixed, we have

$$I_r = \int_{K_i \cap K_0 \neq \emptyset} f_r \, dK_1 \wedge dK_2 \wedge \cdots \wedge dK_n. \tag{6.66}$$

Thus the required mean value is

$$E(f_r) = \frac{\binom{n}{r} (2\pi F)^r (2\pi F_0 + L_0 L)^{n-r} F_0}{[2\pi(F + F_0) + LL_0]^n}. \tag{6.67}$$

Consider the limit case in which $n \to \infty$, leaving constant the total area of the sets K_i, that is, $nF = \alpha =$ constant. Putting $F = \alpha/n$ and taking the limit as $n \to \infty$, we get

(a) If the sets K_i tend to line segments of length s, then $L \to 2s$ and we have

$$E(f_r) \to \frac{F_0}{r!} \left(\frac{\pi\alpha}{\pi\alpha + sL_0} \right)^r \exp\left(- \frac{\pi\alpha}{\pi F_0 + sL_0} \right). \tag{6.68}$$

(b) If the perimeter of the sets K_i tends to zero as $F \to 0$, then

$$E(f_r) \to \frac{F_0}{r!} \left(\frac{\alpha}{F_0} \right)^r \exp\left(- \frac{\alpha}{F_0} \right). \tag{6.69}$$

For $r = 0$ these formulas give the mean value of the area of the part of K_0 that remains uncovered by the sets K_i.

From (6.65) it follows that if a point P and n congruent convex sets K_i are chosen at random so as to meet a bounded convex set K_0, the probability that P is covered exactly by r sets K_i ($r = 0, 1, 2, \ldots, n$) is

$$p_r = \frac{\binom{n}{r} (2\pi F)^r (2\pi F_0 + L_0 L)^{n-r}}{[2\pi(F + F_0) + LL_0]^n}. \tag{6.70}$$

More difficult and not yet solved is the problem of finding the probability that *every* point of K_0 is covered exactly by r sets K_i. The so-called coverage problems are of the following form. Let S be a fixed set and A_i a sequence of random sets in the same space that contains S. Then, for fixed N we want to find the probability that $S \subset \bigcup_1^N A_i$ (coverage probability). Cooke [122,

ISBN 0-201-13500-0

123] has given general upper and lower bounds for coverage probabilities. These probabilities are in general difficult to calculate exactly and we seek asymptotic results. We shall give an example.

Let D be a domain contained in the convex set K_0. The probability that a random point in D is covered exactly by r convex sets K_i is given by (6.70). Assume that K_0 expands to the whole plane in such a way that $n/F_0 \to \rho$ (positive constant). Then, independently of the shape of K_0, we have $L_0/F_0 \to 0$ (Chapter 4, Section 4) and (6.70) becomes

$$\lim p_r = \frac{(\rho F)^r}{r!} e^{-\rho F}. \tag{6.71}$$

The countable set of convex sets K_i of area F generated by this process constitutes a field of convex sets of density ρ. According to (6.71) the number of them that cover a point of the plane is a Poisson random variable with parameter ρF. As before, the probability that every point of D is covered exactly by r sets K_i is not known. For the limiting case area of $D \to \infty$, Miles [412] has given the following asymptotic results.

The probability that every point of D is covered by at least r plates K_i is

$$\sim \exp\{- r\rho F^* \exp(- \rho F)[1 + \cdots + ((\rho F)^r/r!)]\}, \qquad r = 0, 1, 2, \ldots, \tag{6.72}$$

and the probability that every point of D is covered by at most r plates K_i is (asymptotically as $F^* \to \infty$)

$$\sim \exp\left[- (r+1)\rho F^* \exp(- \rho F)\left(\frac{(\rho F)^r}{r!} + \frac{(\rho F)^{r+1}}{(r+1)!} + \cdots\right)\right], \qquad r = 0, 1, \ldots, \tag{6.73}$$

where F^* is the area of D.

For a review of the literature on a class of coverage problems see that of W. C. Guenther and P. J. Terragno [251]. For covering problems on the sphere, see Chapter 18, Section 6, Note 4.

8. Notes and Exercises

1. *Two mean values.* Let K_0 be a fixed convex set of area F_0 and perimeter L_0 and let $K_1, K_2, \ldots, K_n$ be n congruent convex sets of area F and perimeter L. Assume that the sets K_i are dropped at random on the plane with the condition of intersecting K_0. Let u_r be the length of the boundary of the part of K_0 that is covered exactly by r sets K_i. This boundary may be composed of several closed curves. Then the mean value of u_r is the ratio of

$$\int u_r \, dK_1 \wedge dK_2 \wedge \cdots \wedge dK_n = \binom{n}{r}(2\pi F)^r(2\pi F_0 + LL_0)^{n-r}L_0$$

ISBN 0-201-13500-0

$$+ n\left[\binom{n-1}{r-1}(2\pi F)^{r-1}(2\pi F_0 + LL_0)^{n-r}2\pi F_0 L\right.$$

$$\left. + \binom{n-1}{r}(2\pi F)^r(2\pi F_0 + LL_0)^{n-r-1}2\pi F_0 L\right] \quad (6.74)$$

to

$$\int_{K_i \cap K_0 \neq \emptyset} dK_1 \wedge dK_2 \wedge \cdots \wedge dK_n = [2\pi(F + F_0) + LL_0]^n. \quad (6.75)$$

The mean value of the number N_r of separate domains that are covered exactly r times (in Fig. 6.11, $N_0 = 2$, $N_1 = 4$, $N_2 = 4$, $N_3 = 1$) is equal to the ratio of

$$\int N_r\, dK_1 \wedge dK_2 \wedge \cdots \wedge dK_n$$

$$= n(2\pi F)^{r-1}(2\pi F_0 + LL_0)^{n-r-1}2\pi F_0\left[\binom{n-1}{r}2\pi F + \binom{n-1}{r-1}(2\pi F_0 + LL_0)\right]$$

$$+ \binom{n}{r}(2\pi F)^r(2\pi F_0 + LL_0)^{n-r} + \binom{n}{2}2\pi L^2 F_0(2\pi F)^{r-2}(2\pi F_0 + LL_0)^{n-r-2}$$

$$\cdot\left[\binom{n-2}{r-2}(2\pi F_0 + LL_0)^2 + \binom{n-1}{r-1}4\pi F(2\pi F_0 + LL_0) + \binom{n-2}{r}(2\pi F)^2\right]$$

$$+ LL_0 n(2\pi F)^{r-1}(2\pi F_0 + LL_0)^{n-r-1}\left[\binom{n-1}{r-1}(2\pi F_0 + LL_0) + \binom{n-1}{r}2\pi F\right]$$

(6.76)

to (6.75). In (6.74) and (6.76) when the binomial coefficients make no sense, they should be replaced by zeros [560]. For related mean values and their generalization to three-dimensional space see [658].

2. *Number of clumps in a random distribution of convex sets.* Let N small convex laminae be placed at random on an area A. A *clump* of laminae is defined as a set of laminae each of which overlaps or is overlapped by at least one other member of the set. For convenience, a single lamina (i.e., nonoverlapped) will also be considered as a clump. Armitage [12] and Mack [385] have considered the problem of finding approximate formulas for the expected number of clumps, a problem that has importance in certain techniques of particle counting on a sample plate, where overlapping particles are difficult to distinguish from single particles, owing to their small size. We will consider the simple case in which all N laminae $K_1, K_2, \ldots, K_N$ are congruent and have area f and perimeter u.

Let P_i, ϕ_i be the point and the direction that define the position of K_i, so that the kinematic density for K_i is $dK_i = dP_i \wedge d\phi_i$. Let α be a domain

ISBN 0-201-13500-0

contained in the area A such that every K_i that intersects α has the point P_i in A. Given at random a lamina K_r, the probability that $P_r \in \alpha$ is α/A and therefore the mean number of laminae with this property is $N\alpha/A$. On the other hand, if K_r is fixed, the probability that $K_i \cap K_r = \emptyset$ for a random K_i is $1 - (4\pi f + u^2)/2\pi A$ (using (6.48) and the probability that $K_i \cap K_r = \emptyset$ for all $i \neq r$) is $[1 - (4\pi f + u^2)/2\pi A]^{N-1}$. Therefore the mean number of laminae K_r that do not overlap (mean number c_1 of clumps with a single lamina) is

$$E(c_1) = \frac{N\alpha}{A}\left[1 - \frac{4\pi f + u^2}{2\pi A}\right]^{N-1}. \tag{6.77}$$

To find the mean number of clumps (including clumps with a single lamina), we note that each clump can be determined by the lamina K_r for which the point P_r has the greatest abscissa x_r. Then, with K_r fixed, the probability that K_i $(i \neq r)$ intersects K_r is $(4\pi f + u^2)/2\pi A$ and by symmetry, the probability that K_i intersects K_r and the abscissas of P_i satisfy $x_i \geqslant x_r$ is $(4\pi f + u^2)/4\pi A$ and the probability that none of the K_i $(i \neq r)$ intersects K_r and has $x_i \geqslant x_r$ will be $[1 - (4\pi f + u^2)/4\pi A]^{N-1}$. Thus the mean number of K_r with such a property (i.e., the mean number of clumps c, including those composed of a single lamina) is

$$E(c) = \frac{N\alpha}{A}\left[1 - \frac{4\pi f + u^2}{4\pi A}\right]^{N-1}. \tag{6.78}$$

Assuming that $N \to \infty$, $A \to \infty$ in such a way that $N/A \to \lambda$ (mean number of laminae per unit area), we have

$$E(c_1) \to \lambda\alpha \exp[-\lambda(2f + u^2/2\pi)], \tag{6.79}$$

$$E(c) \to \lambda\alpha \exp[-\lambda(f + u^2/4\pi)]. \tag{6.80}$$

These formulas are due to Mack [385], who also considers the case of noncongruent laminae. For details and numerous examples of the theory of random clumping, see Roach's book [511].

3. *Expected number of aggregates in a random distribution of points.* Suppose that n points are independently and uniformly distributed over a domain of the plane. A k-aggregate with respect to a given domain D is defined as a set of k points that can be covered by such a domain D without covering any other point. An interesting problem is that of finding the mean value of the number of such aggregates for domains D of different size and shape. The problem has been considered by Mack [382, 383].

Ambarcumjan [10] uses the term "cluster" for the finite set of n points and assuming that it is contained in a basic circle of radius R centered at the origin, determines the probability $P_k(r)$ of finding k points of the cluster in a random circle $C(r)$ of constant radius r with center distributed uniformly in the interior of the basic circle. If the cluster M is centrally symmetric, with O as the center of symmetry, and the number of points n tends to infinity, then the mean length h_m of the perimeter of the convex hull of a randomly chosen m-subset of the cluster satisfies the inequality $h_{m+2} \geqslant 4\rho(1 - 2^{-m-1})$,

ISBN 0-201-13500-0

where ρ denotes the mean distance of the points of M from O. Another asymptotic result given by Ambarcumjan is $\rho^*/H \leqslant \frac{1}{4}$, where ρ^* denotes the mean distance between the pairs of points in M and H denotes the perimeter of the convex hull of M.

4. *Probability of detecting a convex domain by a linear search.* Let K_0 be a convex domain within which lies another convex domain K. The process of making a linear random cut of length s in K_0 is called a *linear search.* The problem arises of detecting K, that is, of the cut of length s intersecting the

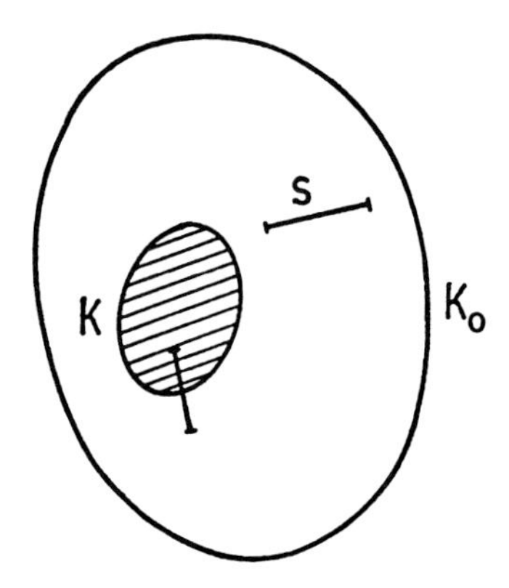

Figure 6.12.

domain K (Fig. 6.12). Assume that all the segments S of length s that cut K are contained in K_0. According to (6.33) the probability of detecting K is

$$p = \frac{m(S; S \cap K \neq \emptyset)}{m(S \subset K_0)} = \frac{2\pi F + 2sL}{m(S \subset K_0)} \tag{6.81}$$

where F and L are, respectively, the area and the perimeter of K. If K_0 is a circle of radius $R > s/2$, according to (6.34) we have

$$p(S \cap K \neq \emptyset) = \frac{4\pi F + 4sL}{\pi[4\pi R^2 - 8R^2 \arcsin(s/2R) - 2s(4R^2 - s^2)^{1/2}]}. \tag{6.82}$$

If K_0 is a rectangle of sides a, b ($a > s$, $b > s$), according to (6.35) we have

$$p(S \cap K \neq \emptyset) = \frac{\pi F + sL}{\pi ab - 2(a + b)s + s^2}. \tag{6.83}$$

(See [260].) Vinogradov and Zaragradsi [703] discussed the search for random objects uniformly positioned in a convex domain K_0, where the search is conducted by tracing a curve C of fixed length L, contained in K_0, with constant velocity. See also [468]; on search by networks, see [80]. For applications in ecology, see [404]; other results are given in [219].

ISBN 0-201-13500-0

5. *Self-intersections in continuous random walk.* An n-step random walk ($n \geqslant 3$) is a sequence of n straight segments, called steps, in the plane, such that each step is of length 1, the first step starts at the origin, and each successive step starts at the end of the previous one. Every step is in a random direction with a uniform distribution angle. Neglecting certain events of probability 0, a self intersection is defined as the event when for some i and j, with $1 \leqslant i < j \leqslant n$ and $j - 1 > 1$, the ith step and the jth step have in common exactly one point interior to each step.

Let $E(n)$ be the expected number of self-intersections. Then, for large n, the following asymptotic result is known.

$$E(n) \sim (2/\pi^2) n \log n.$$

See the article by Melzak [406], which also gives the exact value of $E(n)$ (rather involved).

6. *A new proof of the isoperimetric inequality.* Assume that K_0 and K_1 are two congruent convex sets. Then (6.47) gives $m(K_1; K_1 \cap K_0 \neq \emptyset) = 4\pi F_0 + L_0^2$. Using expression (6.32) for dK_1, we easily obtain

$$m(K_1; K_1 \cap K_0 \neq \emptyset) = 4\pi F_0 + 4 \int_{P \notin K_0} \left(\frac{\phi}{2} - \sin \frac{\phi}{2} \right) dP$$

where ϕ is the angle that K_0 can rotate with center $P \notin K_0$ without being exterior to its initial position. Therefore we have

$$\int_{P \notin K_0} \left(\frac{\phi}{2} - \sin \frac{\phi}{2} \right) dP = \frac{L_0^2}{4}.$$

Comparison with Crofton's formula (4.23) gives the isoperimetric inequality $L_0^2 - 4\pi F_0 \geqslant 0$ [531].

7. *Random figures not uniformly distributed.* The kinematic measure gives equal weight to all points P and all orientations ϕ of geometrical figures. This is a consequence of the assumed invariance under translation and rotation. If we assume only invariance under translation, we have a density of the form $dK^* = F(\phi)\, dP \wedge d\phi$ and can calculate integral formulas and probabilities for random figures whose orientations are not uniformly distributed. This nonisotropic case has been investigated by S. W. Dufour [155].

Exercise 1. *Some integral formulas on the intersection of convex sets.* Let K_0 and K_1 be two bounded convex sets in the plane, of area F_0, F_1 and perimeter L_0, L_1 respectively. We denote by F_{i} the area of $K_0 \cap K_1$; $F_{0\mathrm{e}}$ the area of $K_0 - K_0 \cap K_1$; $F_{1\mathrm{e}}$ the area of $K_1 - K_0 \cap K_1$; $L_{1\mathrm{i}}$ the length of the arc of ∂K_1 that is interior to K_0; $L_{0\mathrm{i}}$ the length of the arc of ∂K_0 that is interior to K_1; $L_{1\mathrm{e}}$ the length of the arc of ∂K_1 that is exterior to K_0; $L_{0\mathrm{e}}$ the length of the arc of ∂K_0 that is exterior to K_1. Prove as an exercise the following integral formulas:

$$\int L_{0\mathrm{i}}\, dK_1 = 2\pi F_1 L_0, \qquad \int L_{1\mathrm{i}}\, dK_1 = 2\pi F_0 L_1,$$

ISBN 0-201-13500-0

$$\int L_{1e}\, dK_1 = 2\pi L_1 F_1 + L_0 L_1^2, \qquad \int L_{0e}\, dK_1 = 2\pi L_0 F_0 + L_1 L_0^2,$$

$$\int (L_{1e}L_{0e} - L_{1i}L_{0i})\, dK_1 = L_0^2 L_1^2, \qquad \int (F_{1e}F_{0e} - F_i^2)\, dK_1 = F_0 F_1 L_0 L_1$$

where the integrals are taken over all positions of K_1 in which $K_1 \cap K_0 \neq \emptyset$.

EXERCISE 2. *Problems in geometric probability.* We shall give some examples of problems in geometric probability that can be solved using the results of this chapter. Throughout these problems, we denote by F_i, L_i the area and the perimeter, respectively, of the bounded convex set K_i.

(a) *Let K_0, K_1 be two convex sets in the plane such that $K_1 \subset K_0$. The convex set K_2 is placed at random in the plane so as to meet K_0. Find the probability that it intersects K_1.*

Solution. From (6.48) we deduce

$$p = \frac{2\pi(F_1 + F_2) + L_1 L_2}{2\pi(F_0 + F_2) + L_0 L_2}.$$

(b) *A point P and a convex set K_1 are chosen at random such that $P \in K_0$ and $K_0 \cap K_1 \neq \emptyset$. Find the probability that $P \in K_1 \cap K_0$.*

Solution.

$$p = \frac{2\pi F_1}{2\pi(F_0 + F_1) + L_0 L_1}.$$

(c) *A line G and a convex set K_1 are placed at random in the plane; both intersect K_0. Find the probability that $G \cap K_1 \cap K_0 \neq \emptyset$.*

Solution.

$$p = \frac{2\pi(F_0 L_1 + F_1 L_0)}{L_0[2\pi(F_0 + F_1) + L_0 L_1]}.$$

In particular, if K_1 is a line segment of length s we have $F_1 = 0$, $L_1 = 2s$ and the required probability becomes

$$p = \frac{2\pi F_0 s}{L_0(\pi F_0 + L_0 s)}.$$

For $s \to \infty$ we get that the probability that two random chords of a convex set K_0 will intersect inside K_0 is $p = 2\pi F_0/L_0^2$.

(d) *Let K_1 and K_2 be two random convex sets that intersect a fixed convex set K_0. Find the probability that $K_0 \cap K_1 \cap K_2 \neq \emptyset$.*

Solution. The solution is the ratio of (6.63) to the product $[2\pi(F_1 + F_0) + L_1 L_0][2\pi(F_2 + F_0) + L_2 L_0]$. In particular, if K_1 and K_2 are two segments of length s_1, s_2, respectively, we have

ISBN 0-201-13500-0

$$p = \frac{2\pi s_1 s_2 F_0}{(\pi F_0 + L_0 s_1)(\pi F_0 + L_0 s_2)}.$$

(e) *Assume n random congruent convex sets K_i that intersect a fixed convex set K_0. Show that the probability that all K_i have a point in common interior to K_0 is*

$$p = \frac{(2\pi)^n(F^n + nF_0F^{n-1}) + (2\pi)^{n-1}\left(nLL_0F^{n-1} + \binom{n}{2}F_0LF^{n-2}\right)}{[2\pi(F + F_0) + LL_0]^n}. \tag{6.84}$$

The proof follows by straightforward induction from Problem (d), F and L denoting, respectively, the area and the perimeter of the congruent sets K_i.

(f) *Assume that n points are chosen at random inside a circle K of radius R. Find the probability that they may be enclosed in a circle of radius r ($r \leqslant R$) contained in K.*

Solution. The problem is equivalent to that of finding the probability that n circles of radius r whose centers are chosen in K have a nonvoid intersection with the circle of radius $R - r$ concentric with K. Thus, the solution is given by (6.84) with $F = \pi r^2$, $L = 2\pi r$, $F_0 = \pi(R - r)^2$, $L_0 = 2\pi(R - r)$. The result is

$$p = \frac{r^{2n-3}}{\pi R^{2n}}\left[\pi r^3 + n\pi r(R - r)^2 + 2\pi n r^2(R - r) + \binom{n}{2}(R - r)^2\right].$$

(g) *Let K_0 be a fixed convex polygon. An oriented segment S^* such that it cannot intersect two nonconsecutive sides of K_0 is placed at random with the condition of intersecting K_0. Find* (i) *the probability p_0 that S^* is interior to K_0;* (ii) *the probability p_1 that S^* has one and only one point in common with ∂K_0;* (iii) *the probability that S^* has two intersection points with ∂K_0.*

The solutions are $p_i = m_i/2(\pi F_0 + L_0 s)$ where s is the length of S and m_i are the measures (6.44).

We shall give an application of this result. Assume a rectangle R of sides a, vb divided in v partial rectangles R_i by parallel lines to the base a at a distance b apart (Fig. 6.13). The measure of the set of segments S^* of length $s \leqslant b$ that are contained in R is, according to (6.35),

$$m(S^*; S^* \subset R) = 2\pi vab - 4s(a + vb) + 2s^2 \tag{6.85}$$

and the sum of the measures of the sets of those segments that are contained in the rectangles R_i ($i = 1, 2, \ldots$) is

$$vm(S^*; S^* \subset R_i) = [2\pi ab - 4s(a + b) + 2s^2]v. \tag{6.86}$$

Consequently, we can state that if a rectangle R of sides a, vb ($a > b$) is ruled with equidistant lines parallel to the base a at a distance b apart, then the probability that a needle of length s ($s < b$) chosen at random in the interior of R will not cross one of the parallel lines is

ISBN 0-201-13500-0

$$p = \frac{(\pi ab - 2s(a + b) + s^2)v}{\pi vab - 2s(a + vb) + s^2}. \tag{6.87}$$

If $a \to \infty$, $v \to \infty$ so that the entire plane is ruled by parallel lines a distance b apart, we again obtain the result of the Buffon needle problem (Chapter 5, Section 2).

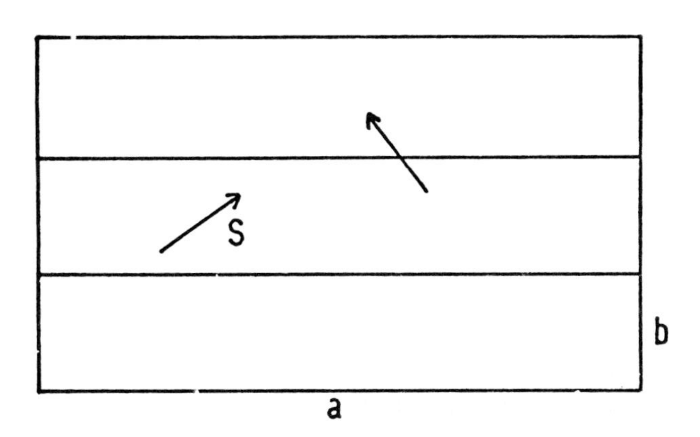

Figure 6.13.

(h) *Let K_0 be a convex polygon and let h be the maximal length of a segment which cannot intersect two nonconsecutive sides of K_0. Let P_1, P_2 be two random points chosen in the interior of K_0. Find the probability that the distance $|P_1P_2|$ is equal to or less than h.*

Solution. Let $|P_1P_2| = r$. From the expression of the area element in polar coordinates we have $dP_1 \wedge dP_2 = r\, dP_1 \wedge dr \wedge d\phi$ where ϕ is the angle between the line P_1P_2 and a reference direction in the plane. If we leave r fixed, then $dP_1 \wedge d\phi$ is the kinematic density for the segment P_1P_2 of length r, and using (6.44) we have

$$m(P_1, P_2; |P_1P_2| < h) = \int_{r \leqslant h} dP_1 \wedge dP_2 = \int_0^h m_0 r\, dr$$

$$= \pi F_0 h^2 - \tfrac{2}{3} L_0 h^3 + \tfrac{1}{8} h^4 \sum_{A_i} [1 + (\pi - A_i) \cot A_i]$$

and the desired probability is

$$p(r \leqslant h) = \frac{1}{F_0{}^2} \left\{ \pi F_0 h^2 - \frac{2}{3} L_0 h^3 + \frac{1}{8} h^4 \sum_{A_i} [1 + (\pi - A_i) \cot A_i] \right\}.$$

(i) *Let P_1, P_2 be two points chosen at random in the interior of a circle of diameter D. Find the probability that the distance $|P_1P_2|$ is equal to or less than $h \leqslant D$.*

ISBN 0-201-13500-0

Solution.

$$p(|P_1P_2| \leqslant h) = \frac{4}{\pi D^2}[\pi h^2 + \alpha(D^2/4 - h^2) - (D^2/4 + h^2/2)\sin\alpha]$$

where $\alpha = 2 \operatorname{arc} \sin(h/D)$.

Analogous results when P_1, P_2 are random points in the interior of triangles, squares, and polygons in general have been computed by E. Borel [64].

In general the distribution function prob$(r \leqslant x)$ of the distance r between two points of a convex set K is given by the integral with respect to r, from 0 to x, of $rm(r)/F^2$, where F is the area of K and $m(r)$ denotes the measure of all oriented segments of length r that lie inside K. This follows immediately from the equality $dP_1 \wedge dP_2 = r\,dr \wedge d\phi \wedge dP_2$. For the generalization to E_n, see [217].

ISBN 0-201-13500-0

CHAPTER 7

Fundamental Formulas of Poincaré and Blaschke

1. A New Expression for the Kinematic Density

Let $(O; x, y)$ be a fixed frame and let Γ_0 denote a fixed rectifiable curve composed of a finite number of arcs with continuous turning tangent at

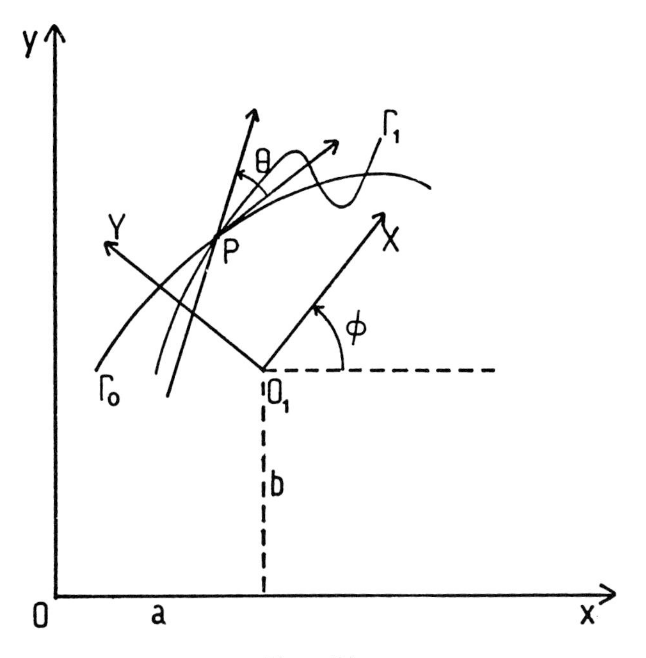

Figure 7.1.

ENCYCLOPEDIA OF MATHEMATICS and Its Applications, Gian-Carlo Rota (ed.).
1, Luis A. Santaló, Integral Geometry and Geometric Probability

ISBN 0-201-13500-0

every point. Let

$$x = x(s_0), \qquad y = y(s_0) \tag{7.1}$$

be the equations of Γ_0 referred to the arc length s_0 as parameter and to the coordinate system $(O; x, y)$ (Fig. 7.1).

Let $(O_1; X, Y)$ be a moving frame and Γ_1 a moving curve, also piecewise smooth, whose equations with respect to the frame $(O_1; X, Y)$ are

$$X = X(s_1), \qquad Y = Y(s_1) \tag{7.2}$$

where s_1 denotes the arc length of Γ_1. With respect to the coordinate system $(O; x, y)$ the equations of Γ_1 become

$$x = a + X \cos\phi - Y \sin\phi, \qquad y = b + X \sin\phi + Y \cos\phi \tag{7.3}$$

where a, b are the coordinates of O_1 and ϕ is the angle between Ox and O_1X.

The intersection points of Γ_0 and Γ_1 are given by the system

$$\begin{aligned} x(s_0) &= a + X(s_1)\cos\phi - Y(s_1)\sin\phi, \\ y(s_0) &= b + X(s_1)\sin\phi + Y(s_1)\cos\phi \end{aligned} \tag{7.4}$$

with the unknowns s_0, s_1. From (7.4) we deduce

$$\begin{aligned} da &= x'\,ds_0 - (X'\cos\phi - Y'\sin\phi)\,ds_1 + (X\sin\phi + Y\cos\phi)\,d\phi \\ db &= y'\,ds_0 - (X'\sin\phi + Y'\cos\phi)\,ds_1 - (X\cos\phi - Y\sin\phi)\,d\phi \end{aligned} \tag{7.5}$$

where the prime symbols denote derivatives. By exterior multiplication we get

$$da \wedge db \wedge d\phi = [(X'y' - x'Y')\cos\phi - (Y'y' + X'x')\sin\phi]\,ds_0 \wedge ds_1 \wedge d\phi. \tag{7.6}$$

If α_0 denotes the angle between the tangent to Γ_0 at the point $P \in \Gamma_0 \cap \Gamma_1$ and the x axis, and α_1 denotes the angle between the tangent to Γ_1 at the same point and the X axis, we have

$$x' = \cos\alpha_0, \qquad y' = \sin\alpha_0, \qquad X' = \cos\alpha_1, \qquad Y' = \sin\alpha_1, \tag{7.7}$$

and the kinematic density can be written $dK_1 = da \wedge db \wedge d\phi = \sin(\alpha_0 - \alpha_1 - \phi)\,ds_0 \wedge ds_1 \wedge d\phi$. If θ is the angle between Γ_0 and Γ_1 at P, then $|\theta| = |\alpha_0 - \alpha_1 - \phi|$ and since α_0 and α_1 are functions only of s_0 and s_1, we conclude that

$$dK_1 = |\sin\theta|\,ds_0 \wedge ds_1 \wedge d\theta. \tag{7.8}$$

ISBN 0-201-13500-0

This is a very useful expression for dK_1. It says that the probability that the angle between Γ_0 and a random curve Γ_1 at any intersection point lies between θ and $\theta + d\theta$ for $0 \leqslant \theta \leqslant \pi$ is $\frac{1}{2} \sin \theta \, d\theta$. The expected value of θ is thus

$$E(\theta) = \pi/2. \tag{7.9}$$

2. Poincaré's Formula

Suppose that we want to calculate the measure of the set of curves congruent to Γ_1 that have common point with the fixed curve Γ_0 or, in other words, the measure of the set of locations of Γ_1 in which it intersects Γ_0. This measure is equal to the integral of dK_1 over all values of the parameters a, b, ϕ for which $\Gamma_0 \cap \Gamma_1 \neq \emptyset$ and in general it lacks a simple form, depending on the shape of the curves Γ_0, Γ_1. However, if we integrate both sides of expression (7.8) over all values of s_0, s_1, θ (assuming $|\theta| \leqslant \pi$), we get, on the right-hand side,

$$\int_0^{L_0} ds_0 \int_0^{L_1} ds_1 \int_{-\pi}^{\pi} |\sin \theta| \, d\theta = 4L_0L_1. \tag{7.10}$$

On the left-hand side of (7.8) each position of Γ_1 has been counted as many times as it has intersection points with Γ_0, so that we can write

$$\int_{\Gamma_0 \cap \Gamma_1 \neq \emptyset} n \, dK_1 = 4L_0L_1 \tag{7.11}$$

where n is the number of intersection points of Γ_0 and Γ_1 and L_0, L_1 are the lengths of Γ_0 and Γ_1.

Formula (7.11) is known in integral geometry as *Poincaré's formula.* To be precise, Poincaré only considered formula (7.11) on the sphere as the mean value of the number of intersection points between spherical curves of given length [484, p. 143]. In such a form it was also known by Barbier [21]. The foregoing proof assumes that the curves are piecewise smooth [523]. However, the formula holds for the case in which the curves are only assumed rectifiable [377, 378, 455]. For a different approach, see [351].

If A denotes the area of the region covered by the points O_1 (origin of the frame attached to Γ_1) for all positions of Γ_1 such that $\Gamma_0 \cap \Gamma_1 \neq \emptyset$, then Poincaré's formula says that the expected number of intersections of Γ_1 and Γ_0 is $2L_0L_1/\pi A$. Obviously each curve can be composed by a separate set of curves of total length L_0 or L_1 (see [434]).

ISBN 0-201-13500-0

Note that multiplying both sides of (7.8) by $|\theta|$ and performing the integration over all values of s_0, s_1, θ $(-\pi \leqslant \theta \leqslant \pi)$, we obtain

$$\int_{\Gamma_0 \cap \Gamma_1 \neq \emptyset} \sum_1^n |\theta^i| \, dK_1 = 2\pi L_0 L_1 \tag{7.12}$$

where θ^i are the values of θ at the intersection points $\Gamma_0 \cap \Gamma_1$.

A particular case. If Γ_1 is a circle of radius r, we may choose the expression $dK_1 = dP \wedge d\phi$, P being the center of the circle. Then for each position of P the angle ϕ can vary from 0 to 2π without the change of n and Poincaré's formula gives

$$\int n \, dP = 4rL_0, \tag{7.13}$$

the integral being extended over the whole plane. From (7.13) we have $L_0 = (1/4r) \int n \, dP$, which may be taken as the definition of the length of a continuum of points Γ_0. This definition may be extended to curves in n-dimensional euclidean space (as has been done by Santaló [535]).

3. Total Curvature of a Closed Curve and of a Plane Domain

Let Γ be a closed oriented curve in the plane with continuous curvature κ. The *total curvature* of Γ is defined as the integral

$$c(\Gamma) = \int_\Gamma \kappa \, ds \tag{7.14}$$

where s is the arc length of Γ. Since $\kappa = d\tau/ds$, where τ is the angle of the oriented tangent to Γ with the x axis, it follows that c can be defined as

$$c(\Gamma) = \int_\Gamma d\tau, \tag{7.15}$$

that is, the total curvature is equal to the total variation of τ when the tangent describes Γ in the sense given by the orientation of Γ.

If Γ is the union of a finite number of arcs $a_1 = A_1A_2$, $a_2 = A_2A_3, \ldots,$ $a_m = A_mA_1$ of class C^2 and we denote by $\theta(A_i)$ the exterior angles at the corners (i.e., $\theta(A_i)$ is the angle of the tangent to a_{i-1} and the tangent to a_i at the point A_i, $-\pi \leqslant \theta \leqslant \pi$), the total curvature is defined by

$$c(\tau) = \sum_{i=1}^m \int_{a_i} d\tau + \sum_{i=1}^m \theta(A_i). \tag{7.16}$$

ISBN 0-201-13500-0

A closed curve is called *simple* if it has no double points. Then the so-called theorem of turning tangents asserts that the total curvature of a simple closed curve is $\pm 2\pi$ where the sign depends on the orientation of the curve [113].

The definition of total curvature can also be applied to plane domains. Let D be a domain of the plane bounded by a finite number of closed piecewise smooth curves without double points. We suppose these closed curves oriented so that the domain lies on the left over each contour. Then the total curvature of D is the total curvature of its boundary ∂D. For instance, assuming the positive orientation to be counterclockwise, the total curvature of the shaded domain in Fig. 7.2a is -2π and the total curvature of the shaded domain in Fig. 7.2b is 2π.

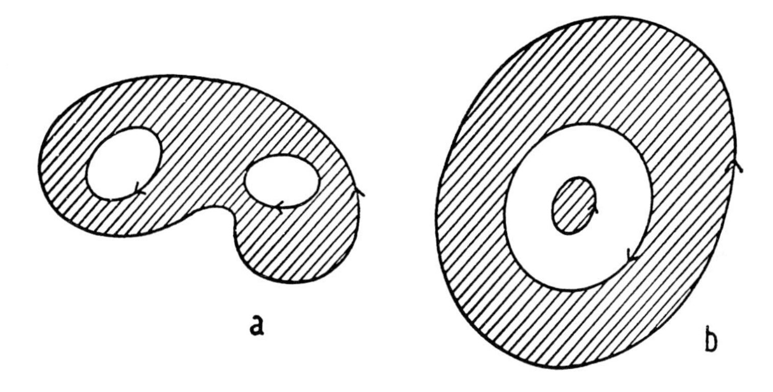

Figure 7.2.

Note that the total curvature of a domain D is always an integer multiple of 2π. Therefore we can write

$$c(D) = 2\pi\chi(D) \tag{7.17}$$

where $\chi(D)$ is an integer called the *Euler characteristic* of D. The Euler characteristic of a domain bounded by a single simple curve is $+1$. In general, if D can be decomposed into f faces or cells (sets homeomorphic to a disk) with a edges and v vertices, it is well known that (see, e.g., the results of Coxeter [128])

$$\chi(D) = v - a + f. \tag{7.18}$$

ISBN 0-201-13500-0

4. Fundamental Formula of Blaschke

Assume a domain D_0 bounded by a finite number of oriented closed piecewise smooth curves without double points. Let c_0 be the total curvature,

F_0 the area, and L_0 the perimeter of D_0. Let D_1 be another domain of the same kind of total curvature c_1, area F_1, and perimeter L_1. We assume D_1 moving in the plane with kinematic density dK_1. Calling c_{01} the total curvature of the domain $D_0 \cap D_1$, we wish to prove the following *fundamental kinematic formula of Blaschke*

$$\int_{D_0 \cap D_1 \neq \emptyset} c_{01} \, dK_1 = 2\pi(F_0 c_1 + F_1 c_0 + L_0 L_1). \tag{7.19}$$

Proof. Denote by a_i^j ($j = 0, 1$) the arcs with continuous tangent of ∂D_j and by A_i^j the corner points. According to (7.16) the total curvature of D_j is (Fig. 7.3)

$$c_j = \sum_i \int_{a_i^j} d\tau + \sum_i \theta(A_i^j), \qquad j = 0, 1. \tag{7.20}$$

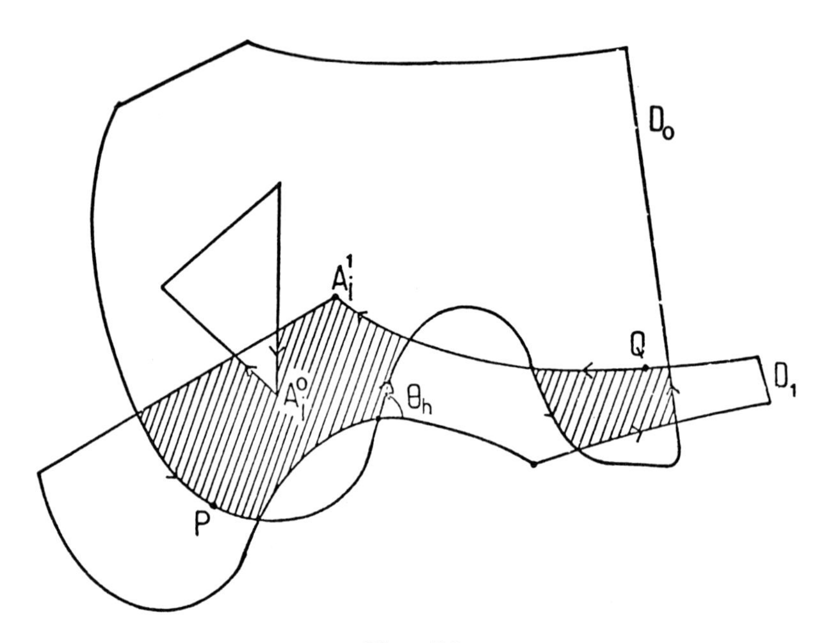

Figure 7.3.

If $P(s_0)$ denotes a point of ∂D_0 and $Q(s_1)$ denotes a point of ∂D_1 (s_0, s_1 being the respective arc lengths), the total curvature of $D_0 \cap D_1$ is

$$c_{01} = \int_{P \in D_1} d\tau(s_0) + \int_{Q \in D_0} d\tau(s_1) + \sum_{A_i^0 \in D_1} \theta(A_i^0) + \sum_{A_i^1 \in D_0} \theta(A_i^1) + \sum_h \theta_h \tag{7.21}$$

ISBN 0-201-13500-0

where θ_h are the angles between ∂D_0 and ∂D_1 at their intersection points, assuming these boundaries oriented so that D_0, D_1, and thus $D_0 \cap D_1$, lie on the left; then, except for positions of D_1 that belong to a set of measure zero, we have $0 \leqslant \theta_h \leqslant \pi$, so that $|\theta_h| = \theta_h$. Let us consider

$$I_1 = \int_{P \in D_1} d\tau(s_0) \wedge dK_1 \tag{7.22}$$

where dK_1 is the kinematic density referring to D_1 as a moving set. We have

$$I_1 = \int_{\partial D_0} d\tau(s_0) \int_{P \in D_1} dK_1 = 2\pi F_1 \int_{\partial D_0} d\tau(s_0) = 2\pi F_1 \sum_i \int_{a_i^0} d\tau(s_0) \tag{7.23}$$

and also

$$I_1 = \int_{D_0 \cap D_1 \neq \emptyset} dK_1 \int_{P \in D_1} d\tau(s_0) = \int_{D_0 \cap D_1 \neq \emptyset} \left(\int_{P \in D_1} d\tau(s_0) \right) dK_1. \tag{7.24}$$

Consequently we have

$$\int_{D_0 \cap D_1 \neq \emptyset} \left(\int_{P \in D_1} d\tau(s_0) \right) dK_1 = 2\pi F_1 \sum_i \int_{a_i^0} d\tau(s_0). \tag{7.25}$$

On the other hand, we have

$$\int_{D_0 \cap D_1 \neq \emptyset} \left(\sum_{A_i^0 \in D_1} \theta(A_i^0) \right) dK_1 = \sum_i \int_{A_i^0 \in D_1} \theta(A_i^0)\, dK_1 = 2\pi F_1 \sum_i \theta(A_i^0). \tag{7.26}$$

From (7.25) and (7.26), using that we may set $dK_1 = dK_0$ (invariance of kinematic density under inversion of the motion), it follows that

$$\int_{D_0 \cap D_1 \neq \emptyset} \left(\int_{Q \in D_0} d\tau(s_1) \right) dK_1 = 2\pi F_0 \sum_i \int_{a_i^1} d\tau(s_1) \tag{7.27}$$

and

$$\int_{D_0 \cap D_1 \neq \emptyset} \left(\int_{A_i^1 \in D_0} \theta(A_i^1) \right) dK_1 = 2\pi F_0 \sum_i (A_i^1) \tag{7.28}$$

and by virtue of (7.12), since $0 \leqslant \theta_h \leqslant \pi$, we also have

$$\int_{D_0 \cap D_1 \neq \emptyset} \left(\sum_i \theta_i \right) dK_1 = 2\pi L_0 L_1. \tag{7.29}$$

ISBN 0-201-13500-0

Summing the last five equalities and applying (7.21) and (7.20), we get the stated formula (7.19).

Particular cases. 1. If D_0 and D_1 are convex domains, we have $c_0 = c_1 = c_{01} = 2\pi$ and formula (7.19) reduces to (6.48), as expected.

2. If each domain D_0, D_1 is bounded by a single simple curve, then $c_0 = c_1 = 2\pi$ and $D_0 \cap D_1$ is composed of a certain number ν of separate domains,

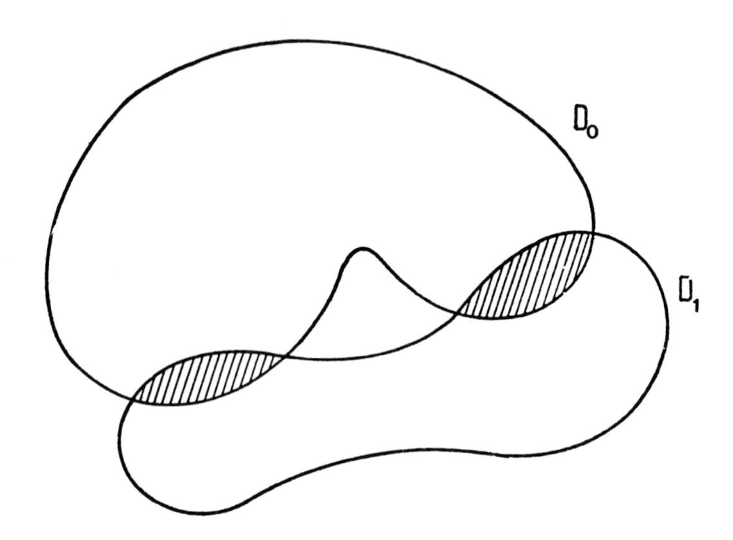

Figure 7.4.

each of total curvature 2π; hence $c_{01} = 2\pi\nu$. For instance, for the case in Fig. 7.4 we have $\nu = 2$. The fundamental formula (7.19) then takes the form

$$\int_{D_0 \cap D_1 \neq \emptyset} \nu \, dK_1 = 2\pi(F_0 + F_1) + L_0 L_1. \tag{7.30}$$

3. If D_0 is the union of a finite number m of separate convex sets $D_0{}^i$ $(i = 1, 2, \ldots, m)$ of total area F_0 and total perimeter L_0, we have $c_0 = 2\pi m$ and assuming that D_1 is bounded by a single simple curve, so that $c_1 = 2\pi$, we have

$$\int_{D_0 \cap D_1 \neq \emptyset} \nu \, dK_1 = 2\pi(F_0 + mF_1) + L_0 L_1 \tag{7.31}$$

where ν is the number of sets $D_0{}^i$ that are met by D_1 (e.g., in Fig. 7.5 $\nu = 5$, $m = 7$).

ISBN 0-201-13500-0

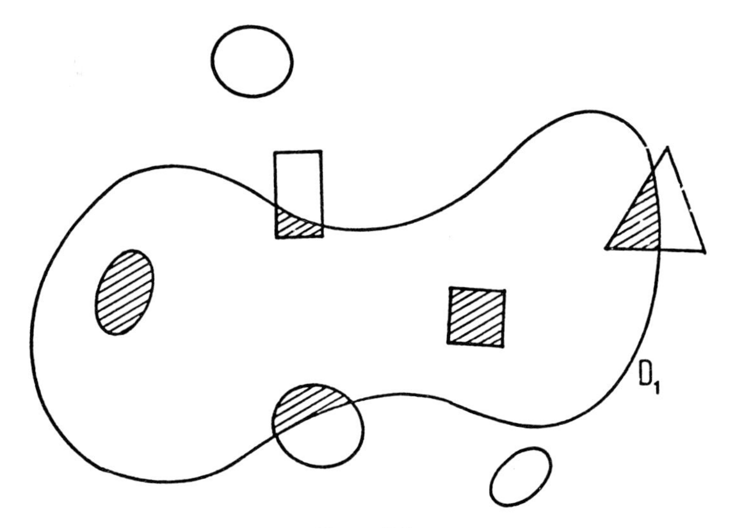

Figure 7.5.

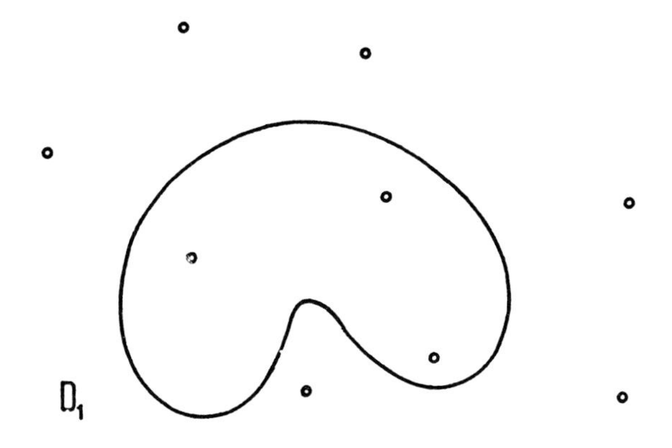

Figure 7.6.

In particular, if D_0 reduces to m points, we get

$$\int \nu \, dK_1 = 2\pi m F_1 \tag{7.32}$$

where ν is the number of points that belong to D_1 for each position of this domain (in Fig. 7.6 $m = 9$, $\nu = 3$) and the integral is extended over the whole of plane.

4. Assume that D_1 is a rectangle of sides a, b $(a > b)$. We have $F_1 = ab$, $L_1 = 2(a + b)$, $c_1 = 2\pi$. For the kinematic density we may take the form $dK_1 = dG_1^* \wedge dt$ (6.30), where G_1^* is the line through the center of the rectangle that is parallel to the sides of length a. Then, for each position of

ISBN 0-201-13500-0

G_1, the integral on the left-hand side of (7.19) can be divided into two parts: part I, corresponding to values of t for which the sides of length b do not intersect D_0; and part II, corresponding to values of t for which one or both sides of length b meet D_0. For the positions of part I, c_{01} does not depend on t and the integral of dt gives a factor of the form $a - h_1$, where h_1 is less than or equal to the diameter of D_0. If $a \to \infty$ and b remains fixed, then for any G_1 the positions of part II have finite measure, so that by dividing both sides of (7.19) by a and taking the limit as $a \to \infty$, we have

$$\int_{B \cap D_0 \neq \emptyset} c_{01}\, dG_1{}^* = 2\pi(bc_0 + 2L_0) \tag{7.33}$$

where B denotes the strip of breadth b that has G_1 as its midparallel. If we put $dG_1{}^* = 2dB$, where dB denotes the density for nonoriented strips (Chapter 5, Section 1) we get

$$\int_{B \cap D_0 \neq \emptyset} c_{01}\, dB = \pi(bc_0 + 2L_0), \tag{7.34}$$

which is the fundamental formula for strips of breadth b. For instance, in Fig. 7.7 $c_0 = -2\pi$, $c_{01} = 6\pi$.

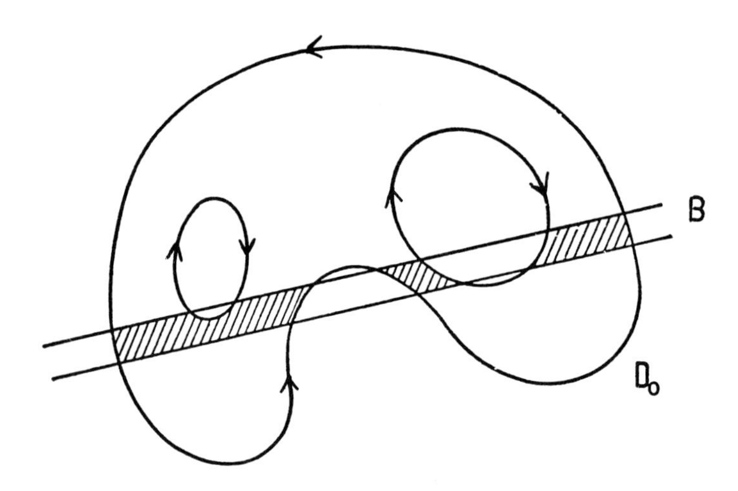

Figure 7.7.

The fundamental formula (7.19) may be extended to more general domains than those considered here: see the work of Blaschke [51] and Nöbeling [454]; for another kind of proof, see that of Hadwiger [270, 274].

ISBN 0-201-13500-0

5. The Isoperimetric Inequality

Let K_i ($i = 0, 1$) be two bounded congruent convex sets of area F and perimeter L. Blaschke's and Poincaré's formulas applied to K_0, K_1, and their boundaries give, respectively,

$$\int_{K_0 \cap K_1 \neq \emptyset} dK_1 = 4\pi F + L^2, \qquad \int_{K_0 \cap K_1 \neq \emptyset} n \, dK_1 = 4L^2. \tag{7.35}$$

Denoting by m_i the measure of the set of positions of K_1 in which ∂K_1 has i points in common with ∂K_0, (7.35) can be written (since each position of K_1 with i odd belongs to a set of measure zero)

$$m_2 + m_4 + \cdots = 4\pi F + L^2, \qquad 2m_2 + 4m_4 + 6m_6 + \cdots = 4L^2.$$

From these equations it follows that

$$L^2 - 4\pi F = m_4 + 2m_6 + 3m_8 + \cdots \tag{7.36}$$

and since the measures m_i are always $\geqslant 0$, we get the classical isoperimetric inequality

$$L^2 - 4\pi F \geqslant 0. \tag{7.37}$$

Note that (7.36) gives a geometrical interpretation of the *isoperimetric deficit* $\Delta = L^2 - 4\pi F$ of a convex set. If $K \equiv K_0 \equiv K_1$ is not convex, we consider the convex hull K^*. Between the areas F, F^* of K and K^* and the lengths L, L^* of ∂K and ∂K^* there exist the relations $F \leqslant F^*$, $L^* \leqslant L$, so that $L^2 - 4\pi F \geqslant L^{*2} - 4\pi F^* \geqslant 0$. Therefore the isoperimetric inequality holds good for any domain of the plane, not necessarily convex.

Stronger isoperimetric inequalities. Let K be a bounded convex set. Consider the maximal circle that can be contained in K (let r_i be its radius) and the minimal circle that can contain K (let r_e be its radius). Obviously $r_i \leqslant r_e$, r_i being equal to r_e if and only if K is a circle. Let K_1 be a circle of radius r satisfying the condition

$$r_i \leqslant r \leqslant r_e. \tag{7.38}$$

As before, let m_i be the measure of positions of K_1 in which its boundary has exactly i points in common with the boundary of K. In the same way as above, we have

$$m_2 + m_4 + \cdots = 2\pi F + 2\pi^2 r^2 + 2\pi r L, \qquad 2m_2 + 4m_4 + \cdots = 8\pi r L,$$

and thus $m_4 + 2m_6 + \cdots = 2\pi(rL - F - r^2)$. Since $m_i \geqslant 0$, we obtain

ISBN 0-201-13500-0

Bonnesen's inequality [62]

$$rL - F - \pi r^2 \geqslant 0, \tag{7.39}$$

which holds for any r within the limits of (7.38).

If we consider the identity

$$Lr - F - \pi r^2 = \frac{L^2}{4\pi} - F - \pi\left(\frac{L}{2\pi} - r\right)^2 \tag{7.40}$$

and call s_i and s_e to the expressions $(m_4 + 2m_6 + \cdots)/2\pi$ corresponding to the circles of radius r_i and r_e, respectively, we have

$$\frac{L^2}{4\pi} - F = \pi\left(\frac{L}{2\pi} - r_i\right)^2 + s_i, \qquad \frac{L^2}{4\pi} - F = \pi\left(\frac{L}{2\pi} - r_e\right)^2 + s_e.$$

Consequently

$$\frac{L^2}{4\pi} - F = \frac{1}{2}\left[\pi\left(\frac{L}{2\pi} - r_i\right)^2 + \pi\left(\frac{L}{2\pi} - r_e\right)^2 + s_i + s_e\right]$$

and therefore

$$L^2 - 4\pi F \geqslant \tfrac{1}{2}[(L - 2\pi r_i)^2 + (2\pi r_e - L)^2]. \tag{7.41}$$

This inequality is stronger than (7.37) and shows that the equality $L^2 - 4\pi F = 0$ holds only when $r_i = r_e = L/2\pi$, that is, when K is a circle. In other words, circles are the only sets that enclose the maximum area for a specified perimeter. This result and (7.36) give the following theorem (due to Fujiwara [206] and Bol [61]).

If a bounded convex set K is not a circle, then there are positions of kinematic measure > 0 in which the boundary of K has at least four points in common with the boundary of another convex set congruent to K.

Taking the general inequality $2(x^2 + y^2) \geqslant (x + y)^2$ into account, we find that from (7.41) it follows that

$$L^2 - 4\pi F \geqslant \pi^2(r_e - r_i)^2, \tag{7.42}$$

which is another isoperimetric inequality due to Bonnesen [62].

An upper limit for the isoperimetric deficit. Assume now that the boundary ∂K of the convex set K has a continuous radius of curvature ρ. Let ρ_m and ρ_M be the smallest and the greatest values, respectively, of ρ. Let K_1 be a circle of radius r such that either $r \leqslant \rho_m$ or $r \geqslant \rho_M$. According to (6.53), the measure of the set of circles K_1 that are contained in K or that contain K satisfies

ISBN 0-201-13500-0

$$\frac{m_0}{2\pi} = F + \pi r^2 - Lr = \pi\left(\frac{L}{2\pi} - r\right)^2 - \left(\frac{L^2}{4\pi} - F\right). \tag{7.43}$$

Since $m_0 \geqslant 0$ and (7.43) holds for $r = \rho_m$ and $r = \rho_M$, we have

$$\frac{L^2}{4\pi} - F \leqslant \pi\left(\frac{L}{2\pi} - \rho_m\right)^2, \qquad \frac{L^2}{4\pi} - F \leqslant \pi\left(\rho_M - \frac{L}{2\pi}\right)^2. \tag{7.44}$$

Using that $2\pi\rho_m \leqslant L \leqslant 2\pi\rho_M$, we can multiply these inequalities and we obtain

$$\frac{L^2}{4\pi} - F \leqslant \pi\left(\frac{L}{2\pi} - \rho_m\right)\left(\rho_M - \frac{L}{2\pi}\right). \tag{7.45}$$

The general inequality $4xy \leqslant (x + y)^2$ then gives

$$L^2 - 4\pi F \leqslant \pi^2(\rho_M - \rho_m)^2, \tag{7.46}$$

which is an upper limit for the isoperimetric deficit due to Bottema [66]. The equality sign holds if and only if $\rho_m = \rho_M$, that is, if K is a circle.

Pleijel [480] has shown that $L^2 - 4\pi F \leqslant \pi(4 - \pi)(\rho_M - \rho_m)^2$, which is an improvement of (7.46).

6. Hadwiger's Conditions for a Domain to Be Able to Contain Another

Following Hadwiger [263, 264], we wish to obtain some sufficient conditions in order that a given domain K_1 of area F_1 bounded by a simple closed curve of length L_1 can be contained in another domain K_0 of area F_0 bounded by a simple closed curve of length L_0. Since we will apply the formulas of Poincaré (7.11) and Blaschke (7.19), we first assume that the boundaries ∂K_0 and ∂K_1 are piecewise smooth, but then we shall see that the result remains valid under more general conditions.

Assume K_0 fixed and K_1 moving in the plane. Let n be the number of points of $\partial K_0 \cap \partial K_1$ and let ν be the number of pieces of which the intersection $K_0 \cap K_1$ is composed. If m_0 denotes the measure of all positions of K_1 in which either $K_1 \subset K_0$ or $K_1 \supset K_0$, formula (7.30) can be written

$$m_0 + \int_{\partial K_0 \cap \partial K_1 \neq \emptyset} \nu \, dK_1 = 2\pi(F_0 + F_1) + L_0 L_1. \tag{7.47}$$

Each piece of $K_0 \cap K_1$, when $\partial K_0 \cap \partial K_1 \neq \emptyset$, is bounded at least by an arc of ∂K_0 and an arc of ∂K_1. Thus $\nu \leqslant n/2$ and (7.11) and (7.47) give

$$m_0 \geqslant 2\pi(F_0 + F_1) - L_0 L_1. \tag{7.48}$$

ISBN 0-201-13500-0

Hence, if the right-hand side exceeds zero, then it is sure that there are positions of K_1 in which either it contains K_0 or it is contained in K_0. To remove the condition that ∂K_0 and ∂K_1 are piecewise smooth it is enough to consider domains $K_0^{(m)}$, $K_1^{(m)}$ whose boundaries are polygonal lines inscribed in ∂K_0 and ∂K_1. If we have $2\pi(F_0 + F_1) - L_0 L_1 > 0$ and K_0, K_1 are rectifiable curves, we can take $K_0^{(m)}$, $K_1^{(m)}$ such that $2(F_0^{(m)} + F_1^{(m)}) - L_0^{(m)} L_1^{(m)} > 0$ for any $m > \mu$. This means that for any $m > \mu$ one of the domains $K_0^{(m)}$, $K_1^{(m)}$ can be contained in the other. In the limit the same property will be true for K_0 and K_1. Observe that we say that K_0 is contained in K_1 when no point of K_0 is outside of K_1 (the boundary curves may have a common point). Therefore we have the following theorem:

In order that a domain K_1 bounded by a simple closed rectifiable curve can be contained in, or contain, a domain K_0 of the same kind, it is sufficient that

$$2\pi(F_0 + F_1) - L_0 L_1 > 0. \tag{7.49}$$

This condition does not really distinguish between the cases $K_0 \subset K_1$ and $K_1 \subset K_0$. To make this point clear, let us consider the inequality

$$L_0 L_1 - 4\pi F_1 > (L_0^2 L_1^2 - 16\pi^2 F_0 F_1)^{1/2} \tag{7.50}$$

where the expression under the root is always nonnegative, in accordance with the isoperimetric inequality (7.37), which holds for K_0 and K_1. If (7.50) holds, then (7.49) is also true and thus either K_0 can be contained in K_1 or K_1 can be contained in K_0. Furthermore, if (7.50) holds, then $L_0 L_1 - 4\pi F_1 > 0$, and by adding this inequality to (7.49) we get $2\pi(F_0 - F_1) > 0$, that is, $F_0 > F_1$. This means that the case in which $K_0 \subset K_1$ must be excluded and K_0 must be capable of containing K_1.

If instead of (7.50) we consider the inequality

$$4\pi F_0 - L_0 L_1 > (L_0^2 L_1^2 - 16\pi^2 F_0 F_1)^{1/2}, \tag{7.51}$$

we note that if it is satisfied, then (7.49) is also satisfied; furthermore, the inequality $4\pi F_0 - L_0 L_1 > 0$ or $16\pi^2 F_0^2 - L_0^2 L_1^2 > 0$, which follows from (7.51), together with $L_0^2 L_1^2 - 16\pi^2 F_0 F_1 > 0$, gives $F_0 > F_1$, that is, if (7.51) holds, then $K_1 \subset K_0$. Thus we can state:

In order that a domain K_1 bounded by a simple closed rectifiable curve can be contained in another domain K_0 of the same kind, it is sufficient that either condition (7.50) *or condition* (7.51) *be satisfied.*

Of course, these *sufficient* conditions are not necessary. (They were obtained by Hadwiger [263, 264].)

ISBN 0-201-13500-0

A particular case. Suppose that K_1 is a circle of radius r. From (7.50) we deduce the following result: in order that a circle of radius r can be contained in a domain K_0 of area F_0 bounded by a simple closed curve of length L_0 it is *sufficient* that

$$r < \frac{L_0 - (L_0{}^2 - 4\pi F_0)^{1/2}}{2\pi} = r_0. \tag{7.52}$$

Similarly, from (7.51) we have that in order for a circle of radius R to be able to contain a domain K_0 of area F_0 bounded by a simple curve of length L_0, it is *sufficient* that

$$R > \frac{L_0 + (L_0{}^2 - 4\pi F_0)^{1/2}}{2\pi} = R_0. \tag{7.53}$$

These inequalities show that for any ε there are circles of radius $r_0 - \varepsilon$ ($R_0 + \varepsilon$) that are contained in K_0 (that contain K_0) and thus taking the limit as $\varepsilon \to 0$ we get the theorem:

For a domain K_0 of area F_0 bounded by a simple curve of length L_0 there exists a circle of radius

$$r \geqslant \frac{L_0 - (L_0{}^2 - 4\pi F_0)^{1/2}}{2\pi} \tag{7.54}$$

contained in K_0 and a circle of radius

$$R \leqslant \frac{L_0 + (L_0{}^2 - 4\pi F)^{1/2}}{2\pi} \tag{7.55}$$

that contains K_0.

Since $L_0 - (L_0{}^2 - 4\pi F)^{1/2} \geqslant 2\pi(F_0/L_0)$, the right-hand side of (7.54) may be replaced by F_0/L_0. This result is the work of Grünwald and Turan [249].

7. Notes

1. *More isoperimetric inequalities*. If K_0 and K_1 are domains bounded by a single simple curve of finite length such that neither of them can be contained in the other, then conditions (7.50) and (7.51) cannot hold and we have

$$(L_0{}^2 L_1{}^2 - 16\pi^2 F_0 F_1)^{1/2} \geqslant L_0 L_1 - 4\pi F_1, \tag{7.56}$$

$$(L_0{}^2 L_1{}^2 - 16\pi^2 F_0 F_1)^{1/2} \geqslant 4\pi F_0 - L_0 L_1 \tag{7.57}$$

and by permuting K_0 and K_1, we obtain

ISBN 0-201-13500-0

$$(L_0^2 L_1^2 - 16\pi^2 F_0 F_1)^{1/2} \geqslant L_0 L_1 - 4\pi F_0, \tag{7.58}$$

$$(L_0^2 L_1^2 - 16\pi^2 F_0 F_1)^{1/2} \geqslant 4\pi F_1 - L_0 L_1. \tag{7.59}$$

If r_i is the radius of the largest circle K_1 that is contained in K_0, the measure of the positions in which $K_1 \subset K_0$ is zero. Thus (7.56) and (7.57) hold and we have

$$L_0^2 - 4\pi F_0 \geqslant (L_0 - 2\pi r_i)^2, \qquad L_0^2 - 4\pi F_0 \geqslant \left(\frac{2F_0}{r_i} - L_0\right)^2. \tag{7.60}$$

Similarly, if r_e is the radius of the smallest circle that contains K_0, we have

$$L_0^2 - 4\pi F_0 \geqslant \left(L_0 - \frac{2F_0}{r_e}\right)^2, \qquad L_0^2 - 4\pi F_0 \geqslant (2\pi r_e - L_0)^2. \tag{7.61}$$

From these inequalities, using that $x^2 + y^2 \geqslant \frac{1}{2}(x + y)^2$, we find that

$$L_0^2 - 4\pi F_0 \geqslant r_i^{-2}(F_0 - \pi r_i^2)^2, \qquad L_0^2 - 4\pi F_0 \geqslant F_0^2\left(\frac{1}{r_i} - \frac{1}{r_e}\right)^2. \tag{7.62}$$

Inequalities of this kind were given by Bonnesen [62] and Hadwiger [264]. Other sharpened forms of the isoperimetric inequality have been given by D. C. Benson [26].

2. *Minkowski's inequality for convex sets.* Let K_0, K_1 be two convex sets with support functions p_0, p_1, respectively. Assuming K_0 fixed, we know that the measure of the set of translates of K_1 that meet K_0 is $F_0 + F_1 + 2F_{01}^*$ where F_{01}^* is the mixed area of K_0 and the set K_1^* obtained by reflecting K_1 in a point (Chapter 6, Section 5). That is, if P_1 is a point attached to K_1, we have

$$\int_{K_0 \cap K_1 \neq \emptyset} dP_1 = F_0 + F_1 + 2F_{01}^*. \tag{7.63}$$

Consider now the convex curves ∂K_0, ∂K_1. We want to find an analogue of Poincaré's formula (7.11) for ∂K_0, ∂K_1 when only translations are taken into account. Let s_0, s_1 denote the arc lengths of the boundaries ∂K_0, ∂K_1 and let P_1 be a point of $\partial K_0 \cap \partial K_1$. Since the corner points of ∂K_0 and ∂K_1 form a countable set (Chapter 1, Section 1), we may assume that both curves are differentiable at P_1 and thus the area element at P_1 can be written $dP_1 = |\sin\theta|\, ds_0 \wedge ds_1$ where θ is the angle between ∂K_0 and ∂K_1 at P_1. If s_0 is fixed, integration of $|\sin\theta|\, ds_1$ gives the projection of ∂K_1 on the line perpendicular to the tangent to ∂K_0, which is equal to $2[p_1(\phi) + p_1(\phi + \pi)]$ where ϕ denotes the angle that makes the normal of ∂K_0 at P_1 with a reference direction. Therefore, if n is the number of intersections of ∂K_1 and ∂K_0, taking into account (1.12), we can write

$$\int_{\partial K_0 \cap \partial K_1 \neq \emptyset} n\, dP_1 = 2\int_{\partial K_0} [p_1(\phi) + p_1(\phi + \pi)]\, ds_0 = 4(F_{01} + F_{01}^*) \tag{7.64}$$

ISBN 0-201-13500-0

where F_{01} is the mixed area of K_0 and K_1. From this equality and from (7.63) we get the average of n,

$$E(n) = \frac{4(F_{01} + F_{01}^*)}{F_0 + F_1 + 2F_{01}^*}. \tag{7.65}$$

Assuming that ∂K_0 and ∂K_1 always intersect, we have $E(n) \geqslant 2$. Then (7.65) gives

$$2F_{01} \geqslant F_0 + F_1 \tag{7.66}$$

and since $(F_0 + F_1)^2 \geqslant 4F_0F_1$, we have

$$F_{01}^2 \geqslant F_0F_1 \tag{7.67}$$

It can be proved that this Minkowski inequality holds good for any pair of convex sets. Note that if K_1 is the unit circle, we have $F_1 = \pi$, $F_{01} = \frac{1}{2}\int_0^{2\pi} p_0 \, d\phi = L_0/2$ and (7.67) specializes to $L_0{}^2 - 4\pi F_0 \geqslant 0$, which is the isoperimetric inequality.

A stronger form of (7.67) is the following. The inradius of K_0 relative to K_1 is the largest real number r_0 such that a translate of r_0K_1 is in K_0 (r_0K_1 denotes the transform of K_1 by a homothety of ratio r_0) and the circumradius of K_0 relative to K_1 is the smallest real number R_0 such that a translate of R_0K_1 contains K_0. Obviously, $R_0 \geqslant r_0$ with equality if and only if K_1 and K_0 are similar and similarly placed. Note that if K_1 is the unit circle, then r_0 and R_0 are the radii r_i and r_e of Section 5. With these definitions the strongest inequality

$$F_{01}^2 - F_0F_1 \geqslant (F_1{}^2/4)(R_0 - r_0)^2 \tag{7.68}$$

can be proved. This inequality is due to Bonnesen [62]. For a proof in the style of Section 5, see that of Blaschke [51]; an elementary and detailed proof has been given by Flanders [201].

3. *Difference set and bisected chords of a convex set.* Following Chakerian and Stein [101], we are going to give an application of (7.64). Let ∂K_0 be a convex curve and let $\partial K_0{}^*$ be the reflection of ∂K_0 in the point $Q \in K_0$ (Fig. 7.8). The curve $\partial K_0{}^*$, up to a translation, is independent of Q. If P_1 is the image of a fixed point $P_0 \in K_0$ by the reflection in Q, then P_1 is fixed in $K_0{}^*$ and formula (7.64) applies. Noting that $dP_1 = 4dQ$, we have

$$\int_{Q\in K_0} n(Q)\, dQ = F_{00}^* + F_{00} \tag{7.69}$$

where $F_{00} = F_0$ (mixed area of K_0 and K_0) and F_{00}^* denotes the mixed area of K_0 and its reflection $K_0{}^*$; $n(Q)$ is the number of intersection points of ∂K_0 and $\partial K_0{}^*$.

Let us consider the so-called *difference set* DK_0, which has the support function $p(\phi) = p_0(\phi) + p_0(\phi + \pi)$. According to (1.9) the area of DK_0 is $F(DK_0) = 2F_0 + 2F_{00}^*$ and Minkowski's inequality (7.66) applied to K_0 and $K_0{}^*$ gives $F(DK_0) \geqslant 4F_0$ where equality holds if and only if K_0 is

ISBN 0-201-13500-0

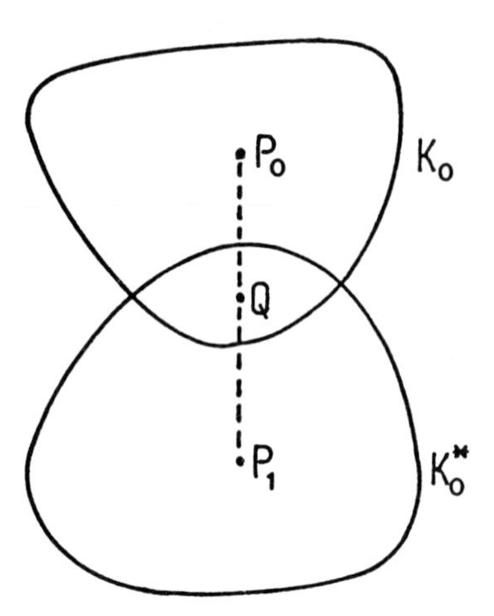

Figure 7.8.

centrally symmetric. Furthermore, it can be shown (as has been done by Bonnesen and Fenchel [63, p. 105]) that $F(DK_0) \leqslant 6F_0$ (with equality if and only if K_0 is a triangle), so that we have

$$F_0 \leqslant F_{00}^* \leqslant 2F_0 \tag{7.70}$$

where equality holds on the left for centrally symmetric sets and on the right for triangles.

Let $\nu(Q)$ denote the number of chords of K_0 bisected by Q. Clearly $2\nu(Q) = n(Q)$ and (7.69) and (7.70) give

$$F_0 \leqslant \int_{Q \in K_0} \nu(Q)\, dQ \leqslant \tfrac{3}{2}F_0 \tag{7.71}$$

with equality holding for the same cases as in (7.70). Since $\nu(Q) \geqslant 1$, (7.71) implies that K_0 is centrally symmetric if and only if the set of points of K_0 bisecting more than one chord has measure zero. This generalizes a result of Viet [701].

Put $M_k = \{Q \in K_0;\ \nu(Q) = k\}$ and call $F(M_k)$ = area of M_k. Then, from the foregoing formulas, Chakerian and Stein [101] have deduced the following results:

(a) If k is even, $F(M_k) = 0$;
(b) $3F_0 \leqslant 4F(M_1) \leqslant 4F_0$;
(c) If K_0 is of constant breadth, then $1 \geqslant F(M_1)/F_0 \geqslant F(M_1(R)/F(R) \sim 0.943$, where R is the Reuleaux triangle (Chapter 1, Section 4). Equality holds on the right if and only if K_0 is a Reuleaux triangle and on the left if and only if K_0 is a circle.

4. *Close packing of segments.* Let $n(Q)$ convex plates be placed at random inside a square Q of area $A(Q)$ in such a way that the distance between two of them is at least $2r$. Let $F(Q)$, $L(Q)$ denote the average area and the average

ISBN 0-201-13500-0

ISBN 0-201-13500-0

perimeter of these plates. Assume that Q increases infinitely in such a way that $F(Q)$, $L(Q)$, and $n(Q)/A(Q)$ tend to the limits F, L, N, respectively. Then it is easy to prove that $N \leqslant (F + Lr + \pi r^2)^{-1}$. If the plates reduce to line segments with an average length a, this result can be improved to $N \leqslant (2ar + 2\sqrt{3}\, r^2)^{-1}$ [181].

CHAPTER 8

Lattices of Figures

1. Definitions and Fundamental Formula

A set of points in the plane is called a *domain* if it is open and connected. A set of points is called a *region* if it is the union of a domain with some, none, or all its boundary points.

By a *lattice of fundamental regions* in the plane we understand a sequence of congruent regions $\alpha_1, \alpha_2, \ldots$ that satisfies the following conditions:

(i) Every point P of the plane belongs to one and only one region α_i;

(ii) Every α_i can be superposed on α_0 by a motion t_i that superposes on every α_h an α_s, that is, by a motion that takes the whole lattice onto itself.

The set of motions $\{t_i\}$ such that $\alpha_0 = t_i\alpha_i$ is a discrete subgroup of the group of motions. Such groups are called crystallographic groups. There are seventeen classes of nonisomorphic crystallographic groups, but for any given group there are infinitely many possible fundamental regions. It is not our purpose to present details on these groups, which are explored, for instance, in the books of Coxeter [127] and Guggenheimer [254]. Figures 8.1 to 8.5 are examples of lattices whose fundamental regions are squares, parallelograms, hexagons, or figures of more complicated shape.

Let D_0 be a figure in the plane, that is, a set of points, which can be a region bounded by a finite number of closed curves without double points, a set of rectifiable curves, a finite number of points, etc. Suppose that D_0 is contained in the fundamental region α_m of a given lattice. Let D_1 be another figure without the condition of being contained in a fundamental region. We assume D_0 fixed and D_1 moving with kinematic density $dK_1 = dP \wedge d\phi$, where P is a point of D_1 and ϕ denotes the angle that a fixed direction attached

ENCYCLOPEDIA OF MATHEMATICS and Its Applications, Gian-Carlo Rota (ed.).
1, Luis A. Santaló, Integral Geometry and Geometric Probability

ISBN 0-201-13500-0

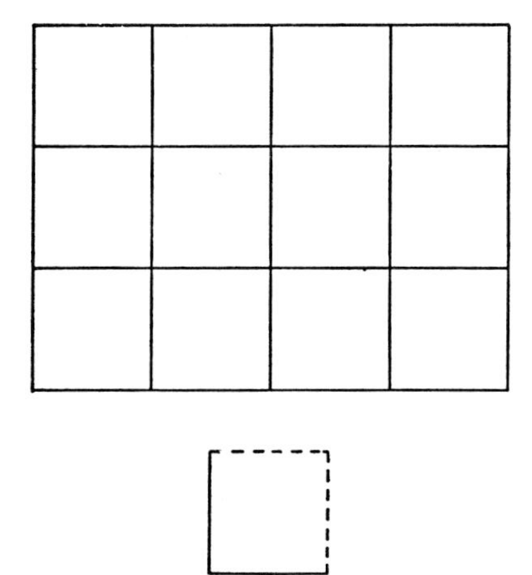

Figure 8.1.

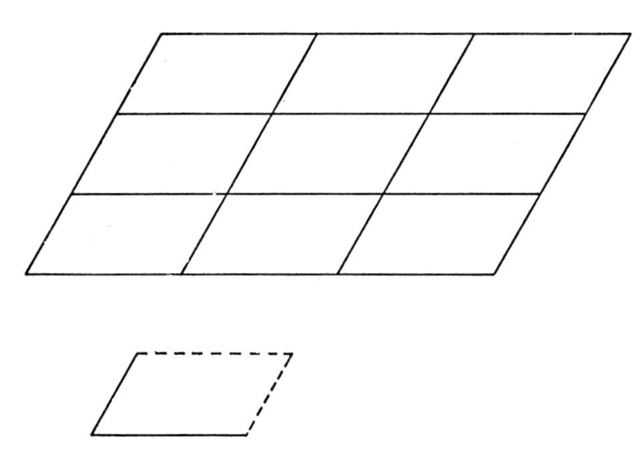

Figure 8.2.

to D_1 makes with a reference direction. Consider the integral

$$I = \int_{D_0 \cap D_1 \neq \emptyset} f(D_0 \cap D_1)\, dK_1 \tag{8.1}$$

where f denotes a real-valued function of the intersection $D_0 \cap D_1$ such that $f(\emptyset) = 0$. We can write

$$I = \sum_i \int_{\alpha_i} f(D_0 \cap D_1)\, dK_1 \tag{8.2}$$

ISBN 0-201-13500-0

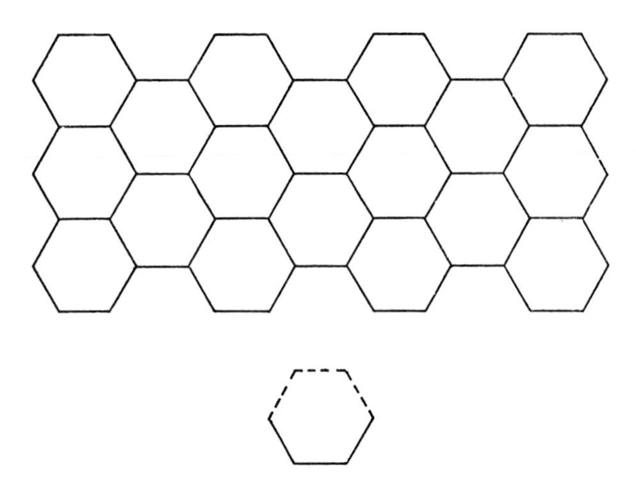

Figure 8.3.

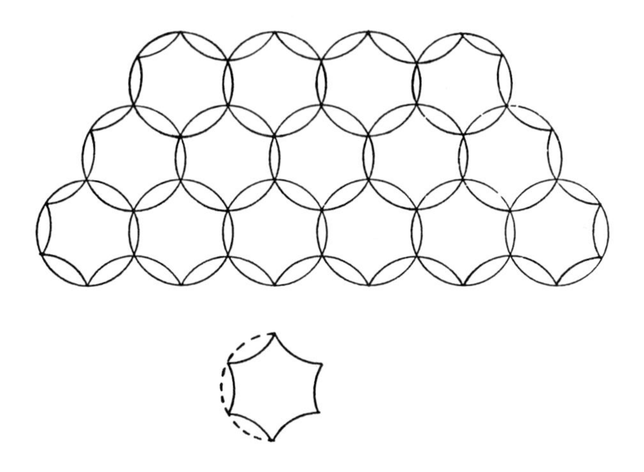

Figure 8.4.

where the sum is extended over all fundamental regions, and for each i the integral is extended over all $P \in \alpha_i$ and $0 \leqslant \phi \leqslant 2\pi$.

By the motion t_i the fundamental region α_i is carried over into α_0, that is, $t_i\alpha_i = \alpha_0$. Consequently, by the change of variables $K_1 \to t_iK_1$ (i.e., the moving frame P, ϕ is translated by t_i), taking the invariance of the kinematic density into account (i.e., $dK_1 = d(t_iK_1)$), we have

$$I = \sum_i \int_{\alpha_0} f(D_0 \cap t_iD_1)\, dK_1 \tag{8.3}$$

ISBN 0-201-13500-0

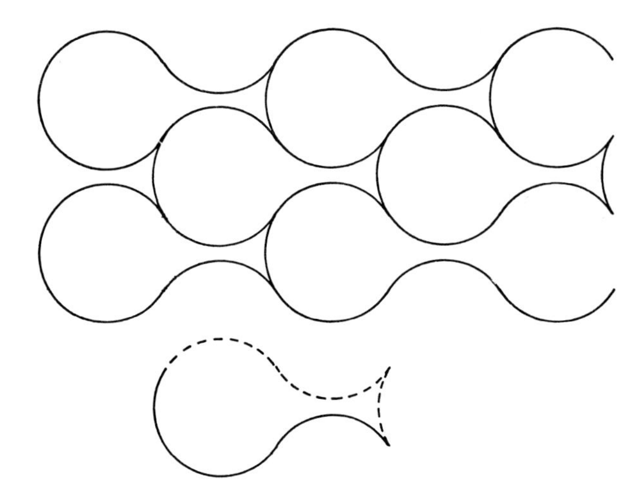

Figure 8.5.

and since the intersection $D_0 \cap t_i D_1$ is congruent with $t_i^{-1} D_0 \cap D_1$, we can write

$$I = \int_{\alpha_0} \left(\sum_i f(t_i^{-1} D_0 \cap D_1) \right) dK_1. \tag{8.4}$$

Therefore, if we draw on the plane the set of all sets $t_i^{-1} D_0$ $(i = 0, 1, 2, \ldots)$ and for each position of D_1 with $P \in \alpha_0$, $0 \leqslant \phi \leqslant 2\pi$, we carry out the sum $\sum_i f(t_i^{-1} D_0 \cap D_1)$, the value of integral (8.4) coincides with the value of integral (8.1).

2. Lattices of Domains

Suppose that D_0 and D_1 are closed domains bounded by a finite number of simple, piecewise smooth closed curves. Let F_i, L_i, c_i be the area, perimeter, and total curvature, respectively, of D_i $(i = 0, 1)$. Assuming D_0 contained in the fundamental region α_0, we consider the lattice of all translates $t_i^{-1} D_0$ $(i = 0, 1, 2, \ldots)$. Then, taking $f(D_0 \cap D_1) =$ total curvature of $D_0 \cap D_1$, from the fundamental formula (7.19) and from (8.4) we deduce

$$\int_{\alpha_0} c_{01}\, dK_1 = 2\pi(F_0 c_1 + F_1 c_0) + L_0 L_1 \tag{8.5}$$

where c_{01} means now the total curvature of the intersection of D_1 with all figures $t_i^{-1} D_0$, that is, with the lattice of figures generated by the reproduction in each α_i of D_0. The integration in (8.5) is over the field $P \in \alpha_0$, $0 \leqslant \phi \leqslant 2\pi$.

ISBN 0-201-13500-0

For instance, if D_0 and D_1 are both bounded by a single closed curve, we have $c_0 = c_1 = 2\pi$, $c_{01} = 2\pi\nu$, where ν denotes the number of pieces of the intersection $\sum_i (t_i^{-1}D_0) \cap D_1$ and we have

$$\int_{\alpha_0} \nu \, dK_1 = 2\pi(F_0 + F_1) + L_0 L_1. \tag{8.6}$$

For the case in Fig. 8.6 we have $\nu = 8$.

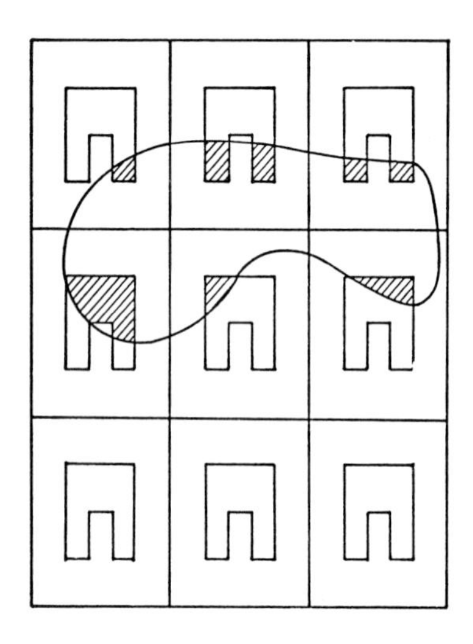

Figure 8.6.

The case in which D_0 coincides with the fundamental region α_0 together with the necessary boundary points in order that D_0 be a closed set is of special interest. Then, formula (8.6) holds and ν means the number of pieces into which D_1 is divided by the lattice. For instance, in the case of Fig. 8.7, $\nu = 8$.

Since the volume of the field of integration is $2\pi\alpha_0$, where α_0 also denotes the area of the fundamental region α_0, from (8.6) we deduce

The mean value of the number of pieces into which a closed domain D_1 of area F_1 bounded by a single curve of length L_1 is divided when it is put at random on a lattice whose fundamental regions have area α_0 and contour of length L_0 is

$$E(\nu) = \frac{2\pi(\alpha_0 + F_1) + L_0 L_1}{2\pi\alpha_0}. \tag{8.7}$$

ISBN 0-201-13500-0

The number N of fundamental regions that have a common point with D_1 is always $N \leqslant \nu$ (e.g., in the case of Fig. 8.7 we have $N = 6$, $\nu = 8$). Consequently, $E(N) \leqslant E(\nu)$ and we get

Any closed domain D_1 of area F_1 bounded by a single curve of length L_1 can be covered by a number μ of fundamental regions, of area α_0 and contour L_0, that satisfies the inequality $\mu \leqslant E(\nu)$ where $E(\nu)$ has the value (8.7).

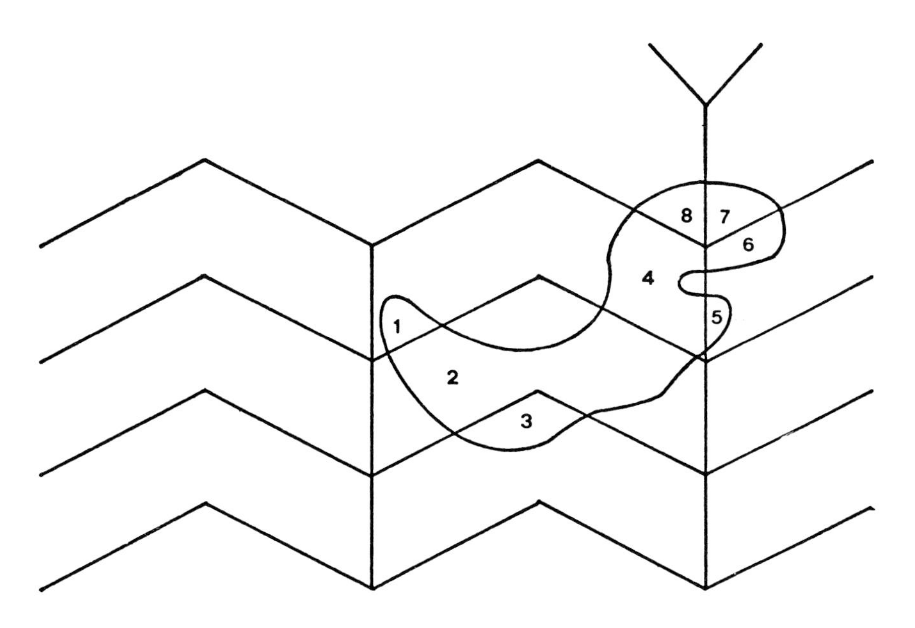

Figure 8.7.

Application of this result to the case of the lattice of squares of side a ($\alpha_0 = a^2$, $L_0 = 4a$) gives the result that every domain D_1 bounded by a single curve can be covered by a number of squares that is less than or equal to

$$\mu_s = 1 + \frac{2L_1}{\pi a} + \frac{F_1}{a^2}. \tag{8.8}$$

If the lattice is that of regular hexagons of side a (Fig. 8.3), we find that D_1 can be covered by a number of hexagons not exceeding

$$\mu_h = 1 + \frac{2L_1}{\sqrt{3}\,\pi a} + \frac{2F_1}{3\sqrt{3}\,a^2}. \tag{8.9}$$

If we consider the circles circumscribed about the regular hexagons of side a, we get the result that D_1 can be covered by a number of circles of

ISBN 0-201-13500-0

radius a that is less than or equal to the same value as μ_h. These results are due to Hadwiger [263]. For the generalization to n dimensions, see Chapter 15, Section 9. Problems on lattices in Riemannian two-dimensional space and in three-dimensional euclidean space have been considered by Trandafir [675, 677, 677a]; see also [464, 465].

3. Lattices of Curves

Let D_0 and D_1 be piecewise smooth curves of lengths L_0 and L_1, respectively. From (7.11) and (8.4), taking $f(D_0 \cap D_1)$ = number of points of the intersection $D_0 \cap D_1$, it follows that

$$\int_{\alpha_0} n \, dK_1 = 4L_0L_1 \tag{8.10}$$

where n means the number of intersection points of D_1 with the lattice of curves $t_i^{-1}D_0$ $(i = 0, 1, 2, \ldots)$. In Fig. 8.8 we have $n = 4$. Consequently, we have

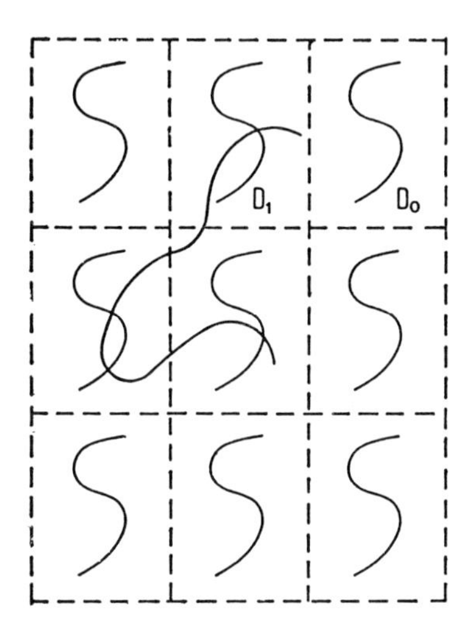

Figure 8.8.

If the fundamental regions of a given lattice have area α_0 and each contains a curve of length L_0, then the mean value of the number of intersection points of these curves with a curve D_1 of length L_1 placed at random on the plane is

$$E(n) = \frac{2L_0L_1}{\pi\alpha_0}. \tag{8.11}$$

ISBN 0-201-13500-0

For instance, consider the lattice of rectangles of sides a, b and let D_0 be composed of two consecutive sides of these rectangles (Fig. 8.9). We have $\alpha_0 = ab$, $L_0 = a + b$, and thus

$$E(n) = \frac{2(a + b)L_1}{\pi ab}. \tag{8.12}$$

If $a \to \infty$, the lattice becomes the lattice of parallel lines a distance b apart and we get $E(n) = 2L_1/\pi b$. In particular, if D_1 is a line segment of length

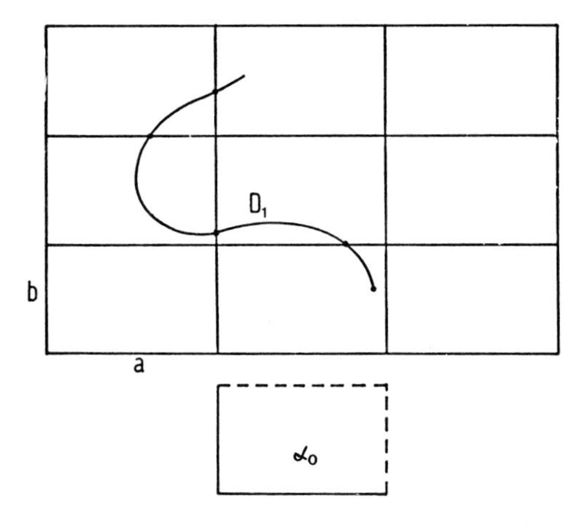

Figure 8.9.

$L_1 \leqslant b$, then n can take only the values 0, 1 and $E(n)$ is the probability that a random segment of length L_1 will cross one of the parallel lines. We again have the result of Buffon (Chapter 5, Section 2).

4. Lattices of Points

Let D_0 be composed of a finite number, say m, of points. Taking $f(D_0 \cap D_1)=$ number of points of D_0 that are contained in D_1, from (7.32) and (8.4) we find that

$$\int_{\alpha_0} n \, dK_1 = 2\pi m F_1 \tag{8.13}$$

where n is the number of points of the lattice that belong to D_1. For instance, in the case of Fig. 8.10 we have $m = 3$, $n = 4$ and in Fig. 8.11 $m = 1$, $n = 2$. The result may be stated as follows.

ISBN 0-201-13500-0

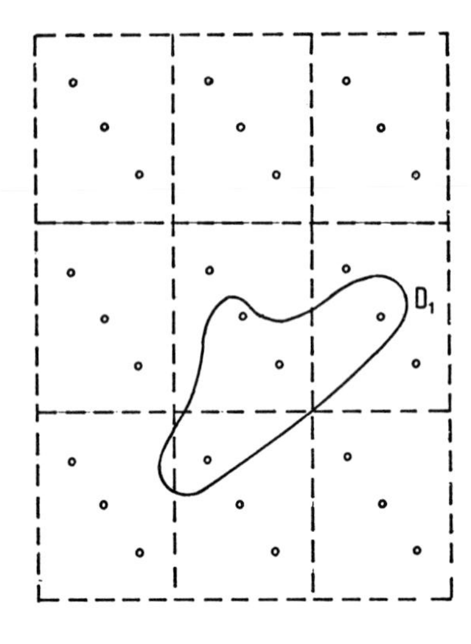

Figure 8.10.

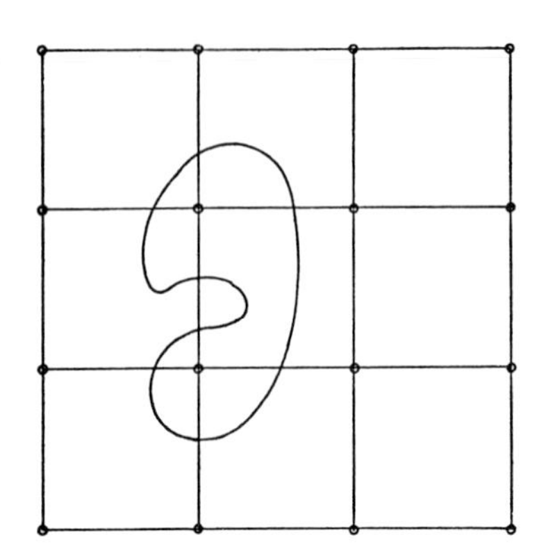
Figure 8.11.

If each fundamental region contains m points, the mean number of points contained in a domain D_1 of area F_1 placed at random in the plane is

$$E(n) = \frac{mF_1}{\alpha_0}. \tag{8.14}$$

We will give three applications of this mean value.

1. Consider the lattice of points formed by the vertices of a lattice of equilateral triangles of side a. The fundamental regions are parallelograms

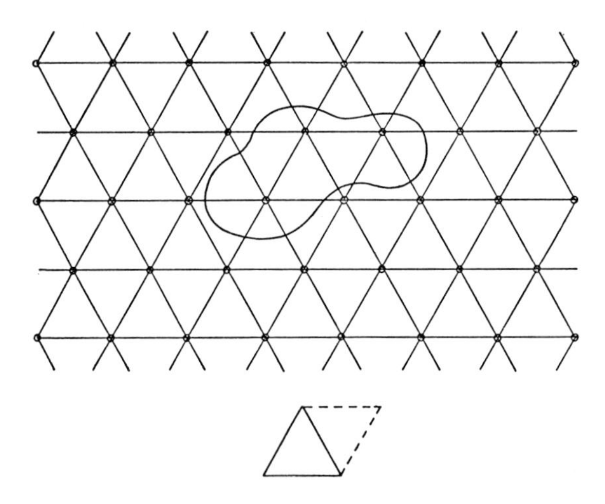
Figure 8.12.

ISBN 0-201-13500-0

formed by two triangles and we have $\alpha_0 = (\sqrt{3}/2)a^2$, $m = 1$ (Fig. 8.12). Thus we have $E(n) = 2F_1(\sqrt{3}\,a^2)^{-1}$ and we get the theorem:

It is always possible to put n points inside a given domain of area F_1 so that the minimal distance a between two of them satisfies the condition $a^2 \geqslant 2F_1(\sqrt{3}\,n)^{-1}$ [179].

2. If only translations of D_1 are taken into account, formula (7.32) holds good without the factor 2π on the right-hand side, so that the mean value (8.14) holds good without change. We will prove that if D_1 is bounded and closed, there are positions of D_1 in which it contains at least $[mF_1/\alpha_0] + 1$ lattice points, where [ ] means "integral part." Indeed, if there exists a set of positions of positive measure for which $n < [mF_1/\alpha_0]$, to compensate it there will be a set of positions of positive measure for which $n > [mF_1/\alpha_0]$ and therefore for some positions n will be equal to $[mF_1/\alpha_0] + 1$. Suppose that, except for locations of D_1 that belong to a set of measure zero, we have $n = [mF_1/\alpha_0]$. Take a position of D_1 in which some of these points are in the boundary ∂D_1. Let ε be the minimal distance to D_1 from the lattice points that do not belong to D_1; since D_1 is bounded and closed, we have $\varepsilon > 0$. By translations of length $< \varepsilon$ we can ascertain that some of the lattice points that belong to ∂D_1 are left out of D_1 and thus we will have a set of positions of positive measure for which $n < [mF_1/\alpha_0]$, contrary to the hypothesis. Consequently we have proved Blichfeldt's theorem [53]:

If D_1 is a bounded, closed domain of area F_1 in the plane of a lattice of points that has m points in each fundamental region, assumed of area α_0, there are always translates of D_1 that contain at least $[mF_1/\alpha_0] + 1$ points of the lattice.

3. Given a lattice of points, a difficult problem is that of finding the shape of the domain of minimal area such that it contains at least a given number of lattice points for any position of it in the plane.

Consider, for instance, the lattice of points of integral coordinates with reference to a pair of rectangular axes, and ask for the closed convex set of minimal area such that, however it is displaced in the plane, a point of the integral lattice is covered. In this case, from (8.14) it follows that $E(n) = F_1$ and the required condition implies $E(n) \geqslant 1$, so that we have $F_1 \geqslant 1$. If we consider translations only, the solution is given obviously by the unit square, with $F_1 = 1$. When all motions are taken into account, the area $F_1 = 1$ is not enough. D. B. Sawyer ([593], for the case of convex sets having a center) and J. J. Schäffer ([594], for the general case) have shown that the solution is given by the domain consisting of a unit square and two parabolic segments, the angle formed by the tangents to the parabola and the side of the square being $\pi/4$ (Fig. 8.13). The area of this minimal set is $F_1 = 4/3$. (See also [332].)

ISBN 0-201-13500-0

A similar problem is the following (devised by Scott [603]). Let K be a bounded convex set in the euclidean plane. Let $\delta(K)$ be the width of K. If $\delta(K) \geqslant \frac{1}{2}(2 + \sqrt{3})$, then K contains a point of the integral lattice. Equality is necessary when and only when K is the equilateral triangle of side length $(2 + \sqrt{3})/\sqrt{3}$.

Figure 8.13.

A domain D not necessarily convex is said to be admissible relative to the lattice of points in the plane with integer coordinates if no points of the lattice are interior to D. Then, Bender [25] proved that $2F \leqslant L$, where F is the area and L the perimeter of D. See also the results of Silver [612], Hadwiger [277, 278], and Wills [726, 727]. The following general result was obtained by Bokowski, Hadwiger, and Wills [60]: let K be a convex body in euclidean n-space with volume V and surface area F, and let N be the number of lattice points (points with integer coordinates) interior to K. Then $N > V - F/2$ and the factor $\frac{1}{2}$ cannot be replaced by a smaller number. (See also [598a].)

The problem of finding in the plane the convex region of smallest area that, after suitable translations and rotations, will cover any given arc of length L has been considered by Poole and Gerriets [491], who assert that a closed rhombus R with diagonals L and $3^{-1/2}L$ will cover every such arc. They also mention that R may be "truncated" to obtain an even smaller region of area less than $0.2861L^2$ with the required property. Related problems have been considered by Chakerian and Klamkin [100] and Wetzel [714a].

The problem of determining the variance of the number of lattice points covered by a random region is in general difficult. Kendall and Rankin [334] have treated the case of lattice points inside a random sphere in E_n. (For related results see [335].)

5. Notes and Exercise

1. *Probabilities on lattices.* Consider a lattice whose fundamental regions are convex polygons of area α_0. The boundaries of the fundamental regions form a linear lattice (or infinite graph) generated by the transforms by t_i of a certain arc of the boundary of α_0. Let u_0 denote the length of this arc. For instance, for the lattice of parallelograms of sides a, b and angle θ in Fig. 8.2,

ISBN 0-201-13500-0

we have $\alpha_0 = ab \sin \theta$, $u_0 = a + b$; for the lattice of regular hexagons in Fig. 8.3, we have $\alpha_0 = (3\sqrt{3}/2)a^2$, $u_0 = 3a$. In this section, the term lattice will indicate the lattice of the boundaries of the fundamental regions.

Let S^* be an oriented segment of length r such that it cannot have more than two points in common with the lattice. The measure m_0 of the set of locations of S^* in which it is contained in a fundamental domain is given by (6.44). On the other hand, calling m_i $(i = 0, 1, 2)$ the measure of the locations of S^* in which it has i points in common with the lattice, according to (8.10) we have $m_1 + 2m_2 = 4ru_0$ and the measure of all locations of S^* that are not equivalent under the group of motions $\{t_i\}$ is $m_0 + m_1 + m_2 = 2\pi\alpha_0$. If m_0 is known, these formulas give m_1 and m_2 and we have solved the problem of finding the probability that S^* has 0, 1, or 2 common points with the lattice.

Examples. (i) Consider the lattice of parallelograms in Fig. 8.2. According to (6.44) we have

$$m_0 = 2\pi ab \sin \theta - 4r(a + b) + r^2[2 + (\pi - 2\theta) \cot \theta], \tag{8.15}$$

and making use of the results above we obtain

$$m_1 = 4r(a + b) - 2r^2[2 + (\pi - 2\theta) \cot \theta],$$

$$m_2 = r^2[2 + (\pi - 2\theta) \cot \theta]. \tag{8.16}$$

Consequently, we can state

A needle of length r is thrown at random on the plane containing a lattice of congruent parallelograms of sides a, b and angle θ. We assume that the needle cannot intersect the lattice in more than two points. Then the probabilities of 0, 1, *or* 2 *intersection points are*

$$p_0 = 1 - \frac{2r(a + b)}{\pi ab \sin \theta} + \frac{r^2}{2\pi ab \sin \theta}[2 + (\pi - 2\theta) \cot \theta],$$

$$p_1 = \frac{2r(a + b)}{\pi ab \sin \theta} - \frac{r^2(2 + (\pi - 2\theta) \cot \theta}{\pi ab \sin \theta},$$

$$p_2 = \frac{r^2[2 + (\pi - 2\theta) \cot \theta]}{2\pi ab \sin \theta}.$$

(ii) For a lattice of regular hexagons of side a and a needle of length $r \leqslant a$, the probabilities are

$$p_0 = 1 - \frac{4\sqrt{3}\, r}{3\pi a} + \left(\frac{\sqrt{3}}{3\pi} - \frac{1}{9}\right)\frac{r^2}{a^2},$$

$$p_1 = \frac{4\sqrt{3}\, r}{3\pi a} - \left(\frac{2\sqrt{3}}{3\pi} - \frac{2}{9}\right)\frac{r^2}{a^2},$$

$$p_2 = \left(\frac{\sqrt{3}}{3\pi} - \frac{1}{9}\right)\frac{r^2}{a^2}.$$

ISBN 0-201-13500-0

These problems have been considered by Santaló [532].

2. *Lattices of convex sets.* Suppose that each fundamental region contains a single convex set D_0 of area F_0 and perimeter L_0. Let D_1 be another convex set of area F_1 and perimeter L_1 such that it cannot intersect more than one

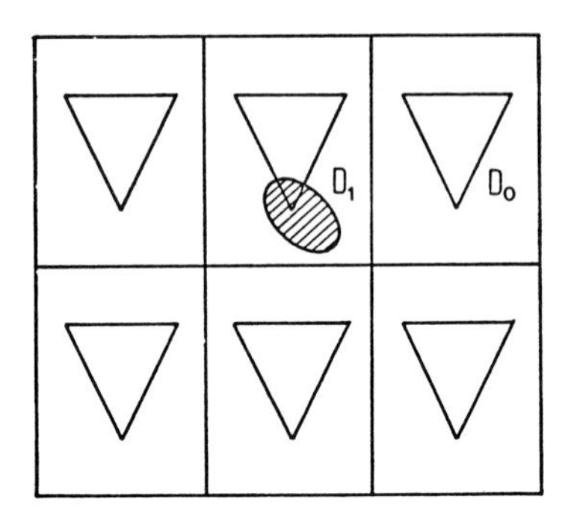

Figure 8.14.

of the sets D_0 (Fig. 8.14). Then, according to (7.19) we have $m(D_1; D_0 \cap D_1 \neq \emptyset) = 2\pi(F_1 + F_0) + L_0 L_1$ and since the measure of all positions of D_1 (up to a motion t_i) is equal to $2\pi\alpha_0$, we have

The probability that a convex set D_1 placed at random on the plane of a lattice of convex sets D_0 will intersect one of them (under the hypothesis that D_1 can meet only one of the sets D_0) is

$$p = \frac{2\pi(F_0 + F_1) + L_0 L_1}{2\pi\alpha_0}.$$

EXERCISE. Let D_1 be a segment of length r, greater than or equal to the diameter of D_0 and such that it cannot intersect more than one of the convex sets D_0 of the lattice. Show that the probabilities of cutting the boundaries ∂D_0 in 0, 1, or 2 points are

$$p_0 = 1 - \frac{F_0}{\alpha_0} - \frac{rL_0}{\pi\alpha_0}, \qquad p_1 = \frac{2F_0}{\alpha_0}, \qquad p_2 = \frac{rL_0}{\pi\alpha_0} - \frac{F_0}{\alpha_0}.$$

3. *Lattices of equilateral triangles.* Consider the lattice of equilateral triangles of side a (Fig. 8.12). A segment of length $r \leqslant (\sqrt{3}/2)a$ placed at random on the lattice can cut the lattice in 0, 1, 2, or 3 points. The respective probabilities are [532]

$$p_0 = 1 - \frac{4\sqrt{3}\,r}{\pi a} + 2\left(\frac{\sqrt{3}}{2\pi} + \frac{1}{3}\right)\frac{r^2}{a^2},$$

$$p_1 = \frac{4\sqrt{3}\,r}{\pi a} - \left(\frac{\sqrt{3}}{\pi} + \frac{5}{3}\right)\frac{r^2}{a^2},$$

ISBN 0-201-13500-0

$$p_2 = \left(\frac{4}{3} - \frac{\sqrt{3}}{\pi}\right)\frac{r^2}{a^2}, \qquad p_3 = \left(\frac{\sqrt{3}}{\pi} - \frac{1}{3}\right)\frac{r^2}{a^2}.$$

4. *Convex sets of particular shape and lattices.* (a) Consider a lattice of rectangles of sides a, b ($a \geqslant b$) (Fig. 8.15) and let K_1 be a convex set of

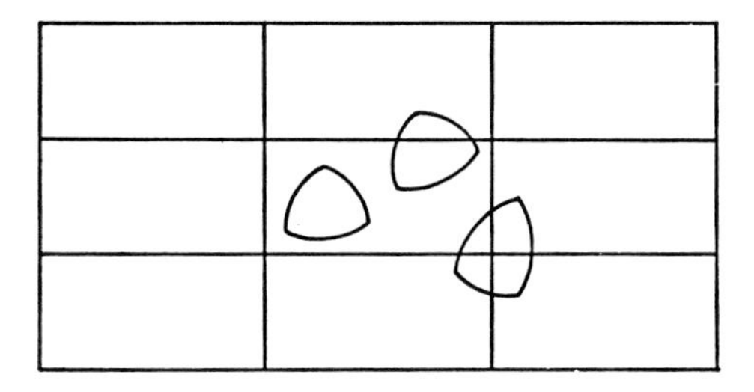

Figure 8.15.

constant breadth $h \leqslant b$. Assuming K_1 placed at random on the plane, the probabilities that it meets 0, 1, or 2 lines of the lattice are

$$p_0 = 1 - \frac{(a+b)h}{ab} + \frac{h^2}{ab}, \qquad p_1 = \frac{(a+b)h}{ab} - \frac{2h^2}{ab}, \qquad p_2 = \frac{h^2}{ab}.$$

(b) Consider the lattice of equilateral triangles of side a (Fig. 8.16) and a triangular convex set K_1 whose circumscribed equilateral triangles have side

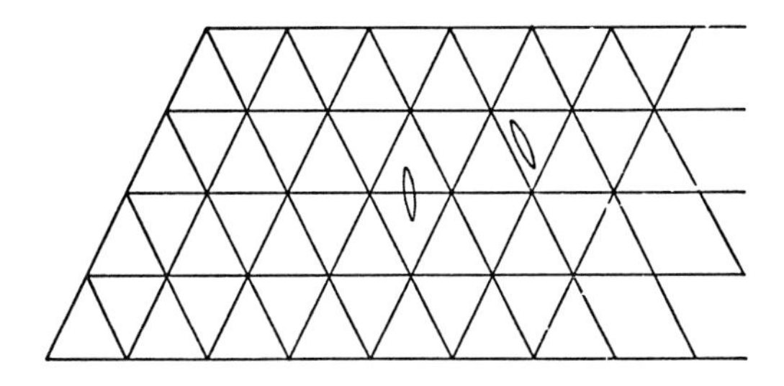

Figure 8.16.

$a_1 \leqslant a$ (Chapter 1, Section 4). If we assume that K_1 is placed at random in the plane, the probability that it intersects the lattice is $p = 2a_1/a - (a_1/a)^2$ and the probability that K_1 falls entirely inside a triangle of the lattice is $p_0 = 1 - (2a_1/a) + (a_1/a)^2$ [551].

(c) Consider the lattice of squares of side a and with center at each vertex of the lattice draw a circle of radius $a/4$ (Fig. 8.17). Show, as an exercise, that (i) the mean number of intersections of the circles of the lattice with a random curve of length L is $E(n) = L/a$. (ii) If a lamina K of arbitrary shape with 16 small holes in it is dropped at random on the plane, the mean number of holes that fall inside a circle of the lattice is equal to π.

ISBN 0-201-13500-0

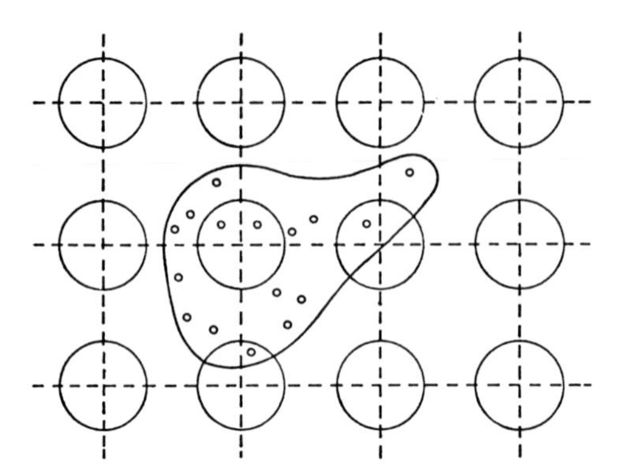

Figure 8.17.

5. *Measuring the length of a curve.* Applying formula (8.12) to a lattice of squares of side a, we obtain $L_1 = (\pi/4)aE(n)$. This result can be used to provide a practical method of measuring the length of curves. Suppose that we cover the curve with a transparent grid of squares of side a. If we count the number of intersections of the curve with the lines of the grid and take the average by rotating the grid through successive multiples of π/n, then $(\pi/4)a$ times this average is an estimate for L_1. Estimates of the errors involved in this method have been analyzed by Moran [427].

ISBN 0-201-13500-0

PART II

General Integral Geometry

CHAPTER 9

Differential Forms and Lie Groups

1. Differential Forms

In Part I we treated integral geometry in the plane from an elementary point of view. The densities for sets of points and sets of lines and the kinematic density were obtained by particular devices and their invariance properties were proved directly and independently in each case.

We want to include the foregoing results and ideas in a general context. In order to do this we need (i) a base space M that contains certain geometric objects (in the preceding part, M was the euclidean plane and the objects were points, lines, or congruent figures); (ii) a group of transformations $\mathfrak{G}$ operating on M (in the preceding part $\mathfrak{G}$ was the group of motions in the plane). Once we have the space M and the group $\mathfrak{G}$, the next step is to define a measure for sets of geometric objects invariant under $\mathfrak{G}$ and then to compute this measure for some specific sets in order to obtain some results of geometric interest.

We will assume known the usual definitions and fundamental theorems about differentiable manifolds and Lie groups, as contained for instance in the books of Sterneberg [628], Kobayashi and Nomizu [341a], or Bishop and Goldberg [32]. In this chapter we will give only a résumé of the principal properties of differential systems and invariant measures on Lie groups and homogeneous spaces.

ENCYCLOPEDIA OF MATHEMATICS and Its Applications, Gian-Carlo Rota (ed.).
1, Luis A. Santaló, Integral Geometry and Geometric Probability

ISBN 0-201-13500-0

Let M be a differentiable manifold of class C^∞ and dimension n. Let T_p denote the tangent vector space at the point p and let $T(M)$ denote the union of all tangent spaces T_p. A vector field X is a function $X: M \to T(M)$ that assigns to each point $p \in M$ a vector $X_p \in T_p$. If X is a vector field and f belongs to the ring of functions of class C^∞ defined on M, then Xf is the function $p \to X_p(f)$ or $Xf(p) = X_pf$. If x_i is a local coordinate system, a vector field is given by the expression

$$X = \sum_1^n a^i(x) \frac{\partial}{\partial x_i} \tag{9.1}$$

where we will always assume that the functions $a^i(x)$ are of class C^∞.

If T_p^* denotes the dual vector space of T_p (called the cotangent space of M at p) and $T^*(M)$ is the union of all spaces T_p^*, a 1-form on M is a function $M \to T^*(M)$ that assigns to each point $p \in M$ a covector $\omega_p \in T_p^*$. The expression of a 1-form in a local coordinate system is

$$\omega = \sum_1^n \alpha_i(x)\, dx_i \tag{9.2}$$

where $\alpha_i(x)$ are functions of class C^∞. If $\bigwedge T_p^*(M)$ denotes the exterior algebra over $T_p^*(M)$, a q-form $\omega^{(q)}$ is an assignment of an element of degree q in $\bigwedge T_p^*(M)$ to each point $p \in M$. In terms of a local coordinate system x_i, $\omega^{(q)}$ can be expressed uniquely as

$$\omega^{(q)} = \sum_{i_1 < \cdots < i_q} \alpha_{i_1 \ldots i_q}\, dx_{i_1} \wedge dx_{i_2} \wedge \cdots \wedge dx_{i_q}$$

where $\alpha_{i_1 \ldots i_q}$ are functions on M of class C^∞. There are no q-forms $\neq 0$ when $q > n$. The q-forms are called differential forms of degree q and we adopt the convention that functions are differential forms of degree 0. Clearly the set of all q-forms on M is a module on the ring of functions of class C^∞ defined on M. A basis for this module is the set of exterior products $dx_{i_1} \wedge \cdots \wedge dx_{i_q}$ $(i_1 < \cdots < i_q)$. More generally, if $\omega_1, \omega_2, \ldots, \omega_n$ is a basis for the 1-forms, then the set of exterior products $\omega_{i_1} \wedge \cdots \wedge \omega_{i_q}$ $(i_1 < \cdots < i_q)$ is a basis for the q-forms.

The exterior derivative or exterior differential of $\omega^{(q)}$ is defined by

$$\begin{aligned} d\omega^{(q)} &= \sum_i d\alpha_{i_1 \ldots i_q} dx_{i_1} \wedge \cdots \wedge dx_{i_q} \\ &= \sum_{i,k} \frac{\partial \alpha_{i_1 \ldots i_q}}{\partial x_k} dx_k \wedge dx_{i_1} \wedge \cdots \wedge dx_{i_q} \end{aligned} \tag{9.3}$$

where the sums are extended over all permutations of the indices i_h such that $i_1 < \cdots < i_q$ and $1 \leqslant k \leqslant n$. Note the properties

ISBN 0-201-13500-0

$$d(d\omega^{(q)}) = 0, \qquad d(\omega^{(q)} \wedge \omega^{(s)}) = d\omega^{(q)} \wedge \omega^{(s)} + (-1)^q \omega^{(q)} \wedge d\omega^{(s)} \quad (9.4)$$

for each integer $q, s \geqslant 1$.

For the case of 1-forms it is sometimes useful to have an expression of the exterior differential in terms of ordinary differentials. To this end note that the exterior product $dx_1 \wedge dx_2$ can be considered as the expression of the element of signed area in R^2. Thus, if (dx_1, dx_2) and $(\delta x_1, \delta x_2)$ are two independent infinitesimal vectors, we have $dx_1 \wedge dx_2 = dx_1\, \delta x_2 - dx_2\, \delta x_1$ and the exterior differential of the 1-form ω (9.2) can be written

$$d\omega = \sum_{i,k}^{n} \frac{\partial \alpha_i}{\partial x_k} dx_k \wedge dx_i = \sum_{i,k} \frac{\partial \alpha_i}{\partial x_k} (dx_k\, \delta x_i - \delta x_k\, dx_i). \quad (9.5)$$

Using that $d\, \delta x_i = \delta\, dx_i$ (which is the statement of the commutativity of the second-order partial derivatives), we have [90, p. 36]

$$d\omega = d\omega(\delta) - \delta\omega(d). \quad (9.6)$$

This formula is invariant under change of coordinates and will prove useful later on. Note that the symbol d on the left-hand side denotes exterior differentiation and that on the right-hand side denotes ordinary differentiation. However, since (9.6) is not used often, there is no danger in retaining this somewhat misleading notation.

Finally, note that given r 1-forms $\phi_i = \sum \alpha_{ij}\, dx_j$ $(i = 1, 2, \ldots, r)$, we have

$$\phi_1 \wedge \phi_2 \wedge \cdots \wedge \phi_r = \sum_{i_1 < \cdots < i_r} \det(\alpha_{i_1} \cdots \alpha_{i_r})\, dx_{i_1} \wedge \cdots \wedge dx_{i_r} \quad (9.7)$$

where $\det(\alpha_{i_1} \cdots \alpha_{i_r})$ denotes the determinant formed by the columns of order $i_1, \ldots, i_r$ of the matrix (α_{ij}). In particular, if $r = n$, we have

$$\phi_1 \wedge \cdots \wedge \phi_n = \det(\alpha_{ij})\, dx_1 \wedge dx_2 \wedge \cdots \wedge dx_n. \quad (9.8)$$

The 1-forms $\phi_1, \ldots, \phi_r$ are independent if the rank of the matrix (α_{ij}) is r at every point of M, that is, according to (9.7), if and only if

$$\phi_1 \wedge \cdots \wedge \phi_r \neq 0. \quad (9.9)$$

2. Pfaffian Differential Systems

Let ω be a 1-form on the differentiable manifold M and let X be a vector field on M. The equation $\omega X = 0$ means that at each point $p \in M$, the covector $\omega(p)$ maps the vector X_p into zero. We then say that X is a solution of the equation $\omega = 0$ and such an equation is called a Pfaffian equation.

More generally, if $\omega_1, \omega_2, \ldots, \omega_r$ are r independent 1-forms on M, the system

ISBN 0-201-13500-0

$$\omega_1 = 0, \quad \omega_2 = 0, \quad \ldots, \quad \omega_r = 0 \tag{9.10}$$

is called a Pfaffian system on M. Two Pfaffian systems $\omega_i = 0\,(i = 1, 2,\ldots, r)$ and $\phi_j = 0$ $(j = 1, 2,\ldots, m)$ are called equivalent on M (or on an open set of M) if there exist functions λ_{ij}, μ_{ij} of class C^∞ such that

$$\omega_i = \sum_{j=1}^{m} \lambda_{ij}\phi_j, \qquad \phi_j = \sum_{i=1}^{r} \mu_{ij}\omega_i. \tag{9.11}$$

It is clear that if the 1-forms ω_i are independent and the 1-forms ϕ_j are independent, conditions (9.11) can only be verified if $m = r$ and the matrix (μ_{ij}) is the inverse of the matrix (λ_{ji}). If we assume that $\omega_i = \sum \alpha_{ik}\,dx_k$, then a vector field $X = \sum a^j(\partial/\partial x_j)$ is a solution of system (9.10) if and only if

$$\sum_{k=1}^{n} \alpha_{ik}a^k = 0, \qquad i = 1, 2,\ldots, r. \tag{9.12}$$

If the 1-forms ω_i are assumed independent, the rank of the matrix (α_{ik}) is equal to r and then at each point $p \in M$ the vectors X_p whose components $a^k(p)$ satisfy linear equations (9.12) form a linear subspace of dimension $n - r$, or an $(n - r)$-plane, of the tangent space T_p. Thus, with each Pfaffian system (9.10) there is associated a field of $(n - r)$-planes, that is, a function that assigns to each $p \in M$ an $(n - r)$-plane of T_p. Such a function is called an r-dimensional *distribution* on M. We say that this field of $(n - r)$-planes is the *annihilator* or the *solution* of system (9.10). A submanifold M' of M is called an *integral manifold* of system (9.10) if for every $p \in M'$ the tangent space T_p' of M' at p is contained in the annihilator of (9.10). Conversely, if we have a field of $(n - r)$-planes on M, spanned by $n - r$ independent vector fields $X_i' = \sum a_i^k(\partial/\partial x_k)$ $(i = 1,\ldots, n - r;\ k = 1, 2,\ldots, n)$, the system of linear equations

$$\sum_{h=1}^{n} \alpha_h a_i^h = 0 \qquad (i = 1, 2,\ldots, n - r) \tag{9.13}$$

defines an r-plane of the cotangent space $T_p{}^*$ at each point $p \in M$. That is, to each field of $(n - r)$-planes there is associated a function (called an r-dimensional *codistribution*) that assigns to each $p \in M$ an r-plane of $T_p{}^*$, that is, a set of r independent 1-forms defined up to a linear combination.

A Pfaffian system (9.10) is called *completely integrable* on M (or on an open set of M) if there exists a set of r functions $y_i(x_1, x_2,\ldots, x_n)\,(i = 1, 2,\ldots, r)$ such that system (9.10) is equivalent to the system

$$dy_1 = 0, \quad dy_2 = 0, \quad \ldots, \quad dy_r = 0. \tag{9.14}$$

ISBN 0-201-13500-0

In other words, system (9.10) is completely integrable if there exist r functions $y_i(x_1,\ldots,x_n)$ such that

$$\omega_i = \sum_{h=1}^{r} a_{ih}\, dy_h \tag{9.15}$$

where (a_{ih}) is an invertible matrix whose elements a_{ij} are of class C^∞. If system (9.10) is completely integrable, then there is an integral submanifold of dimension $n - r$ through each point $p \in M$. In fact, in the neighborhood of each point p, the equations $y_i(x) = y_i(p)$ define a submanifold of dimension $n - r$ that obviously is an integral submanifold (since (9.14) is equivalent to (9.10)). Conversely, given a family of $(n - r)$-dimensional differentiable submanifolds with the property that through each point of M there passes one and only one submanifold, then they are integral submanifolds of a completely integrable Pfaffian system. Indeed, in a local coordinate system, the given manifolds are represented by

$$F_i(x_1, x_2,\ldots, x_n; \xi_1, \xi_2,\ldots, \xi_r) = 0, \qquad i = 1, 2,\ldots, r, \tag{9.16}$$

where each point p determines one and only one set of parameters $\xi_1,\ldots, \xi_r$. The stated Pfaffian system is

$$\sum_{j=1}^{n} \frac{\partial F_i}{\partial x_j}\, dx_j = 0, \qquad i = 1, 2,\ldots, r \tag{9.17}$$

where the parameters ξ_i are replaced by functions of x_i deduced from (9.16).

Finally, note that if system (9.10) is completely integrable, that is, the 1-forms ω_i can be written in the form of (9.15), then we have

$$d\omega_i = \sum_{j=1}^{n} \sum_{h=1}^{r} \frac{\partial a_{ih}}{\partial x_j}\, dx_j \wedge dy_h, \qquad i = 1, 2,\ldots, r, \tag{9.18}$$

and using equations (9.14), we get $d\omega_i = 0$. We can therefore state

A necessary condition for system (9.10) *to be completely integrable is that the exterior differentials $d\omega_i$ be zero when the equations of the system are taken into account.*

In other words, *in order that system* (9.10) *be completely integrable, it is necessary that the conditions*

$$d\omega_i \wedge \omega_1 \wedge \omega_2 \wedge \cdots \wedge \omega_r = 0 \qquad (i = 1, 2,\ldots, r) \tag{9.19}$$

hold.

The preceding conditions are also sufficient (theorem of Frobenius) but we shall need them only as necessary conditions. For a proof of Frobenius' theorem see the work of Flanders [201], Sternberg [628], or H. Cartan [91].

ISBN 0-201-13500-0

3. Mappings of Differentiable Manifolds

Before beginning the study of Lie groups, we need to recall some definitions about mappings of differentiable manifolds. Let $\Phi: M \to M'$ be a mapping of the differentiable manifold M of dimension n into the differentiable manifold M' of dimension m. The mapping Φ is called differentiable if for any $f \in C^\infty$ defined on M', it is $f \circ \Phi \in C^\infty$. If Φ is differentiable, then there is an induced linear mapping $d\Phi: T(M) \to T(M')$, called the differential of Φ, defined by $d\Phi(X_p)f = X_p(f \circ \Phi)$ for any vector $X_p \in T_p$ and any function $f \in C^\infty$ defined in a neighborhood of $p' = \Phi(p)$. In terms of local coordinates, if Φ has the expression $x_\alpha' = x_\alpha'(x_1, x_2, \ldots, x_n)$, $\alpha = 1, 2, \ldots, m$, and $X_p = \sum_1^n a^i(\partial/\partial x_i)_p$, then

$$(d\Phi)X_p = \sum_{i,\alpha} a^i \left(\frac{\partial x_\alpha'}{\partial x_i}\right)_p \left(\frac{\partial}{\partial x_\alpha'}\right)_{p'} \tag{9.20}$$

where the ranges of indices are $i = 1, 2, \ldots, n$; $\alpha = 1, 2, \ldots, m$.

From this expression follows this chain rule: if $\Phi: M \to M'$ and $\Psi: M' \to M''$, then

$$d(\Psi \circ \Phi) = d\Psi \circ d\Phi. \tag{9.21}$$

The differentiable mapping $\Phi: M \to M'$ also induces a mapping $\Phi^*: T^*(M') \to T^*(M)$, called the *dual* of $d\Phi$, defined as follows:

$$\Phi^*(\omega')X_p = \omega' X_{p'} \tag{9.22}$$

where $X_{p'} = (d\Phi)X_p$ and ω' denotes a 1-form of $T_{p'}^*$. If $\omega' = \sum \alpha_\alpha' \, dx_\alpha'$, the expression of $\Phi^*(\omega')$ is

$$\Phi^*\left(\sum_1^m \alpha_\alpha' \, dx_\alpha'\right) = \sum_{i,\alpha} \alpha_\alpha' \frac{\partial x_\alpha'}{\partial x_i} dx_i. \tag{9.23}$$

Note that the matrices of $d\phi$ and Φ^* are transposed matrices.

The mapping Φ^* may be extended to a mapping $\omega'^{(q)} \to \omega^{(q)}$ of q-forms as follows.

$$\Phi^*\left(\sum_\alpha \alpha'_{\alpha_1\alpha_2\ldots\alpha_q} dx'_{\alpha_1} \wedge \cdots \wedge dx'_{\alpha_q}\right)$$

$$= \sum_{\alpha,i} \alpha'_{\alpha_1\ldots\alpha_q} \frac{\partial x'_{\alpha_1}}{\partial x_{i_1}} \cdots \frac{\partial x'_{\alpha_q}}{\partial x_{i_q}} dx_{i_1} \wedge \cdots \wedge dx_{i_q}. \tag{9.24}$$

In particular, if $m = n$ and ω' is an n-form on M', we have

$$\Phi^*\omega_n' = |J|\omega_n = (\det d\Phi)\omega_n \tag{9.25}$$

ISBN 0-201-13500-0

where $|J|$ denotes the determinant of the Jacobian matrix $J = (\partial x_\alpha'/\partial x_i)$ $(i = 1, 2, \ldots, n; \alpha = 1, 2, \ldots, m)$.

From (9.24) it is easy to deduce the following formulas.

$$(\Psi \circ \Phi)^* = \Phi^* \circ \Psi^*, \qquad \Phi^*(\omega \wedge \phi) = \Phi^*(\omega) \wedge \Phi^*(\phi), \qquad \Phi^*(d\omega) = d(\Phi^*(\omega)). \tag{9.26}$$

4. Lie Groups; Left and Right Translations

A Lie group $\mathfrak{G}$ is a group whose elements are points of a differentiable manifold, which we shall represent by the same letter $\mathfrak{G}$, such that the group operation $(a, b) \to ab^{-1}$ $(a, b \in \mathfrak{G})$ is a differentiable mapping of $\mathfrak{G} \times \mathfrak{G}$ into $\mathfrak{G}$. If $e \in \mathfrak{G}$ is the identity, the mapping $x \to x^{-1}$ can be written $(e, x) \to ex^{-1}$, so that it is also a differentiable mapping $\mathfrak{G} \to \mathfrak{G}$.

The general linear group $GL(n)$ (group of all $n \times n$ invertible real matrices) is a Lie group. The map $a_{ij} = x_{n(i-1)+j}$ $(i, j = 1, 2, \ldots, n)$ represents each matrix (a_{ij}) into a point of R^{n^2}. The group manifold of $GL(n)$ is then the n^2-dimensional affine space with the points $\det(a_{ij}) = F(x_1, \ldots, x_{n^2}) = 0$ excluded. Integral geometry deals mainly with Lie groups of matrices, that is, subgroups of $GL(n)$ that are differentiable manifolds, for instance, the special linear group $SL(n)$ (group of all $n \times n$ real matrices of determinant equal to 1) and the orthogonal group $O(n)$ (group of all $n \times n$ real matrices such that $OO^t = I$, where O^t means the transpose of O and I denotes the $n \times n$ unit matrix). Since $OO^t = I$ implies $n(n + 1)/2$ conditions, the dimension of $O(n)$ is $n(n - 1)/2$.

For each fixed $g_0 \in \mathfrak{G}$ we have two differentiable mappings $\mathfrak{G} \to \mathfrak{G}$: (a) the *left translations* $L_{g_0}\colon g \to g_0 g$; and (b) the *right translations* R_{g_0}: $g \to g g_0$.

A vector field X on $\mathfrak{G}$ is said to be *left invariant* if for every $g, g_0 \in \mathfrak{G}$, we have

$$(dL_{g_0})X_g = X_{g_0 g}. \tag{9.27}$$

By the substitution $g_0 \to g$, $g \to e$, we get

$$X_g = (dL_g)X_e. \tag{9.28}$$

Hence, any left-invariant vector field can be generated by left translations from a tangent vector at the identity e.

Similarly, a vector field X on G is called *right invariant* if for every $g, g_0 \in \mathfrak{G}$ we have $(dR_{g_0})X_g = X_{gg_0}$, and therefore, setting $g_0 \to g$, $g \to e$, we have $X_g = (dR_g)X_e$, which proves that any right-invariant vector field can be generated by right translations of a tangent vector at the identity.

ISBN 0-201-13500-0

For all $g_0, g_1 \in \mathfrak{G}$ we have $L_{g_0}R_{g_1} = L_{g_0}(gg_1) = g_0gg_1, R_{g_1}L_{g_0}g = R_{g_1}(g_0g) = g_0gg_1$, so that

$$L_{g_0}R_{g_1} = R_{g_1}L_{g_0}. \tag{9.29}$$

On the other hand, according to (9.21) we have $d(L_{g_0} \circ R_{g_1}) = dL_{g_0} \circ dR_{g_1}$ and thus we have the commutative property

$$dL_{g_0} \circ dR_{g_1} = dR_{g_1} \circ dL_{g_0}. \tag{9.30}$$

We now note that $L_{g_1g_0}: g \to (g_1g_0)g = g_1(g_0g)$, $R_{g_1g_0}: g \to g(g_1g_0) = (gg_1)g_0$ and consequently

$$L_{g_1g_0} = L_{g_1} \circ L_{g_0}, \qquad R_{g_1g_0} = R_{g_0} \circ R_{g_1}. \tag{9.31}$$

Hence, by (9.21),

$$dL_{g_1g_0} = dL_{g_1} \circ dL_{g_0}, \qquad dR_{g_1g_0} = dR_{g_0} \circ dR_{g_1}. \tag{9.32}$$

From (9.31) it follows that $L_{g_0^{-1}} \circ L_{g_0} = R_{g_0} \circ R_{g_0^{-1}} =$ identity and therefore

$$dL_{g_0^{-1}} = (dL_{g_0})^{-1}, \qquad dR_{g_0^{-1}} = (dR_{g_0})^{-1}. \tag{9.33}$$

Adjoint representation. Let T_e denote the tangent space of $\mathfrak{G}$ at the identity e. Each $g_0 \in \mathfrak{G}$ induces a linear transformation of T_e into itself defined by

$$\mathrm{ad}(g_0): X_e \to (dR_{g_0^{-1}} \circ dL_{g_0})X_e. \tag{9.34}$$

This transformation is called the *adjoint transformation* of g_0 and it is linear because dL_{g_0} and $dR_{g_0^{-1}}$ are linear transformations. The mapping $g \to \mathrm{ad}(g)$ is called the adjoint representation of $\mathfrak{G}$ and it is a homomorphism of $\mathfrak{G}$ into the group of automorphisms of T_e (i.e., into $GL(n)$). In fact, using (9.30) and (9.32), we can easily prove that $\mathrm{ad}(g_1g_0): X_e \to \mathrm{ad}(g_1) \circ \mathrm{ad}(g_0)X_e$ and $\mathrm{ad}(g_0{}^{-1}) = (\mathrm{ad}(g_0))^{-1}$.

5. Left-Invariant Differential Forms

Any left translation L_{g_0} induces the dual mapping (9.22)

$$L^*_{g_0}: T^*_{g_0g} \to T_g^* \tag{9.35}$$

of the cotangent space $T^*_{g_0g}$ onto T_g^*. A 1-form $\omega \in T^*(G)$ is called *left invariant* if

$$L^*_{g_0}\omega(g_0g) = \omega(g) \tag{9.36}$$

ISBN 0-201-13500-0

for all $g, g_0 \in \mathfrak{G}$. In terms of local coordinates, denoting by x_i, x_i^0, and $y_i = y_i(x_i; x_i^0)$ the coordinates of g, g_0, and $g_0 g$, respectively, we can write invariance condition (9.36) as

$$\sum_1^n a_i(y)\, dy_i = \sum_{i,h} a_i(y(x)) \frac{\partial y_i}{\partial x_h}\, dx_h = \sum_1^n a_i(x)\, dx_i \tag{9.37}$$

and this expression justifies the symbolic notation

$$\omega(g_0 g) = \omega(g) \tag{9.38}$$

sometimes used for indicating the left invariance of ω. For instance, $\omega = x^{-1}\, dx$ is a left-invariant form for the multiplicative group of real numbers because $\omega(x_0 x) = (x_0 x)^{-1}\, d(x_0 x) = x^{-1}\, dx = \omega(x)$ for any constant x_0.

If ω_1 and ω_2 are left-invariant 1-forms, so are $a\omega_1 + b\omega_2$ for any constants a, b, so that the left-invariant 1-forms are the elements of a vector space on the field of real numbers. From (9.36) it follows that $L_{g^{-1}}^*\omega(e) = \omega(g)$. Hence, all left-invariant 1-forms on $\mathfrak{G}$ can be generated from the 1-forms at the identity and therefore the vector space of the left-invariant 1-forms has the dimension of T_e^*, that is, the dimension n of $\mathfrak{G}$. The 1-forms $\omega_1, \omega_2, \ldots, \omega_n$ of any basis for this space are called *Maurer–Cartan forms* for $\mathfrak{G}$. In other words, a set of Maurer–Cartan 1-forms for $\mathfrak{G}$ is any set of n linearly independent left-invariant 1-forms on $\mathfrak{G}$. Each left-invariant 1-form can then be written

$$\omega = \sum_1^n a_i \omega_i \tag{9.39}$$

where a_i are constants. For a direct proof of this fact, note that if ω is a left-invariant 1-form, we have

$$L_{g_0}^* \omega(g_0 g) = \sum_1^n a_i(g_0 g)\omega_i(g_0 g) = \omega(g) = \sum_1^n a_i(g)\omega_i(g)$$

and since $\omega_i(g_0 g) = \omega_i(g)$, we get $\sum_i (a_i(g_0 g) - a_i(g))\omega_i(g) = 0$. Since the 1-forms ω_i are assumed independent, the last equation proves that $a_i(g_0 g) = a_i(g)$ and thus $a_i = \text{constant}$ $(i = 1, 2, \ldots, n)$.

In general we have the following important theorem:

Any left-invariant h-form can be written

$$\omega^{(h)} = \sum_{i_1, \ldots, i_h} a_{i_1 i_2 \ldots i_h}\, \omega_{i_1} \wedge \cdots \wedge \omega_{i_h} \tag{9.40}$$

where the 1-forms ω_{i_s} belong to a set of Maurer–Cartan forms and the coefficients $a_{i_1 \ldots i_h}$ are constants.

The proof is the same as above, using that the $\binom{n}{h}$ products $\omega_{i_1} \wedge \omega_{i_2} \wedge \cdots \wedge \omega_{i_h}$ form a basis for the module of the h-forms.

ISBN 0-201-13500-0

COROLLARY. If $\mathfrak{G}$ is of dimension n, all left-invariant n-forms are equal, except for a constant factor.

6. Maurer-Cartan Equations

Let $\{\omega_i\}$ $(i = 1, 2, \ldots, n)$ be a set of Maurer–Cartan 1-forms, that is, a basis for the vector space of the left-invariant 1-forms on the group $\mathfrak{G}$. Because of the last of equations (9.26), we have $L_g{}^* \, d\omega = d(L_g{}^*\omega)$. Thus the exterior derivatives $d\omega_i$ are left-invariant 2-forms, and according to the last theorem in the preceding section we can write

$$d\omega_i = \sum_{j,k=1}^{n} C^i_{jk} \, \omega_j \wedge \omega_k \tag{9.41}$$

where C^i_{jk} are constants. Setting $C^i_{jk} - C^i_{kj} = C'^i_{jk}$ and again writing C^i_{jk} instead of C'^i_{jk}, we can write equations (9.41) as

$$d\omega_i = \tfrac{1}{2} \sum_{j,k=1}^{n} C^i_{jk} \, \omega_j \wedge \omega_k \tag{9.42}$$

where C^i_{jk} are constants such that

$$C^i_{jk} + C^i_{kj} = 0. \tag{9.43}$$

Equations (9.42) are known as the *Maurer–Cartan equations* or *structure equations for the group* $\mathfrak{G}$. The constants C^i_{jk} are called the *structure constants* for $\mathfrak{G}$ with respect to the basis $\{\omega_i\}$.

Sometimes it is useful to have structure equations (9.42) in terms of ordinary differential calculus. Taking (9.6) into account, we can write equations (9.42) as

$$\begin{aligned} d\omega_i(\delta) - \delta\omega_i(d) &= \tfrac{1}{2} \sum_{j,k} C^i_{jk}(\omega_j(d)\omega_k(\delta) - \omega_j(\delta)\omega_k(d)) \\ &= \sum_{j,k} C^i_{jk} \, \omega_j(d)\omega_k(\delta), \end{aligned} \tag{9.44}$$

where on the right-hand side we have an ordinary product of differential forms.

Properties of the structure constants. The structure constants satisfy the conditions of skew symmetry (9.43). Moreover, exterior differentiation of the Maurer–Cartan equations, using that $d(d\omega_i) = 0$, gives

$$\begin{aligned} 0 &= \tfrac{1}{4} \sum C^i_{jk}(C^j_{pq} \, \omega_p \wedge \omega_q \wedge \omega_k - C^k_{pq} \, \omega_j \wedge \omega_p \wedge \omega_q) \\ &= \tfrac{1}{2} \sum C^i_{jk} C^j_{pq} \, \omega_p \wedge \omega_q \wedge \omega_k \\ &= \sum_{j} \sum_{p<q<k} (C^i_{jk} C^j_{pq} + C^i_{jq} C^j_{kp} + C^i_{jp} C^j_{qk}) \omega_p \wedge \omega_q \wedge \omega_k. \end{aligned}$$

ISBN 0-201-13500-0

Since the products $\{\omega_p \wedge \omega_q \wedge \omega_k\}$ $(p < q < k)$ form a basis for the vector space of left-invariant 3-forms and are independent, we have

$$\sum_{j=1}^{n} (C^i_{jk} C^j_{pq} + C^i_{jq} C^j_{kp} + C^i_{jp} C^j_{qk}) = 0 \tag{9.45}$$

for all values of $i, p, q, k \leqslant n$.

The structure constants defined by equations (9.41) depend on the basis $\{\omega_i\}$. If we replace this basis with another $\{\phi_i\}$ such that

$$\phi_i = \sum_{h=1}^{n} u_{ih}\omega_h, \qquad \omega_i = \sum_{h=1}^{n} U_{ih}\phi_h \tag{9.46}$$

where u_{ih}, U_{ih} are constants such that

$$\sum_{h=1}^{n} u_{ih} U_{hj} = \delta_{ij}, \tag{9.47}$$

we have the following new set of structure constants.

$$C'^j_{sk} = \sum_{i,m,h} C^i_{mh} U_{ms} U_{hk} u_{ji}. \tag{9.48}$$

These equations show that:

(a) The conditions $C^i_{mh} = 0$ are independent of the basis. They give an intrinsic property of $\mathfrak{G}$. We can prove that they are the necessary and sufficient conditions that $\mathfrak{G}$ be commutative.

(b) From (9.47) and (9.48) it follows that

$$\sum_{j=1}^{n} C'^j_{sj} = \sum_{i,m} C^i_{mi} U_{ms}$$

and hence

$$\sum_{i=1}^{n} C^i_{mi} = 0, \qquad m = 1, 2, \ldots, n, \tag{9.49}$$

is a set of equations that is independent of the basis and thus it gives another intrinsic property of $\mathfrak{G}$. We shall see later that system (9.49) characterizes the so-called unimodular groups.

Construction of the Maurer–Cartan forms for matrix groups. Let $\mathfrak{G}$ be a group of matrices $g = (g_{ij})$. Consider the matrix

$$\Omega = g^{-1}\, dg \tag{9.50}$$

whose elements are the 1-forms

ISBN 0-201-13500-0

$$\omega_{ij} = \sum_{h=1}^{n} \gamma_{ih}\, dg_{hj} \tag{9.51}$$

where γ_{ih} denote the elements of the inverse matrix g^{-1}. The matrix Ω is left invariant, that is, the forms ω_{ij} are left invariant. Indeed, we have

$$L_{g_0}^{*}\Omega(g_0 g) = (g_0 g)^{-1}\, d(g_0 g) = g^{-1} g_0^{-1} g_0\, dg = g^{-1}\, dg = \Omega(g), \tag{9.52}$$

which is the condition (9.36) of left invariance.

If the matrices $g \in \mathfrak{G}$ are $r \times r$ matrices and $\mathfrak{G}$ is of dimension n, we want to show that among the r^2 1-forms ω_{ij} there exist exactly n that are linearly independent. In fact, at the identity we have $\Omega(e) = (dg)_e = (dg_{ij}(e))$ and because $dg_{ij}(e)$ form a basis for the cotangent space to $GL(r)$ at the identity, they contain a basis for the cotangent space to $\mathfrak{G}$ at e and the covectors that form this basis generate n linearly independent 1-forms. Hence, from the forms (9.51) we can select a basis for the left-invariant 1-forms, that is, a set of Maurer–Cartan forms for $\mathfrak{G}$.

Exterior differentiation of (9.50) gives $d\Omega = dg^{-1} \wedge dg = -\, g^{-1}\, dg\, g^{-1} \wedge dg$; that is,

$$d\Omega = -\, \Omega \wedge \Omega. \tag{9.53}$$

Among the r^2 equations contained in this matrix equation are the Maurer–Cartan equations for $\mathfrak{G}$.

Example. Consider the three-dimensional matrix group

$$g = \begin{pmatrix} x & 0 & y \\ x \log x & x & z \\ 0 & 0 & 1 \end{pmatrix}, \qquad x \neq 0. \tag{9.54}$$

The group manifold is R^3 except for the plane $x = 0$. We have

$$\Omega = g^{-1}\, dg = \begin{pmatrix} dx/x & 0 & dy/x \\ dx/x & dx/x & -x^{-1} \log x\, dy + dz/x \\ 0 & 0 & 0 \end{pmatrix}. \tag{9.55}$$

Thus a set of Maurer–Cartan forms is

$$\omega_1 = \frac{dx}{x}, \qquad \omega_2 = \frac{dy}{x}, \qquad \omega_3 = -\frac{\log x}{x}\, dy + \frac{dz}{x}. \tag{9.56}$$

Since

$$\Omega \wedge \Omega = \begin{pmatrix} 0 & 0 & \omega_1 \wedge \omega_2 \\ 0 & 0 & \omega_1 \wedge \omega_2 + \omega_1 \wedge \omega_3 \\ 0 & 0 & 0 \end{pmatrix},$$

ISBN 0-201-13500-0

the structure equations for $\mathfrak{G}$ are

$$d\omega_1 = 0, \qquad d\omega_2 = -\,\omega_1 \wedge \omega_2, \qquad d\omega_3 = -\,\omega_1 \wedge \omega_2 - \omega_1 \wedge \omega_3 \tag{9.57}$$

and the structure constants that are not zero are

$$C^2_{12} = -\,C^2_{21} = -\,1, \qquad C^3_{12} = -\,C^3_{21} = -\,1, \qquad C^3_{13} = -\,C^3_{31} = -\,1. \tag{9.58}$$

Right-invariant 1-forms. All that we have said about left-invariant 1-forms can be repeated almost without change for right-invariant 1-forms. A 1-form ϖ is called *right invariant* if

$$R^*_{g_0}\varpi(gg_0) = \varpi(g) \tag{9.59}$$

for all $g, g_0 \in \mathfrak{G}$. Briefly we shall write this condition

$$R_g{}^*\varpi = \varpi \qquad \text{or} \qquad \varpi(gg_0) = \varpi(g), \tag{9.60}$$

understanding that the true meaning of (9.60) is the relation (9.59), since strictly speaking $\varpi(gg_0)$ is an element of the cotangent space $T^*_{gg_0}$ and $\varpi(g)$ is an element of $T_g{}^*$.

As in the case of left-invariant forms, the right-invariant 1-forms are elements of a vector space of dimension n over the real field. For matrix groups, the right-invariant 1-forms are given by the matrix

$$\bar{\Omega} = dg\, g^{-1}, \tag{9.61}$$

which is right invariant since $R^*_{g_0}\bar{\Omega} = d(gg_0)(gg_0)^{-1} = \bar{\Omega}$. A basis for the vector space of the right-invariant 1-forms is obtained by choosing n linearly independent elements of the matrix $\bar{\Omega}$. From (9.61) and (9.50) we get

$$\Omega(g^{-1}) = g\, dg^{-1} = -\, dg\, g^{-1} = -\,\bar{\Omega}(g). \tag{9.62}$$

Therefore it is always possible to choose basis $\{\omega_i\}$, $\{\varpi_i\}$ such that

$$\varpi_i(g) = -\,\omega_i(g^{-1}), \qquad i = 1, 2, \ldots, n. \tag{9.63}$$

In a manner like that by which we obtained (9.42) we arrive at

$$d\varpi_i = \tfrac{1}{2} \sum_{j,k=1}^{n} \bar{C}^i_{jk}\, \varpi_j \wedge \varpi_k \tag{9.64}$$

where $\bar{C}^i_{jk}$ are constants. Applying (9.64) to g^{-1} and taking (9.63) into account, we get

$$-\, d\omega_i = \tfrac{1}{2} \sum_{j,k=1}^{n} \bar{C}^i_{jk}\, \omega_j \wedge \omega_k \tag{9.65}$$

ISBN 0-201-13500-0

and comparison with (9.42) yields $\bar{C}^i_{jk} = -C^i_{jk}$. Thus, the structure equations for right-invariant 1-forms can be written

$$d\bar{\omega}_i = -\tfrac{1}{2}\sum_{j,k=1}^{n} C^i_{jk}\,\bar{\omega}_j \wedge \bar{\omega}_k. \tag{9.66}$$

For commutative groups, left- and right-invariant forms are the same and (9.66) and (9.42) give $C^i_{jk} = 0$. That is, *commutative groups have structure constants equal to zero.* We can prove that this condition is also sufficient for the commutativity of $\mathfrak{G}$.

As for (9.44), structure equations (9.66) can be written

$$d\bar{\omega}_i(\delta) - \delta\bar{\omega}_i(d) = -\sum_{j,k} C^i_{jk}\,\bar{\omega}_j(d)\bar{\omega}_k(\delta). \tag{9.67}$$

Example. Consider the group of matrices (9.54). We have

$$\bar{\Omega} = dg\,g^{-1} = \begin{pmatrix} x^{-1}\,dx & 0 & -yx^{-1}\,dx + dy \\ x^{-1}\,dx & x^{-1}\,dx & -yx^{-1}\,dx - zx^{-1}\,dx + dz \\ 0 & 0 & 0 \end{pmatrix},$$

so that a system of right-invariant 1-forms is

$$\bar{\omega}_1 = \frac{dx}{x}, \qquad \bar{\omega}_2 = -\frac{y}{x}\,dx + dy, \qquad \bar{\omega}_3 = -\frac{y+z}{x}\,dx + dz. \tag{9.68}$$

The structure equations are

$$d\bar{\omega}_1 = 0, \qquad d\bar{\omega}_2 = \bar{\omega}_1 \wedge \bar{\omega}_2, \qquad d\bar{\omega}_3 = \bar{\omega}_1 \wedge \bar{\omega}_2 + \bar{\omega}_1 \wedge \bar{\omega}_3. \tag{9.69}$$

The structure constants are the same as (9.58) with opposite sign.

7. Invariant Volume Elements of a Group: Unimodular Groups

Let $\{\omega_i\}$ be a set of Maurer–Cartan 1-forms of the group $\mathfrak{G}$. The exterior product

$$d_L\mathfrak{G} = \omega_1 \wedge \omega_2 \wedge \cdots \wedge \omega_n \tag{9.70}$$

is a left-invariant n-form that, according to the corollary in Section 5, is unique up to a constant factor. We shall write the left invariance of $d_L\mathfrak{G}$ symbolically as follows.

$$d_L(g_0\mathfrak{G}) = d_L\mathfrak{G}, \qquad g_0 \in \mathfrak{G}. \tag{9.71}$$

The n-form $d_L\mathfrak{G}$ is called the *left-invariant volume element* of $\mathfrak{G}$. In integral geometry it is also known as the *kinematic density* of $\mathfrak{G}$. The kinematic

ISBN 0-201-13500-0

density considered in Chapter 6 is the left-invariant volume element of the group of motions on the plane.

In like manner, the right-invariant volume element of $\mathfrak{G}$ is defined as

$$d_{\mathrm{R}}\mathfrak{G} = \overline{\omega}_1 \wedge \overline{\omega}_2 \wedge \cdots \wedge \overline{\omega}_n \tag{9.72}$$

where $\{\overline{\omega}\}$ is a basis for the space of the right-invariant 1-forms. The right invariance of $d_{\mathrm{R}}\mathfrak{G}$ can be written symbolically as

$$d_{\mathrm{R}}(\mathfrak{G}g_0) = d_{\mathrm{R}}\mathfrak{G}, \qquad g_0 \in \mathfrak{G}. \tag{9.73}$$

By integration of the left-invariant (right-invariant) volume element we have a left-invariant (right-invariant) measure on $\mathfrak{G}$. This invariant measure exists on any locally compact group and is called the left (right) Haar measure on the group. Most of the properties of the left-invariant (right-invariant) measure that we shall prove for Lie groups are valid for the more general case of locally compact groups (see [443, 500, 710]).

A Lie group is called *unimodular* if its left-invariant volume element is also right invariant. In this case $d_{\mathrm{L}}\mathfrak{G}$ and $d_{\mathrm{R}}\mathfrak{G}$ coincide except for a constant factor and we can choose bases $\{\omega_i\}$, $\{\overline{\omega}_i\}$ such that $d_{\mathrm{L}}\mathfrak{G} = d_{\mathrm{R}}\mathfrak{G}$. If $\mathfrak{G}$ is unimodular, from (9.63) it follows that (up to the sign)

$$d_{\mathrm{L}}\mathfrak{G}(g) = d_{\mathrm{L}}\mathfrak{G}(g^{-1}); \tag{9.74}$$

that is, for unimodular groups, the invariant volume element is also invariant under the mapping $g \to g^{-1}$. This invariance corresponds to the so-called invariance under inversion of the motion (see Chapter 6, Section 3).

We shall now give some conditions for a group $\mathfrak{G}$ to be unimodular. Since the n 1-forms $\overline{\omega}_i$ are independent and the module of 1-forms on $\mathfrak{G}$ is of dimension n, we can write

$$\omega_i = \sum_{k=1}^{n} a_{ik}\overline{\omega}_k \tag{9.75}$$

where a_{ik} are functions on $\mathfrak{G}$. If we consider the structure equations in the form (9.67) and assume that δ denotes an elementary left translation and d an elementary right translation, using the right invariance of $\overline{\omega}_i$, we have $d\overline{\omega}_i(\delta) = 0$ and (9.67) becomes

$$\delta\overline{\omega}_i(d) = \sum_{j,k=1}^{n} C^i_{jk}\, \overline{\omega}_j(d)\overline{\omega}_k(\delta). \tag{9.76}$$

On the other hand, by the left invariance of $\omega_i(d)$ we have $\delta\omega_i(d) = 0$ and from (9.75) we get

$$\sum_{k=1}^{n} (\delta a_{ik}\, \overline{\omega}_k(d) + a_{ik}\, \delta\overline{\omega}_k(d)) = 0. \tag{9.77}$$

ISBN 0-201-13500-0

From (9.76) and (9.77) we find that

$$\sum_{k=1}^{n}\left(\delta a_{ik} - \sum_{m,j=1}^{n} C_{jk}^{m} a_{im}\bar{\omega}_j(\delta)\right)\bar{\omega}_k(d) = 0 \tag{9.78}$$

and since the 1-forms $\bar{\omega}_k(d)$ are independent, using d instead of δ, we have

$$da_{ik} - \sum_{j,m=1}^{n} C_{jk}^{m} a_{im}\bar{\omega}_j = 0. \tag{9.79}$$

On the other hand, calling Δ the determinant $|a_{ik}|$ of system (9.75), up to the sign, we have

$$d_{\mathrm{L}}\mathfrak{G} = \Delta(g)\, d_{\mathrm{R}}\mathfrak{G}. \tag{9.80}$$

The derivative of a determinant is given by

$$d\Delta = \sum_{i,k=1}^{n} \alpha_{ik}\, da_{ik} \tag{9.81}$$

where α_{ik} are defined by

$$\sum_{i=1}^{n} \alpha_{ik} a_{ih} = \Delta\delta_{kh}. \tag{9.82}$$

Multiplying both sides of (9.79) by α_{ik} and summing for i, k, we have

$$\frac{d\Delta}{\Delta} = \sum_{j,k=1}^{n} C_{jk}^{k}\bar{\omega}_j. \tag{9.83}$$

If $\mathfrak{G}$ is unimodular, then Δ is a constant and $d\Delta = 0$. Conversely, if Δ is constant, then the volume elements can be normalized so that $\Delta = 1$ and from (9.80) it follows that $\mathfrak{G}$ is unimodular. Hence we have that *in order for $\mathfrak{G}$ to be unimodular it is necessary and sufficient that*

$$\sum_{k=1}^{n} C_{jk}^{k} = 0 \quad \textit{for} \quad j = 1, 2, \ldots, n. \tag{9.84}$$

Properties of $\Delta(g)$. The determinant $\Delta(g)$ defined by (9.80) has important properties, which we shall now investigate. Applying the right translation R_{g_0} to both sides of (9.80) and using the right invariance of $d_{\mathrm{R}}\mathfrak{G}$, we get

$$R_{g_0}^{*}\, d_{\mathrm{L}}\mathfrak{G} = \Delta(gg_0)\, R_{g_0}^{*}\, d_{\mathrm{R}}\mathfrak{G} = \Delta(gg_0)\, d_{\mathrm{R}}\mathfrak{G} = \frac{\Delta(gg_0)}{\Delta(g)}\, d_{\mathrm{L}}\mathfrak{G}. \tag{9.85}$$

According to (9.29), $R_{g_0}^{*}$ commutes with any left-invariant n-form and hence $R_{g_0}^{*}\, d_{\mathrm{L}}\mathfrak{G}$ is again left invariant. Thus, from the Corollary in Section 5, it

ISBN 0-201-13500-0

follows that $R_{g_0}^* d_L\mathfrak{G}$ and $d_L\mathfrak{G}$ differ by a constant factor, and using (9.85) we have

$$\frac{\Delta(gg_0)}{\Delta(g)} = c \tag{9.86}$$

where c is constant with respect to g, but can depend on g_0. To find the value of c we consider the case in which $g = e$. Then, using (9.63) we have $\bar{\omega}_i(e) = -\omega_i(e)$ and hence, up to the sign, $(d_L\mathfrak{G})_e = (d_R\mathfrak{G})_e$, that is, $\Delta(e) = 1$. Thus $c = \Delta(g_0)$ and we can write

$$\Delta(gg_0) = \Delta(g)\,\Delta(g_0), \qquad \Delta(g^{-1}) = \frac{1}{\Delta(g)} \tag{9.87}$$

for all $g, g_0 \in \mathfrak{G}$. These equalities show that $\Delta(g)$ is a homomorphism of $\mathfrak{G}$ into the multiplicative group of positive real numbers.

From (9.85) and (9.87) we deduce $R_{g_0}^* d_L\mathfrak{G} = \Delta(g_0)\, d_L\mathfrak{G}$, which can be written

$$d_L(\mathfrak{G}g_0) = \Delta(g_0)\, d_L\mathfrak{G}. \tag{9.88}$$

If $\mathfrak{G}$ has a finite volume, the integral of $d_L(\mathfrak{G}g_0)$ and the integral of $d_L\mathfrak{G}$ over the whole group have the same value and thus $\Delta(g_0) = 1$ for all $g_0 \in \mathfrak{G}$. Hence, *the groups of finite volume are unimodular*. The converse is not true. For instance, although $GL(n)$ is not of finite volume, it is unimodular, as we shall see later.

A Lie group is said to be *compact* if the underlying differentiable manifold is compact. Since any compact group has finite volume, *compact groups are unimodular*.

Another condition for $\mathfrak{G}$ to be unimodular can be found in terms of the adjoint representation. Indeed, $\mathrm{ad}(g_0)$ induces on T_e^* the linear transformation

$$(\mathrm{ad}(g_0))^*\colon \omega(e) \to (R_{g_0^{-1}} \circ L_{g_0})^*\omega(e) = L_{g_0}^* \circ R_{g_0^{-1}}^*\omega(e) \tag{9.89}$$

for all 1-forms $\omega(e) \in T_e^*$. Note that the matrix of this linear transformation is the transpose of the matrix of (9.34); in consequence, they have the same determinant, which we will indicate by $\det \mathrm{ad}(g_0)$. Applying transformation (9.89) to both sides of (9.80) and using that $R_{g_0^{-1}}^* d_R\mathfrak{G} = d_R\mathfrak{G}$, we obtain

$$L_{g_0}^* \circ R_{g_0^{-1}}^*\, d_L\mathfrak{G} = \Delta(g_0 g g_0^{-1}) L_{g_0}^*\, d_R\mathfrak{G} = \Delta(g_0 g g_0^{-1})\, d_R(g_0\mathfrak{G}). \tag{9.90}$$

Using (9.26) we have

$$L_{g_0}^* \circ R_{g_0^{-1}}^*\, d_L\mathfrak{G} = L_{g_0}^* \circ R_{g_0^{-1}}^*\omega_1 \wedge \cdots \wedge L_{g_0}^* \circ R_{g_0^{-1}}^*\omega_n = \det \mathrm{ad}(g_0)\, d_L\mathfrak{G}$$

and consequently

ISBN 0-201-13500-0

$$\det \operatorname{ad}(g_0)\, d_L\mathfrak{G} = \Delta(g_0 g g_0^{-1})\, d_R(g_0\mathfrak{G}) = \frac{\Delta(g_0 g g_0^{-1})}{\Delta(g_0 g)}\, d_L\mathfrak{G}. \tag{9.91}$$

Putting $g = e$ we have

$$\det \operatorname{ad}(g_0) = \frac{\Delta(e)}{\Delta(g_0)}. \tag{9.92}$$

If $\mathfrak{G}$ is unimodular, then $\Delta =$ constant and from (9.92) it follows that $\det \operatorname{ad}(g_0) = 1$. Conversely, if $\det \operatorname{ad}(g_0) = 1$, we have $\Delta(g_0) = \Delta(e) =$ constant, and $\mathfrak{G}$ is unimodular. Hence we have

A necessary and sufficient condition that the group $\mathfrak{G}$ be unimodular is that

$$\det \operatorname{ad}(g_0) = 1 \tag{9.93}$$

for all $g, g_0 \in \mathfrak{G}$.

8. Notes and Exercises

1. *The group of motions and the group of similitudes in the plane.* The group of motions in the plane was considered in detail in Chapter 6. It is isomorphic to the group of matrices

$$g = \begin{pmatrix} \cos\phi & -\sin\phi & a \\ \sin\phi & \cos\phi & b \\ 0 & 0 & 1 \end{pmatrix}. \tag{9.94}$$

A system of Maurer–Cartan forms is $\omega_1 = \cos\phi\, da + \sin\phi\, db$, $\omega_2 = -\sin\phi\, da + \cos\phi\, db$, $\omega_3 = d\phi$. The structure equations are $d\omega_1 = \omega_3 \wedge \omega_2$, $d\omega_2 = \omega_1 \wedge \omega_3$, $d\omega_3 = 0$ and the nonzero structure constants are $C^1_{23} = -C^1_{32} = -1$, $C^2_{13} = -C^2_{31} = 1$. Hence, conditions (9.84) are satisfied and we have that *the group of motions in the plane is unimodular.* The invariant volume element is $d_L\mathfrak{G} = d_R\mathfrak{G} = da \wedge db \wedge d\phi$. These results agree with those of Chapter 6.

The group of similitudes in the plane is the four-dimensional transformation group

$$x' = \rho(x\cos\phi + y\sin\phi) + a, \qquad y' = \rho(-x\sin\phi + y\cos\phi) + b. \tag{9.95}$$

The parameters are ρ, a, b, ϕ, with $\rho \neq 0$. This group is isomorphic to the group of matrices

$$g = \begin{pmatrix} \rho\cos\phi & \rho\sin\phi & a \\ -\rho\sin\phi & \rho\cos\phi & b \\ 0 & 0 & 1 \end{pmatrix}, \qquad \rho \neq 0. \tag{9.96}$$

ISBN 0-201-13500-0

Using (9.50) we get the set of Maurer–Cartan 1-forms:

$$\omega_1 = d\rho/\rho, \quad \omega_2 = d\phi, \quad \omega_3 = \rho^{-1}\cos\phi\, da - \rho^{-1}\sin\phi\, db,$$
$$\omega_4 = \rho^{-1}\sin\phi\, da + \rho^{-1}\cos\phi\, db. \tag{9.97}$$

The structure equations are

$$d\omega_1 = 0, \qquad d\omega_2 = 0, \qquad d\omega_3 = -\,\omega_1 \wedge \omega_3 - \omega_2 \wedge \omega_4,$$
$$d\omega_4 = \omega_2 \wedge \omega_3 - \omega_1 \wedge \omega_4,$$

and the nonzero structure constants are

$$C^3_{13} = -\,C^3_{31} = -\,1, \qquad C^3_{24} = -\,C^3_{42} = -\,1, \qquad C^4_{23} = -\,C^4_{32} = 1,$$
$$C^4_{14} = -\,C^4_{41} = -\,1.$$

Conditions (9.84) are not satisfied; hence, *the group of similitudes is not unimodular.* The left-invariant volume element is

$$d_{\rm L}\mathfrak{G} = \omega_1 \wedge \omega_2 \wedge \omega_3 \wedge \omega_4 = \frac{d\rho \wedge d\phi \wedge da \wedge db}{\rho^3}.$$

Similarly, the right-invariant volume element is

$$d_{\rm R}\mathfrak{G} = \frac{d\rho \wedge d\phi \wedge da \wedge db}{\rho}.$$

For instance, a set C of circles defined by the center (a, b) and the radius ρ has the following left- and right-invariant measures under the group of similitudes.

$$m_{\rm L}(C) = 2\pi \int_C \frac{da \wedge db \wedge d\rho}{\rho^3}, \qquad m_{\rm R}(C) = 2\pi \int_C \frac{da \wedge db \wedge d\rho}{\rho}.$$

These measures are defined up to a constant factor (see [647]).

2. *Direct product of groups.* A group $\mathfrak{G}$ is said to be the direct product $\mathfrak{G}_1 \times \mathfrak{G}_2$ of its subgroups $\mathfrak{G}_1$ and $\mathfrak{G}_2$ if it is the set of all pairs (g_1, g_2), where $g_1 \in \mathfrak{G}_1$, $g_2 \in \mathfrak{G}_2$ with the multiplication law $(g_1, g_2)(g_1', g_2') = (g_1 g_1', g_2 g_2')$. If $d_{\rm L}\mathfrak{G}_1$ and $d_{\rm L}\mathfrak{G}_2$ are the left-invariant volume elements of $\mathfrak{G}_1$ and $\mathfrak{G}_2$, the left-invariant volume element of $\mathfrak{G}$ is $d_{\rm L}\mathfrak{G} = d_{\rm L}\mathfrak{G}_1 \wedge d_{\rm L}\mathfrak{G}_2$ and similarly, $d_{\rm R}\mathfrak{G} = d_{\rm R}\mathfrak{G}_1 \wedge d_{\rm R}\mathfrak{G}_2$. Consequently, we have that *if $\mathfrak{G}_1$ and $\mathfrak{G}_2$ are unimodular, then $\mathfrak{G}$ is also unimodular.*

EXERCISE 1. Prove the relations

$$d_{\rm L}(\mathfrak{G}g_0) = \det \operatorname{ad}(g_0^{-1})\, d_{\rm L}\mathfrak{G}, \qquad d_{\rm R}(g_0\mathfrak{G}) = \det \operatorname{ad}(g_0)\, d_{\rm R}\mathfrak{G},$$
$$d_{\rm L}(\mathfrak{G}g^{-1}) = \det \operatorname{ad}(g)\, d_{\rm L}\mathfrak{G}. \tag{9.98}$$

ISBN 0-201-13500-0

EXERCISE 2. If C^i_{jk} are the structure constants and a_{ij} are defined by equations (9.75), establish the identities

$$\sum_{p,q=1}^{n} C^i_{pq} a_{qk} a_{ps} = \sum_{r=1}^{n} C^r_{sk} a_{ir} \tag{9.99}$$

for any $i, s, k \leqslant n$.

Example 1. Consider the matrix group

$$g = \begin{pmatrix} a & b \\ 0 & 1 \end{pmatrix}, \qquad a \neq 0.$$

Using (9.50) and (9.61), we get

$$\omega_1 = da/a, \qquad \omega_2 = db/a, \qquad \bar{\omega}_1 = da/a, \qquad \bar{\omega}_2 = -(b/a)\,da + db \tag{9.100}$$

and therefore

$$d_{\mathrm{L}}\mathfrak{G} = \frac{da \wedge db}{a^2}, \qquad d_R\mathfrak{G} = \frac{da \wedge db}{a}, \qquad \Delta = \frac{1}{a}. \tag{9.101}$$

The following mappings are easily obtained:

$$L_{g_0}\colon a \to aa_0, \quad b \to a_0 b + b_0;$$

$$R_{g_0}\colon a \to aa_0, \quad b \to ab_0 + b;$$

$$R_{g_0^{-1}} \circ L_{g_0}\colon a \to a, \qquad b \to -ab_0 + ba_0 + b_0;$$

$$\mathrm{ad}(g_0)\colon \frac{\partial}{\partial a} \to \frac{\partial}{\partial a} - b_0 \frac{\partial}{\partial b}, \quad \frac{\partial}{\partial b} \to a_0 \frac{\partial}{\partial b};$$

so that $\det \mathrm{ad}(g_0) = a_0$, in accordance with (9.92). We can easily check relations (9.98).

Example 2. Consider the group $\mathfrak{G}$ of matrices (9.54). From (9.56) and (9.68) it follows that

$$d_{\mathrm{L}}\mathfrak{G} = x^{-3}\, dx \wedge dy \wedge dz, \qquad d_{\mathrm{R}}\mathfrak{G} = x^{-1}\, dx \wedge dy \wedge dz. \tag{9.102}$$

The group is not unimodular. Comparing with (9.80), we get $\Delta(g) = x^{-2}$, $d\Delta/\Delta = -(2/x)\,dx$. Using (9.58) it is easy to verify that relation (9.83) holds. The group element $g' = gg_0$ is the matrix of the form (9.54) whose elements are $x' = xx_0$, $y' = xy_0 + y$, $z' = xy_0 \log x + xz_0 + z$, so that we have $\Delta(gg_0) = (x^2 x_0^{\,2})^{-1} = \Delta(g)\,\Delta(g_0)$ and $d_{\mathrm{L}}(\mathfrak{G}g_0) = x'^{-3}\, dx' \wedge dy' \wedge dz' = (xx_0)^{-3} x_0\, dx \wedge dy \wedge dz = \Delta(g_0)\, d_{\mathrm{L}}\mathfrak{G}$, as it should be according to (9.88).
The mapping $g \to g_0 g g_0^{\,-1}$ is

$$x' = x, \qquad y' = -y_0 x + x_0 y + y_0,$$

$$z' = -xy_0 \log x + x_0 y \log x_0 - z_0 x + x_0 z + z_0,$$

ISBN 0-201-13500-0

and therefore the 1-form $\omega' = a'\,dx' + b'\,dy' + c'\,dz'$ goes to

$$\omega = [(a' - y_0 b' - (y_0 \log x + y_0 + z_0)c']\,dx$$
$$+ [b'x_0 + x_0 \log x_0 c']\,dy + c'x_0\,dz.$$

The identity e corresponds to $x = 1$, $y = 0$, $z = 0$; hence, the transformation $(\mathrm{ad}(g_0))^*$ can be written

$$a = a' - y_0 b' - (y_0 + z_0)c', \qquad b = x_0 b' + x_0 \log x_0 c', \qquad c = x_0 c',$$

and we have $\det \mathrm{ad}(g_0) = x_0{}^2$. With this value we can easily verify equation (9.92) and identities (9.98).

Example 3. Consider the group $GL(2)$, that is, the four-dimensional group of 2×2 invertible matrices

$$g = \begin{pmatrix} x & y \\ z & t \end{pmatrix}, \qquad xt - yz = D \neq 0. \tag{9.103}$$

The matrix $\Omega = g^{-1}\,dg$ has the elements

$$\omega_1 = \frac{1}{D}(t\,dx - y\,dz), \qquad \omega_2 = \frac{1}{D}(t\,dy - y\,dt),$$

$$\omega_3 = \frac{1}{D}(-\,z\,dx + x\,dz), \qquad \omega_4 = \frac{1}{D}(-\,z\,dy + x\,dt).$$

The structure equations, deduced from (9.53), are

$$d\omega_1 = -\,\omega_2 \wedge \omega_3, \qquad d\omega_2 = -\,\omega_1 \wedge \omega_2 - \omega_2 \wedge \omega_4,$$

$$d\omega_3 = \omega_1 \wedge \omega_3 + \omega_3 \wedge \omega_4, \qquad d\omega_4 = \omega_2 \wedge \omega_3,$$

so that the structure constants that are not zero are

$$C^1_{23} = -\,C^1_{32} = -\,1, \qquad C^2_{12} = -\,C^2_{21} = -\,1, \qquad C^2_{24} = -\,C^2_{42} = -\,1,$$

$$C^3_{13} = -\,C^3_{31} = 1, \qquad C^3_{34} = -\,C^3_{43} = 1, \qquad C^4_{23} = -\,C^4_{32} = 1.$$

Therefore conditions (9.84) are verified and we can state that *the group $GL(2)$ is unimodular*. We shall see later that the same is true for $GL(n)$. The invariant volume element is

$$d_L\mathfrak{G} = d_R\mathfrak{G} = \omega_1 \wedge \omega_2 \wedge \omega_3 \wedge \omega_4 = \frac{dx \wedge dy \wedge dz \wedge dt}{D^2}. \tag{9.104}$$

Example 4. Let $\mathfrak{G}$ be the group of matrices

$$g = \begin{pmatrix} x & y \\ 0 & x \end{pmatrix}, \qquad x \neq 0.$$

It is a commutative group, so that it is unimodular. The general method gives

ISBN 0-201-13500-0

$$\omega_1 = \bar{\omega}_1 = dx/x, \qquad \omega_2 = \bar{\omega}_2 = dy/x - (y/x^2)\,dx.$$

The structure equations are $d\omega_1 = 0$, $d\omega_2 = 0$ and the invariant volume element is $d_L\mathfrak{G} = d_R\mathfrak{G} = x^{-2}\,dx \wedge dy$.

Example 5. Let $\mathfrak{G}$ be the matrix group

$$g = \begin{pmatrix} x & y \\ 0 & 1/x \end{pmatrix}, \qquad x \neq 0.$$

By the general method we have

$$\omega_1 = dx/x, \quad \omega_2 = dy/x + (y/x^2)\,dx, \quad \bar{\omega}_1 = dx/x, \quad \bar{\omega}_2 = -\,y\,dx + x\,dy.$$

The structure equations are $d\omega_1 = 0$, $d\omega_2 = -\,2\omega_1 \wedge \omega_2$ and the structure constants that are not zero are $C_{12}^2 = -\,C_{21}^2 = -\,2$. The group is not unimodular. The volume elements are $d_L\mathfrak{G} = x^{-2}\,dx \wedge dy$ and $d_R\mathfrak{G} = dx \wedge dy$. Hence we have $\Delta = 1/x^2$ and $d\Delta/\Delta = -\,2(dx/x)$, in accordance with (9.83). The adjoint mapping is $A\,dx + B\,dy \to (A - 2x_0y_0B)\,dx + Bx_0{}^2\,dy$. Hence, det ad$(g_0) = x_0{}^2$.

ISBN 0-201-13500-0

CHAPTER 10

Density and Measure in Homogeneous Spaces

1. Introduction

In this chapter we shall use only left invariance, so that, for simplicity, we shall say invariance instead of left invariance. The left-invariant volume element $d_L\mathfrak{G}$ will be represented simply by $d\mathfrak{G}$.

Let $\mathfrak{G}$ be a Lie group of dimension n and let $\mathfrak{H}$ be a closed subgroup of $\mathfrak{G}$ of dimension $n - m$ (a subgroup $\mathfrak{H}$ is called closed if the set of points $\mathfrak{H}$ is closed in the underlying differentiable manifold of $\mathfrak{G}$). The set of left cosets $g\mathfrak{H}$, $g \in \mathfrak{G}$, is the homogeneous space $\mathfrak{G}/\mathfrak{H}$, which, as is known, admits a differentiable manifold structure of dimension m (see, e.g., [628, p. 230]). We want to find the conditions for the existence of a nonzero m-form on $\mathfrak{G}/\mathfrak{H}$ invariant under $\mathfrak{G}$. Such an m-form is called a *density* on $\mathfrak{G}/\mathfrak{H}$ and it gives rise, by integration, to an *invariant measure* on $\mathfrak{G}/\mathfrak{H}$. Since $\mathfrak{G}$ acts transitively on $\mathfrak{G}/\mathfrak{H}$, the invariant density, if it exists, is unique up to a constant factor.

Note that $\mathfrak{H}$ and its left cosets $g\mathfrak{H}$ ($g \in \mathfrak{G}$) are differentiable submanifolds of the differentiable manifold $\mathfrak{G}$, in such a way that through each point of $\mathfrak{G}$ there passes one and only one submanifold $g\mathfrak{H}$. Hence, according to Section 2 of Chapter 9, they are integral manifolds of a completely integrable Pfaffian system

$$\omega_1 = 0, \quad \omega_2 = 0, \quad \ldots, \quad \omega_m = 0 \tag{10.1}$$

where ω_i are 1-forms on $\mathfrak{G}$.

Because $\mathfrak{H}$ and its left cosets as a whole are invariant under $\mathfrak{G}$, the left-hand members of (10.1) can be chosen to be invariant 1-forms and therefore to be

---

ENCYCLOPEDIA OF MATHEMATICS and Its Applications, Gian-Carlo Rota (ed.).
1, Luis A. Santaló, Integral Geometry and Geometric Probability

ISBN 0-201-13500-0

linear combinations with constant coefficients of a system of Maurer–Cartan forms for $\mathfrak{G}$ (Chapter 9, Section 5). Because these forms are defined up to a linear combination with constant coefficients, we may assume that 1-forms (10.1) are the m first forms of a Maurer–Cartan system for $\mathfrak{G}$. The differential m-form

$$d(\mathfrak{G}/\mathfrak{H}) = \omega_1 \wedge \omega_2 \wedge \cdots \wedge \omega_m \tag{10.2}$$

is then invariant under $\mathfrak{G}$ and, up to a constant factor, it is the unique m-form with this property. Indeed, any other invariant m-form can be written $f(g)\, d(\mathfrak{G}/\mathfrak{H})$ and by the property of invariance we have $f(g_0 g)\, d(\mathfrak{G}/\mathfrak{H}) = f(g)\, d(\mathfrak{G}/\mathfrak{H})$; hence, $f(g_0 g) = f(g)$ for any g_0 and since $\mathfrak{G}$ is transitive on $\mathfrak{G}/\mathfrak{H}$, it follows that $f(g) =$ constant. However, $d(\mathfrak{G}/\mathfrak{H})$ is not always a density for $\mathfrak{G}/\mathfrak{H}$ because its value can change when the points $g \in \mathfrak{G}$ displace on the submanifold $g\mathfrak{H}$. We shall prove the following theorem.

A necessary and sufficient condition for the m-form $d(\mathfrak{G}/\mathfrak{H})$ to be a density for $\mathfrak{G}/\mathfrak{H}$ is that its exterior differential vanish, that is,

$$d(d(\mathfrak{G}/\mathfrak{H})) = 0. \tag{10.3}$$

Proof. Let x_i be the coordinates of the point $g \in \mathfrak{G}$ in a local coordinate system. We can make a change of coordinates $(x_1, x_2, \ldots, x_n) \to (\xi_1, \xi_2, \ldots, \xi_m, y_{m+1}, \ldots, y_n)$ such that the submanifolds $g\mathfrak{H}$ be represented by equations $\xi_i =$ constant $(i = 1, 2, \ldots, m)$ and y_j $(j = m+1, \ldots, n)$ be local coordinates on $g\mathfrak{H}$. Then system (10.1) is equivalent to $d\xi_1 = 0$, $d\xi_2 = 0, \ldots, d\xi_m = 0$ and we have

$$d(\mathfrak{G}/\mathfrak{H}) = F(\xi_1, \ldots, \xi_m, y_{m+1}, \ldots, y_n)\, d\xi_1 \wedge \cdots \wedge d\xi_m. \tag{10.4}$$

Under a variation δ on $g\mathfrak{H}$, the coordinates ξ_i remain constant and we have

$$\delta(d(\mathfrak{G}/\mathfrak{H})) = \sum_{j=m+1}^{n} \frac{\partial F}{\partial y_j}\, dy_j \wedge d\xi_1 \wedge \cdots \wedge d\xi_m. \tag{10.5}$$

On the other hand, by exterior differentiation of (10.4) we get

$$d(d(\mathfrak{G}/\mathfrak{H})) = \sum_{j=1}^{m} \frac{\partial F}{\partial \xi_j}\, d\xi_j \wedge d\xi_1 \wedge \cdots \wedge d\xi_m$$

$$+ \sum_{j=m+1}^{n} \frac{\partial F}{\partial y_j}\, dy_j \wedge d\xi_1 \wedge \cdots \wedge d\xi_m = \delta(d(\mathfrak{G}/\mathfrak{H})) \tag{10.6}$$

because the first sum vanishes. Consequently, in order that $\delta(d(\mathfrak{G}/\mathfrak{H})) = 0$, that is, for $d(\mathfrak{G}/\mathfrak{H})$ to be invariant under displacements on the manifolds $g\mathfrak{H}$, it is necessary and sufficient that $d(d(\mathfrak{G}/\mathfrak{H})) = 0$. This proves the theorem.

ISBN 0-201-13500-0

Alternative approach. We shall give other forms for the conditions of existence of an invariant density on $\mathfrak{G}/\mathfrak{H}$ (we always refer to "nonzero" invariant densities). Let $\xi_1, \xi_2, \ldots, \xi_m, y_{m+1}, \ldots, y_n$ be the local coordinates introduced above. The elements of $\mathfrak{H}$, which we will indicate by g_{H}, have coordinates $(e_1, e_2, \ldots, e_m, y_{m+1}, \ldots, y_n)$ where $(e_1, e_2, \ldots, e_n)$ are the coordinates of the identity. Each element of $\mathfrak{G}$ belongs to a coset $g\mathfrak{H}$ and we may take as representative of the coset the element h of coordinates $(\xi_1, \xi_2, \ldots, \xi_m, e_{m+1}, e_{m+2}, \ldots, e_n)$. Since the forms (10.1) are the first m forms of a Maurer–Cartan system of $\mathfrak{G}$ and they are equivalent to $d\xi_1 = 0, d\xi_2 = 0, \ldots, d\xi_m = 0$, it follows that they have the form $\omega_i(\xi, y, d\xi) = 0$ $(i = 1, 2, \ldots, m)$, that is, they do not involve the differentials $dy_{m+1}, \ldots, dy_n$. Moreover, if in the remaining forms of the Maurer–Cartan system of $\mathfrak{G}$ we set $\xi_1 = e_1, \xi_2 = e_2, \ldots, \xi_m = e_m$, the resulting forms will be invariant under $\mathfrak{H}$ and therefore they constitute a Maurer–Cartan system of $\mathfrak{H}$. Hence we have

The Maurer–Cartan forms of $\mathfrak{G}$ can be written

$$\omega_i = \omega_i(\xi, y, d\xi), \qquad i = 1, 2, \ldots, m,$$

$$\omega_s = \omega_s{}^*(\xi, y, d\xi) + \omega_s{}^{\mathrm{H}}(y, dy), \qquad s = m + 1, \ldots, n \tag{10.7}$$

where the 1-forms $\omega_s{}^{\mathrm{H}}(y, dy)$ are a set of Maurer–Cartan forms for the group $\mathfrak{H}$.

By exterior multiplication of the forms (10.7) we get

$$d\mathfrak{G}(g) = d(\mathfrak{G}/\mathfrak{H})(h) \wedge d\mathfrak{H}(g_{\mathrm{H}}) \tag{10.8}$$

where $d\mathfrak{H}(g_{\mathrm{H}}) = \omega_{m+1}^{\mathrm{H}} \wedge \cdots \wedge \omega_n{}^{\mathrm{H}}$ is the invariant volume element of $\mathfrak{H}$ at the element g_{H} and $g = hg_{\mathrm{H}}$.

In order that $d(\mathfrak{G}/\mathfrak{H})(h)$ be a density it must be invariant under displacements on the cosets $g\mathfrak{H}$; that is, it does not change under $h \to hg_{\mathrm{H}}$. Hence we must have

$$d(\mathfrak{G}/\mathfrak{H})(hg_{\mathrm{H}}) = d(\mathfrak{G}/\mathfrak{H})(h) \qquad \text{for all} \quad g_{\mathrm{H}} \in \mathfrak{H}. \tag{10.9}$$

Recall equation (9.88), which we can now write $d(\mathfrak{G}g_{\mathrm{H}}) = \Delta(g_{\mathrm{H}})\, d\mathfrak{G}$ and the corresponding equation $d(\mathfrak{H}g_{\mathrm{H}}) = \delta(g_{\mathrm{H}})\, d\mathfrak{H}$ for the group $\mathfrak{H}$. By the translation $g \to gg_{\mathrm{H}}$, (10.8) gives $d(\mathfrak{G}g_{\mathrm{H}}) = d(\mathfrak{G}/\mathfrak{H})(hg_{\mathrm{H}}) \wedge d(\mathfrak{H}g_{\mathrm{H}})$, which can be written

$$\Delta(g_{\mathrm{H}})\, d\mathfrak{G} = d(\mathfrak{G}/\mathfrak{H})(hg_{\mathrm{H}}) \wedge \delta(g_{\mathrm{H}})\, d\mathfrak{H}. \tag{10.10}$$

From this equation and (10.8) it follows that

$$\frac{\Delta(g_{\mathrm{H}})}{\delta(g_{\mathrm{H}})} = \frac{d(\mathfrak{G}/\mathfrak{H})(hg_{\mathrm{H}})}{d(\mathfrak{G}/\mathfrak{H})(h)} \tag{10.11}$$

ISBN 0-201-13500-0

for each $g_H \in \mathfrak{H}$. Thus, in order that condition (10.9) be satisfied we must have $\Delta(g_H) = \delta(g_H)$. Since we are always considering densities in absolute value, we can state the following condition of Weil [710]:

A necessary and sufficient condition for the existence of an invariant density on the homogeneous space $\mathfrak{G}/\mathfrak{H}$ is that

$$|\Delta(g_H)| = |\delta(g_H)| \tag{10.12}$$

for all $g_H \in \mathfrak{H}$.

According to (9.92), condition (10.12) implies that

$$|\det \mathrm{ad}_G(g_H)| = |\det \mathrm{ad}_H(g_H)|. \tag{10.13}$$

Consequences. (a) If $\mathfrak{G}$ is unimodular, we have $\det \mathrm{ad}_G(g) = 1$ for all $g \in \mathfrak{G}$ and from (10.13) it follows that *if $\mathfrak{G}/\mathfrak{H}$ has an invariant density and $\mathfrak{G}$ is unimodular, then $\mathfrak{H}$ is also unimodular.*

(b) If (10.13) holds, then by (9.92) we have $|\Delta(g_H)/\delta(g_H)| = |\Delta(e)/\delta(e)| =$ constant, and since $d\mathfrak{G}$ and $d\mathfrak{H}$ are defined up to a constant factor, they can be normalized so that $|\Delta(g_H)| = |\delta(g_H)|$. Then we have that *condition* (10.13) *is a necessary and sufficient condition for the existence of an invariant density on $\mathfrak{G}/\mathfrak{H}$.*

(c) If $\mathfrak{G}$ and $\mathfrak{H}$ are unimodular, then condition (10.12) holds and we have *if $\mathfrak{G}$ and $\mathfrak{H}$ are unimodular, then $\mathfrak{G}/\mathfrak{H}$ has an invariant density.*

(d) Recall that as a consequence of (9.87) the map $g \to \Delta(g)$ is a homomorphism of $\mathfrak{G}$ into the multiplicative group of positive real numbers. If $\mathfrak{H}$ is compact, then its image under Δ or δ is a compact subgroup of the multiplicative group of positive real numbers and therefore it must reduce to unity, that is, $\Delta(g_H) = \delta(g_H) = 1$. Hence we have *if a closed subgroup $\mathfrak{H}$ of $\mathfrak{G}$ is compact, then there is an invariant measure on $\mathfrak{G}/\mathfrak{H}$.*

That a compact subgroup of the multiplicative group of positive real numbers consists only of the unity element can be proved as follows: if it contains any real number $a \neq 1$, then it will contain a^m for any m, and this is not possible since $a^m \to \infty$ or $a^m \to 0$ as $m \to \infty$; both cases are impossible since a^m must remain in a bounded closed interval not including 0.

Note that $|\det \mathrm{ad}_G(g_H)| = 1$ does not imply that $|\det \mathrm{ad}_G(g)| = 1$ for any $g \in \mathfrak{G}$. Therefore it is possible that $\mathfrak{H}$ be unimodular and $\mathfrak{G}/\mathfrak{H}$ have an invariant density without $\mathfrak{G}$ being unimodular. For instance, let $\mathfrak{G}$ be the group of matrices

$$g = \begin{pmatrix} a & b \\ 0 & 1 \end{pmatrix}, \qquad a \neq 0$$

and $\mathfrak{H}$ be the subgroup of matrices

ISBN 0-201-13500-0

$$g_H = \begin{pmatrix} 1 & b \\ 0 & 1 \end{pmatrix}.$$

Then $\mathfrak{H}$ is unimodular because it is commutative. The Maurer–Cartan forms for $\mathfrak{G}$ are $\omega_1 = da/a$, $\omega_2 = db/a$, so that the cosets $g\mathfrak{H}$ are defined by $\omega_1 = 0$ and since $d\omega_1 = 0$, it follows that by criterion (10.3), $\mathfrak{G}/\mathfrak{H}$ has an invariant density. However, $\mathfrak{G}$ is not unimodular, since $\bar{\omega}_1 = da/a, \bar{\omega}_2 = -(b/a)\,da + db$ and therefore

$$d_L\mathfrak{G} = a^{-2}\, da \wedge db, \qquad d_R\mathfrak{G} = a^{-1}\, da \wedge db.$$

The linear transformation $\mathrm{ad}_G(g_0)$ is $A' = A$, $B' = -b_0A + a_0B$, so that $\det \mathrm{ad}_G(g_0) = a_0$. We have $\det \mathrm{ad}_G(g_0) = 1$ if $g_0 = g_H \in \mathfrak{H}$ and $\det \mathrm{ad}_G(g_0) \neq 1$ otherwise.

2. Invariant Subgroups and Quotient Groups

A subgroup $\mathfrak{H}$ of a Lie group is called *invariant* or *normal* if for all $g_H \in \mathfrak{H}$ and all $g \in \mathfrak{G}$ the relation $gg_Hg^{-1} \in \mathfrak{H}$ holds. Then, if $\mathfrak{H}$ is closed, it is well known that with the definition of the product $(x\mathfrak{H})(y\mathfrak{H}) = xy\mathfrak{H}$, the homogeneous space $\mathfrak{G}/\mathfrak{H}$ becomes a Lie group, called the *quotient group*, which is also denoted by $\mathfrak{G}/\mathfrak{H}$.

Let $d_L(\mathfrak{G}/\mathfrak{H})$ denote the left-invariant density of this group $\mathfrak{G}/\mathfrak{H}$. If $h \in \mathfrak{G}$ is a representative of the coset $g\mathfrak{H}$, from the relation $hg_Hh^{-1} \in \mathfrak{H}$ it follows that $hg_H = g_H{}^*h$, where $g_H{}^* \in \mathfrak{H}$ and we can write symbolically $(\mathfrak{G}/\mathfrak{H})g_H = g_H{}^*(\mathfrak{G}/\mathfrak{H})$. Therefore the left invariance $d_L(\mathfrak{G}/\mathfrak{H}) = d_L(g_H{}^*(\mathfrak{G}/\mathfrak{H}))$ implies that $d_L(\mathfrak{G}/\mathfrak{H}) = d_L((\mathfrak{G}/\mathfrak{H})g_H)$, and then from (10.9) it follows that $d_L(\mathfrak{G}/\mathfrak{H})$ is an invariant density for the homogeneous space $\mathfrak{G}/\mathfrak{H}$. Hence we have

If $\mathfrak{H}$ is a closed invariant subgroup of $\mathfrak{G}$, the homogeneous space $\mathfrak{G}/\mathfrak{H}$ has an invariant density under $\mathfrak{G}$ that is equal to the left-invariant density for the quotient group $\mathfrak{G}/\mathfrak{H}$.

From this theorem and consequence (a) of the preceding section we deduce that *if $\mathfrak{G}$ is unimodular and $\mathfrak{H}$ is a closed invariant subgroup of $\mathfrak{G}$, then $\mathfrak{H}$ is unimodular.*

The condition of $\mathfrak{H}$ being an invariant subgroup cannot be dropped. For instance, $GL(2)$ is unimodular and its closed subgroup of matrices

$$\begin{pmatrix} x & y \\ 0 & z \end{pmatrix}$$

with $xz = 1$ is not unimodular. The converse is not always true. For instance, the group of matrices

ISBN 0-201-13500-0

$$g_{\mathrm{H}} = \begin{pmatrix} 1 & h \\ 0 & 1 \end{pmatrix}$$

is a closed unimodular invariant subgroup of the group of matrices

$$\begin{pmatrix} x & y \\ 0 & 1 \end{pmatrix}, \qquad x \neq 0;$$

this group, however, is not unimodular. Of course, if $\mathfrak{H}$ is unimodular and $\mathfrak{G}/\mathfrak{H}$ is compact, then $\mathfrak{G}$ is unimodular.

3. Other Conditions for the Existence of a Density on Homogeneous Spaces

Applying transformation (9.89) to both sides of (10.8), we obtain

$$L^*_{g_{\mathrm{H}}} \circ R^*_{g_{\mathrm{H}}^{-1}} d\mathfrak{G}(e) = L^*_{g_{\mathrm{H}}} \circ R^*_{g_{\mathrm{H}}^{-1}} d((\mathfrak{G}/\mathfrak{H})(e)) \wedge L^*_{g_{\mathrm{H}}} \circ R^*_{g_{\mathrm{H}}^{-1}} d\mathfrak{H}(e). \quad (10.14)$$

The diffeomorphism $\tau(g_{\mathrm{H}})\colon g\mathfrak{H} \to g_{\mathrm{H}}g\mathfrak{H}$ of $\mathfrak{G}/\mathfrak{H}$ onto itself is equivalent to $g\mathfrak{H} \to g_{\mathrm{H}}g\mathfrak{H}g_{\mathrm{H}}^{-1}$ (since $g_{\mathrm{H}} \in \mathfrak{H}$); thus the first factor on the right-hand side in (10.14) is the mapping $\tau^*(g_{\mathrm{H}})_e$. Equating the determinants of the matrices of linear transformation (10.14) and using that the matrices of τ^* and $d\tau$ are transposed matrices, we have

$$\det \mathrm{ad}_{\mathrm{G}}(g_{\mathrm{H}}) = \det \mathrm{ad}_{\mathrm{H}}(g_{\mathrm{H}}) \cdot \det(d\tau(g_{\mathrm{H}}))_e. \quad (10.15)$$

The linear group of transformations $(d\tau(g_{\mathrm{H}}))_e$, which acts on the tangent space to $\mathfrak{G}/\mathfrak{H}$ at e (i.e., the point represented by the coset $\mathfrak{H}$), is called the *linear isotropy group* of $\mathfrak{G}$. From (10.13) and (10.15) we have the theorem (see [353]):

A necessary and sufficient condition for the existence of an invariant density on the homogeneous space $\mathfrak{G}/\mathfrak{H}$ is

$$|\det d\tau(g_{\mathrm{H}})_e| = 1, \quad (10.16)$$

that is, that the linear isotropy group have determinant equal to unity.

Finally, we shall give a last condition for the existence of invariant density on $\mathfrak{G}/\mathfrak{H}$. From (10.7), writing the structure equations $d\omega_s = \frac{1}{2}\sum C^s_{ik}\, \omega_i \wedge \omega_k$ $(s = m+1, \ldots, n)$ and equating the terms that do not involve the differentials $d\xi_i$, we get $d\omega_s{}^{\mathrm{H}} = \frac{1}{2}\sum C^s_{ik}\, \omega_i{}^{\mathrm{H}} \wedge \omega_k{}^{\mathrm{H}}$ $(i, k, s = m+1, \ldots, n)$. Thus, the structure constants C^s_{ik} for $i, k, s = m+1, \ldots, n$ are the same for the group $\mathfrak{G}$ as for the subgroup $\mathfrak{H}$. On the other hand, according to (10.7), for the points g_{H} (ξ_i = constant) we have $\omega_i(g_{\mathrm{H}}) = 0$ $(i = 1, 2, \ldots, m)$ and $\omega_i(g_{\mathrm{H}}) =$

ISBN 0-201-13500-0

$\omega_i^{\mathrm{H}}(g_{\mathrm{H}})$ for $i = m + 1, \ldots, n$. Hence, using (9.63) we have $\overline{\omega}_i(g_{\mathrm{H}}) = -\omega_i(g_{\mathrm{H}}^{-1}) = 0\ (i = 1, 2, \ldots, m)$. Therefore, the condition $\Delta = \delta$ or $d\Delta/\Delta = d\delta/\delta$ becomes, in view of (9.83),

$$\sum_{s=m+1}^{n} \sum_{k=1}^{n} C_{sk}^{k} \overline{\omega}_s = \sum_{s=m+1}^{n} \sum_{k=m+1}^{n} C_{sk}^{k} \overline{\omega}_s,$$

which gives

$$\sum_{s=m+1}^{n} \left(\sum_{k=1}^{m} C_{sk}^{k} \right) \overline{\omega}_s = 0.$$

Since the 1-forms $\overline{\omega}_s$ are independent, we can state the following criterion of Chern [105].

A necessary and sufficient condition for the existence of an invariant density on the homogeneous space $\mathfrak{G}/\mathfrak{H}$ is that

$$\sum_{k=1}^{m} C_{sk}^{k} = 0 \qquad \text{for} \quad s = m + 1, \ldots, n. \tag{10.17}$$

A method of computing invariant densities based on the construction of models of homogeneous spaces as varieties in affine spaces, has been given by Gaeta [211].

4. Examples

1. Let $\mathfrak{G}$ be the group of matrices

$$g = \begin{pmatrix} 1 & 0 & \log x \\ y & x & z \\ 0 & 0 & 1 \end{pmatrix}, \qquad x \neq 0, \tag{10.18}$$

and let $\mathfrak{H}$ be the one-dimensional subgroup

$$g_{\mathrm{H}} = \begin{pmatrix} 1 & 0 & 0 \\ y & 1 & 0 \\ 0 & 0 & 1 \end{pmatrix}. \tag{10.19}$$

A set of Maurer–Cartan 1-forms for $\mathfrak{G}$ is

$$\omega_1 = dx/x, \qquad \omega_2 = dy/x, \qquad \omega_3 = -yx^{-2}\,dx + x^{-1}\,dz, \tag{10.20}$$

and the structure equations are

$$d\omega_1 = 0, \quad d\omega_2 = -\omega_1 \wedge \omega_2, \quad d\omega_3 = \omega_1 \wedge \omega_2 - \omega_1 \wedge \omega_3. \tag{10.21}$$

The subgroup $\mathfrak{H}$ is defined by $x = 1$, $z = 0$. Hence system (10.1) is now $dx = 0$, $dz = 0$; that is, $\omega_1 = 0$, $\omega_3 = 0$ and the invariant density, if it exists, must be $d(\mathfrak{G}/\mathfrak{H}) = \omega_1 \wedge \omega_3$. Because $d(\omega_1 \wedge \omega_3) = d\omega_1 \wedge \omega_3 - \omega_1 \wedge d\omega_3 = 0$,

ISBN 0-201-13500-0

condition (10.3) is fulfilled and hence the space $\mathfrak{G}/\mathfrak{H}$ has the invariant density $d(\mathfrak{G}/\mathfrak{H}) = \omega_1 \wedge \omega_3$. The cosets $g\mathfrak{H}$ (points of $\mathfrak{G}/\mathfrak{H}$) are determined by the x, z coordinates (ξ_1, ξ_2 coordinates of Section 1) and in terms of these coordinates the invariant density becomes

$$d(\mathfrak{G}/\mathfrak{H}) = x^{-2}\, dx \wedge dz. \tag{10.22}$$

We want to verify, as an exercise, that the other conditions for the existence of an invariant density on $\mathfrak{G}/\mathfrak{H}$ are satisfied. Indeed, from (10.21) it follows that the nonzero structure constants are $C_{12}^2 = -C_{21}^2 = -1$, $C_{12}^3 = -C_{21}^3 = 1$, $C_{13}^3 = -C_{31}^3 = -1$ and therefore conditions (10.17) are fulfilled. The mapping $g \to g_0 g g_0^{-1}$ is $x' = x$, $y' = -y_0 x + x_0 y + y_0$, $z' = (y_0 \log x_0 - z_0)\, x + y_0 \log x - x_0 y \log x_0 + x_0 z - y_0 \log x_0 + z_0$ and $|\det \mathrm{ad}_G(g_0)| = x_0^2$. If $g_0 \in \mathfrak{H}$, we have $x_0 = 1$ and $|\det \mathrm{ad}_G(g_H)| = 1$. The mapping $g_H \to g_{H_0} g_H g_{H_0}^{-1}$ can be written $y' = y$ and therefore $|\det \mathrm{ad}_H(g_H)| = 1$. Hence, condition (10.13) is fulfilled. The coset $g\mathfrak{H}$ is the set of matrices

$$\begin{pmatrix} 1 & 0 & \log x \\ y + xy_0 & x & z \\ 0 & 0 & 1 \end{pmatrix}$$

for all $y_0 \in R^1$. Thus, if $g_H{}^1$ is the element $x = 1$, $y = y_1$, $x = 0$, the mapping $\tau(g_H{}^1)$ is $x' = x$, $y' = y + y_1$, $z' = y_1 \log x + z$, so that the mapping $(d\tau(g_H{}^1))_e$ may be written $x' = x$, $y' = y$, $z' = y_1 + z$. Thus $|\det(d\tau(g_H{}^1))_e| = 1$, according to (10.16).

2. Consider the matrix groups

$$\mathfrak{G}: g = \begin{pmatrix} x & 0 & 0 \\ 0 & y & z \\ 0 & 0 & 1 \end{pmatrix}, \quad xy \neq 0; \qquad \mathfrak{H}: g_H = \begin{pmatrix} 1 & 0 & 0 \\ 0 & y & 0 \\ 0 & 0 & 1 \end{pmatrix}, \quad y \neq 0, \tag{10.23}$$

such that $\mathfrak{H} \subset \mathfrak{G}$. A system of Maurer–Cartan forms for $\mathfrak{G}$ is $\omega_1 = dx/x$, $\omega_2 = dy/y$, $\omega_3 = dz/y$, and the structure equations are $d\omega_1 = 0$, $d\omega_2 = 0$' $d\omega_3 = -\omega_2 \wedge \omega_3$. The cosets $g\mathfrak{H}$ are defined by $x =$ constant, $z =$ constant, so that system (10.1) becomes $\omega_1 = 0$, $\omega_3 = 0$. We must see whether $d(\mathfrak{G}/\mathfrak{H}) = \omega_1 \wedge \omega_3$ is or is not an invariant density for $\mathfrak{G}/\mathfrak{H}$. We have $d(d(\mathfrak{G}/\mathfrak{H})) = d\omega_1 \wedge \omega_3 - \omega_1 \wedge d\omega_3 = \omega_1 \wedge \omega_2 \wedge \omega_3 \neq 0$. Hence, the homogeneous space $\mathfrak{G}/\mathfrak{H}$ does not have an invariant density under $\mathfrak{G}$. The same result yields the other criteria of Sections 2 and 3. For instance, from the structure equations it follows that $C_{21}^1 + C_{22}^2 + C_{23}^3 = -1$, and hence condition (10.17) is not fulfilled. A set of right-invariant 1-forms is $\bar{\omega}_1 = dx/x$, $\bar{\omega}_2 = dy/y$, $\bar{\omega}_3 = -(z/y)\, dy + dz = -z\omega_2 + y\omega_3$, so that according to the definition of $\Delta(g)$ (9.80), we have $\Delta(g) = 1/y$. Thus the group $\mathfrak{G}$ is not unimodular. For the group $\mathfrak{H}$ we have $\omega_1{}^H = \bar{\omega}_1{}^H = dy/y$, $\delta(g_H) = 1$ and hence $\mathfrak{H}$ is unimodular.

ISBN 0-201-13500-0

Condition (10.12) is not fulfilled. The reader can easily compute $|\det \mathrm{ad}_G(g_1)| = y_1$, $|\det \mathrm{ad}_H(g_1)| = 1$, $|\det(d\tau(g_H{}^1))_e| = y_1$.

5. Lie Transformation Groups

Let $\mathfrak{G}$ be a Lie group and let M be a differentiable manifold. The group $\mathfrak{G}$ is called a *Lie transformation group* of M if to each $x \in \mathfrak{G}$ there is associated a homeomorphism $p \to xp$ $(p \in M)$ of M onto itself, such that (a) $(x_1x_2)p = x_1(x_2p)$ for $x_1, x_2 \in \mathfrak{G}$, $p \in M$; (b) the mapping $(x, p) \to xp$ is a differentiable mapping of $\mathfrak{G} \times M$ onto M. It follows that for each $x \in \mathfrak{G}$ the mapping $p \to xp$ is a diffeomorphism of M onto itself. We call xp the transform of p by x. The group of motions on the plane is an example of a Lie transformation group.

A Lie transformation group is called *transitive* if for every pair of points $p_1, p_2 \in M$ there is at least one $x \in \mathfrak{G}$ such that $xp_1 = p_2$. If for every pair of points $p_1, p_2 \in M$ there is one unique x such that $xp_1 = p_2$, then $\mathfrak{G}$ is called *simply transitive*. If the identity e is the only element in $\mathfrak{G}$ that leaves all of M fixed, then $\mathfrak{G}$ is called *effective*.

Assume that $\mathfrak{G}$ acts transitively on M. The elements s of $\mathfrak{G}$ such that $sp_0 = p_0$ for a given point $p_0 \in M$ form a subgroup $\mathfrak{G}_0$ of $\mathfrak{G}$ called the *isotropy group* at p_0 (or the *stability subgroup* of $\mathfrak{G}$ at p_0). This group $\mathfrak{G}_0$ is closed in $\mathfrak{G}$. If $\mathfrak{G}$ is effective, the homogeneous space $\mathfrak{G}/\mathfrak{G}_0$ is mapped in a continuous one-to-one way onto M by the map $\psi: x\mathfrak{G}_0 \to xp_0$; that is, to the coset $x\mathfrak{G}_0$ (an element of $\mathfrak{G}/\mathfrak{G}_0$) there corresponds the point xp_0 of M and, conversely, to the point $p \in M$ such that $p = xp_0$ corresponds the coset $x\mathfrak{G}_0$. Let $\{y\}$ be a set of representatives of $\mathfrak{G}/\mathfrak{G}_0$, that is, a set of elements $y \in \mathfrak{G}$ such that $y\mathfrak{G}_0$ $(y \in \{y\})$ are all cosets of $\mathfrak{G}/\mathfrak{G}_0$ and if $y_1 \neq y_2$, then $y_1\mathfrak{G}_0 \neq y_2\mathfrak{G}_0$. We call $\{y\}$ a set of translations of the group $\mathfrak{G}$. By the mapping ψ above, the set $\{y\}$ is mapped in a continuous one-to-one way onto the points of M.

If $\mathfrak{G}/\mathfrak{G}_0$ has an invariant density under $\mathfrak{G}$, the mapping ψ gives rise to an invariant density for points of M, the invariant measure of a point set $X \subset M$ being

$$m(X) = \int_{\mathfrak{G}/\mathfrak{G}_0} \phi(y)\, d(\mathfrak{G}/\mathfrak{G}_0) \tag{10.24}$$

where $\phi(y) = 1$ if $yp_0 \in X$ and $\phi(y) = 0$ otherwise. The density $d(\mathfrak{G}/\mathfrak{G}_0)$ is called the *density for points* in M that is invariant under $\mathfrak{G}$ and we set

$$dP = d(\mathfrak{G}/\mathfrak{G}_0). \tag{10.25}$$

From (9.74) and (10.8) it follows that if $\mathfrak{G}_0$ and $\mathfrak{G}$ are unimodular, then $dP = d(\mathfrak{G}/\mathfrak{G}_0)$ is invariant under the mapping $y \to y^{-1}$ and (10.24) can be

ISBN 0-201-13500-0

written

$$m(X) = \int_{\mathfrak{G}/\mathfrak{G}_0} \phi(y^{-1})\, dP \tag{10.26}$$

where $\phi(y^{-1}) = 1$ if $p_0 \in yX$ and $\phi(y^{-1}) = 0$ otherwise.

We give two applications of these formulas.

1. Let K_0 be a set of N fixed points $p_1, p_2, \ldots, p_N$ in M. Writing (10.26) for each point p_i and summing, we get

$$\int_{\mathfrak{G}/\mathfrak{G}_0} \nu(K_0 \cap yX)\, dP = Nm(X) \tag{10.27}$$

where $\nu(K_0 \cap yX)$ denotes the number of points of K_0 that belong to yX. If $m(M) = m(\mathfrak{G}/\mathfrak{G}_0)$ denotes the volume of M, it follows that the mean value of the number of points of K_0 covered by the set yX, for all y, is

$$E(\nu) = \frac{m(X)}{m(M)} N. \tag{10.28}$$

Since the mean value $E(\nu)$ lies between the greatest and the smallest values of ν, we can state the following theorem.

Let M be a differentiable manifold on which the group $\mathfrak{G}$ acts transitively. Assume that the homogeneous space $\mathfrak{G}/\mathfrak{G}_0$ has an invariant density ($\mathfrak{G}_0$ being the isotropy group at $p_0 \in M$) and that $\mathfrak{G}$ and $\mathfrak{G}_0$ are unimodular. Let X be a given point set in M of volume $m(X)$ and assume that M has a finite volume $m(M)$ (both $m(X)$ and $m(M)$ measured with the point density $d(\mathfrak{G}/\mathfrak{G}_0) = dP$). Then, given N points $p_i \in M$ $(i = 1, 2, \ldots, N)$, there exists a translation $y \in \mathfrak{G}$ such that yX contains at least (at most) $Nm(X)/m(M)$ of the given points. It certainly contains a number greater (less) than $Nm(X)/m(M)$ if X is closed in M and $m(X) < m(M)$.

The proof of the last part of the theorem is the same as that in Chapter 8, Section 4.

2. Consider the complex variable $z = \xi + i\eta$. The group $SL(2)$ (group of all 2×2 matrices of determinant equal to 1) acts on the upper half plane $\eta > 0$ as the transformation group

$$z' = \frac{az + b}{pz + q}, \qquad aq - bp = 1, \tag{10.29}$$

where a, b, p, q are real numbers. The Maurer–Cartan forms for $SL(2)$ are

$$\omega_1 = q\, da - b\, dp, \qquad \omega_2 = q\, db - b\, dq, \qquad \omega_3 = -p\, da + a\, dp, \tag{10.30}$$

ISBN 0-201-13500-0

and using that $d(aq - bp) = 0$, we have

$$da = a\omega_1 + b\omega_3, \qquad db = -b\omega_1 + a\omega_2, \qquad dp = p\omega_1 + q\omega_3,$$
$$dq = -q\omega_1 + p\omega_2. \tag{10.31}$$

The structure equations are

$$d\omega_1 = -\omega_2 \wedge \omega_3, \qquad d\omega_2 = -2\omega_1 \wedge \omega_2, \qquad d\omega_3 = -2\omega_3 \wedge \omega_1. \tag{10.32}$$

The isotropy group $\mathfrak{G}_0$ at $z = i$ is characterized by the conditions $b = -p$, $a = q$, which according to (10.31) are equivalent to $\omega_1 = 0$, $\omega_2 + \omega_3 = 0$. Since $d(\omega_1 \wedge (\omega_2 + \omega_3)) = 0$, it follows from criterion (10.3) that $\mathfrak{G}/\mathfrak{G}_0$ has the invariant density $d(\mathfrak{G}/\mathfrak{G}_0) = dP = \omega_1 \wedge (\omega_2 + \omega_3)$. Noting that $dz = (\omega_2 + \omega_3 + 2i\omega_1)/(p + iq)^2$, $z - \bar{z} = 2i(p^2 + q^2)^{-1}$, we get

$$dP = i\frac{dz \wedge d\bar{z}}{(z - \bar{z})^2} = -\frac{d\xi \wedge d\eta}{2\eta^2}. \tag{10.33}$$

Consider now the subgroup $\mathfrak{G}_1$ of the elements of $\mathfrak{G} \equiv SL(2)$ that preserves the half circle $|z| = 1$ $(\eta > 0)$. They are characterized by the condition $1 = (a^2 + 2ab\xi + b^2)(p^2 + 2pq\xi + q^2)^{-1}$ for any ξ. Thus we have the conditions $a^2 + b^2 = p^2 + q^2$, $ab = pq$, from which we deduce $a = q$, $b = p$, or $a = -q$, $b = -p$. These conditions are equivalent to $\omega_1 = 0$, $\omega_2 - \omega_3 = 0$. Since $d(\omega_1 \wedge (\omega_2 - \omega_3)) = 0$, it follows that the homogeneous space $\mathfrak{G}/\mathfrak{G}_1$ has the invariant density $d(\mathfrak{G}/\mathfrak{G}_1) = \omega_1 \wedge (\omega_2 - \omega_3)$. Element (10.29) of $\mathfrak{G}$ transforms the half circle $|z| = 1$ $(\eta > 0)$ onto the half circle with center $(\xi, 0)$ and radius r given by

$$\xi = (bq - ap)(q^2 - p^2)^{-1}, \qquad r = (q^2 - p^2)^{-1}. \tag{10.34}$$

Using (10.30), we find by an easy computation that

$$d(\mathfrak{G}/\mathfrak{G}_1) = \omega_1 \wedge (\omega_2 - \omega_3) = -(d\xi \wedge dr)/2r^2. \tag{10.35}$$

This is the invariant density under the group (10.29) for half circles whose centers are on the ξ axis.

The upper half plane $\eta > 0$ with the transformation group (10.29) is the classical Poincaré model for noneuclidean hyperbolic geometry. The density for points (10.33) is then, up to a constant factor, the noneuclidean area element and (10.35) is the invariant density for sets of lines (geodesics) in the hyperbolic plane (see, e.g., [360]).

6. Notes and Exercises

1. *An analogue to Blichfeldt's and Minkowski's theorems.* Let M be a differentiable manifold on which a Lie transformation group $\mathfrak{G}$ acts transitively.

ISBN 0-201-13500-0

Assume that the homogeneous space $\mathfrak{G}/\mathfrak{G}_0$ ($\mathfrak{G}_0$ being the isotropy group at p_0) has an invariant density $d(\mathfrak{G}/\mathfrak{G}_0)$ and that there exists a partition of M into fundamental regions D_h ($h = 0, 1, \ldots$), $D_h \cap D_m = 0$ if $h \neq m$, and a discrete subgroup $\mathfrak{F}$ of $\mathfrak{G}$ such that (a) each D_h is the transform of D_0 by the transformation $x_h \in \mathfrak{F}$ (x_0 = identity of $\mathfrak{F}$), that is, $D_h = x_h D_0$ ($x_h \in \mathfrak{F}$, $h = 0, 1, 2, \ldots$); (b) each x_h transforms a fundamental region D_j into a fundamental region D_m with $j \neq m$ if $h \neq 0$; (c) the fundamental regions have a finite volume $0 < m(D_0) = m(D_h)$. Then we can prove the following theorem:

Let p_i be N points contained in the fundamental region D_0. We consider the lattice of all points $x_h p_i$ ($x_h \in \mathfrak{F}$) and a function $f(p_i)$ defined on the points p_i and their translates, such that $f(p_i) = f(x_h p_i)$. Then, for every point set X of M there are translates yX ($\{y\}$ = set of representatives of $\mathfrak{G}/\mathfrak{G}_0$) for which the sum $\sum f(p_i)$ extended over the lattice points contained in yX is not less than

$$\frac{m(X)}{m(D_0)} \sum_1^N f(p_i) \tag{10.36}$$

and there are translates yX for which that sum is not greater than (10.36). *If X is closed in M, then there are translates for which the sum $\sum f(p_i)$ is certainly less and translates for which it is certainly greater than* (10.36).

If $f(p_i) = 1$, we get the mean value of the number ν of lattice points that are covered by yX, namely, $E(\nu) = Nm(X)/m(D_0)$, and the theorem gives bounds for the number of lattice points contained in suitable translates of X. If M is the euclidean n-dimensional space E_n, $\mathfrak{G}$ is the group of motions in it, and $\mathfrak{F}$ the subgroup of translations that leaves unchanged the lattice of points with integral coordinates, then $\mathfrak{G}/\mathfrak{G}_0$ is the group of translations in E_n and the theorem coincides with the classical Blichfeldt theorem:

Consider in E_n the lattice of points with integral coordinates; let X be a measurable closed set such that $m(X) \geqslant k$; then there are translations y such that yX contains at least k lattice points. For $n = 2$, see Chapter 8, Section 5.

If M is the unit disk $|z| \leqslant 1$ of the complex plane, $\mathfrak{G}$ is the group of linear transformations that maps M onto itself, and $\mathfrak{F}$ is a real discontinuous subgroup of $\mathfrak{G}$ (Fuchsian group), the theorem gives some results of Tsuji [680]. Consider the case in which $N = 1$. Given a domain X that contains p_0, we shall say that the domain X^* contained in X is an m-domain of X with respect to the group $\mathfrak{G}$ if for each pair $x, y \in \mathfrak{G}$ the following conditions are satisfied: (a) if $xp_0 \in X^*$, then $x^{-1}p_0 \in X^*$; (b) if $xp_0 \in X^*$ and $yp_0 \in X^*$, then $xyp_0 \in X$. With these conditions, we can prove that *if the domain X that contains the point p_0 possesses an m-domain X^* such that $m(X^*) > m(D_0)$, then X contains at least one lattice point distinct from p_0.*

If M is E_n and $\mathfrak{G}$, $\mathfrak{F}$ are the group of translations and the subgroup that preserves the points with integral coordinates, this theorem is a classical Minkowski theorem [361]. For applications to the unit disk of the complex plane and to Fuchsian groups, see the work of Tsuji [679] and Santaló [571].

ISBN 0-201-13500-0

2. *Measurable groups according to* M. I. Stoka. Let $\mathfrak{G}$ be a Lie group that acts transitively as a transformation group of a differentiable manifold M. Let $\mathfrak{G}_0$ be the isotropy group at $p_0 \in M$. If the homogeneous space $\mathfrak{G}/\mathfrak{G}_0$ has an invariant density $d(\mathfrak{G}/\mathfrak{G}_0)$ (density for points in M), then $\mathfrak{G}$ is said to be a measurable group in the sense of Stoka [646]. In other words, a transitive group $\mathfrak{G}$ is called measurable if there exists an invariant density under $\mathfrak{G}$ for sets of points. The following theorems can be proved:

(i) All simply transitive groups on M are measurable (since $d(\mathfrak{G}/\mathfrak{G}_0)$ is then the invariant density of $\mathfrak{G}$).

(ii) If M has dimension n, any group of dimension $n + 1$ acting on M is either measurable or has a measurable subgroup of dimension n.

(iii) If $\mathfrak{G}$ has a transitive subgroup $\mathfrak{G}_1 \subset \mathfrak{G}$ that is not measurable, then $\mathfrak{G}$ is not measurable.

(iv) If $\mathfrak{G}$ possesses two simply transitive subgroups $\mathfrak{G}_1$, $\mathfrak{G}_2$, a necessary and sufficient condition for $\mathfrak{G}$ to be measurable is that $\mathfrak{G}_1$ and $\mathfrak{G}_2$ have the same invariant density. For instance, the affine group $\mathfrak{G}: x' = ax + b$ $(a \neq 0)$ that acts on the real line has the subgroup $\mathfrak{G}_1: x' = x + b$, which is measurable with invariant density equal to dx, and the subgroup of homotheties $x' = ax$, $a \neq 0$, which is also measurable with dx/x as invariant density. Hence, since the two densities are different, it follows that the group $\mathfrak{G}$ is not measurable; that is, the sets of points on the real line do not have an invariant density under the affine group $\mathfrak{G}$. A fortiori, the projective group $x' = (ax + b)/(cx + d)$, $ad - bc \neq 0$, is not measurable, that is, it lacks an invariant density for sets of points, as can also be seen by direct computation.

Other properties of measurable groups and the necessary and sufficient conditions for a group to be measurable in terms of the infinitesimal operators and the structure constants of the group were given by Stoka [646]; see also the work of Chebotarev [102], Drinfel'd [152, 153], and Vranceanu [706].

3. *The chain rule.* (a) Let $\mathfrak{S} \subset \mathfrak{F}$ be closed subgroups of the Lie group $\mathfrak{G}$. If we assume that the homogeneous spaces $\mathfrak{G}/\mathfrak{F}$ and $\mathfrak{G}/\mathfrak{S}$ have nonzero invariant densities $d(\mathfrak{G}/\mathfrak{F})$ and $d(\mathfrak{G}/\mathfrak{S})$, respectively, then $\mathfrak{F}/\mathfrak{S}$ also has an invariant density, which with a suitable normalization can be written

$$d(\mathfrak{G}/\mathfrak{S}) = d(\mathfrak{G}/\mathfrak{F}) \wedge d(\mathfrak{F}/\mathfrak{S}). \tag{10.37}$$

See [298, p. 449]; on homogeneous spaces with finite invariant measure, see [436].

(b) Let $\mathfrak{H}$, $\mathfrak{K}$ be closed subgroups of the Lie group $\mathfrak{G}$. Assume that $\mathfrak{G}/\mathfrak{H}$ and $\mathfrak{G}/\mathfrak{K}$ have invariant density under $\mathfrak{G}$. Then a necessary and sufficient condition for $\mathfrak{H}/\mathfrak{H} \cap \mathfrak{K}$ and $\mathfrak{K}/\mathfrak{H} \cap \mathfrak{K}$ to have an invariant density is that $\mathfrak{G}/\mathfrak{H} \cap \mathfrak{K}$ have invariant density [661].

4. *Relative invariant density in homogeneous spaces.* Let $\mathfrak{G}$ be a Lie group. We know that a differential form Ω defined on $\mathfrak{G}$ is called invariant under $\mathfrak{G}$ (recall that in this chapter invariance means "left invariance") if $\Omega(g_0 g) = \Omega(g)$ for all $g, g_0 \in \mathfrak{G}$. More generally, the form Ω is called relatively invariant under $\mathfrak{G}$ if there exists a positive real-valued function $\phi: \mathfrak{G} \to R^+$ (called the multiplier of Ω) such that $\Omega(g_0 g) = \phi(g_0)\Omega(g)$, $g, g_0 \in \mathfrak{G}$.

ISBN 0-201-13500-0

Let $\mathfrak{H}$ be a closed subgroup of $\mathfrak{G}$ and consider the homogeneous space $\mathfrak{G}/\mathfrak{H}$. Assume that $\mathfrak{G}/\mathfrak{H}$ is of dimension m and let Ω be an m-form on $\mathfrak{G}$. Then Ω will be an m-form on $\mathfrak{G}/\mathfrak{H}$ if it remains invariant under translations on the cosets $g\mathfrak{H}$, that is, if $\Omega(gg_{\mathrm{H}}) = \Omega(g)$ for $g \in \mathfrak{G}$, $g_{\mathrm{H}} \in \mathfrak{H}$. We have seen criteria for finding the m-forms on $\mathfrak{G}/\mathfrak{H}$ that are invariant under $\mathfrak{G}$. We shall now give a criterion for finding the m-forms on $\mathfrak{G}/\mathfrak{H}$ that are relatively invariant under $\mathfrak{G}$.

Let $\omega_1 = 0, \omega_2 = 0, \ldots, \omega_m = 0$ be the Pfaffian system that determines the cosets $g\mathfrak{H}$ of $\mathfrak{G}/\mathfrak{H}$. Put $\omega = \omega_1 \wedge \omega_2 \wedge \cdots \wedge \omega_m$. As a generalization of criterion (10.3) it can be shown that a necessary and sufficient condition for the existence of a relatively invariant density for $\mathfrak{G}/\mathfrak{H}$ of multiplier ϕ is $d(\phi\omega) = 0$, provided that $\phi(g_0 g) = \phi(g_0)\phi(g)$. Then, the desired relatively invariant m-form is $\Omega = \phi\omega$. Moreover, the multiplier ϕ restricted to $\mathfrak{H}$ satisfies the equation $\phi(g_{\mathrm{H}}) = \det \mathrm{ad}_{\mathrm{H}}(g_{\mathrm{H}})/\det \mathrm{ad}_{\mathrm{G}}(g_{\mathrm{H}})$, with the same notation as in (10.13). See [521]. From $\phi(g_1 g_0) = \phi(g_1)\phi(g_0)$ it follows that, up to a constant factor, $\phi(g_{\mathrm{H}}) = \delta(g_{\mathrm{H}})/\Delta(g_{\mathrm{H}})$. Any strictly positive function ϕ on $\mathfrak{G}$ satisfying these conditions gives rise to a relatively invariant positive density on $\mathfrak{G}/\mathfrak{H}$. More general results are discussed by Bourbaki [69] and Reiter [500].

EXERCISE 1. *Triangular groups*. Let $ST(n)$, $n \geqslant 2$, be the group of all $n \times n$ matrices g with determinant equal to unity and such that $g_{ij} = 0\,(1 \leqslant j < i \leqslant n)$ and $g_{ii} > 0$ $(1 \leqslant i \leqslant n)$, called the special triangular group. Show that it is not unimodular. The group $ST_1(n)$ of all $n \times n$ matrices such that $g_{ij} = 0$ $(1 \leqslant j < i \leqslant n)$ and $g_{ii} = 1$ $(1 \leqslant i \leqslant n)$ is a closed invariant subgroup of $ST(n)$. Show that it is unimodular (see [500]).

ISBN 0-201-13500-0

CHAPTER 11

The Affine Groups

1. The Groups of Affine Transformations

Let R^n denote the space of n-tuples of real numbers $(x_1, x_2, \ldots, x_n)$ with the usual topology. We shall call it Cartesian n-dimensional space. The r-dimensional linear spaces of R^n will be called r-planes (for $r = 1$, lines; for $r = 2$, planes; and for $r = n - 1$, hyperplanes).

The results of the preceding chapters will now be applied to the case of the affine groups of R^n. We shall represent by x the point of coordinates $(x_1, x_2, \ldots, x_n)$ as well as the $n \times 1$ matrix whose elements these coordinates are. An *affinity* or an *affine transformation* is a transformation of R^n onto itself, represented by a matrix equation of the form

$$x' = ax + b, \qquad \det a \neq 0, \tag{11.1}$$

where $a = (a_{ij})$ and $b = (b_i)$ are $n \times n$ and $n \times 1$ matrices, respectively. A group of affinities is a group of transformations of the form (11.1). These groups are isomorphic to the corresponding groups of $(n + 1) \times (n + 1)$ matrices

$$g = \begin{pmatrix} a & b \\ 0 & 1 \end{pmatrix} \tag{11.2}$$

with the customary rules

$$g_2 g_1 = \begin{pmatrix} a_2 a_1 & a_2 b_1 + b_2 \\ 0 & 1 \end{pmatrix}, \qquad g^{-1} = \begin{pmatrix} a^{-1} & -a^{-1} b \\ 0 & 1 \end{pmatrix}. \tag{11.3}$$

ENCYCLOPEDIA OF MATHEMATICS and Its Applications, Gian-Carlo Rota (ed.).
1, Luis A. Santaló, Integral Geometry and Geometric Probability

ISBN 0-201-13500-0

The Maurer–Cartan forms for a group of affinities are those of the corresponding group of matrices, that is, a set of independent 1-forms of the matrix $g^{-1}\,dg$, which in this case splits up into the matrices

$$\Omega_1 = a^{-1}\,da, \qquad \Omega_2 = a^{-1}\,db. \tag{11.4}$$

By exterior differentiation and using the fact that $da^{-1} = -\,a^{-1}\,da\,a^{-1}$, we get the following structure equations.

$$d\Omega_1 = -\,\Omega_1 \wedge \Omega_1, \qquad d\Omega_2 = -\,\Omega_1 \wedge \Omega_2. \tag{11.5}$$

In explicit form, if we set

$$a = (a_{ij}), \qquad b = (b_i), \qquad a^{-1} = (\alpha_{ij}), \qquad \Omega_1 = (\omega_{ij}), \qquad \Omega_2 = (\omega_i), \tag{11.6}$$

equations (11.4) can be written in the form

$$\omega_{ij} = \sum_{h=1}^{n} \alpha_{ih}\,da_{hj}, \qquad \omega_i = \sum_{h=1}^{n} \alpha_{ih}\,db_h, \tag{11.7}$$

and the structure equations become

$$d\omega_{ij} = -\sum_{h=1}^{n} \omega_{ih} \wedge \omega_{hj}, \qquad d\omega_i = -\sum_{h=1}^{n} \omega_{ih} \wedge \omega_h. \tag{11.8}$$

If the group of affinities has dimension r, among the $n^2 + n$ 1-forms ω_{ij}, ω_i there are exactly r of them that are linearly independent and they form a system of Maurer–Cartan forms for the group. For instance, for the general affine group $\mathfrak{A}(n)$ (a is any $n \times n$ nonsingular matrix and b any $n \times 1$ matrix) we have $r = n^2 + n$ and the Maurer–Cartan forms are all the forms ω_{ij}, ω_i. For the *special affine group* $\mathfrak{A}_s(n)$ ($\det a = 1$) we have $r = n^2 + n - 1$ and therefore there is a relation between the 1-forms ω_{ij}, ω_i. Indeed, developing the equation $d(\det a) = 0$, we get

$$\sum_{i,h} \alpha_{hi}\,da_{ih} = 0 \qquad \text{or} \qquad \omega_{11} + \omega_{22} + \cdots + \omega_{nn} = 0 \tag{11.9}$$

and the Maurer–Cartan forms of $\mathfrak{A}_s(n)$ are all the forms ω_{ij}, ω_i except one of the forms ω_{ii}.

The moving frames of E. Cartan. In order to give a geometric meaning to the Pfaffian forms ω_{ij}, ω_h, let us consider in R^n a fixed frame $(p_0; \mathbf{e}_1{}^0, \mathbf{e}_2{}^0, \ldots, \mathbf{e}_n{}^0)$, composed of a point p_0 and n independent vectors $\mathbf{e}_i{}^0$, and a moving frame $(p; \mathbf{e}_1, \mathbf{e}_2, \ldots, \mathbf{e}_n)$, which results from the fixed frame by the affinity represented by (11.1). Let us assume that p_0 is the origin $(0, 0, \ldots, 0)$ and that $\mathbf{e}_i{}^0$ are the coordinate vectors

$$\mathbf{e}_1{}^0(1, 0, \ldots, 0), \ \mathbf{e}_2{}^0(0, 1, 0, \ldots, 0), \ \ldots, \ \mathbf{e}_n{}^0(0, \ldots, 0, 1). \tag{11.10}$$

ISBN 0-201-13500-0

If we introduce the vector matrices

$$\mathbf{e}^0 = (\mathbf{e}_1{}^0\ \mathbf{e}_2{}^0\ \cdots\ \mathbf{e}_n{}^0), \qquad \mathbf{e} = (\mathbf{e}_1\ \mathbf{e}_2\ \cdots\ \mathbf{e}_n), \tag{11.11}$$

we can write

$$p - p_0 = \mathbf{e}^0 b, \qquad \mathbf{e} = \mathbf{e}^0 a, \tag{11.12}$$

and therefore

$$dp = \mathbf{e}^0\, db = \mathbf{e}a^{-1}\, db = \mathbf{e}\Omega_2, \qquad d\mathbf{e} = \mathbf{e}^0\, da = \mathbf{e}a^{-1}\, da = \mathbf{e}\Omega_1, \tag{11.13}$$

which are the so-called moving frame equations for affine groups. They can be written

$$dp = \sum_{i=1}^{n} \omega_i \mathbf{e}_i, \qquad d\mathbf{e}_i = \sum_{j=1}^{n} \omega_{ji}\mathbf{e}_j. \tag{11.14}$$

Note that from the equation $\mathbf{e} = \mathbf{e}^0 a$ it follows that the components of the vectors $\mathbf{e}_i$ of the moving frame with respect to the fixed frame are the columns $a_{1i}, a_{2i}, \ldots, a_{ni}$ of the matrix a. Thus, det a can be written $|\mathbf{e}_1\ \mathbf{e}_2\ \cdots\ \mathbf{e}_n|$ where each vector $\mathbf{e}_i$ must be replaced by the column of its components with respect to the fixed frame. With this notation, the special affinities are those such that $|\mathbf{e}_1\ \mathbf{e}_2\ \cdots\ \mathbf{e}_n| = 1$, and using (11.14) we have

$$\omega_i = |\mathbf{e}_1\, \mathbf{e}_2 \cdots \mathbf{e}_{i-1}\, dp\, \mathbf{e}_{i+1} \cdots \mathbf{e}_n|, \quad \omega_{ji} = |\mathbf{e}_1 \cdots \mathbf{e}_{j-1}\, d\mathbf{e}_i\, \mathbf{e}_{j+1} \cdots \mathbf{e}_n|. \tag{11.15}$$

Invariant volume of affine groups. We have seen that the left-invariant 1-forms for affine groups are given by the matrices Ω_1, Ω_2 of (11.4). The right-invariant 1-forms are given by the matrix $dg\, g^{-1}$, which now splits up into

$$\bar{\Omega}_1 = da\, a^{-1}, \qquad \bar{\Omega}_2 = -\, da\, a^{-1} b + db. \tag{11.16}$$

In order to obtain the explicit form for the invariant volume elements of affine groups, we shall consider separately the cases in which $b = 0$ and $b \neq 0$.

When the column matrix b is the zero matrix, we have the so-called *homogeneous affinities* (or *central affinities*)

$$x' = ax, \qquad \det a \neq 0. \tag{11.17}$$

This group, which we shall represent by $\mathfrak{A}^0(n)$, is isomorphic to $GL(n)$. According to (11.7), the left-invariant volume element is $d_{\mathrm{L}}\mathfrak{A}^0(n) = d_{\mathrm{L}}GL(n) = \bigwedge_k (\det \alpha)^n\, da_{1k} \wedge da_{2k} \wedge \cdots \wedge da_{nk}$ where $\alpha = a^{-1}$. The right-invariant 1-forms are $\varpi_{ih} = \sum_1^n da_{ik}\alpha_{kh}$, as follows from (11.16), and thus $d_{\mathrm{R}}\mathfrak{A}^0(n) = d_{\mathrm{R}}GL(n) = \bigwedge_i (\det \alpha)^n\, da_{i1} \wedge da_{i2} \wedge \cdots \wedge da_{in}$. Since $\det \alpha = (\det a)^{-1}$, we can state

ISBN 0-201-13500-0

The group $\mathfrak{A}^0(n)$ *(or the group* $GL(n)$*) is unimodular. The left- and right-invariant volume element can be written*

$$d\mathfrak{A}^0(n) = dGL(n) = \frac{1}{(\det a)^n} \bigwedge_{i,k} da_{ik}. \tag{11.18}$$

The special homogeneous affine group in R^n is the group, which we shall represent by $\mathfrak{A}_s{}^0(n)$,

$$x' = ax, \qquad \det a = 1, \tag{11.19}$$

that is isomorphic to $SL(n)$. This group is a closed invariant subgroup of $GL(n)$. For if $g \in GL(n)$, $g_H \in SL(n)$, we have $\det(gg_Hg^{-1}) = 1$ and hence $gg_Hg^{-1} \in SL(n)$. Consequently, since $GL(n)$ is unimodular, from Chapter 10, Section 3, it follows that $SL(n)$ is also unimodular. Its left- and right-invariant volume element is (recall that we always consider densities in absolute value)

$$d\mathfrak{A}_s{}^0(n) = dSL(n) = \sum_{i=1}^{n} (-1)^{n+i} a_{in} \bigwedge_{h,k} da_{hk}(da_{in}) \tag{11.20}$$

where (da_{in}) indicates that the factor da_{in} is to be omitted. For instance, for $n = 2$, we have

$$d\mathfrak{A}_s{}^0(2) = dSL(2) = \frac{da_{11} \wedge da_{12} \wedge da_{21}}{a_{11}}.$$

For the general affine group (11.1), according to (11.7) we have

$$d_L\mathfrak{A}(n) = (\det a)^{-1}\, d_L GL(n) \bigwedge_h db_h. \tag{11.21}$$

The right-invariant volume element is the exterior product of all elements of the matrices (11.16) and it has the value

$$d_R\mathfrak{A}(n) = d_R GL(n) \bigwedge_h db_h. \tag{11.22}$$

Hence we have that *the general affine group is not unimodular*. According to (9.80), we have $\Delta(g) = (\det a)^{-1}$.

The nonhomogeneous special affine group $\mathfrak{A}_s(n)$: $x' = ax + b$, $\det a = 1$, is unimodular and its invariant volume element is

$$d\mathfrak{A}_s(n) = dSL(n) \bigwedge_h db_h \tag{11.23}$$

where $dSL(n)$ is given by (11.20).

ISBN 0-201-13500-0

2. Densities for Linear Spaces with Respect to Special Homogeneous Affinities

Consider the group $\mathfrak{A}_s{}^0(n)$, or simply $\mathfrak{A}_s{}^0$, (11.19) of special homogeneous affinities acting on the n-dimensional space R^n. This group is transitive and effective on $R^n - O$. Let L_r be a fixed r-plane that does not go through the origin and let $\mathfrak{H}_r$ denote the subgroup of $\mathfrak{A}_s{}^0$ consisting of the elements that leave L_r fixed. It is clear that $\mathfrak{H}_r$ is closed in $\mathfrak{A}_s{}^0$ and that there is a one-to-one correspondence between the set of r-planes and the points of the homogeneous space $\mathfrak{A}_s{}^0/\mathfrak{H}_r$. The invariant density for $\mathfrak{A}_s{}^0/\mathfrak{H}_r$ under $\mathfrak{A}_s{}^0$, if it exists, is called the invariant density for r-planes. In order to see whether such an invariant density exists or not, we assume that L_r is the r-plane that goes through the end point of $\mathbf{e}_1$ and is parallel to the r-plane spanned by $\mathbf{e}_2, \mathbf{e}_3, \ldots, \mathbf{e}_{r+1}$, where $(\mathbf{e}_1, \mathbf{e}_2, \ldots, \mathbf{e}_n)$ is a moving frame of origin O that satisfies the condition $|\mathbf{e}_1\, \mathbf{e}_2 \cdots \mathbf{e}_n| = 1$. The elements of $\mathfrak{A}_s{}^0$ that leave L_r fixed are those for which $d\mathbf{e}_i$ ($i = 1, 2, \ldots, r + 1$) are linear combinations of $\mathbf{e}_2, \mathbf{e}_3, \ldots, \mathbf{e}_{r+1}$. Thus, according to the second set of equations (11.14), they are characterized by the conditions

$$\omega_{1i} = 0 \qquad \text{for} \quad i = 1, 2, \ldots, r + 1;$$

$$\omega_{ji} = 0 \qquad \text{for} \quad j = r + 2, \ldots, n,\ i = 1, 2, \ldots, r + 1. \tag{11.24}$$

This is system (10.1) of the general theory. From (10.3) it follows that a necessary and sufficient condition for the sets of L_r to have an invariant density under $\mathfrak{A}_s{}^0$ is that

$$d(\bigwedge \omega_{1i} \bigwedge \omega_{ji}) = 0 \tag{11.25}$$

where the exterior products have the range of indices of (11.24). The structure equations of $\mathfrak{A}_s{}^0$ are the first set of equations (11.8) with condition (11.9). When these equations are taken into account, (10.25) becomes

$$\bigwedge \omega_{1i} \bigwedge \omega_{ji} \wedge \left(\sum_{h=1}^{r+1} \omega_{hh} \right) = 0, \qquad i = 1, \ldots, r + 1, \qquad j = r + 2, \ldots, n. \tag{11.26}$$

This condition is satisfied if and only if

$$\omega_{11} \wedge (\omega_{22} + \cdots + \omega_{r+1,r+1}) = 0 \tag{11.27}$$

and since (11.9) is the only relation between the Maurer–Cartan 1-forms, it follows that either $r = 0$ or $r = n - 1$.

For $r = 0$ the density for points becomes $dL_0 = \omega_{11} \wedge \omega_{21} \wedge \cdots \wedge \omega_{n1} = da_{11} \wedge da_{21} \wedge \cdots \wedge da_{n1}$ and since $a_{11}, a_{21}, \ldots, a_{n1}$ are the coordinates of

ISBN 0-201-13500-0

the end point of $\mathbf{e}_1$, we have that the density for points is equal, up to a constant factor, to the volume element of the space, which is an obvious result, since the special affinities are volume preserving. For $r = n - 1$, the density for hyperplanes becomes

$$dL_{n-1} = \omega_{11} \wedge \omega_{12} \wedge \cdots \wedge \omega_{1n}. \tag{11.28}$$

This expression corresponds to the hyperplane that goes through the end point of $\mathbf{e}_1$ and is parallel to $\mathbf{e}_2, \mathbf{e}_3, \ldots, \mathbf{e}_n$. In order to give a metric interpretation of this density, we proceed as follows. Let $\mathbf{b}$ be the unit normal vector to L_{n-1} through the origin O and let $\mathbf{b}_2, \mathbf{b}_3, \ldots, \mathbf{b}_n$ be $n - 1$ orthonormal unit vectors of origin O that are contained in the hyperplane $(O, \mathbf{e}_2, \ldots, \mathbf{e}_n)$ parallel to L_{n-1}. The scalar product $\mathbf{b} \cdot d\mathbf{b}_i$ is equal to the arc element on the $(n - 1)$-dimensional unit sphere U_{n-1} at the end point of $\mathbf{b}$ in the direction of $\mathbf{b}_i$, so that, denoting by du_{n-1} the $(n - 1)$-dimensional volume element of U_{n-1} at the end point of $\mathbf{b}$, we have

$$du_{n-1} = \bigwedge_i (\mathbf{b} \cdot d\mathbf{b}_i), \qquad i = 2, 3, \ldots, n. \tag{11.29}$$

On the other hand, since vectors $\mathbf{b}_i$ and $\mathbf{e}_i$ belong to the same hyperplane, we may write

$$\mathbf{e}_i = \sum_{k=2}^{n} \lambda_{ik} \mathbf{b}_k, \qquad i = 2, 3, \ldots, n, \tag{11.30}$$

and therefore

$$\bigwedge (\mathbf{b} \cdot d\mathbf{e}_i) = \det(\lambda_{ik}) \bigwedge (\mathbf{b} \cdot d\mathbf{b}_i) = \det(\lambda_{ik})\, du_{n-1}. \tag{11.31}$$

Since $\mathbf{b} \bullet \mathbf{e}_i = 0\ (i = 2, \ldots, n)$, from (11.14) we have $(\mathbf{b} \cdot d\mathbf{e}_i) = \omega_{1i}(\mathbf{b} \bullet \mathbf{e}_1) = \omega_{1i}\rho_1 \cos\theta\ (i = 1, 2, \ldots, n)$ where ρ_1 is the length of $\mathbf{e}_1$ and θ is the angle between $\mathbf{e}_1$ and $\mathbf{b}$. That is, if p denotes the distance from O to L_{n-1}, we have $p = \rho_1 \cos\theta$ and from (11.31) we deduce

$$\left(\bigwedge_{i=2}^{n} \omega_{1i} \right) p^{n-1} = \det(\lambda_{ik})\, du_{n-1}. \tag{11.32}$$

Moreover, from $(\mathbf{b} \cdot d\mathbf{e}_1) = \omega_{11}(\mathbf{b} \bullet \mathbf{e}_1) = \omega_{11} p$ and $(\mathbf{b} \cdot d\mathbf{e}_1) = dp - (\mathbf{e}_1 \cdot d\mathbf{b})$, using that $(\mathbf{e}_1 \cdot d\mathbf{b}) \wedge du_{n-1} = 0$, which is a consequence of (11.29) and the relations $\mathbf{b} \bullet \mathbf{b}_i = 0$, $\mathbf{b}^2 = 1$, $\mathbf{b} \cdot d\mathbf{b} = 0$, we get

$$(\omega_{11} \wedge du_{n-1}) p = dp \wedge du_{n-1}. \tag{11.33}$$

On the other hand, since the parallelepiped spanned by $\mathbf{e}_1, \mathbf{e}_2, \ldots, \mathbf{e}_n$ has volume 1, according to the condition $|\mathbf{e}_1\ \mathbf{e}_2 \cdots \mathbf{e}_n| = 1$, and the parallelepiped

ISBN 0-201-13500-0

spanned by $\mathbf{e}_2, \mathbf{e}_3, \ldots, \mathbf{e}_n$ has its volume equal to $\det(\lambda_{ik})$, we have $\det(\lambda_{ik}) \cdot p = 1$ and thus we finally get

$$dL_{n-1} = \frac{dp \wedge du_{n-1}}{p^{n+1}}, \tag{11.34}$$

which is the wanted metric interpretation for dL_{n-1}. We may summarize the foregoing results in the following theorem.

The points and the hyperplanes that do not pass through the origin are the only linear subspaces that have an invariant density under the special homogeneous affine group $\mathfrak{A}_s^0$. *The density for points equals the volume element. The density for hyperplanes may be written in the metric form* (11.34) *where* p *is the distance from the origin to the hyperplane and* du_{n-1} *denotes the volume element of the* $(n - 1)$*-dimensional unit sphere corresponding to the direction of the normal to the hyperplane* (*i.e., the solid angle element corresponding to the unit normal vector to the hyperplane*).

An application to convex bodies. Let $p = p(u_{n-1})$ denote the support function of the convex body K in E_n with respect to the origin O contained in K. That is, p is equal to the distance from O to the support hyperplane of K that is normal to the direction u_{n-1}. According to (11.34) the measure of the set of hyperplanes exterior to K, which is invariant under $\mathfrak{A}_s^0$, is

$$I(K, O) = m(L_{n-1}; L_{n-1} \cap K = \emptyset) = \frac{1}{n} \int_{U_{n-1}} \frac{du_{n-1}}{p^n} \tag{11.35}$$

where the integral is extended over the whole $(n - 1)$-dimensional unit sphere U_{n-1}. This integral $I(K, O)$ is an affine invariant that depends on K and on the point $O \in K$. There exists a unique point $O_M \in K$ such that $I(K, O_M) = \min I(K, O)$ $(O \in K)$. Consequences of this fact have been given by Petty [476]. Note that $I(K, O)$ is the volume of K^*, the polar reciprocal body of K with respect to O (defined by the vector radius $1/p$). Properties of this invariant were reported in Chapter 1, Section 6, Note 2.

Linear subspaces through the origin. The group $\mathfrak{A}_s^0$ leaves invariant the origin O and hence it maps linear subspaces through the origin into linear subspaces through the origin. We shall now see if the r-planes $L_{r[0]}$ through the origin have an invariant density under $\mathfrak{A}_s^0$ $(0 < r < n)$.

Assume $L_{r[0]}$ determined by the vectors $\mathbf{e}_1, \mathbf{e}_2, \ldots, \mathbf{e}_r$. Then, in order that $L_{r[0]}$ remain fixed under the mapping (11.19) the differentials $d\mathbf{e}_1, d\mathbf{e}_2, \ldots, d\mathbf{e}_r$ must be linear combinations of $\mathbf{e}_1, \mathbf{e}_2, \ldots, \mathbf{e}_r$ and (11.14) gives

$$\omega_{ji} = 0 \qquad \text{for} \quad i = 1, 2, \ldots, r; \qquad j = r + 1, \ldots, n. \tag{11.36}$$

ISBN 0-201-13500-0

According to the general theory, in order that the sets of r-planes have an invariant density it is necessary and sufficient that $d(\bigwedge \omega_{ji}) = 0$ where the exterior product refers to the range of indices in (11.36). Taking (11.8) and (11.9) into account, we have

$$d(\bigwedge \omega_{ji}) = \bigwedge \omega_{ji} \wedge \left(\sum_{r+1}^{n} \omega_{jj} - \sum_{1}^{r} \omega_{ii} \right) = -2 \bigwedge \omega_{ji} \wedge \left(\sum_{1}^{r} \omega_{ii} \right). \quad (11.37)$$

Since (11.9) is the only relation between the 1-forms ω_{ji} and $r < n$, differential form (11.37) cannot be zero. Thus we have that *the linear subspaces through the origin do not have an invariant density under* $\mathfrak{A}_s^0$.

Let us now consider as elements sets of m-tuples of linear subspaces $(L_{r_1[0]}, L_{r_2[0]}, \ldots, L_{r_m[0]})$ of dimensions $r_1, r_2, \ldots, r_m$ passing through the origin and having no other common point. Assume that these m-tuples transform transitively under $\mathfrak{A}_s^0$. We take the moving frame so that $L_{r_h[0]}$ is determined by the vectors $\mathbf{e}_{r_1+\ldots+r_{h-1}+1}, \mathbf{e}_{r_1+\ldots+r_{h-1}+2}, \ldots, \mathbf{e}_{r_1+\ldots+r_{h-1}+r_h}$ $(h = 1, 2, \ldots, m)$. The elements of $\mathfrak{A}_s^0$ that leave the m-tuple fixed form a closed subgroup $\mathfrak{H}(r_1, \ldots, r_m)$ characterized by the conditions that the differentials $d\mathbf{e}_i (r_1 + \cdots + r_{h-1} + 1 \leqslant i \leqslant r_1 + \cdots + r_h)$ are linear combination of $\mathbf{e}_s$ for $r_1 + \cdots + r_{h-1} + 1 \leqslant s \leqslant r_1 + \cdots + r_h$. Consequently, according to (11.14) we have $\omega_{ji} = 0$ for all pairs i, j between the ranges

$$r_1 + \cdots + r_{h-1} + 1 \leqslant i \leqslant r_1 + \cdots + r_h, \quad 1 \leqslant j \leqslant r_1 + \cdots + r_{h-1},$$

$$r_1 + \cdots + r_h + 1 \leqslant j \leqslant n,$$

for $h = 1, 2, \ldots, m$ (with $r_0 = 0$).

By exterior multiplication of all ω_{ij} corresponding to these ranges of indices we get a differential form Φ which will be an invariant density for the homogeneous space $\mathfrak{A}_s^0/\mathfrak{H}(r_1 + \cdots + r_m)$ and consequently for m-tuples $(L_{r_1[0]}, \ldots, L_{r_m[0]})$ if and only if $d\Phi = 0$ (according to condition (10.3)). Using structure equations (11.8), we see by an easy calculation that $d\Phi = \Phi \wedge \sum \omega_{hh}$ $(h = 1, 2, \ldots, r_1 + r_2 + \cdots + r_m)$. Since (11.9) is the only relation between the forms ω_{ij}, the condition $d\Phi = 0$ holds if and only if

$$r_1 + r_2 + \cdots + r_m = n \quad (11.38)$$

and we have the theorem

The m-tuples $(L_{r_1[0]}, \ldots, L_{r_m[0]})$ *of linear spaces of dimensions* $r_1, r_2, \ldots, r_m$ *passing through the origin and having no other common point have an invariant density under* $\mathfrak{A}_s^0$ *if and only if condition* (11.38) *holds.*

In this case, the density for m-tuples $(L_{r_1[0]}, \ldots, L_{r_m[0]})$ is the differential form Φ of degree $(n - r_1)r_1 + \cdots + (n - r_m)r_m$. For example, on the plane,

ISBN 0-201-13500-0

$n = 2$, no invariant density exists under $\mathfrak{A}_s{}^0(2)$ for sets of lines through the origin; such a density does exist, however, for pairs of lines (L_1, L_1'). If the lines are determined by the angles ϕ, ϕ' that they form with the x axis (assuming an orthogonal Cartesian system), the invariant density takes the form

$$d(L, L') = \sin^{-2}(\phi - \phi')\, d\phi \wedge d\phi'.$$

3. Densities for Linear Subspaces with Respect to the Special Nonhomogeneous Affine Group

We now consider the group $\mathfrak{A}_s$ or $\mathfrak{A}_s(n)$

$$x' = ax + b, \qquad \det a = 1. \tag{11.39}$$

Each element of $\mathfrak{A}_s$ is defined by the position of an n-frame $(p; \mathbf{e}_1, \mathbf{e}_2, \ldots, \mathbf{e}_n)$ such that $|\mathbf{e}_1\, \mathbf{e}_2 \cdots \mathbf{e}_n| = 1$ and the Maurer–Cartan forms are given by (11.4). The structure equations are (11.8) with condition (11.9). Let L_r be the r-plane defined by the point p and the vectors $\mathbf{e}_1, \ldots, \mathbf{e}_r$ $(0 \leqslant r < n)$. From (11.13) it follows that the subgroup $\mathfrak{H}_r$ of the affinities that leave L_r fixed is characterized by the conditions

$$\omega_i = 0 \qquad \text{for} \quad i = r + 1, \ldots, n;$$

$$\omega_{jh} = 0 \qquad \text{for} \quad h = 1, \ldots, r;\ j = r + 1, \ldots, n. \tag{11.40}$$

This is system (10.1) of the general theory. The homogeneous space $\mathfrak{A}_s/\mathfrak{H}_r$ will have an invariant density under $\mathfrak{A}_s$ if and only if $d(\bigwedge \omega_i \bigwedge \omega_{jh}) = 0$ where the indices have the ranges indicated in (11.40). Making use of structure equations (11.8) we easily get, up to the sign,

$$d(\textstyle\bigwedge \omega_i \bigwedge \omega_{jh}) = \bigwedge \omega_i \bigwedge \omega_{jh} \wedge (2(\omega_{r+1,r+1} + \cdots + \omega_{nn})). \tag{11.41}$$

Since the forms ω_i, ω_{jh} are only related by equation (11.9), the right-hand side of (11.41) is zero if and only if $r = 0$. Hence we have that *except for points, the linear subspaces L_r have no invariant density under $\mathfrak{A}_s$*. For points, $r = 0$, the invariant density is $dL_0 = \omega_1 \wedge \cdots \wedge \omega_n = db_1 \wedge db_2 \wedge \cdots \wedge db_n$, which is equal to the volume element of the space.

For sets of m-tuples of linear spaces $(L_{r_1}, \ldots, L_{r_m})$ without common point, proceeding as in the case of the homogeneous group $\mathfrak{A}_s{}^0$, we prove the following theorem.

The sets of m-tuples of linear subspaces $(L_{r_1}, \ldots, L_{r_m})$ without common point, whose dimensions r_i satisfy the condition $r_1 + r_2 + \cdots + r_m + m \leqslant n + 1$ and transform transitively under $\mathfrak{A}_s(n)$, have an invariant density

ISBN 0-201-13500-0

under $\mathfrak{A}_s(n)$ if and only if either all r_i are equal to zero or the condition

$$r_1 + r_2 + \cdots + r_m + m = n + 1 \tag{11.42}$$

holds.

For instance, the sets of (point L_0, hyperplane L_{n-1}) such that L_0 is not in L_{n-1} have an invariant density whose metric expression is $d(L_0, L_{n-1}) = p^{-(n+1)}\, dL_0 \wedge du_{n-1} \wedge dp$, where dL_0 is the volume element at the point L_0, du_{n-1} is the solid angle element corresponding to the direction perpendicular to L_{n-1}, and p is the distance from the point to the hyperplane.

Density for sets of parallel hyperplanes. We will prove that sets of parallel hyperplanes have an invariant density under $\mathfrak{A}_s$. Since pairs of parallel hyperplanes transform transitively under $\mathfrak{A}_s$, we can take the hyperplane determined by the point p and the vectors $\mathbf{e}_1, \mathbf{e}_2, \ldots, \mathbf{e}_{n-1}$ and its parallel through the end point of $\mathbf{e}_n$. According to (11.14), the system whose integral varieties correspond to these parallel hyperplanes is

$$\omega_n = 0, \qquad \omega_{nh} = 0 \qquad \text{for} \quad h = 1, 2, \ldots, n. \tag{11.43}$$

In view of (10.3), the condition that the exterior product $d\mathscr{P} = \omega_n \wedge \omega_{n1} \wedge \cdots \wedge \omega_{nn}$ be a density is that $d(d\mathscr{P}) = 0$. Taking the structure equations into account we have, up to the sign,

$$d(d\mathscr{P})) = d\mathscr{P} \wedge \left(\sum_1^n \omega_{ii} \right) = 0.$$

Therefore we have

The sets of parallel hyperplanes have an invariant density with respect to the group $\mathfrak{A}_s(n)$ that is given by

$$d\mathscr{P} = \omega_n \wedge \omega_{n1} \wedge \cdots \wedge \omega_{nn}. \tag{11.44}$$

It is not difficult to obtain a metric expression for $d\mathscr{P}$. If p_1, p_2 denote the distances from the origin to the hyperplanes and du_{n-1} denotes the volume element on the $(n-1)$-dimensional unit sphere corresponding to the direction perpendicular to the hyperplanes, we can prove that [577]

$$d\mathscr{P} = \frac{du_{n-1} \wedge dp_1 \wedge dp_2}{|p_1 - p_2|^{n+2}}. \tag{11.45}$$

Application to convex bodies. Let K be a given convex body in the euclidean space E_n and let $\Delta = \Delta(u_{n-1})$ be the breadth of K corresponding to the direction u_{n-1}. The measure of all pairs of parallel hyperplanes that contain K has the value

ISBN 0-201-13500-0

$$\int \frac{du_{n-1} \wedge dp_1 \wedge dp_2}{|p_2 - p_1|^{n+2}} = \frac{1}{n(n+1)} \int_{U_{n-1}/2} \frac{du_{n-1}}{\Delta^n}. \tag{11.46}$$

This measure gives rise to the affine invariant of K

$$J = \int_{U_{n-1}/2} \frac{du_{n-1}}{\Delta^n} \tag{11.47}$$

where the integral is extended over half of the $(n-1)$-dimensional unit sphere.

If K possesses a center of symmetry, then J is related to the affine invariant $I(K, O)$ (11.35) by the obvious relation $nI(K, O) = 2^{n+1}J$. If K does not have a center of symmetry, J and I are not trivially related. Using the inequalities

$$x^{-n} + y^{-n} \geqslant 2(xy)^{-n/2}, \quad (xy)^{n/2} \leqslant [(x+y)/2]^n, \quad x^{-n} + y^{-n} \geqslant 2^{n+1}(x+y)^{-n},$$

valid for $x > 0$, $y > 0$ and where equality occurs only for $x = y$, and denoting by p_1, p_2 the values of the support function p (with respect to a point O inside K) at opposite points, we have

$$2J = \int_{U_{n-1}} \frac{du_{n-1}}{\Delta^n} = \int_{U_{n-1}} \frac{du_{n-1}}{(p_1 + p_2)^n} \leqslant \frac{1}{2^{n+1}} \int_{U_{n-1}} \left(\frac{du_{n-1}}{p_1{}^n} + \frac{du_{n-1}}{p_2{}^n} \right) = \frac{n}{2^n} I(K, O)$$

and therefore we have

$$J \leqslant \frac{n}{2^{n+1}} I(K, O) \tag{11.48}$$

valid for any point $O \in K$, with equality if and only if K is centrally symmetric about O. We can prove the inequalities

$$\frac{2^{n-1}}{n!(n-1)!} < JV \leqslant \frac{4\pi^n}{2^{n+1}(\Gamma(n/2))^2 n} \tag{11.49}$$

where V is the volume of K [577]. The upper bound is attained if and only if K is an ellipsoid. The lower bound is not the best possible and its best value is not known for $n > 2$. For $n = 2$, Eggleston [161] established that $JV \geqslant 3/2$ and that equality characterizes triangles.

4. Notes and Exercises

1. *Sets of lattices.* Let $\mathbf{e}_1, \mathbf{e}_2, \ldots, \mathbf{e}_n$ be linearly independent vectors in R^n. The set of all points $x = u_1\mathbf{e}_1 + \cdots + u_n\mathbf{e}_n$ with integral $u_1, u_2, \ldots, u_n$ is called the lattice with basis $\mathbf{e}_1, \mathbf{e}_2, \ldots, \mathbf{e}_n$. The half-open parallelepiped defined by $x = x_1\mathbf{e}_1 + \cdots + x_n\mathbf{e}_n$ where x_i are real numbers such that

ISBN 0-201-13500-0

$0 \leqslant x_i < 1$ is a fundamental region of the lattice (in the sense of Section 1 of Chapter 8) and the volume of this fundamental region is the value of the determinant $|\mathbf{e}_1\mathbf{e}_2 \cdots \mathbf{e}_n|$, called the determinant of the lattice. Consider the lattice $\mathscr{L}_0$ of all points with basis $\mathbf{e}_i(0, \ldots, 0, 1, 0, \ldots, 0)$, where 1 is the ith component, and the group $\mathfrak{A}_s^{\,0}(n)$ of special homogeneous affine transformations. The lattices $\mathscr{L} = a\mathscr{L}_0$ $(a \in \mathfrak{A}_s^{\,0})$ have determinant 1 and conversely any lattice with determinant 1 is the transform of $\mathscr{L}_0$ by a suitable $a \in \mathfrak{A}_s^{\,0}$. Let $\mathscr{L}$ denote the set of all lattices with determinant 1. If Γ denotes the subgroup of $\mathfrak{A}_s^{\,0}$ whose elements leave $\mathscr{L}_0$ invariant, the set $\mathscr{L}$ can be identified with the homogeneous space $\mathfrak{A}_s^{\,0}/\Gamma$. Since Γ is a discrete group, the invariant density $d\mathscr{L}$ for lattices of determinant 1 coincides with the density $d\mathfrak{A}_s^{\,0} = dSL(n)$ given by (11.20). We shall set $d\mathscr{L} = d\mathfrak{A}_s^{\,0}$.

With this density we can prove that the total measure of $\mathscr{L}$ is $m(\mathscr{L}) = \zeta(2)\zeta(3) \cdots \zeta(n)$, where $\zeta(n) = 1 + 2^{-n} + 3^{-n} + \cdots$. Moreover, given a fixed domain D of volume $v(D)$ in R^n and denoting by N the number of lattice points of $\mathscr{L}$ that belong to D and have coordinates prime to each other (called primitive lattice points) and by N_t the total number of lattice points of $\mathscr{L}$ that are contained in D, we have

$$\int_{\mathbf{L}} N \, d\mathscr{L} = v(D)\zeta(2)\zeta(3) \cdots \zeta(n-1), \qquad \int_{\mathbf{L}} N_t \, d\mathscr{L} = v(D)\zeta(2) \cdots \zeta(n-1)\zeta(n).$$

Therefore we have

(a) The mean value of the number of primitive lattice points contained in a fixed domain D of volume $v(D)$ is $E(N) = v(D)/\zeta(n)$.

(b) The mean value of the number of lattice points contained in D is $E(N_t) = v(D)$.

These mean values are taken over the set of all lattices with determinant 1. As a consequence we have the following theorem of Minkowski and Hlawka:

If D is an n-dimensional domain in R^n that is star-shaped relative to the origin (i.e., a domain that contains with any point x the whole segment λx, $0 \leqslant \lambda \leqslant 1$) of volume $v(D) < \zeta(n)$, then there exists a lattice with determinant 1 such that D does not contain any lattice point other than the origin.

This theorem was conjectured by Minkowski and first proved by Hlawka in 1944. The proof based on the mean values $E(N)$ and $E(N_t)$ is due to Siegel [610a] and Weil [712]. A simplified proof and more general results have been given by Macbeath and Rogers [381]. Other kinds of proofs and interesting related results are quoted in the books of Cassels [92] and Lekkerkerker [361]; see also [557].

2. *Sets of parallel subspaces in affine space.* The result of Section 3 about sets of parallel hyperplanes is a particular case of the following theorem (due to Santaló [578]):

In order that the sets of q-tuples $H(L_{h_1}, \ldots, L_{h_q})$ of parallel linear subspaces of dimensions $h_1, \ldots, h_q$ that transform transitively under the special affine group $\mathfrak{A}_s(n)$ have an invariant density with respect to $\mathfrak{A}_s$, it is necessary and sufficient that the dimensions h_i all be equal, $h_1 = h_2 = \cdots = h_q = h$, and that $q = n + 1 - h$.

ISBN 0-201-13500-0

The case of two parallel hyperplanes corresponds to the case in which $h = n - 1$, $q = 2$.

For $n = 3$ we have the case of two parallel planes and the case of sets of three parallel lines (noncoplanar). In this case, if du_2 denotes the area element of the unit sphere corresponding to the direction of the lines and dP_0, dP_1, dP_2 denote the area elements on a plane perpendicular to the lines at the corresponding intersection points, the measure of the set of three parallel lines whose convex hull contains a convex set K gives the following affine invariant of K:

$$m_1(K) = \frac{1}{64} \int_{U_n/2} du_2 \int \frac{dP_0 \wedge dP_1 \wedge dP_2}{T^6} \tag{11.50}$$

where T denotes the area of the triangle $P_0P_1P_2$. The first integral is extended over all triangles $P_0P_1P_2$ that contain the projection of K on the plane perpendicular to the direction u_2 and the second integral is extended over half of the unit sphere. It should be interesting to find inequalities between the invariant (11.50) and other affine invariants of K.

3. *Sets of parabolas, ellipses, and hyperbolas.* Consider the affine plane. The parabolas transform transitively under the special affine group $\mathfrak{A}_s(2)$. Therefore, it makes sense to seek a density for sets of parabolas invariant under $\mathfrak{A}_s(2)$. Assuming the parabolas given by the equation

$$x^2 + 2bxy + b^2y^2 + px + qy + r = 0,$$

we can show that the invariant density is [104]

$$dP = \frac{db \wedge dp \wedge dq \wedge dr}{(q - bp)^{8/3}}.$$

Similarly, assuming the ellipses given by the equation

$$x^2 + 2axy + (a^2 + b^2)y^2 + 2px + 2qy + r = 0,$$

we have, as the invariant density for sets of ellipses under the group $\mathfrak{A}(2)$,

$$dE = \frac{b}{\Delta^2} da \wedge db \wedge dp \wedge dq \wedge dr, \qquad \Delta = \begin{vmatrix} 1 & a & p \\ a & a^2 + b^2 & q \\ p & q & r \end{vmatrix}.$$

For hyperbolas given by the equation

$$x^2 + 2axy + (a^2 - b^2)y^2 + 2px + 2qy + r = 0$$

the invariant density under $\mathfrak{A}(2)$ is

$$dH = \frac{b}{\Delta^2} da \wedge db \wedge dp \wedge dq \wedge dr, \qquad \Delta = \begin{vmatrix} 1 & a & p \\ a & a^2 - b^2 & q \\ p & q & r \end{vmatrix}.$$

These densities are the work of Stoka [634, 647].

ISBN 0-201-13500-0

4. *The real projective group.* Let V^{n+1} be an $(n + 1)$-dimensional vector space over the real numbers. Let $V_*^{n+1} = V^{n+1} - \{O\}$ denote the set of all nonzero vectors in V^{n+1}. Two vectors $\mathbf{x}$, $\mathbf{y}$ of V_*^{n+1} are said to be in the relation $\rho: \mathbf{x} \sim \mathbf{y}$ if they are linearly dependent, that is, if there exists a real number $c \neq 0$ such that $\mathbf{y} = c\mathbf{x}$. This relation is an equivalence relation (proportionality relation). The quotient V_*^{n+1}/ρ, the set of all equivalence classes, is called the *projective n-dimensional space* P_n. If L_{r+1}^* is an $(r + 1)$-dimensional subspace of V^{n+1} $(r = 0, 1, \ldots, n)$, then $(L_{r+1}^* - \{O\}/\rho$ is a real projective r-dimensional space contained in P_n that is called an r-plane of P_n. If $\mathbf{x}$ is a vector of V_*^{n+1}, then all vectors of the form $c\mathbf{x}$ $(c \neq 0)$ are representatives of the same point x. If the nonzero vectors $\mathbf{x}^0, \mathbf{x}^1, \ldots, \mathbf{x}^r$ define an $(r + 1)$-dimensional subspace in V_*^{n+1}, the corresponding points define an r-plane in P_n.

Let $\mathbf{A}_0, \mathbf{A}_1, \ldots, \mathbf{A}_n$ be a basis for V^{n+1}. Then every vector $\mathbf{x}$ can be represented in the form $\mathbf{x} = x_0\mathbf{A}_0 + x_1\mathbf{A}_1 + \cdots + x_n\mathbf{A}_n$. The numbers $x_0, x_1, \ldots, x_n$ are called the *homogeneous coordinates* of the point x corresponding to the class $c\mathbf{x}$ $(c \neq 0)$. Two points are coincident if their homogeneous coordinates are proportional. If $x_0 \neq 0$, the ratios $x_1/x_0, x_2/x_0, \ldots, x_n/x_0$ are the nonhomogeneous coordinates of the point x. The vector space V^{n+1} is called the vector space associated with P_n and the bases $\mathbf{A}_0, \mathbf{A}_1, \ldots, \mathbf{A}_n$ constitute a system of homogeneous coordinates.

Let x denote the point $x \in P_n$ and the column matrix $(n + 1) \times 1$ whose elements are the homogeneous coordinates of x. A projective transformation or a collineation of P_n into itself is a transformation defined by the matrix equation $x' = Ax$, $\det A \neq 0$, where $A = (a_{ij})$ is an $(n + 1) \times (n + 1)$ nonsingular matrix. Since x and cx represent the same point, we can always assume that $\det A = 1$. The collineations of P_n into itself form the projective group $\mathfrak{P}_n$. A system of Maurer–Cartan forms of $\mathfrak{P}_n$ is a set of independent elements of the matrix $\Omega = A^{-1}\,dA$ and the structure equations are $d\Omega = -\,\Omega \wedge \Omega$. In explicit form, if $A = (a_{ij})$, $A^{-1} = (\alpha_{ij})$, $\Omega = (\omega_{ij})$, we have

$$\omega_{ij} = \sum_{h=0}^{n} \alpha_{ih}\,da_{hj} = -\sum_{h=0}^{n} a_{hj}\,d\alpha_{ih}, \qquad d\omega_{ij} = -\sum_{h=0}^{n} \omega_{ih} \wedge \omega_{hj},$$

and differentiating $\det A = 1$, we get $\omega_{00} + \omega_{11} + \cdots + \omega_{nn} = 0$.

Calling $\mathbf{A}_i$ the vectors of V_*^{n+1} whose components are $(a_{0i}, a_{1i}, \ldots, a_{ni})$, $i = 0, 1, \ldots, n$, we can write the equation $dA = A\Omega$ as

$$d\mathbf{A}_i = \sum_{j=0}^{n} \omega_{ji}\,\mathbf{A}_j \tag{11.51}$$

and hence we have

$$\omega_{ji} = |\mathbf{A}_0 \cdots \mathbf{A}_{j-1}\,d\mathbf{A}_i\,\mathbf{A}_{j+1} \cdots \mathbf{A}_n|. \tag{11.52}$$

The kinematic density for $\mathfrak{P}_n$ has the following metric form. A collineation in P_n is determined by the $n + 2$ points $Q_0, Q_1, \ldots, Q_{n+1}$, which are the transforms of the points $(1, 0, \ldots, 0), (0, 1, 0, \ldots, 0), \ldots, (0, 0, \ldots, 0, 1)$, $(1, 1, \ldots, 1)$. Let S_h denote the volume of the simplex whose vertices are the

ISBN 0-201-13500-0

points $Q_0, Q_1, \ldots, Q_{n+1}$ (Q_h excluded) and let dQ_i denote the volume element at Q_i. Then the invariant volume element of $\mathfrak{P}_n$ (kinematic density of $\mathfrak{P}_n$) can be written [370]

$$d\mathfrak{P}_n = \frac{dQ_0 \wedge dQ_1 \wedge \cdots \wedge dQ_{n+1}}{S_0 S_1 \cdots S_{n+1}}. \tag{11.53}$$

The problem of determining whether sets of m-tuples of linear spaces have an invariant measure under $\mathfrak{P}_n$ can be solved in like manner as for the affine group [557]. The result is the following theorem:

In order that the sets of m-tuples $(L_{h_1}, \ldots, L_{h_m})$ *of linear subspaces without common point, whose dimensions* h_i *satisfy the condition* $h_1 + h_2 + \cdots + h_m + m \leqslant n + 1$, *have an invariant density with respect to the projective group, it is necessary and sufficient that*

$$h_1 + h_2 + \cdots + h_m + m = n + 1. \tag{11.54}$$

In this case the invariant density is given by the exterior product $\bigwedge \omega_{ij}$ *where the forms are given by* (11.52) *and the indices have the ranges*

$$0 \leqslant i \leqslant h_1, \qquad h_1 + 1 \leqslant j \leqslant n,$$

$$h_1 + 1 \leqslant i \leqslant h_1 + h_2 + 1, \qquad 0 \leqslant j \leqslant h_1, \qquad h_1 + h_2 + 2 \leqslant j \leqslant n,$$

$$\cdots$$

$$h_1 + \cdots + h_{m-1} + m - 1 \leqslant i \leqslant h_1 + h_2 + \cdots + h_m + m - 1,$$

$$0 \leqslant j \leqslant h_1 + \cdots + h_{m-1} + m - 2, \qquad h_1 + \cdots + h_m + m \leqslant j \leqslant n.$$

For $m = 2$ this result is due to Varga [686].

Example 1. For the projective line ($n = 1$) the only possible case is that in which $h_1 = h_2 = 0$, that is, pairs of points. The invariant density for the points (A_0, A_1) is $d(A_0, A_1) = \omega_{01} \wedge \omega_{10} = a_{00}a_{11}\, da_{10} \wedge da_{01} - a_{00}a_{01}\, da_{10} \wedge da_{11} - a_{10}a_{11}\, da_{00} \wedge da_{01} + a_{10}a_{01}\, da_{00} \wedge da_{11}$. On the other hand, the nonhomogeneous coordinates of A_0 and A_1 are $\xi = a_{10}/a_{00}$, $\eta = a_{11}/a_{01}$, so that an easy calculation gives $d\xi \wedge d\eta = (a_{00}a_{01})^{-2}\omega_{10} \wedge \omega_{01}$ and $(\eta - \xi)^2 = (a_{01}a_{00})^{-2}$. Hence we have: *The projective invariant density for pairs of points on the line is* $d(A_0, A_1) = (\xi - \eta)^{-2}\, d\xi \wedge d\eta$ *where* ξ, η *are the nonhomogeneous coordinates of the points.* For instance, the measure of the set of pairs of points (A_0, A_1) such that A_0 lies in the interval (ξ_0, ξ_1) and A_1 lies in the interval (η_0, η_1), assuming that both intervals do not intersect, is $m(\xi, \eta; \xi_0 \leqslant \xi \leqslant \xi_1, \eta_0 \leqslant \eta \leqslant \eta_1) = \log(\eta_2, \eta_1, \xi_2, \xi_1)$, where $(\eta_2, \eta_1, \xi_2, \xi_1)$ denotes the cross ratio of the four points $\eta_2, \eta_1, \xi_2, \xi_1$.

Example 2. For the projective plane ($n = 2$) the cases in which condition (11.54) holds are (a) $h_1 = h_2 = h_3 = 0$, $m = 3$ (triples of noncollinear points); (b) $h_1 = 0$, $h_2 = 1$, $m = 2$ (point and line that are not incident).

(a) In order that the points A_0, A_1, A_2 be fixed, using (11.51) we have the conditions $\omega_{10} = \omega_{20} = 0$, $\omega_{01} = \omega_{21} = 0$, $\omega_{02} = \omega_{12} = 0$, and the invariant density becomes $d(A_0, A_1, A_2) = \omega_{10} \wedge \omega_{20} \wedge \omega_{01} \wedge \omega_{21} \wedge \omega_{02} \wedge \omega_{12}$.

ISBN 0-201-13500-0

The nonhomogeneous coordinates of A_0, A_1, A_2 are $A_0(\xi_0 = a_{10}/a_{00}, \eta_0 = a_{20}/a_{00})$, $A_1(\xi = a_{11}/a_{01}, \eta_1 = a_{21}/a_{01})$, $A_2(\xi_2 = a_{12}/a_{02}, \eta_2 = a_{22}/a_{02})$, and with the aid of (11.52) we get $d(A_0, A_1, A_2) = a_{00}^3 a_{01}^3 a_{02}^3 \, d\xi_0 \wedge d\eta_0 \wedge d\xi_1 \wedge d\eta_1 \wedge d\xi_2 \wedge d\eta_2$. Taking into account that the area of the triangle $A_0A_1A_2$ is $T = \frac{1}{2}(a_{00}a_{01}a_{02})^{-1}$, we get, up to a constant factor,

$$d(A_0, A_1, A_2) = T^{-3} \, dA_0 \wedge dA_1 \wedge dA_2$$

where $dA_i = d\xi_i \wedge d\eta_i$ is the area element at the point A_i.

(b) For the point A_0 and the line $G \equiv A_1A_2$ the invariant density is $d(A_0, G) = \omega_{10} \wedge \omega_{20} \wedge \omega_{01} \wedge \omega_{02}$. The metric form of this density is $d(A_0, G) = \delta^{-3} \, dA_0 \wedge dG$, where dA_0 and dG are the metric densities for A and G and δ is the distance from A_0 to G.

Sets of hyperquadrics. A nonsingular hyperquadric C in P_n has a matrix equation of the form $x^t Q x = 0$, where Q is a nonsingular $(n + 1) \times (n + 1)$ symmetric matrix (q_{ij}) and x is the $(n + 1) \times 1$ matrix whose elements are the homogeneous coordinates of the point x. Then the density for sets of nonsingular hyperquadrics invariant under $\mathfrak{P}_n$ can be written $dC = (q^{nn})^{-1} \, dq_{00} \wedge dq_{01} \wedge \cdots \wedge dq_{n-1,n}$, where q^{nn} is the element (n, n) of the inverse matrix Q^{-1} and we have assumed Q normalized in such a way that $\det Q = 1$.

If the equation of the hyperquadric is normalized so that it has the form $\sum \overset{*}{q}_{ij} x_i x_j = 0$ with $\overset{*}{q}_{nn} = 1$ $(i, j = 0, 1, \ldots, n)$, then we have

$$dC = \frac{d\overset{*}{q}_{00} \wedge d\overset{*}{q}_{01} \wedge \cdots \wedge d\overset{*}{q}_{n-1,n}}{(n + 1) \, \Delta^{(n+2)/2}}$$

where $\Delta = \det(\overset{*}{q}_{ij})$.

For instance, the invariant density for sets of conics written in the form $q_{00}x_0^2 + 2q_{01}x_0x_1 + 2q_{02}x_0x_2 + q_{11}x_1^2 + 2q_{12}x_1x_2 + x_2^2 = 0$ is (see [646, 647, 577])

$$dC = \frac{dq_{00} \wedge dq_{01} \wedge dq_{02} \wedge dq_{11} \wedge dq_{12}}{3\Delta^2}.$$

The case of singular hyperquadrics has been considered by Luccioni [371]. The integral geometry of the group of collineations in P_{2n+1} that leave invariant two planes has been investigated by Stanilow [622].

EXERCISE 1. With respect to the general affine group $\mathfrak{A}(2)$, parallelograms transform transitively and have the invariant density $F^{-3} \, dP_0 \wedge dP_1 \wedge dP_2$, where P_0, P_1, P_2 are three vertices of the parallelogram and F denotes its area.

EXERCISE 2. Under the special affine group $\mathfrak{A}_s(2)$ the pairs of lines (G_1, G_2) have an invariant density given by $d(G_0, G_1) = \sin^{-3} \theta \, dG_0 \wedge dG_1$, where dG_0, dG_1 denote the metric density for lines and θ is the angle between G_0 and G_1.

EXERCISE 3. With respect to the group of similitudes (9.95), the pairs of points (P_0, P_1) have the density $d(P_0, P_1) = \delta^{-4} \, dP_0 \wedge dP_1$ where δ is the

ISBN 0-201-13500-0

distance between the points. The pairs of parallel lines (G_0, G_1) have the density $d(G_0, G_1) = \Delta^{-2}\, dG_0 \wedge dp_1$ where Δ is the distance between the two lines and p_1 is the distance from the origin to G_1.

Other examples are given by Stoka [634] and Santaló [581]. The inverse problem of finding the groups when the density of point pairs or pairs of lines is given has been considered by Lucenko and Jurtova [374].

EXERCISE 4. Check the following densities for points and lines in the plane, under the indicated subgroup of the projective group $\mathfrak{P}_2$. Where the densities are missing, it means that they do not exist.

(a) $x' = ax,\ y' = by;\ dP = (xy)^{-1}\, dx \wedge dy,\ dG = (p \sin\phi \cos\phi)^{-1}\, dp \wedge d\phi$.
(b) $x' = ax,\ y' = y + h;\ dP = x^{-1}\, dx \wedge dy,\ dG = (\sin^2\phi \cos\phi)^{-1}\, dp \wedge d\phi$.
(c) $x' = x + c,\ y' = cx + y + h;\ dP = dx \wedge dy,\ dG = \sin^{-3}\phi\, dp \wedge d\phi$.
(d) $x' = ax,\ y' = ay + h;\ dP = x^{-2}\, dx \wedge dy$.
(e) $x' = x + \log a,\ y' = ay + h;\ dP = e^{-x}\, dx \wedge dy,\ dG = (\cos^2\phi \sin\phi)^{-1} \cdot dp \wedge d\phi$.
(f) $x' = x,\ y' = ay + b;\ dG = (\cos^2\phi \sin\phi)^{-1}\, dp \wedge d\phi$.
(g) $x' = (ax + b)(cx + h)^{-1},\ y' = (cx + h)^{-1} y,\ ah - bc = 1;\ dP = y^{-3} \cdot dx \wedge dy,\ dG = \sin^{-3}\phi\, dp \wedge d\phi$.
(h) $x' = ax + b,\ y' = cx + a^{1/2} y + h,\ dG = \sin^3\phi\, dp \wedge d\phi$.

The projective groups in the plane for which there exists a measure for n-tuples of points have been considered by Lucenko [372].

ISBN 0-201-13500-0

CHAPTER 12

The Group of Motions in E_n

1. Introduction

Let $x(x_1,\ldots, x_n)$ and $y(y_1,\ldots, y_n)$ be two points of R^n. If the distance between x and y is defined by $d(x, y) = [(x_1 - y_1)^2 + \cdots + (x_n - y_n)^2]^{1/2}$, then R^n is a metric space called euclidean n-dimensional space and denoted by E_n. An affine transformation of E_n onto itself that preserves distance is called a *motion*. The motions form a subgroup $\mathfrak{M}$ of the affine group, defined by the equations

$$x' = ax + b, \qquad aa^t = e \tag{12.1}$$

where a^t denotes the transpose of the $n \times n$ matrix a and e is the unit $n \times n$ matrix. The condition $aa^t = e$ says that a is an orthogonal matrix. If det $a = + 1$, we have a subgroup of $\mathfrak{M}$, called the group of *special motions*.

The study of $\mathfrak{M}$ can be easily done by the method of moving frames (Chapter 11, Section 1). Let $(p_0; \mathbf{e}_1{}^0,\ldots, \mathbf{e}_n{}^0) \equiv (p_0; \mathbf{e}_i{}^0)$ be an orthonormal fixed frame ($\mathbf{e}_i{}^0 \cdot \mathbf{e}_j{}^0 = \delta_{ij}$). To each motion $g \in \mathfrak{M}$ corresponds a moving frame $(p; \mathbf{e}_i) = g(p_0; \mathbf{e}_i{}^0)$, the transform of $(p_0; \mathbf{e}_i{}^0)$ by g, and conversely, to each orthonormal frame $(p; \mathbf{e}_i)$ corresponds the motion g such that $(p; \mathbf{e}_i) = g(p_0; \mathbf{e}_i{}^0)$. From equations (11.14)

$$dp = \sum_{i=1}^{n} \omega_i \mathbf{e}_i, \qquad d\mathbf{e}_i = \sum_{j=1}^{n} \omega_{ji} \mathbf{e}_j \tag{12.2}$$

using the relations $\mathbf{e}_i \cdot \mathbf{e}_j = \delta_{ij}$, $d\mathbf{e}_i \cdot \mathbf{e}_j + \mathbf{e}_i \cdot d\mathbf{e}_j = 0$, we have

$$\omega_i = dp \cdot \mathbf{e}_i, \qquad \omega_{ji} = d\mathbf{e}_i \cdot \mathbf{e}_j, \qquad \omega_{ji} + \omega_{ij} = 0. \tag{12.3}$$

ENCYCLOPEDIA OF MATHEMATICS and Its Applications, Gian-Carlo Rota (ed.).
1, Luis A. Santaló, Integral Geometry and Geometric Probability

ISBN 0-201-13500-0

In terms of the matrices a, b these 1-forms are given by (11.7), which can now be written (since $\alpha_{ih} = a_{hi}$ for orthogonal matrices)

$$\omega_{ij} = \sum_{h=1}^{n} a_{hi}\, da_{hj}, \qquad \omega_i = \sum_{h=1}^{n} a_{hi}\, db_h. \tag{12.4}$$

The structure equations are

$$d\omega_{ij} = -\sum_{h=1}^{n} \omega_{ih} \wedge \omega_{hj}, \qquad d\omega_i = \sum_{h=1}^{n} \omega_h \wedge \omega_{ih}. \tag{12.5}$$

We proceed to find the invariant volume elements of $\mathfrak{M}$ and some of its subgroups.

The group of translations. Is the subgroup $\mathfrak{T} \subset \mathfrak{M}$ defined by $x' = x + b$. The dimension of $\mathfrak{T}$ is n and the 1-forms db_i form a set of Maurer–Cartan forms. Since $\mathfrak{T}$ is commutative, it is unimodular and its invariant volume element is

$$d\mathfrak{T} = db_1 \wedge db_2 \wedge \cdots \wedge db_n. \tag{12.6}$$

The integral geometry of the group of translations has been investigated by Berwald and Varga [30], Blaschke [44] and Miles [422].

Rotations about a point. The group $\mathfrak{M}_{[0]}$ of rotations about the origin is defined by $x' = ax$ (a orthogonal). It is isomorphic to the orthogonal group $O(n)$. From the condition $aa^t = e$ it follows that $a_{ih}^2 \leqslant 1$ for all i, h and thus the group manifold of $\mathfrak{M}_{[0]}$ is bounded (and closed). Therefore $\mathfrak{M}_{[0]}$ is a compact group and we have, according to Section 6 of Chapter 9, that *the group $\mathfrak{M}_{[0]}$ of rotations about a point in E_n (orthogonal group $O(n)$) is unimodular*. The invariant volume element is given by

$$d\mathfrak{M}_{[0]} = \bigwedge_{i<j} \omega_{ij}. \tag{12.7}$$

As a consequence of (9.63) we have $d\mathfrak{M}_{[0]}(a) = d\mathfrak{M}_{[0]}(a^{-1})$, that is, the invariant volume element $d\mathfrak{M}_{[0]}$ is not changed by replacement of the rotation by its inverse.

We shall compute the total volume of $\mathfrak{M}_{[0]}$. By rotation about the origin O, the end points of the vectors $\mathbf{e}_i$ move on the unit sphere U_{n-1} centered at O. The scalar product $d\mathbf{e}_i \cdot \mathbf{e}_j = \omega_{ji}$ is equal to the arc element on U_{n-1} at the end point of $\mathbf{e}_i$ in the direction of $\mathbf{e}_j$, so that, denoting by du_{n-1} the $(n-1)$-dimensional volume element of U_{n-1} at the end point of $\mathbf{e}_1$, we have

$$du_{n-1} = \omega_{21} \wedge \omega_{31} \wedge \cdots \wedge \omega_{n1}. \tag{12.8}$$

ISBN 0-201-13500-0

By similar reasoning, if U_{n-2} is the $(n-2)$-dimensional unit sphere contained in the $(n-1)$-plane orthogonal to $\mathbf{e}_1$, its $(n-2)$-dimensional volume element at the end point of $\mathbf{e}_2$ is

$$du_{n-2} = \omega_{32} \wedge \omega_{42} \wedge \cdots \wedge \omega_{n2} \tag{12.9}$$

and by induction we get

$$d\mathfrak{M}_{[0]} = du_{n-1} \wedge du_{n-2} \wedge \cdots \wedge du_1. \tag{12.10}$$

Note that the last vector $\mathbf{e}_n$ has two possible senses, $\mathbf{e}_n$ and $-\mathbf{e}_n$, both giving rise to an orthonormal n-frame $(O; \mathbf{e}_i)$. The group $\mathfrak{M}_{[0]}$ includes both possibilities, so that integration of (12.10) over the whole group gives

$$m(\mathfrak{M}_{[0]}) = 2O_{n-1}\, O_{n-2} \cdots O_1 \tag{12.11}$$

where O_i denotes the surface area of the i-dimensional unit sphere (1.22). If only special rotations are considered, then the measure (12.11) should be divided by 2.

For $n = 3$, the kinematic density about the origin O takes the form $d\mathfrak{M}_{[0]} = du_2 \wedge du_1 = \sin\theta\, d\theta \wedge d\phi \wedge d\alpha$, where θ, ϕ are the spherical coordinates on the unit sphere U_2 of the end point of $\mathbf{e}_3$ and α denotes the rotation about this vector $(0 \leqslant \theta \leqslant \pi, 0 \leqslant \phi \leqslant 2\pi, 0 \leqslant \alpha \leqslant 2\pi)$. If the rotation is defined by the axis of rotation with spherical coordinates θ_1, ϕ_1 and the rotation α_1 about it, then a change of coordinates gives $d\mathfrak{M}_{[0]} = 2\sin^2(\alpha_1/2)\sin\theta_1\, d\theta_1 \wedge d\phi_1 \wedge d\alpha_1$. This shows that the guess of choosing uniformly at random both the axis of rotation and the angle of rotation is wrong. (See [144, 335, 408, 670].)

The group of motions. Since $\mathfrak{M}_{[0]}$ and $\mathfrak{T}$ are unimodular, the group of motions $\mathfrak{M}$ is also unimodular and its invariant volume element can be written

$$d\mathfrak{M} = d\mathfrak{M}_{[0]} \wedge d\mathfrak{T}. \tag{12.12}$$

Assuming the motion defined by the frame $(p; \mathbf{e}_i)$, taking into account (12.6) and (12.10), and noting that the volume element of E_n at p is $dP = db_1 \wedge db_2 \wedge \cdots \wedge db_n$, we can write the invariant volume element (12.12) as

$$d\mathfrak{M} = dP \wedge d\mathfrak{M}_{[0]} = dP \wedge du_{n-1} \wedge du_{n-2} \wedge \cdots \wedge du_1. \tag{12.13}$$

Rotations about a q-plane. The motions in E_n that leave invariant all points of a fixed q-plane L_q form the group $\mathfrak{M}_{[q]}$ of rotations about L_q. If we cut L_q by an orthogonal $(n-q)$-plane L_{n-q}, the group $\mathfrak{M}_{[q]}$ is clearly isomorphic to the group $\mathfrak{M}_{[0]}^{(n-q)}$ of rotations about $O = L_q \cap L_{n-q}$ in the euclidean space L_{n-q}. Thus $\mathfrak{M}_{[q]}$ is unimodular and its total volume, according

ISBN 0-201-13500-0

to (12.11), has the value

$$m(\mathfrak{M}_{[q]}) = 2O_{n-q-1}O_{n-q-2}\cdots O_1. \tag{12.14}$$

In all the cases above, if the moving frame $(p; \mathbf{e}_i)$ has the same orientation as the original frame $(p_0; \mathbf{e}_i^0)$, then $\det a = 1$ and we have the groups of "special" rotations or "special" motions. The total measures (12.11) or (12.14) are twice the total measures of the corresponding special groups.

2. Densities for Linear Spaces in E_n

We want to define a density for r-planes L_r in E_n, invariant under the group of motions. Let $\mathfrak{H}_r$ denote the group of all motions that leave invariant a fixed L_r^0. Clearly, $\mathfrak{H}_r$ is closed in $\mathfrak{M}$. If $g \in \mathfrak{M}$, to each coset $g\mathfrak{H}_r$ there corresponds the r-plane gL_r^0 and, conversely, to each L_r there corresponds a coset $g\mathfrak{H}_r$, where g is a motion such that $L_r = gL_r^0$. That is, there is a one-to-one correspondence between the r-planes of E_n and the elements of the homogeneous space $\mathfrak{M}/\mathfrak{H}_r$. The problem of finding an invariant density for sets of r-planes is equivalent to that of finding an invariant density on $\mathfrak{M}/\mathfrak{H}_r$. We will apply the general methods of Chapter 10.

The group $\mathfrak{H}_r$ is the direct product of the group of rotations about L_r^0, say $\mathfrak{M}_{[r]}$, and the group of motions in L_r^0, say $\mathfrak{M}^{(r)}$. Since $\mathfrak{M}_{[r]}$ and $\mathfrak{M}^{(r)}$ are unimodular groups, the product $\mathfrak{H}_r = \mathfrak{M}_{[r]} \times \mathfrak{M}^{(r)}$ is also unimodular. Thus, by consequence (c) of Section 1, Chapter 10, we have that *the homogeneous space* $\mathfrak{M}/\mathfrak{H}_r$ $(r = 0, 1, \ldots, n-1)$ *has an invariant density*. This invariant density is called the density for r-planes and it will be indicated by dL_r. We shall compute the explicit form of dL_r. Let $(p; \mathbf{e}_i)$ be a moving frame in E_n. Assume that L_r^0 is the r-plane spanned by the point p and the r unit vectors $\mathbf{e}_1, \mathbf{e}_2, \ldots, \mathbf{e}_r$. If L_r^0 is kept fixed, then p and the vectors $\mathbf{e}_1, \ldots, \mathbf{e}_r$ can vary only on L_r^0 and therefore we have $dp \cdot \mathbf{e}_\alpha = 0$ $(\alpha = r+1, \ldots, n)$ and $d\mathbf{e}_i \cdot \mathbf{e}_\beta = 0$ $(\beta = r+1, \ldots, n;\ i = 1, 2, \ldots, r)$. Consequently, by (12.3) we have

$$\omega_\alpha = 0, \qquad \omega_{i\beta} = 0, \tag{12.15}$$

with the range of indices

$$\alpha, \beta = r+1, r+2, \ldots, n, \qquad i = 1, 2, \ldots, r. \tag{12.16}$$

System (12.15) is Pfaffian system (10.1) of the general theory. The invariant density for $\mathfrak{M}/\mathfrak{H}_r$ becomes

$$d(\mathfrak{M}/\mathfrak{H}_r) = \bigwedge \omega_{i\beta} \bigwedge \omega_\alpha \tag{12.17}$$

with the range of indices in (12.16). Though we know that this differential

ISBN 0-201-13500-0

form is an invariant density for $\mathfrak{M}/\mathfrak{H}_r$, we can verify as an exercise, using structure equations (12.5), that condition (10.3) is fulfilled. Thus we can state

There exists a density for sets of r-planes in E_n that is given by

$$dL_r = \bigwedge_{i,\beta} \omega_{i\beta} \bigwedge_{\alpha} \omega_\alpha \tag{12.18}$$

where ω_α, $\omega_{i\beta}$ are the forms in (12.3) *and the range of indices is given by* (12.16).

The density dL_r is defined up to a constant factor and it is a form of degree $(n - r)(r + 1)$, as expected, since the space of r-planes of E_n has dimension $(n - r)(r + 1)$. For a direct proof of the invariance properties of form (12.18) see [36, 664, 274].

Example. Consider the case in which $n = 2$, $r = 1$. We have $dL_1 = \omega_{12} \wedge \omega_2 = (d\mathbf{e}_1 \cdot \mathbf{e}_2) \wedge (dp \cdot \mathbf{e}_2)$. If x, y are the coordinates of p and $\mathbf{e}_1(\cos\theta, \sin\theta)$, $\mathbf{e}_2(-\sin\theta, \cos\theta)$, we have $\omega_{12} = d\theta$, $\omega_2 = -\sin\theta\, dx + \cos\theta\, dy$ and $dL_1 = -\sin\theta\, d\theta \wedge dx + \cos\theta\, d\theta \wedge dy$, which is form (3.11) of the density for lines in the plane.

3. A Differential Formula

Let $p, p_1, \ldots, p_r$ be $r + 1$ points in L_r. Assuming that p is the origin of the moving frame $(p; \mathbf{e}_i)$ and that L_r is the r-plane spanned by $(p; \mathbf{e}_1, \mathbf{e}_2, \ldots, \mathbf{e}_r)$, we can write

$$p_i - p = \sum_{j=1}^{r} \lambda_{ij}\mathbf{e}_j, \quad dp_i - dp = \sum_{j=1}^{r} d\lambda_{ij}\mathbf{e}_j + \sum_{j=1}^{r} \lambda_{ij}\, d\mathbf{e}_j \qquad (i = 1, \ldots, r).$$

Multiplying by $\mathbf{e}_\alpha$ $(\alpha = r + 1, \ldots, n)$, we have

$$dp_i \cdot \mathbf{e}_\alpha - dp \cdot \mathbf{e}_\alpha = \sum_{j=1}^{r} \lambda_{ij}\mathbf{e}_\alpha \cdot d\mathbf{e}_j$$

or

$$dp_i \cdot \mathbf{e}_\alpha = \omega_\alpha + \sum_{j=1}^{r} \lambda_{ij}\omega_{\alpha j} \qquad (\alpha = r + 1, \ldots, n).$$

Exterior multiplication of these equalities for $i = 1, 2, \ldots, r$ and by $dp \cdot \mathbf{e}_\alpha = \omega_\alpha$ gives

$$(dp \cdot \mathbf{e}_\alpha) \bigwedge_i (dp_i \cdot \mathbf{e}_\alpha) = r!S\omega_\alpha \bigwedge_j \omega_{\alpha j} \qquad (i, j = 1, 2, \ldots, r) \tag{12.19}$$

where

ISBN 0-201-13500-0

$$r!S = \begin{vmatrix} 1 & 0 & 0 & \cdots & 0 \\ 1 & \lambda_{11} & \lambda_{12} & \cdots & \lambda_{1r} \\ \cdot & \cdot & \cdot & \cdots & \cdot \\ 1 & \lambda_{r1} & \lambda_{r2} & \cdots & \lambda_{rr} \end{vmatrix}; \tag{12.20}$$

that is, S is equal to the volume of the simplex of vertices $p, p_1, \ldots, p_r$ contained in L_r. Exterior multiplication of equalities (12.19) for $\alpha = r + 1, \ldots, n$ gives

$$\bigwedge_{\alpha} (dp \cdot \mathbf{e}_\alpha) \bigwedge_{i,\alpha} (dp_i \cdot \mathbf{e}_\alpha) = (r!S)^{n-r} \, dL_r. \tag{12.21}$$

The volume element of E_n at p_i can be written $dP_i(E_n) = \bigwedge_h (dp_i \cdot \mathbf{e}_h)$ $(h = 1, 2, \ldots, n)$ and the volume element of L_r at p_i is $dP_i(L_r) = \bigwedge_j (dp_i \cdot \mathbf{e}_j)$ $(j = 1, 2, \ldots, r)$. Thus, by exterior multiplication of (12.21) by $\bigwedge_h (dp \cdot \mathbf{e}_h) \bigwedge_{i,j} \cdot$ $(dp_i \cdot \mathbf{e}_j)$ $(h, i, j = 1, 2, \ldots, r)$ we get

$$dP(E_n) \wedge dP_1(E_n) \wedge \cdots \wedge dP_r(E_n) = (r!S)^{n-r} \, dP(L_r) \, dP_1(L_r) \wedge \cdots \wedge dP_r(L_r) \wedge dL_r. \tag{12.22}$$

This is a very useful formula due to Blaschke [37]. Similar formulas have been obtained by Petkantschin [471] and Kingman [337]. Miles [420] applies (12.22) to problems in random geometry, for instance, to determine all the moments of the random volume of various isotropic random r-dimensional simplexes in E_n $(r = 1, 2, \ldots, n)$.

As particular cases of (12.22) we have:

(a) $r = 1$. Let P, P_1 be two points of the line L_1 of E_n. If t, t_1 are their coordinates on L_1, we have $dP(L_1) = dt$, $dP_1(L_1) = dt_1$, $S = t_1 - t$, and (12.22) becomes (taking the absolute value of S)

$$dP(E_n) \wedge dP_1(E_n) = |t_1 - t|^{n-1} \, dt \wedge dt_1 \wedge dL_1, \tag{12.23}$$

which for $n = 2$ is formula (4.2).

(b) $r = n - 1$. If $p, p_1, \ldots, p_{n-1}$ are n independent points of the hyperplane L_{n-1}, we have

$$dP(E_n) \wedge dP_1(E_n) \wedge \cdots \wedge dP_{n-1}(E_n)$$
$$= (n-1)!S \, dP(L_{n-1}) \wedge dP_1(L_{n-1}) \wedge \cdots \wedge dP_{n-1}(L_{n-1}) \wedge dL_{n-1}. \tag{12.24}$$

4. Density for r-Planes about a Fixed q-Plane

Assume a q-plane L_q^0 fixed in E_n. We seek a density for r-planes L_r $(r > q)$ that contain L_q^0. Let L_q^0 be determined by the point p and the unit vectors $\mathbf{e}_1, \mathbf{e}_2, \ldots, \mathbf{e}_q$ and consider the r-plane L_r^0 determined by L_q^0 and the unit

ISBN 0-201-13500-0

vectors $\mathbf{e}_{q+1}, \ldots, \mathbf{e}_r$. The group $\mathfrak{H}_{r[q]}$ of all motions that kept L_r^0 fixed, considered as a subgroup of the group $\mathfrak{H}_q$ of all motions that kept L_q^0 fixed, is defined by

$$\omega_{hi} = 0, \qquad i = q+1, \ldots, r, \quad h = r+1, \ldots, n, \tag{12.25}$$

for $\mathbf{e}_{q+1}, \ldots, \mathbf{e}_r$ can only vary in L_r^0. Consequently the invariant volume element of $\mathfrak{H}_q/\mathfrak{H}_{r[q]}$, which is equal to the density for r-planes about L_q^0, reads

$$dL_{r[q]} = \bigwedge_{h,i} \omega_{hi} \qquad (i = q+1, \ldots, r;\ h = r+1, \ldots, n). \tag{12.26}$$

From the structure equations (12.5) for $\mathfrak{M}$, it follows that $d(dL_{r[q]}) = 0$. Hence, according to (10.3) the space $\mathfrak{H}_q/\mathfrak{H}_{r[q]}$ has the invariant density (12.26) and we can state

The r-planes about a q-plane in E_n ($q < r \leqslant n - 1$) have an invariant density given by (12.26).

For $q = 0$ we get the density for r-planes about a fixed point

$$dL_{r[0]} = \bigwedge_{i,h} \omega_{hi} \qquad (i = 1, 2, \ldots, r;\ h = r+1, \ldots, n), \tag{12.27}$$

which is the invariant volume element of the Grassmann manifold $G_{r,n-r}$. Recall that the Grassmann manifold $G_{n,p}$ is the set of n-planes through the origin of R^{n+p} with a suitable differentiable manifold structure (see, e.g., [341a]).

Between the r-planes $L_{r[0]}$ and their orthogonal complement $L_{n-r[0]}$ we have the duality

$$dL_{r[0]} = dL_{n-r[0]}, \tag{12.28}$$

as follows immediately from (12.27) and (12.3).

The set of all L_r about L_q^0 has a finite measure. In order to compute this measure we consider first the case in which $q = 0$. Let $L_{n-1[0]}$ be the $(n-1)$-plane perpendicular to $\mathbf{e}_1$ through the origin p of the moving frame $(p; \mathbf{e}_i)$. Calling $L_{r-1[0]}$ to the $(r-1)$-plane $L_{r[0]} \cap L_{n-1[0]}$, we have that the density for $L_{r-1[0]}$ in $L_{n-1[0]}$ is

$$dL_{r-1[0]}^{(n-1)} = \bigwedge_{h,i} \omega_{hi} \qquad (i = 2, \ldots, r;\ h = r+1, \ldots, n). \tag{12.29}$$

From (12.27) and (12.29), using that the volume element on U_{n-1} at the end point of $\mathbf{e}_1$ is $du_{n-1} = \omega_{21} \wedge \omega_{31} \wedge \cdots \wedge \omega_{n1}$, we get

$$dL_{r[0]} \wedge \omega_{21} \wedge \omega_{31} \wedge \cdots \wedge \omega_{r1} = dL_{r-1[0]}^{(n-1)} \wedge du_{n-1}. \tag{12.30}$$

ISBN 0-201-13500-0

If du_{r-1} denotes the volume element of the unit $(r-1)$-sphere in $L_{r[0]}$, (12.30) can be written

$$dL_{r[0]} \wedge du_{r-1} = dL_{r-1[0]}^{(n-1)} \wedge du_{n-1}. \tag{12.31}$$

Exterior multiplication by du_{r-2} using (12.31) gives (up to the sign)

$$\begin{aligned} dL_{r[0]} \wedge du_{r-1} \wedge du_{r-2} &= dL_{r-1[0]}^{(n-1)} \wedge du_{r-2} \wedge du_{n-1} \\ &= dL_{r-2[0]}^{(n-2)} \wedge du_{n-1} \wedge du_{n-2}. \end{aligned} \tag{12.32}$$

Successive exterior multiplication by $du_{r-3}, \ldots, du_{r-q}$, taking (12.31) into account each time, gives

$$dL_{r[0]} \wedge du_{r-1} \wedge \cdots \wedge du_{r-q} = dL_{r-q[0]}^{(n-q)} \wedge du_{n-1} \wedge \cdots \wedge du_{n-q}. \tag{12.33}$$

For $q = r - 1$, since $dL_{1[0]}^{(n-r+1)} = du_{n-r}$, we have

$$dL_{r[0]} \wedge du_{r-1} \wedge \cdots \wedge du_1 = du_{n-1} \wedge \cdots \wedge du_{n-r}. \tag{12.34}$$

Integrating over the hemispheres of the unit spheres $U_{n-1}, U_{n-2}, \ldots, U_1$ and dividing by $O_0 = 2$, since we consider unoriented $L_{r[0]}$, we get

The total measure of the unoriented r-planes of E_n through a fixed point (i.e., the volume of the Grassmann manifold $G_{r,n-r}$) is

$$m(G_{r,n-r}) = m(G_{n-r,r}) = \int_{G_{r,n-r}} dL_{r[0]} = \frac{O_{n-1}O_{n-2}\cdots O_{n-r}}{O_{r-1}O_{r-2}\cdots O_1 O_0} \tag{12.35}$$

where O_i is the surface area of the i-dimensional unit sphere (1.22).

The measure of all $L_{r[q]}$, that is, the measure of all r-planes of E_n that contain a given L_q^0, can be deduced from (12.35). Let $L_{n-q[0]}$ be an $(n-q)$-plane perpendicular to L_q^0 and let O be the intersection point. Each $L_{r[q]}$ can be defined by the intersection $L_{r[q]} \cap L_{n-q[0]}$, which is an $(r-q)$-plane through O, and consequently the measure of all $L_{r[q]}$ in E_n is equal to the measure of all $L_{r-q[0]}$ in $L_{n-q[0]}$. Formula (12.35) then gives

$$\int_{\text{Total}} dL_{r[q]} = \frac{O_{n-q-1}O_{n-q-2}\cdots O_{n-r}}{O_{r-q-1}O_{r-q-2}\cdots O_1 O_0}. \tag{12.36}$$

Example 1. The measure of all the lines through a point in E_3 corresponds to $n = 3$, $r = 1$, $q = 0$ and (12.36) gives the value 2π.

Example 2. The measure of all the planes through a line in E_3 corresponds to $n = 3$, $r = 2$, $q = 1$ and (12.36) gives the value π.

ISBN 0-201-13500-0

Note that we have considered unoriented r-planes. If they were considered oriented, the measures above would be multiplied by 2.

5. Another Form of the Density for r-Planes in E_n

Formula (12.18) gives dL_r in terms of a point p and r orthonormal vectors $\mathbf{e}_i$ contained in L_r. It is sometimes useful to introduce the elements of the $(n - r)$-plane perpendicular to L_r through a fixed point O. To this end, let $L_{n-r[0]}$ be the $(n - r)$-plane perpendicular to L_r through O and let p be the point $L_r \cap L_{n-r[0]}$. Let ρ be the distance Op. Choose the orthonormal frame $(p; \mathbf{e}_i)$ in such a way that $p, \mathbf{e}_1, \ldots, \mathbf{e}_r$ are in L_r and $\mathbf{e}_{r+1}$ have the direction of Op. The 1-form $\omega_{r+h} = dp \cdot \mathbf{e}_{r+h}$ is the arc element of $L_{n-r[0]}$ at p in the direction $\mathbf{e}_{r+h}$. Thus $\omega_{r+1} \wedge \omega_{r+2} \wedge \cdots \wedge \omega_n$ is the volume element of $L_{n-r[0]}$ at p, which we will denote by $d\sigma_{n-r}$, so that

$$d\sigma_{n-r} = \omega_{r+1} \wedge \omega_{r+2} \wedge \cdots \wedge \omega_n. \tag{12.37}$$

From (12.27) and (12.28) it follows that the product $\bigwedge \omega_{hj}$ $(h = r + 1, \ldots, n;$ $j = 1, 2, \ldots, r)$ is equal to $dL_{n-r[0]}$ and hence we have

$$dL_r = d\sigma_{n-r} \wedge dL_{n-r[0]}. \tag{12.38}$$

Example 1. $n = 2$, $r = 1$. This corresponds to lines on the plane. In this case $L_{1[0]}$ is the line perpendicular to L_1 through O and $d\sigma_1$ is the arc length on $L_{1[0]}$. On the other hand, $dL_{1[0]}$ is the density for lines about O, that is, $dL_{1[0]} = d\phi$, where ϕ is the angle of $L_{1[0]}$ with a fixed reference direction. With these notations, (12.38) gives (3.5).

Example 2. $r = 1$. A line L_1 in E_n may be determined by its direction and its intersection point p with the hyperplane $L_{n-1[0]}$ perpendicular to L_1 through O. The density $dL_{n-1[0]}$ is the volume element du_{n-1} of the unit sphere U_{n-1} at the end point of the vector $\mathbf{e}_1$ parallel to L_1 and $d\sigma_{n-1}$ is the volume element of $L_{n-1[0]}$ at p. We have

$$dL_1 = d\sigma_{n-1} \wedge du_{n-1}. \tag{12.39}$$

Example 3. $r = n - 1$. Let ρ be the distance from the origin O to the hyperplane L_{n-1}; then $d\sigma_1 = d\rho$. The density $dL_{1[0]}$ is the volume element du_{n-1} of the unit sphere U_{n-1} at the end point of the vector $\mathbf{e}_n$ perpendicular to L_{n-1} through O. Thus we have

$$dL_{n-1} = d\rho \wedge du_{n-1}. \tag{12.40}$$

For $n = 2$ we again have the density for lines in the plane in the form of (3.5).

ISBN 0-201-13500-0

6. Sets of Pairs of Linear Spaces

1. Let L_{n-1} and L^*_{n-1} be two hyperplanes of E_n. We want to express $dL_{n-1} \wedge dL^*_{n-1}$ in terms of the density dL_{n-2} of the intersection $L_{n-2} = L_{n-1} \cap L^*_{n-1}$ and the densities $d\phi_1$, $d\phi_2$ of the angles of L_{n-1} and L^*_{n-1} about L_{n-2}.

For this purpose let us consider the orthonormal frames $(p; \mathbf{e}_1, \ldots, \mathbf{e}_n)$, $(p; \mathbf{e}_1, \mathbf{e}_2, \ldots, \mathbf{e}_{n-2}, \mathbf{e}^*_{n-1}, \mathbf{e}_n{}^*)$, and $(p; \mathbf{e}_1, \ldots, \mathbf{e}_{n-2}, \mathbf{a}, \mathbf{b})$ such that (a) $p, \mathbf{e}_1, \ldots, \mathbf{e}_{n-2}$ define L_{n-2}; (b) $\mathbf{e}_{n-1}$ is contained in L_{n-1} and $\mathbf{e}_n$ is the unit normal vector to L_{n-1}; (c) $\mathbf{e}^*_{n-1}$ is contained in L^*_{n-1} and $\mathbf{e}_n{}^*$ is the unit normal vector to L^*_{n-1}; (d) $\mathbf{a}$ and $\mathbf{b}$ are constant unit vectors. By (12.18) and (12.3) we have

$$dL_{n-1} = \bigwedge_i (d\mathbf{e}_i \cdot \mathbf{e}_n) \wedge (dp \cdot \mathbf{e}_n), \qquad i = 1, 2, \ldots, n-1,$$

$$dL^*_{n-1} = \bigwedge_i (d\mathbf{e}_i \cdot \mathbf{e}_n{}^*) \wedge (d\mathbf{e}^*_{n-1} \cdot \mathbf{e}_n{}^*) \wedge (dp \cdot \mathbf{e}_n{}^*), \qquad i = 1, 2, \ldots, n-2,$$

$$dL_{n-2} = \bigwedge_i (d\mathbf{e}_i \cdot \mathbf{a})(d\mathbf{e}_i \cdot \mathbf{b})(dp \cdot \mathbf{a})(dp \cdot \mathbf{b}), \qquad i = 1, 2, \ldots, n-2. \tag{12.41}$$

Calling ϕ_1 and ϕ_2 the angles of L_{n-1} and L^*_{n-1} about L_{n-2} that are equal to the angles of $\mathbf{e}_n$ and $\mathbf{e}_n{}^*$ with $\mathbf{b}$, we have

$$\mathbf{e}_{n-1} = \cos\phi_1 \mathbf{a} + \sin\phi_1 \mathbf{b}, \qquad \mathbf{e}_n = -\sin\phi_1 \mathbf{a} + \cos\phi_1 \mathbf{b},$$

$$\mathbf{e}^*_{n-1} = \cos\phi_2 \mathbf{a} + \sin\phi_2 \mathbf{b}, \qquad \mathbf{e}_n{}^* = -\sin\phi_2 \mathbf{a} + \cos\phi_2 \mathbf{b},$$

and therefore

$$(d\mathbf{e}_i \cdot \mathbf{e}_n) \wedge (d\mathbf{e}_i \cdot \mathbf{e}_n{}^*) = \sin(\phi_2 - \phi_1)(d\mathbf{e}_i \cdot \mathbf{a}) \wedge (d\mathbf{e}_i \cdot \mathbf{b})$$

$$(d\mathbf{e}_{n-1} \cdot \mathbf{e}_n) = d\phi_1, \qquad (d\mathbf{e}^*_{n-1} \cdot \mathbf{e}_n{}^*) = d\phi_2,$$

$$(dp \cdot \mathbf{e}_n) \wedge (dp \cdot \mathbf{e}_n{}^*) = \sin(\phi_2 - \phi_1)(dp \cdot \mathbf{a}) \wedge (dp \cdot \mathbf{b}). \tag{12.42}$$

From (12.41) and (12.42) it follows, up to the sign, that

$$dL_{n-1} \wedge dL^*_{n-1} = \sin^{n-1}(\phi_2 - \phi_1)\, d\phi_2 \wedge d\phi_1 \wedge dL_{n-2}. \tag{12.43}$$

This is the formula that we want to obtain. For $n = 2$ we have (4.19).

2. We can extend (12.43) to the case of two subspaces L_r, L_s such that $r + s - n \geqslant 0$. To this end let $L_{r+s-n} = L_r \cap L_s$ and consider the following sets of unit orthogonal vectors through $p \in L_{r+s-n}$:

(a) $\mathbf{e}_1, \mathbf{e}_2, \ldots, \mathbf{e}_{r+s-n}$, which span L_{r+s-n};
(b) $\mathbf{b}_1, \mathbf{b}_2, \ldots, \mathbf{b}_{2n-r-s}$, which span the $(2n-r-s)$-plane L_{2n-r-s} perpendicular to L_{r+s-n} through p;

ISBN 0-201-13500-0

(c) $\mathbf{f}_1, \mathbf{f}_2, \ldots, \mathbf{f}_{n-s}$, perpendicular to L_{r+s-n} and contained in L_r;
(d) $\mathbf{f}_1', \mathbf{f}_2', \ldots, \mathbf{f}_{n-r}'$, perpendicular to L_{r+s-n} and normal to L_r;
(e) $\mathbf{g}_1, \mathbf{g}_2, \ldots, \mathbf{g}_{n-r}$, perpendicular to L_{r+s-n} and contained in L_s;
(f) $\mathbf{g}_1', \mathbf{g}_2', \ldots, \mathbf{g}_{n-s}'$, perpendicular to L_{r+s-n} and normal to L_s.

Set

$$\mathbf{f}_i = \sum_{j=1}^{2n-r-s} \alpha_{ij}\mathbf{b}_j, \qquad f_h' = \sum_{j=1}^{2n-r-s} \alpha_{hj}'\mathbf{b}_j,$$

$$\mathbf{g}_h = \sum_{j=1}^{2n-r-s} \beta_{hj}\mathbf{b}_j, \qquad \mathbf{g}_i' = \sum_{j=1}^{2n-r-s} \beta_{ij}'\mathbf{b}_j, \tag{12.44}$$

where the ranges of indices are

$$i = 1, 2, \ldots, n-s; \qquad h = 1, 2, \ldots, n-r; \quad j = 1, 2, \ldots, 2n-r-s. \tag{12.45}$$

According to (12.18) and (12.3) we have

$$dL_{r+s-n} = \bigwedge_j (dp\cdot\mathbf{b}_j) \bigwedge_{m,j} (d\mathbf{e}_m\cdot\mathbf{b}_j), \qquad m = 1, 2, \ldots, r+s-n,$$

$$dL_{r[r+s-n]} = \bigwedge_{i,h} (d\mathbf{f}_i\cdot\mathbf{f}_h'), \qquad dL_{s[r+s-n]} = \bigwedge_{h,i} (d\mathbf{g}_h\cdot\mathbf{g}_i'),$$

$$dL_r = \bigwedge_h (dp\cdot\mathbf{f}_h') \bigwedge_{m,h} (d\mathbf{e}_m\cdot\mathbf{f}_h') \bigwedge_{i,h} (d\mathbf{f}_i\cdot\mathbf{f}_h'),$$

$$dL_s = \bigwedge_i (dp\cdot\mathbf{g}_i') \bigwedge_{m,i} (d\mathbf{e}_m\cdot\mathbf{g}_i') \bigwedge_{h,i} (d\mathbf{g}_h\cdot\mathbf{g}_i').$$

Substituting the values in (12.44) into the foregoing expressions and introducing the determinant

$$\Delta = \begin{vmatrix} \mathbf{f}_1'\cdot\mathbf{b}_1 & \mathbf{f}_1'\cdot\mathbf{b}_2 & \cdots & \mathbf{f}_1'\cdot\mathbf{b}_{2n-r-s} \\ \mathbf{f}_2'\cdot\mathbf{b}_1 & \mathbf{f}_2'\cdot\mathbf{b}_2 & \cdots & \mathbf{f}_2'\cdot\mathbf{b}_{2n-r-s} \\ \cdot & \cdot & \cdots & \cdot \\ \mathbf{f}_{n-r}'\cdot\mathbf{b}_1 & \mathbf{f}_{n-r}'\cdot\mathbf{b}_2 & \cdots & \mathbf{f}_{n-r}'\cdot\mathbf{b}_{2n-r-s} \\ \mathbf{g}_1'\cdot\mathbf{b}_1 & \mathbf{g}_1'\cdot\mathbf{b}_2 & \cdots & \mathbf{g}_1'\cdot\mathbf{b}_{2n-r-s} \\ \cdot & \cdot & & \cdot \\ \mathbf{g}_{n-s}'\cdot\mathbf{b}_1 & \mathbf{g}_{n-s}'\cdot\mathbf{b}_2 & \cdots & \mathbf{g}_{n-s}'\cdot\mathbf{b}_{2n-r-s} \end{vmatrix}, \tag{12.46}$$

we easily get the desired formula

$$dL_r \wedge dL_s = \Delta^{r+s-n+1}\, dL_{r[r+s-n]} \wedge dL_{s[r+s-n]} \wedge dL_{r+s-n}. \tag{12.47}$$

3. Finally we want to consider sets of pairs of linear spaces $(L_r, L_{i+1}^{(r)})$ such that $L_{i+1}^{(r)} \subset L_r$ $(i+1 \leqslant r)$. Choose the moving frame $(p; \mathbf{e}_1, \mathbf{e}_2, \ldots, \mathbf{e}_n)$ so

ISBN 0-201-13500-0

that $p, \mathbf{e}_1, \ldots, \mathbf{e}_{i+1}$ belong to $L_{i+1}^{(r)}$ and $p, \mathbf{e}_1, \ldots, \mathbf{e}_r$ belong to L_r. The density for the $(i + 1)$-planes $L_{i+1}^{(r)}$ in L_r is, according to (12.18),

$$dL_{i+1}^{(r)} = \bigwedge_{j,\beta} \omega_{j\beta} \bigwedge_{\alpha} \omega_\alpha \tag{12.48}$$

where

$$j = 1, 2, \ldots, i + 1; \qquad \alpha, \beta = i + 2, \ldots, r. \tag{12.49}$$

On the other hand, the density $dL_{r[i+1]}$ of L_r about $L_{i+1}^{(r)}$ is, according to (12.26),

$$dL_{r[i+1]} = \bigwedge_{h,\alpha} \omega_{h\alpha}, \qquad h = r + 1, \ldots, n; \quad \alpha = i + 2, \ldots, r. \tag{12.50}$$

Furthermore, (12.18) gives

$$dL_{i+1} = \bigwedge_{j,m} \omega_{jm} \bigwedge_{h} \omega_h, \qquad m, h = i + 2, \ldots, n; \quad j = 1, 2, \ldots, i + 1. \tag{12.51}$$

From these results we get

$$dL_{i+1}^{(r)} \wedge dL_r^* = dL_{r[i+1]} \wedge dL_{i+1} \tag{12.52}$$

where on the left-hand side the r-planes L_r are oriented because each unoriented L_r that contains $L_{i+1}^{(r)}$ is the support of two $L_{r[i+1]}$ about $L_{i+1}^{(r)}$ that differ in orientation.

If we consider only linear spaces through a fixed point O of E_n, (12.52) still holds and may be written

$$dL_{i+1[0]}^{(r)} \wedge dL_{r[0]}^* = dL_{r[i+1]} \wedge dL_{i+1[0]}. \tag{12.53}$$

Formulas in the style of (12.52) and (12.53) and more general ones have been given by Petkantschin [471].

Example. Let $n = 3$, $r = 2$, $i = 0$. Let dG be the density for lines in E_3, dE the density for planes, $dE(G)$ the density for planes about G, and $dG(E)$ the density for lines in E. Then (12.52) becomes

$$dG(E) \wedge dE^* = dE(G) \wedge dG. \tag{12.54}$$

7. Notes

1. *Stiefel manifolds.* Consider the set of all orthonormal frames $(\mathbf{e}_1, \mathbf{e}_2, \ldots, \mathbf{e}_r)$ through the origin O in E_{r+n}. This set can be made, in a natural manner, into a differentiable manifold called the Stiefel manifold $S_{r,n}$. If the components

ISBN 0-201-13500-0

of the vectors $\mathbf{e}_i$ with respect to a fixed orthonormal $(n + r)$-frame are taken as columns of a matrix, the result is an $(r + n) \times r$ matrix A such that $A^t A = e$ (unit $r \times r$ matrix). The Stiefel manifold $S_{r,n}$ can be considered as the set of such matrices. For instance, $S_{r,0}$ is the set of $r \times r$ orthogonal matrices and can be identified with the orthogonal group $O(r)$. More generally, $S_{r,n}$ can be identified with the homogeneous space $O(r + n)/O(n)$. The density for sets of r-frames invariant under the group of rotations about O is obtained as follows. Let $(\varepsilon_1, \ldots, \varepsilon_n)$ be the orthonormal frame complementary to $(\mathbf{e}_1, \ldots, \mathbf{e}_r)$. The Pfaffian system (10.1) that determines the r-frame $(\mathbf{e}_1, \mathbf{e}_2, \ldots, \mathbf{e}_r)$ is $\omega_{ih} = \mathbf{e}_i \cdot d\mathbf{e}_h = 0$, $\phi_{\alpha h} = \varepsilon_\alpha \cdot d\mathbf{e}_h = 0$ $(1 \leqslant i, h \leqslant r;\ 1 \leqslant \alpha \leqslant n)$. Therefore the volume element of the Stiefel manifold is

$$dS_{r,n} = \bigwedge_{i<h} \omega_{ih} \bigwedge_{\alpha,k} \phi_{\alpha k} \qquad (1 \leqslant i, h, k \leqslant r,\ 1 \leqslant \alpha \leqslant n). \tag{12.55}$$

By reasoning similar to that in Section 1 for the orthogonal group, we prove that (12.55) can be written $dS_{r,n} = du_{r+n-1} \wedge du_{r+n-2} \wedge \cdots \wedge du_n$ and the total volume of the Stiefel manifold $S_{r,n}$ is

$$m(S_{r,n}) = \int_{S_{r,n}} dS_{r,n} = O_{r+n-1} O_{r+n-2} \cdots O_n. \tag{12.56}$$

In particular, for $n = 0$ we get the volume of $O(r) \equiv M_{[0]}$.

2. *Some differential formulas.* (a) Let $L_{r[0]}$ and $L_{s[0]}$ be two orthogonal subspaces in E_n through O. Assume $r + s \geqslant n$. It can be shown that

$$dL_{r[0]} \wedge dL_{s[n-r]} = dL_{s[0]} \wedge dL_{r[n-s]} \tag{12.57}$$

where $dL_{s[n-r]}$ denotes the density of L_s about the $(n - r)$-plane perpendicular to L_r and similarly $dL_{r[n-s]}$.

(b) Let L_{n-p}^0 be a fixed $(n - p)$-plane. Let L_i^0 be the orthogonal projection of a given L_i on L_{n-p}^0 (assuming $i < n - p$) and let L_{i+p} be the $(i + p)$-plane perpendicular to L_{n-p}^0 that contains L_i and L_i^0. Then it can be proved that

$$dL_i(E_n) = dL_i^0(L_{n-p}^0) \wedge dL_i(L_{i+p}) \tag{12.58}$$

where $dL_h(L)$ denotes the density of L_h in L.

(c) Let L_r and L_s be two linear subspaces of E_n with $r + s \geqslant n$. If $L_{r+s-n} = L_r \cap L_s$, consider a $(2n - r - s)$-plane, say L_{2n-r-s}, perpendicular to L_{r+s-n}, and let O be the intersection point $L_{2n-r-s} \cap L_{r+s-n}$. Put $L_{n-s}^* = L_{2n-r-s} \cap L_r$, $L_{n-r}^* = L_{2n-r-s} \cap L_s$, and consider the $(2n - r - s - 1)$-dimensional unit sphere $U_{2n-r-s-1}$ of L_{2n-r-s} with center O. The intersections $U_{2n-r-s-1} \cap L_{n-s}^*$ and $U_{2n-r-s-1} \cap L_{n-r}^*$ are great spheres U_{n-s-1}^* and U_{n-r-1}^* of U_{2n-r-s}. Consider the great circles meeting both U_{n-s-1}^* and U_{n-r-1}^* at right angles and let $\theta_1, \theta_2, \ldots,$ be the spherical distances between the intersection points. These distances $\theta_1, \theta_2, \ldots$ are called the *angles* between L_r and L_s.

Consider now a fixed L_s, say L_s^0, and a variable L_r. We want to express the density dL_r in terms of the density $dL_{r+s-n}^{(s)}$ of $L_{r+s-n} = L_r \cap L_s^0$ as an $(r + s - n)$-plane of L_s^0 and the density $dL_{r[r+s-n]}$ of L_r about L_{r+s-n}. The

ISBN 0-201-13500-0

result is

$$dL_r = (\sin\theta_1 \sin\theta_2 \cdots)^{r+s-n+1}\, dL^{(s)}_{r+s-n} \wedge dL_{r[r+s-n]}. \tag{12.59}$$

Example 1. $r = 1$, $s = 1$, $n = 2$. Then (12.59) coincides with (3.7).

Example 2. $r = 1$, $s = 2$, $n = 3$. Then we have

$$dL_1 = \sin\theta\, dP \wedge du_2 \tag{12.60}$$

where $dP = dL_0^{(2)}$ is the area element of the plane L_2^0 at the intersection point $P = L_1 \cap L_2^0$, du_2 is the density for lines about P, and θ is the angle between L_1 and L_2^0.

These results are the work of Petkantschin [471].

3. *Volume in terms of concurrent cross sections.* Let $P_1, P_2, \ldots, P_{n-1}$ be points in E_n and call $T(P_1, P_2, \ldots, P_{n-1}, O)$ the volume of the simplex with vertices $P_1, \ldots, P_{n-1}, O$ (O being a fixed point taken as origin). Then from (12.22) it follows easily that

$$\begin{aligned} &dP_1(E_n) \wedge dP_2(E_n) \wedge \cdots \wedge dP_{n-1}(E_n) \\ &= (n-1)!\,T(P_1, \ldots, P_{n-1}, O)\, dP_1(L_{n-1}) \wedge \cdots \wedge dP_{n-1}(L_{n-1}) \wedge du_{n-1} \end{aligned} \tag{12.61}$$

where L_{n-1} is the $(n-1)$-plane spanned by $P_1, P_2, \ldots, P_{n-1}, O$ and $du_{n-1} = dL_{n-1}$ denotes the area element of the $(n-1)$-unit sphere U_{n-1} at the end point of the unit normal vector to L_{n-1}.

If K_i $(i = 1, 2, \ldots, n-1)$ are convex bodies in E_n $(n \geqslant 3)$ and $K_i(L_{n-1}) = K_i^*$ denotes the cross sections $K_i \cap L_{n-1}$, by integrating (12.61) over all points $P_i \in K_i$ we get

$$V_1 V_2 \cdots V_{n-1} = \frac{(n-1)!}{2} \int_{U_{n-1}} \left(\int_{K_1^*} \cdots \int_{K^*_{n-1}} T(P_1, \ldots, P_{n-1}, O) \cdot dP_1(L_{n-1}) \wedge \cdots \wedge dP_{n-1}(L_{n-1}) \right) du_{n-1} \tag{12.62}$$

where V_i is the volume of K_i and the factor $\frac{1}{2}$ arises after observing that by integration over U_{n-1} every $K_i(L_{n-1})$ is counted twice.

If we replace the sets $K_{n-h+1}, \ldots, K_{n-1}$ by the unit ball S_n with center O, after some computation (12.62) yields expressions for $V_1 V_2 \cdots V_{n-h}$ in terms of the volume $T(P_1, \ldots, P_{n-h}, O)$. For instance, for $h = n-1$, (12.62) yields the well-known formula

$$V_1 = \frac{1}{n} \int_{U_{n-1}} r^n\, du_{n-1} \tag{12.63}$$

where r is the distance from O to the point of ∂K_1 in the direction u_{n-1}. The computations can be seen in Busemann's article [75]. If $V_i^* = V_i(L_{n-1})$

ISBN 0-201-13500-0

denote the $(n-1)$-dimensional volume of the cross section K_i^*, Busemann has also proved the inequality

$$V_1 V_2 \cdots V_{n-1} \geqslant \frac{1}{n} \frac{\kappa_n^{n-2}}{\kappa_{n-1}^n} \int_{U_{n-1}} (V_1^* V_2^* \cdots V_{n-1}^*)^{n/(n-1)} \, du_{n-1} \tag{12.64}$$

where κ_i denotes the volume of the i-dimensional unit ball (1.22), and the equality sign holds only when K_i are homothetic ellipsoids with center O. In particular, for a convex body K, assuming $n \geqslant 3$, we have

$$V^{n-1} \geqslant \frac{1}{n} \frac{\kappa_n^{n-2}}{\kappa_{n-1}^n} \int_{U_{n-1}} (V^*)^n \, du_{n-1} \tag{12.65}$$

with equality, if $V > 0$, only for ellipsoids with center O.

For a reinterpretation of these results and other relations of the same type, see that of Petty [473].

4. *Mean density in E_n.* Let $\phi(P)$ be a function with which a finite mass is distributed throughout E_n, that is, a real-valued, nonnegative, measurable function on E_n that is bounded over any bounded subset of E_n $(n \geqslant 2)$. The mean density $\delta(\phi)$ of ϕ is defined (provided the limit exists) as

$$\delta(\phi) = \lim_{R \to \infty} \left(\int_{S_R} \phi(P) \, dP \right) \kappa_n^{-1}(R)$$

where $\kappa_n(R)$ means the volume of the n-ball S_R of radius R. Let $H \equiv L_{n-1}$ be a hyperplane through O and let ϕ_H be the restriction of ϕ to H. If the mean density $\delta(\phi_H)$ exists for every H through O, then $\delta(\phi) \geqslant \inf \delta(\phi_H)$ and if, in addition, $\delta(\phi)$ is bounded, then $\delta(\phi) \leqslant \sup \delta(\phi_H)$, where inf and sup refer to the set of hyperplanes through O. The result is due to Groemer [241]. Analyze the situation when instead of hyperplanes we consider the set of all r-planes through O $(r = 1, 2, \ldots, n-2)$.

5. *Densities for planes and lines in E_3.* We wish to describe different forms of the densities for planes and lines in E_3, which are easily obtained from the general expressions (12.18) and (12.38) by the rule for change of variables in a differential form. For simplicity we shall represent the planes by E and the lines by G, instead of the general notations L_2 and L_1 used earlier.

According to (12.40), the density for planes is $dE = dp \wedge du_2$, where p is the distance from the origin to E and $du_2 = \sin\theta \, d\phi \wedge d\theta$ is the area element of the unit sphere U_2 at the end point of the unit vector perpendicular to E (θ, ϕ are the polar coordinates on U_2).

(i) If E is given by the equation $\alpha x + \beta y + \gamma z = 1$, we have $\alpha = p^{-1}\cos\phi\sin\theta$, $\beta = p^{-1}\sin\phi\sin\theta$, $\gamma = p^{-1}\cos\theta$, and $d\alpha \wedge d\beta \wedge d\gamma = p^{-4}\sin\theta \, dp \wedge d\phi \wedge d\theta$, so that the density for planes becomes

$$dE = \frac{d\alpha \wedge d\beta \wedge d\gamma}{(\alpha^2 + \beta^2 + \gamma^2)^2}. \tag{12.66}$$

ISBN 0-201-13500-0

(ii) If a, b, c are the abscissas of the intersection points of E with the coordinate axis of a rectangular system, we have $a = 1/\alpha$, $b = 1/\beta$, $c = 1/\gamma$ and thus

$$dE = \frac{a^2b^2c^2\, da \wedge db \wedge dc}{(b^2c^2 + a^2c^2 + a^2b^2)^2}. \tag{12.67}$$

(iii) Let E_0 be a fixed plane in E_3. A plane E can be determined by the line $G = E \cap E_0$ and the angle ψ between E and E_0. Then

$$dE = \sin^2 \psi \, dG(E_0) \wedge d\psi \tag{12.68}$$

where $dG(E_0)$ denotes the density for lines in E_0. This formula has been applied to obtain the angular distribution of planes in space from observations of the angular distribution of their traces on one or more section planes (see [435]).

According to (12.39), the density for lines G is $dG = d\sigma_2 \wedge du_2$, where $d\sigma_2$ is the area element on a plane perpendicular to G at the intersection point and du_2 is the area element on U_2 at the end point of the unit vector parallel to G.

(i) If G is determined by its direction and its intersection point (x, y) with a fixed plane, denoting by γ the angle between G and the direction perpendicular to the fixed plane, we have

$$dG = |\cos \gamma| \, dx \wedge dy \wedge du_2. \tag{12.69}$$

(ii) If v_1, v_2, v_3 are the direction cosines of G, we have $du_2 = (dv_1 \wedge dv_2)/v_3$ and $v_3 = \cos \gamma$, so that (12.69) can be written

$$dG = dv_1 \wedge dv_2 \wedge dx \wedge dy. \tag{12.70}$$

(iii) If G is defined by the equations $x = az + m$, $y = bz + q$ with respect to a rectangular coordinate system, the direction cosines are $v_1 = a(1 + a^2 + b^2)^{-1/2}$, $v_2 = b(1 + a^2 + b^2)^{-1/2}$, $v_3 = c(1 + a^2 + b^2)^{-1/2}$, and using that the coordinates of the intersection point of G with the (x, y)-plane are $x = m$, $y = q$, we find that (12.70) becomes

$$dG = (1 + a^2 + b^2)^{-2} \, da \wedge db \wedge dm \wedge dq. \tag{12.71}$$

Two parameter families of lines (called *congruences* of lines) have two invariant densities. The first is the area element of the spherical image of the lines. The second density is more important; it can be written

$$dG^* = -\sin \alpha_1 \, dx \wedge d\alpha_1 - \sin \alpha_2 \, dy \wedge d\alpha_2 \tag{12.72}$$

where α_1, α_2, α_3 are the angles of G with the coordinate axis of a rectangular coordinate system (x, y, z) and x, y are the coordinates of the intersection point of G with the (x, y)-plane. This density was introduced by E. Cartan [85]. See also the work of Vidal Abascal [695] and Pohl [481]. Sulanke [663] has studied in detail the integral geometry of such congruences of lines in E_3. For the generalization to E_n, see the work of Stanilov and Sulanke [623]. Relations between the densities of lines and planes in E_3 were given by Blaschke [51] and Varga [685]. For applications to mineralogy, see [54–56].

ISBN 0-201-13500-0

6. *Distribution of distances in an n-ball.* The probability distribution of the distance r between two points P, Q taken at random inside an n-dimensional ball of radius R has been considered by Deltheil [143], Hammersley [288], Lord [369], and Miles [420]; see also [335]. If $f_n(x)\,dx$ denotes the probability that the distance r will assume a value in the interval x, $x + dx$ and we put $r = 2R\lambda$, we prove that

$$f_n(\lambda) = 2^n n \lambda^{n-1} I_{1-\lambda^2}[(n+1)/2, \tfrac{1}{2}] \tag{12.73}$$

where $I_x(p, q)$ is the complete beta function, defined by

$$I_x(p, q) = \frac{\Gamma(p+q)}{\Gamma(p)\Gamma(q)} \int_0^x t^{p-1}(1-t)^{q-1}\,dt.$$

For $n = 1, 2, 3$ we get

$$f_1(\lambda) = 2(1-\lambda), \quad f_2(\lambda) = 16\lambda\pi^{-1}[\arccos\lambda - \lambda(1-\lambda^2)^{1/2}],$$
$$f_3(\lambda) = 12\lambda^2(1-\lambda)^2(2+\lambda),$$

which correspond to the cases of a segment, a disk, and a three-dimensional ball, respectively.

From (12.73) we can obtain the kth moments of r,

$$\mu_k(n) = \left(\frac{n}{n+k}\right)^2 \frac{\Gamma(n+k+1)\Gamma(n/2)}{\Gamma(n+1+k/2)\Gamma((n+k)/2)} R^k. \tag{12.74}$$

In particular, the first moment or mean value of the distance PQ in an n-dimensional ball of radius R becomes

$$\mu_1(n) = \left(\frac{n}{n+1}\right)^2 \frac{\Gamma(n+2)\Gamma(n/2)}{\Gamma(n+3/2)\Gamma((n+1)/2)} R.$$

When n becomes large it can be shown that the distribution tends to the normal distribution with mean $\sqrt{2}\,R$ and variance $(2n)^{-1}R^2$. It follows that for n large, the distance is almost always nearly equal to $\sqrt{2}\,R$, the distance between the extremities of two orthogonal radii.

A more general result has been given by Miles [420]. Let $P_1, P_2, \ldots, P_i$ be independent random points in the interior of the unit n-ball and let $Q_1, \ldots, Q_j$ be independent random points on its bounding $(n-1)$-sphere. Consider the case in which $2 \leqslant i + j \leqslant n + 1$; in it, the convex hull of the $i + j$ points is almost surely an r-simplex ($r = i + j - 1$) having the points as vertices. Then, the r-dimensional volume T of this random r-simplex has moments

$$E(T^k) = \frac{1}{\kappa_n{}^i O_{n-1}^j} \int T^k\, dP_1 \wedge \cdots \wedge dP_i \wedge dQ_1 \wedge \cdots \wedge dQ_j$$
$$= \left(\frac{1}{r!}\right)^k \left(\frac{n}{n+k}\right)^i \frac{\Gamma[\frac{1}{2}(r+1)(n+k) - j + i]}{\Gamma[\frac{1}{2}(rn+n+rk) - j + 1]}$$

ISBN 0-201-13500-0

$$\cdot \left\{ \frac{\Gamma(n/2)}{\Gamma[(n+k)/2]} \right\}^r \prod_{h=1}^{r-1} \frac{\Gamma[(n-r+k+h)/2]}{\Gamma[(n-r+h)/2]} \qquad (12.75)$$

where $dP_1, \ldots, dP_i$ denote the volume elements in E_n and $dQ_1, \ldots, dQ_j$ the volume elements on the unit $(n-1)$-sphere; κ_n and O_{n-1} are given by (1.22).

7. *Random clusters.* Consider points scattered at random in E_3 in such a way that the probability of a volume V containing exactly s points is $p_s(\lambda, V) = e^{-\lambda V}(\lambda V)^s/s!$ (Poisson process of intensity λ). Two points are said to be connected if they lie within a distance $2R$ of each other. A *cluster* of size n is a set of n points, each of which is connected to at least one other point in the set but none of which are connected to points not contained in the set. Although the distribution of the number of points in a cluster and the expected cluster size depend on the two parameters λ and R, dimensional analysis suggests that they are functions of the one composite parameter $t = \lambda R^3$. Let $p_n(t)$ be the probability distribution of cluster sizes and $E(t)$ the expected cluster size. Roberts and Storey [515] consider the problem of finding the critical value $t = t_c$ at which $E(t)$ becomes infinite and that for values of $t > t_c$ infinite clusters occur. They gave the lower bound $t_c \geqslant 0.0434$ and the Monte Carlo estimate $t_c = 0.0889$. Holcomb, Iwasawa, and Roberts [310] estimate t_c to be 0.070 and Domb [149], by series expansions, suggests the value $t_c = 0.081$.

Tate [668] has also considered clusters of random points in E_3. Let B be the boundary of a right-circular cylinder of length T and cross-sectional area A in such units that $T \geqslant 1$ and $A \geqslant 1$. Let Z, Z' be unit cylinders, right-circular cylinders of unit length and unit cross-sectional area, with axes parallel to the axis of B. Let $\psi_n(\lambda, A, T)$ be the probability that no unit cylinder contains n or more points. A set of k points falling within a unit cylinder is called an aggregate of degree k or k-aggregate. Then, the probability for no aggregates of degree n or higher can be written

$$\psi_n(\lambda, A, T) = \sum_{s=0}^{\infty} e^{-\lambda V} \frac{(\lambda V)^s}{s!} f_n(s, A, T)$$

where $V = AT$ and $f_n(s, A, T)$ is the probability for no aggregate of degree n or higher when B contains exactly s randomly placed points. Analytical expressions for $f_n(s, A, T)$ can be obtained easily only if $A = 1$, $n = 2$, namely, $f_2(s, 1, T) = (1 - (s-1)/T)^s$. In general, $f_n(s, 1, T)$ is a polynome of degree s in T^{-1}. For $n > 2$ the analytical calculation of $\psi_n(\lambda, A, T)$ is impracticable and Tate [668] reported some results based on Monte Carlo method.

ISBN 0-201-13500-0

PART III

Integral Geometry in E_n

CHAPTER 13

Convex Sets in E_n

1. Convex Sets and Quermassintegrale

A set of E_n is called convex if and only if it contains, with each pair of its points, the entire line segment joining them. We shall restrict our attention throughout to convex sets that are bounded and closed. A convex set with nonempty interior is called a *convex body*. The boundary ∂K of a convex body is called a *convex surface*.

We can prove the following properties: (a) A convex surface is almost everywhere differentiable; that is, it has almost everywhere a tangent plane [74, p. 13]. (b) Any convex surface can be approximated by a sequence of analytic convex surfaces and by a sequence of convex polyhedral surfaces [63, p. 36; 195].

A support hyperplane of a convex set K (or a support hyperplane of the convex surface ∂K) is a hyperplane that contains points of K but does not separate any two points of K. If K has interior points, then the support hyperplanes may be defined as those hyperplanes which contain points of ∂K but no interior points of K. Through each boundary point of a convex set there passes at least one support hyperplane. If a boundary point admits only one support hyperplane, it is called a regular point.

Let K be a convex set and let O be a fixed point in E_n. Consider all the $(n-r)$-planes $L_{n-r[0]}$ through O and let K'_{n-r} be the orthogonal projection

ENCYCLOPEDIA OF MATHEMATICS and Its Applications, Gian-Carlo Rota (ed.).
1, Luis A. Santaló, Integral Geometry and Geometric Probability

ISBN 0-201-13500-0

of K into $L_{n-r[0]}$. That is, K'_{n-r} denotes the convex set of all intersection points of $L_{n-r[0]}$ with the r-planes perpendicular to $L_{n-r[0]}$ through each point of K. We want to define the mean value of the volume $V(K'_{n-r})$ of these projected sets.

Equation (12.27) gives the densities $dL_{n-r[0]}$ for $(n-r)$-planes about a fixed point O, that is, the invariant volume element of the Grassmann manifold $G_{n-r,r}$. Then the mean value of the projected volumes $V(K'_{n-r})$ is

$$E(V(K'_{n-r})) = \frac{I_r(K)}{m(G_{n-r,r})} = \frac{O_{r-1}\cdots O_1 O_0}{O_{n-1}\cdots O_{n-r}} I_r(K) \tag{13.1}$$

where $m(G_{n-r,r})$ denotes the volume of the Grassmann manifold $G_{n-r,r}$ given by (12.35) and

$$I_r(K) = \int_{G_{n-r,r}} V(K'_{n-r})\, dL_{n-r[0]} = \int_{G_{r,n-r}} V(K'_{n-r})\, dL_{r[0]} \tag{13.2}$$

is an important characteristic of the convex set K, which we proceed to investigate. Equation (13.2) defines $I_r(K)$ for $r = 1, 2, \ldots, n-1$. For completeness we define

$$I_0(K) = V(K) = \text{volume of } K. \tag{13.3}$$

Formula (12.53) for $r = n-1$ and $i+1 = r$ becomes

$$dL^{n-1}_{r[0]} \wedge dL^{*}_{n-1[0]} = dL_{n-1[r]} \wedge dL_{r[0]} \tag{13.4}$$

and the same formula (12.53) for $n \to n-1$, $i+1 \to r-1$ becomes

$$dL^{r}_{r-1[0]} \wedge dL^{*n-1}_{r[0]} = dL^{n-1}_{r[r-1]} \wedge dL^{n-1}_{r-1[0]}. \tag{13.5}$$

Multiplying both sides of (13.4) by $dL^{r}_{r-1[0]}$ and taking (13.5) into account, we have (up to the sign and noting that an oriented plane is equivalent to two unoriented planes)

$$dL_{n-1[0]} \wedge dL^{n-1}_{r-1[0]} \wedge dL^{n-1}_{r[r-1]} = dL_{r[0]} \wedge dL^{r}_{r-1[0]} \wedge dL_{n-1[r]}. \tag{13.6}$$

Let us multiply by $V(K'_{n-r})$ and perform the integration over all pairs $L_{r[0]}$, $L^{r}_{r-1[0]}$. Since K'_{n-r} is also the orthogonal projection onto $L_{n-r[0]}$ of the convex set K'_{n-1}, orthogonal projection of K onto $L_{n-1[0]}$, and $dL_{n-1[0]} = du_{n-1}$ is the area element of the unit sphere U_{n-1}, on the left-hand side we get $\frac{1}{2}O_{n-r-1}\int I'_{r-1}(K'_{n-1})\,du_{n-1}$, where we have applied that the measure of all $L^{n-1}_{r[r-1]}$ is $\frac{1}{2}O_{n-r-1}$ (12.36) and $I'_{r-1}(K'_{n-1})$ refers to K'_{n-1} as a convex set of $L_{n-1[0]}$. On the right-hand side we get $I_r(K)$ times the measure of all $L^{r}_{r-1[0]}$ (which is equal to $\frac{1}{2}O_{r-1}$) and the measure of all $L_{n-1[r]}$ (which is equal to $\frac{1}{2}O_{n-r-1}$). Therefore, we have the recursion formula

ISBN 0-201-13500-0

$$I_r(K) = \frac{2}{O_{r-1}} \int_{\frac{1}{2}U_{n-1}} I'_{r-1}(K'_{n-1})\, du_{n-1}. \tag{13.7}$$

Note that the unit hemisphere $\frac{1}{2}U_n$ is precisely $G_{1,n-1}$, that is, the manifold of all unoriented lines through the origin O.

Instead of the integrals $I_r(K)$, which are related by (13.2) to the expected values of the projected volumes $V(K'_{n-r})$, it is usual to introduce the so-called *quermassintegrale*, or mean cross-sectional measures, introduced by Minkowski and defined by

$$W_r(K) = \frac{(n-r)O_{n-1}}{nO_{n-r-1}} E(V(K'_{n-r})) = \frac{(n-r)O_{r-1}\cdots O_0}{nO_{n-2}\cdots O_{n-r-1}} I_r(K). \tag{13.8}$$

Then (13.7) becomes

$$W_r(K) = \frac{2(n-1)}{nO_{n-2}} \int_{\frac{1}{2}U_{n-1}} W'_{r-1}(K'_{n-1})\, du_{n-1}, \tag{13.9}$$

which is known as *Kubota's formula* ([348]; see also the work of Bonnesen and Fenchel [63, p. 49], and of Hadwiger [274, p. 209]). For completeness, we put

$$W_0(K) = I_0(K) = V(K), \qquad W_n(K) = O_{n-1}/n, \tag{13.10}$$

where the last formula comes from (13.9) if we assume that W_n is a constant that depends only on n.

It is worth observing that from (13.8) it follows that the *mean breadth* of K is

$$E(V(K_1')) = (2n/O_{n-1})W_{n-1}(K). \tag{13.11}$$

It is easy to show that the quermassintegrale are continuous functionals over the space of convex sets ([274, p. 211]; see also [243]). For the generalization of the quermassintegrale to any measurable set, see [460, 461].

2. Cauchy's Formula

Suppose first that K is a convex polyhedral region. Let f_i be the area of a face of ∂K and let f_i' be the area of its orthogonal projection onto $L_{n-1[0]}$. Then we have $f_i' = |\cos\theta_i| f_i$, where θ_i is the angle between the direction u_{n-1} normal to $L_{n-1[0]}$ and the outer normal to ∂K. Since the sum of all projections is twice the area $V(K'_{n-1})$ of the projection of K, we have

ISBN 0-201-13500-0

$$2V(K'_{n-1}) = \sum_{1}^{m} |\cos \theta_i| f_i \tag{13.12}$$

where m is the number of faces of ∂K.

Multiplying by du_{n-1} and integrating over the unit hemisphere we get

$$2 \int_{\frac{1}{2}U_{n-1}} V(K'_{n-1})\, du_{n-1} = \frac{O_{n-2}}{n-1} F \tag{13.13}$$

where F denotes the $(n-1)$-dimensional surface area of ∂K and we have applied that $\int |\cos \theta_i|\, du_{n-1}$ over $\frac{1}{2}U_{n-1}$ is equal to the projection of $\frac{1}{2}U_{n-1}$ onto a diametral hyperplane, that is, the volume of the $(n-1)$-dimensional unit ball $\kappa_{n-1} = O_{n-2}/(n-1)$.

Equation (13.13) can be written

$$F = \frac{2(n-1)}{O_{n-2}} \int_{\frac{1}{2}U_{n-1}} V(K'_{n-1})\, du_{n-1} = nW_1, \tag{13.14}$$

which is the so-called Cauchy formula for convex polyhedrons. Since any convex set can be expressed as the limit of a sequence of convex polyhedral regions, it follows that the formula also holds for arbitrary convex sets [93, 63, 274]. For a generalization to subsets of the unit n-sphere, see [188].

The case in which $n = 3$. For $n = 3$, if F_u denotes the area of the orthogonal projection of K on a plane perpendicular to the direction u_2, Cauchy's formula can be written

$$F = \frac{1}{\pi} \int_{U_2} F_u\, du_2 \tag{13.15}$$

where F is the surface area of ∂K. The expected value of F_u is

$$E(F_u) = F/4. \tag{13.16}$$

This provides a practical method for measuring the surface area of small convex bodies. If E_{12} and E_{20} denote the means of the projections of K on the faces of a dodecahedron and an icosahedron, respectively, then Moran [424] has shown that $0.2294 \leqslant E_{12}/F \leqslant 0.2696$ and $0.2390 \leqslant E_{20}/F \leqslant 0.2618$. A similar result can be proven for curves. If Γ is a curve of length L, the expected value of the length of the projections on a plane is $E(L_u) = (\pi/4)L$ and the mean values of the projections on the faces of a dodecahedron or icosahedron give approximate values of L. If these mean values are L_{12} and L_{20}, we have $0.229 \leqslant L_{12}/\pi L \leqslant 0.269$ and $0.244 \leqslant L_{20}/\pi L \leqslant 0.252$.

The probability density function $f(F_u)$ (i.e., the function such that the probability that F_u lies in the interval from a to b is given by integrating

ISBN 0-201-13500-0

$f(F_u)$ between these limits) has, in general, an involved expression depending on the shape of K. In some particular cases, however, it can be computed. Let, for instance, K be a right-circular cylinder of height h and radius r. The area of the orthogonal projection onto a plane normal to the direction

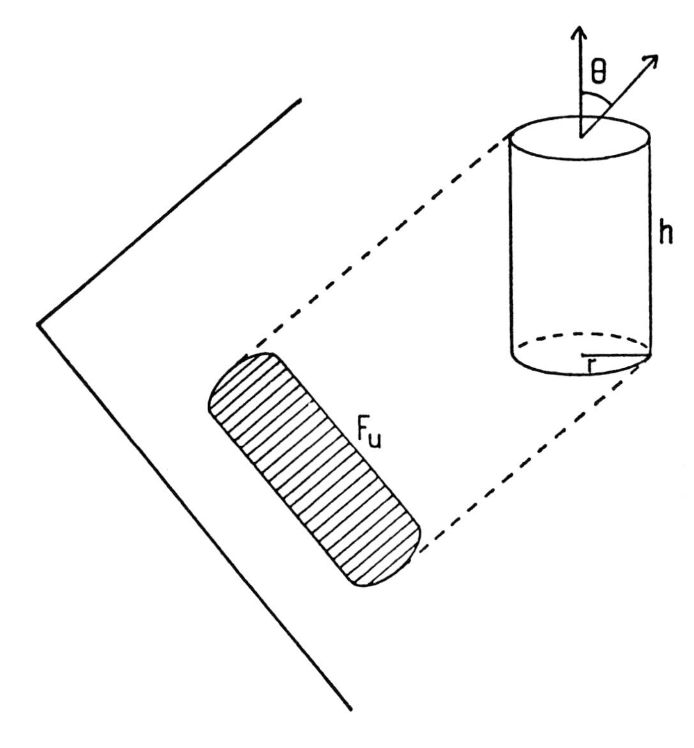

Figure 13.1.

that makes an angle θ with the axis of the cylinder is (Fig. 13.1)

$$F_\theta = \pi r^2 \cos\theta + 2rh \sin\theta. \tag{13.17}$$

The maximum area corresponds to $\theta = \text{arc}\tan(2h/\pi r)$ and has the value $F_M = \pi^2 r^4 + 4h^2 r^2$. The resulting distribution function is

$$f(F_u) = \frac{1}{F_M{}^2}\left[\frac{2rhF_u}{(F_M{}^2 - F_u{}^2)^{1/2}} \pm \pi r^2\right] \tag{13.18}$$

for $\min(\pi r^2, 2rh) < F_u < \max(\pi r^2, 2rh)$, and

$$f(F_u) = \frac{1}{F_M{}^2}\frac{4rhF_u}{(F_M{}^2 - F_u{}^2)^{3/2}} \tag{13.19}$$

for $\max(\pi r^2, 2rh) < F_u < F_M$.

ISBN 0-201-13500-0

The positive sign in (13.18) is used when $\pi r^2 > 2rh$ and the negative sign when $\pi r^2 < 2rh$. Assuming, for instance, $\pi r^2 < 2rh$, we have

$$\int_{\pi r^2}^{F_M} f(F_u)\, dF_u = 1, \qquad E(F_u) = \int_{\pi r^2}^{F_M} F_u f(F_u)\, dF_u = \frac{\pi r^2}{2} + \frac{\pi r h}{2}, \tag{13.20}$$

which is in accord with the result above. The second moment is

$$\int_{\pi r^2}^{F_M} F_u{}^2 f(F_u)\, dF_u = \frac{4rh}{3}(2rh + \pi r^2) + \frac{\pi^2 r^4}{3}. \tag{13.21}$$

These formulas, as well as those for rectangular parallelepipeds, are due to A. G. Walters [707].

3. Parallel Convex Sets; Steiner's Formula

The parallel body K_ρ in the distance ρ of a convex set K is the union of all solid spheres of radius ρ the centers of which are points of K. The boundary ∂K_ρ is called the parallel surface of ∂K in the distance ρ.

We want to prove the following expression for the volume of K_ρ

$$V(K_\rho) = \sum_0^n \binom{n}{i} W_i(K)\rho^i, \tag{13.22}$$

which is valid for any $\rho \geqslant 0$ and is called Steiner's formula for parallel convex sets.

We proceed by induction. For $n = 1$, (13.22) becomes $V(K_\rho) = V(K) + 2\rho$; this is true because K is a line segment of length $V(K)$ and K_ρ is a line segment of length $V(K) + 2\rho$. Assume that (13.22) holds for convex sets of E_{n-1}. Applying it to the convex set K', which is the orthogonal projection of K onto L_{n-1}, we have

$$V'(K_\rho') = \sum_0^{n-1} \binom{n-1}{i} W_i'(K')\rho^i. \tag{13.23}$$

Multiplying both sides of this equality by du_{n-1}, the area element of the unit sphere U_{n-1} at the end point of the unit vector perpendicular to L_{n-1}, and integrating over the whole U_{n-1}, using the formulas of Kubota and Cauchy, we get

$$F(K_\rho) = \sum_0^{n-1} n \binom{n-1}{i} W_{i+1}(K)\rho^i. \tag{13.24}$$

ISBN 0-201-13500-0

On the other hand, we have

$$V(K_\rho) = V(K) + \int_0^\rho F(K_\rho)\, d\rho$$

and thus

$$V(K_\rho) = V(K) + \sum_0^{n-1} \binom{n}{i+1} W_{i+1}(K)\rho^{i+1}.$$

By virtue of (13.10) we can put $W_0(K) = V(K)$ and then the change of indices $i + 1 \to i$ gives the desired formula.

A more general formula can be obtained by simply writing that the parallel body $K_{\rho+\alpha}$ coincides with the parallel body $(K_\rho)_\alpha$ of K_ρ in the distance α. Indeed, we have

$$\sum_0^n \binom{n}{i} W_i(K_\rho)\alpha^i = \sum_0^n \binom{n}{h} W_h(K)(\rho + \alpha)^h \tag{13.25}$$

or

$$\sum_0^n \binom{n}{i} W_i(K_\rho)\alpha^i = \sum_{h=0}^n \binom{n}{h} W_h(K) \sum_{i=0}^h \binom{h}{i} \rho^{h-i}\alpha^i. \tag{13.26}$$

Since this equality holds for any $\alpha \geqslant 0$, we can identify the coefficients of α^i ($i = 0, 1, \ldots, n$) and we get, after substituting $h - i \to j$,

$$W_i(K_\rho) = \sum_{j=0}^{n-i} \binom{n-i}{j} W_{i+j}(K)\rho^j, \tag{13.27}$$

which holds for $i = 0, 1, \ldots, n$.

4. Integral Formulas Relating to the Projections of a Convex Set on Linear Subspaces

Let K be a convex set and let K'_{n-r} be its orthogonal projection onto the $(n - r)$-plane $L_{n-r[0]}$ that passes through the origin O. Formula (13.8) gives the mean volume of K'_{n-r} in terms of $W_r(K)$. We want to find the mean value of the invariants $W_i'(K'_{n-r})$.

To this end we apply (13.8) to the body K_ρ. We have

$$W_r(K_\rho) = \frac{(n-r)O_{r-1}\cdots O_0}{nO_{n-2}\cdots O_{n-r-1}} \int_{G_{r,n-r}} V(K_\rho')_{n-r}\, dL_{r[0]}. \tag{13.28}$$

ISBN 0-201-13500-0

According to (13.27) we have

$$W_r(K_\rho) = \sum_{j=0}^{n-r} \binom{n-r}{j} W_{r+j}(K)\rho^j \tag{13.29}$$

and, applying Steiner's formula to $(K_\rho')_{n-r}$ and using that $(K_\rho')_{n-r} = (K'_{n-r})_\rho$, we get

$$V(K_\rho')_{n-r} = \sum_{j=0}^{n-r} \binom{n-r}{j} W_j'(K'_{n-r})\rho^j. \tag{13.30}$$

Putting (13.30) and (13.29) into (13.28) and equating the coefficients of ρ^j, we have

$$\int_{G_{r,n-r}} W_j'(K'_{n-r})\, dL_{r[0]} = \frac{nO_{n-2}\cdots O_{n-r-1}}{(n-r)O_{r-1}\cdots O_0} W_{r+j}(K), \tag{13.31}$$

which holds for $0 \leqslant j \leqslant n - r \leqslant n - 1$. Consider the cases in which $j = 0, 1$. If $j = 0$, then $W_0' = V(K'_{n-r})$ is the volume of the projection K'_{n-r} and (13.31) coincides with (13.8); that is,

$$\int_{G_{r,n-r}} V(K'_{n-r})\, dL_{r[0]} = \frac{nO_{n-2}\cdots O_{n-r-1}}{(n-r)O_{r-1}\cdots O_0} W_r(K). \tag{13.32}$$

For $j = 1$, according to (13.9) and (13.14) we have $W_1'(K'_{n-r}) = F(K'_{n-r})/(n-r)$, where $F(K'_{n-r})$ denotes the area of $\partial K'_{n-r}$. Then (13.31) becomes

$$\int_{G_{r,n-r}} F(K'_{n-r})\, dL_{r[0]} = \frac{O_{n-2}\cdots O_{n-r-1}}{O_{r-1}\cdots O_0} n W_{r+1}(K). \tag{13.33}$$

5. Integrals of Mean Curvature

If the boundary ∂K of a convex set K is a hypersurface of class C^2, the quermassintegrale W_r can be expressed by means of the integrals of mean curvature of ∂K. It is known that at each point of a hypersurface Σ there are $n-1$ principal directions and $n-1$ principal curvatures $\kappa_1, \kappa_2, \ldots, \kappa_{n-1}$. If $d\sigma$ denotes the area element of Σ, then the rth integral of mean curvature $M_r(\Sigma)$ is defined by

$$M_r(\Sigma) = \binom{n-1}{r}^{-1} \int_\Sigma \{\kappa_{i_1}, \kappa_{i_2}, \ldots, \kappa_{i_r}\}\, d\sigma \tag{13.34}$$

where $\{\kappa_{i_1}, \kappa_{i_2}, \ldots, \kappa_{i_r}\}$ denotes the rth elementary symmetric function of the

ISBN 0-201-13500-0

principal curvatures. The product $\kappa_1\kappa_2\cdots\kappa_{n-1}$ is called the Gauss–Kronecker curvature and is related to the area element du_{n-1} of the spherical image of Σ by the equation

$$\kappa_1\kappa_2\cdots\kappa_{n-1}\,d\sigma = du_{n-1}. \tag{13.35}$$

Recall that the spherical image of Σ is the mapping $\Sigma \to U_{n-1}$, which maps a point P of Σ to the end point of the unit vector through the origin O parallel to the normal vector to Σ at P.

If $R_i = 1/\kappa_i$ $(i = 1, 2, \ldots, n-1)$ are the principal radii of curvature, then (13.34) can be written

$$M_r(\Sigma) = \binom{n-1}{r}^{-1} \int_{U_{n-1}} \{R_{i_1}, R_{i_2}, \ldots, R_{i_{n-r-1}}\}\, du_{n-1} \tag{13.36}$$

where the integration is over the unit sphere U_{n-1}.

If Σ is a compact, orientable hypersurface of class C^2 and $n-1$ is even, then $M_{n-1}(\Sigma)$ is equal to the total volume of the spherical image of Σ and is a topological invariant of Σ that is related to the Euler–Poincaré characteristic $\chi(\Sigma)$ by

$$M_{n-1}(\Sigma) = \tfrac{1}{2}O_{n-1}\chi(\Sigma) \qquad (n-1 \text{ even}). \tag{13.37}$$

If Σ is the boundary of a domain D, for every n we have

$$M_{n-1}(\partial D) = O_{n-1}\chi(D). \tag{13.38}$$

For instance, if D is a topological ball, in particular if D is a convex body, it is known that $\chi(D) = 1$ and we have

$$M_{n-1}(\partial D) = O_{n-1}. \tag{13.39}$$

Finally, recall that if a compact differentiable manifold Σ_m of dimension m has a finite simplicial decomposition and α_i denotes the number of simplexes of dimension i, then

$$\chi(\Sigma_m) = \alpha_0 - \alpha_1 + \alpha_2 - \cdots + (-1)^m\alpha_m. \tag{13.40}$$

Applying this formula to a simplex of dimension n, we easily get that for any topological ball D in E_n we have

$$\chi(D) = 1, \qquad \chi(\partial D) = 1 - (-1)^n. \tag{13.41}$$

For an axiomatic definition of the Euler–Poincaré characteristic see the work of Hadwiger [272, 274], Groemer [243], and Lenz [363].

ISBN 0-201-13500-0

6. Integrals of Mean Curvature and Quermassintegrale

Let K be a convex set and assume that ∂K is a hypersurface of class C^2. If R_i are the principal radii of curvature of ∂K, then the principal radii of curvature of ∂K_ρ are $R_i + \rho$ and the area element $d\sigma_\rho$ of ∂K_ρ becomes (using (13.35))

$$d\sigma_\rho = (R_1 + \rho)(R_2 + \rho)\cdots(R_{n-1} + \rho)\, du_{n-1}. \tag{13.42}$$

Hence, the area of ∂K_ρ is

$$F(K_\rho) = \int_{U_{n-1}} (R_1 + \rho)\cdots(R_{n-1} + \rho)\, du_{n-1}$$

$$= \sum_{r=0}^{n-1} \int \{R_{i_1},\ldots, R_{i_{n-1-r}}\}\rho^r\, du_{n-1} = \sum_{r=0}^{n-1} \binom{n-1}{r} M_r(\partial K)\rho^r. \tag{13.43}$$

For the volume of K_ρ we have

$$V(K_\rho) = V(K) + \int_0^\rho F(K_\rho)\, d\rho = V(K) + \sum_{r=0}^{n-1} \frac{\rho^{r+1}}{r+1}\binom{n-1}{r} M_r(\partial K). \tag{13.44}$$

Comparison of this expression with Steiner's formula (13.22) gives

$$M_r(\partial K) = nW_{r+1}(K). \tag{13.45}$$

Note that $W_{r+1}(K)$ is well defined for any convex set, whereas the definition (13.34) of $M_r(\partial K)$ makes sense only if ∂K is of class C^2. However, we can define $M_r(\partial K)$ by equality (13.45), as we shall do from now on, so that we will have a well-defined integral of mean curvature $M_r(\partial K)$, $r = 0, 1, 2, \ldots, n-1$, for any convex set K. Conversely, equality (13.45) can be used for computing $W_{r+1}(K)$. Indeed, the parallel convex body K_ρ has a smooth boundary ∂K_ρ and $M_r(\partial K_\rho)$ can often be computed directly from (13.34), in which case the limit of $M_r(\partial K_\rho)$ as $\rho \to 0$ yields $nW_{r+1}(K)$. We collect some values of $W_r(K)$ for some particular convex sets of E_n.

(a) *Ball of radius R*

$$W_r(\text{ball}) = (O_{n-1}/n)R^{n-r}, \qquad r = 0, 1, \ldots, n. \tag{13.46}$$

(b) *Right parallelepiped of edges a_i ($i = 1, 2, \ldots, n$)*

$$W_r(\text{right parallelepiped}) = \frac{O_{r-1}}{r\binom{n}{r}}\{a_{i_1}, \ldots, a_{i_{n-r}}\} \tag{13.47}$$

ISBN 0-201-13500-0

where $\{a_{i_1}, a_{i_2}, \ldots, a_{i_{n-r}}\}$ is the $(n-r)$th elementary symmetric function of the a_i. For instance, for a cube of edge a we have

$$W_r(\text{cube}) = (O_{r-1}/r)a^{n-r}. \tag{13.48}$$

(c) *Line segment of length s.* For $n \geqslant 2$ we have

$$W_i = 0 \quad (i = 0, 1, \ldots, n-2), \qquad W_{n-1} = \frac{O_{n-2}}{n(n-1)}\, s, \qquad W_n = \frac{O_{n-1}}{n}. \tag{13.49}$$

For $n = 1$ we have

$$W_0 = s, \qquad W_1 = 2. \tag{13.50}$$

(d) *Convex cylinders.* Let A_p, B_{n-p} be convex domains contained respectively in two orthogonal planes L_p, L_{n-p} of E_n. If $x \in A_p$ and $y \in B_{n-p}$, the set of all points of the form $x + y$ constitutes the convex set $A_p \times B_{n-p}$, called the Minkowski sum of A_p and B_{n-p}. Then we have

$$\begin{aligned} &W_i(A_p \times B_{n-p}) \\ &= \frac{O_{i-1}}{i\binom{n}{i}} \sum_{\nu=1}^{i-1} \frac{\nu(i-\nu)}{O_{\nu-1}O_{i-\nu-1}} \binom{p}{\nu}\binom{n-p}{i-\nu} W_\nu^{(p)}(A_p) W_{i-\nu}^{(n-p)}(B_{n-p}) \\ &\quad + \frac{1}{\binom{n}{i}}\left[\binom{n-p}{i} W_0^{(p)}(A_p) W_i^{(n-p)}(B_{n-p}) + \binom{p}{i} W_i^{(p)}(A_p) W_0^{(n-p)}(B_{n-p})\right] \end{aligned} \tag{13.51}$$

where $W_\nu^{(p)}(A_p)$ denotes the νth quermassintegral of A_p as a convex set of L_p and similarly $W_{i-\nu}^{(n-p)}(B_{n-p})$ denotes the $(i-\nu)$th quermassintegral of B_{n-p} as a convex set of L_{n-p}.

In particular, if $p = 1$ and A_1 is a line segment of length h and B_{n-1} is an $(n-1)$-dimensional ball of radius a, then (13.51) are the quermassintegrale of the cylinder of revolution of radius a and height h; that is,

$$W_i(\text{cylinder}) = \frac{O_{n-2}}{n(n-1)}\left[\frac{(i-1)O_{i-1}}{O_{i-2}}\, a^{n-i} + (n-i)a^{n-i-1}h\right] \tag{13.52}$$

for $i \geqslant 2$ and

$$W_0 = \frac{O_{n-2}}{n-1}\, a^{n-1}h, \qquad W_1 = \frac{O_{n-2}}{n(n-1)}(2a^{n-1} + (n-1)a^{n-2}h). \tag{13.53}$$

ISBN 0-201-13500-0

(e) *Ellipsoid of revolution.* If K is an ellipsoid of revolution of semiaxis a and λa (a = radius of the equator), we have

$$W = (O_{n-1}/n)\lambda^{r+1}a^{n-r}F[(n+1)/2, r/2, n/2; 1-\lambda^2] \tag{13.54}$$

where F is the hypergeometric function. The preceding values are from Hadwiger's book [274, pp. 220–221].

(f) It is worth noting that using (13.45) and (13.27), we obtain, as the integrals of mean curvature of the parallel surface ∂K_ρ to ∂K in the distance ρ,

$$M_i(\partial K_\rho) = \sum_{j=0}^{n-i-1} \binom{n-i-1}{j} M_{i+j}(\partial K)\rho^j, \qquad j = i = 0, 1, \ldots, n-1. \tag{13.55}$$

(g) *The case in which* $n = 3$. For convex bodies K in E_3, the integrals of mean curvature are $M_0 = F$ (area of ∂K); $M_1 = M$ (integral of mean curvature of ∂K); $M_2 = 4\pi$. Then, using (13.45), (13.11), and (13.33), we easily get

$$M = \frac{1}{2\pi}\int_{U_2} L_u\, du_2 = \frac{1}{2}\int_{U_2} \Delta\, du_2 \tag{13.56}$$

where L_u denotes the perimeter of the orthogonal projection of K onto a plane perpendicular to the direction u_2 and Δ is the breadth of K in the direction u_2, that is, the distance between the two support planes of K that are perpendicular to u_2 and that contain K between them. Thus we have the mean values

$$E(L_u) = M/2, \qquad E(\Delta) = M/2\pi. \tag{13.57}$$

It follows from the last equation that M is equal, up to the factor 2π, to the mean distance between parallel support planes and in technical books is also referred to as "the mean caliper diameter" [166].

The method of computing $M(K)$ by first computing $M(K_\rho)$ and then taking the limit as $\rho \to 0$ easily yields the following results.

(a) For a convex polyhedron the edges of which have lengths a_i with corresponding dihedral angles α_i, we have

$$M = \tfrac{1}{2}\sum (\pi - \alpha_i)a_i \tag{13.58}$$

where the sum is extended over all edges of ∂K.

(b) For a right cylinder of revolution of height h and radius r we have

$$M = \pi h + \pi^2 r. \tag{13.59}$$

In particular, for a line segment of length h ($r = 0$) we have $M = \pi h$.

ISBN 0-201-13500-0

(c) For a plane convex set, considered as a flattened convex body of E_3, we have

$$M = (\pi/2)L \tag{13.60}$$

where L is the perimeter of the convex set.

The method of computing M for the parallel surface in the distance ρ and then taking the limit as $\rho \to 0$ can also be applied to the boundary of nonconvex sets. For instance, for a right cylinder Z of height h that has a

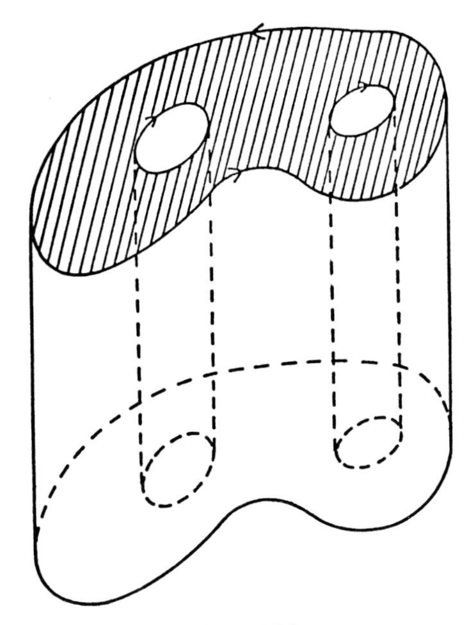

Figure 13.2.

cross section of perimeter L and total curvature c (in Fig. 13.2, $c(Z) = -2\pi$), we have

$$M(Z) = \tfrac{1}{2}(\pi L + ch). \tag{13.61}$$

7. Integrals of Mean Curvature of a Flattened Convex Body

Let K^r be a convex body in the r-plane $L_r \subset E_n$. The boundary ∂K^r, as a set of L_r, is an $(r-1)$-dimensional manifold of L_r that we assume twice differentiable. Let $M_q^{(r)}(\partial K^r)$ $(q = 0, 1, \ldots, r-1)$ be the mean curvature integrals of ∂K^r as a convex surface of L_r. If we consider K^r as a flattened convex body of E_n $(n > r)$, the integrals of mean curvature $M_q^{(n)}(\partial K^r)$ can be evaluated by considering first the mean curvatures of the convex body

ISBN 0-201-13500-0

K_ε^r parallel to K^r in a distance ε (as a convex set of E_n) and then taking the limit as $\varepsilon \to 0$. We state without proof the following result, which is not difficult to obtain (see [574, 274]).

(i) If $q \geqslant n - r$, then

$$M_q^{(n)}(\partial K^r) = \frac{\binom{r-1}{q-n+r}}{\binom{n-1}{q}} \frac{O_q}{O_{q-n+r}} M_{q-n+r}^{(r)}(\partial K^r), \qquad q \geqslant n - r. \tag{13.62}$$

(ii) If $q = n - r - 1$, then

$$M_{n-r-1}^{(n)}(\partial K^r) = \binom{n-1}{n-r-1}^{-1} O_{n-r-1} V_r(K^r) \tag{13.63}$$

where $V_r(K^r)$ denotes the r-dimensional volume of K^r.

(iii) If $q < n - r - 1$, then

$$M_q^{(n)}(\partial K^r) = 0, \qquad q < n - r - 1. \tag{13.64}$$

From these mean curvature integrals we pass to the quermassintegrale by (13.45).

Example 1. For the ordinary space E_3 we have the following possibilities.

(a) $r = 1$. K^1 reduces to a line segment of length s. The mean curvature integrals are

$$M_0^{(3)} = 0, \qquad M_1^{(3)} = \pi s, \qquad M_2^{(3)} = 4\pi. \tag{13.65}$$

(b) $r = 2$. K^2 is a plane convex domain. Let u be its perimeter and f its area. We get

$$M_0^{(3)} = 2f, \qquad M_1^{(3)} = (\pi/2)M_0^{(2)} = (\pi/2)u, \qquad M_2^{(3)} = 2M_1^{(2)} = 4\pi. \tag{13.66}$$

Example 2. For a ball of radius R and dimension r, considered as a convex body of E_n, we have

$$M_q^{(n)} = \frac{\binom{r-1}{q-n+r}}{\binom{n-1}{q}} \frac{O_q}{O_{q-n+r}} \frac{O_{r-1}}{r} R^{n-q-1} \tag{13.67}$$

for $q = n - r, n - r - 1, \ldots, n - 1$ and

ISBN 0-201-13500-0

$$M_q^{(n)} = \frac{1}{r\binom{n-1}{q}} O_{n-r-1} O_{r-1} \qquad \text{for} \quad q = n - r - 1. \tag{13.68}$$

For $q < n - r - 1$ we have $M_q^{(n)} = 0$.

Example 3. For an r-dimensional cube of edges a, as a convex set of E_n, we have

$$M_q^{(n)} = \frac{r\binom{r-1}{q-n+r}}{\binom{n-1}{q}} \frac{O_q}{q-n+r+1} a^{n-q-1} \qquad \text{for} \quad q = n-r, \ldots, n-1, \tag{13.69}$$

and

$$M_{n-r-1}^{(n)} = \binom{n-1}{r}^{-1} O_{n-r-1} a^r. \tag{13.70}$$

For $q < n - r - 1$ we have $M_q^{(n)} = 0$.

Example 4. For a line segment of length s in E_n we have $M_{n-1}^{(n)} = O_{n-1}$, $M_q^{(n)} = 0$ for $q = 0, 1, \ldots, n-3$, and

$$M_{n-2}^{(n)} = \frac{O_{n-2}}{n-1} s. \tag{13.71}$$

8. Notes

1. *Volume, surface area, and integral of mean curvature for convex bodies of various shapes in E_3.* (a) For regular polyhedrons whose circumsphere has the radius R, we have the following values volume V, surface area F, and mean curvature integral M, for the [270].

Tetrahedron

$$V = (8\sqrt{3}/27)R^3, \qquad F = (8\sqrt{3}/3)R^2, \qquad M = 2\sqrt{6}\, R \arccos(-1/3).$$

Cube

$$V = (8\sqrt{3}/9)R^3, \qquad F = 8R^2, \qquad M = 2\sqrt{3}\,\pi R.$$

Octahedron

$$V = (4/3)R^3, \qquad F = 4\sqrt{3}\, R^2, \qquad M = 6\sqrt{2}\, R \arccos(1/3).$$

Dodecahedron

$$V = (2/9)\sqrt{15}(\sqrt{5} + 1)R^3, \qquad F = (200 - 40\sqrt{5})^{1/2} R^2,$$

ISBN 0-201-13500-0

$$M = 5\sqrt{3}(\sqrt{5} - 1)R \text{ arc tan } 2.$$

Icosahedron

$$V = (40 + 8\sqrt{5})^{1/2}R^3/3, \qquad F = \sqrt{3}(10 - 2\sqrt{5})R^2,$$

$$M = (450 - 90\sqrt{5})^{1/2}R \text{ arc sin}(2/3).$$

(b) For the rectangular parallelepiped whose concurrent edges have lengths a_1, a_2, a_3, we have

$$V = a_1a_2a_3, \qquad F = 2(a_1a_2 + a_2a_3 + a_1a_3), \qquad M = \pi(a_1 + a_2 + a_3).$$

(c) For the cone of revolution having height h and radius R we have

$$V = (1/3)\pi R^2 h, \qquad F = \pi R^2 + \pi R(h^2 + R^2)^{1/2},$$

$$M = \pi^2 R + \pi h - \pi R \text{ arc tan}(h/R).$$

(d) For the cylinder of height h and radius R,

$$V = \pi R^2 h, \qquad F = 2\pi R(h + R), \qquad M = \pi(h + \pi R).$$

(e) For the hemisphere of radius R,

$$V = (2/3)\pi R^3, \qquad F = 2\pi R^2, \qquad M = 2\pi R(1 + \pi/4).$$

(f) For the ellipsoid of revolution with semiaxes a, a, λa, we have, for $0 < \lambda < 1$,

$$V = (4/3)\pi\lambda a^3, \quad F = 2\pi[1 + \lambda^2(1 - \lambda^2)^{-1/2} \log\{[1 + (1 - \lambda^2)^{1/2}]\lambda^{-1}\}]a^2,$$

$$M = 2\pi[\lambda + (1 - \lambda^2)^{-1/2} \text{ arc cos } \lambda]a,$$

and for $1 < \lambda < \infty$

$$V = (4/3)\pi\lambda a^3, \qquad F = 2\pi[1 + \lambda^2(\lambda^2 - 1)^{-1/2} \text{ arc cos}(1/\lambda)]a^2,$$

$$M = 2\pi\{\lambda + (\lambda^2 - 1)^{-1/2} \log[\lambda + (\lambda^2 - 1)^{1/2}]\}a.$$

(g) The values of some other geometric quantities relating to convex bodies in E_3 and upper and lower estimates for them may be found in [490].

2. *Integral formulas for convex bodies in* E_3. Let K be a convex body in E_3. Denote by G_s the secant lines of K and by G_e the lines that are exterior to K. Let $d\sigma_1$, $d\sigma_2$ denote the area elements of ∂K at the intersection points P_1 and P_2 of G_s with K. Let s denote the length of the chord $G_s \cap K$, ω the angle between the support planes of K through G_e, and α_1, α_2 the angles between G_s and the outer normals of ∂K at P_1, P_2, respectively. Finally, let t_1, t_2 denote the distances from P_1, P_2 to G_e and k_1, k_2 the Gaussian curvature of ∂K at P_1, P_2. Then the following relations hold.

$$dG_s = \frac{\cos\alpha_1 \cos\alpha_2}{s^2} d\sigma_1 \wedge d\sigma_2, \qquad dG_e = \frac{k_1k_2t_1t_2}{\sin^2\omega} d\sigma_1 \wedge d\sigma_2,$$

ISBN 0-201-13500-0

$$dG_e = \frac{k_1 k_2 t_1 t_2}{\cos\alpha_1 \cos\alpha_2 \sin^2\omega} dG_s, \qquad dG_e = \frac{k_1 k_2 s^4}{\sin^4\omega} dG_s.$$

From these expressions the following integral formulas can be proved, where the integrals are taken over all G_e exterior to K and all G_s that meet K:

$$\int \frac{\sin^m \omega}{t_1 t_2} dG_e = 4\pi^{5/2} \frac{\Gamma(m/2)}{\Gamma((m+1)/2)},$$

$$\int (2\omega - \sin 2\omega)\left(\frac{1}{t_1} + \frac{1}{t_2}\right) dG_e = 16\pi M,$$

$$\int \frac{\sin^2\omega}{k_1 k_2 t_1 t_2} dG_e = \left(\frac{\pi}{2}\right) F^2, \qquad \int \frac{\sin^4\omega}{k_1 k_2 s^4} dG_e = \left(\frac{\pi}{2}\right) F,$$

$$\int \frac{\sin^4\omega}{k_1 k_2 s^3} dG_e = 2\pi V, \qquad \int \frac{\sin^4\omega}{k_1 k_2} dG_e = 6V^2,$$

$$\int \frac{k_1 k_2 s^2}{\cos\alpha_1 \cos\alpha_2} dG_s = 8\pi^2, \qquad \int \frac{s^2}{\cos\alpha_1 \cos\alpha_2} dG_s = \frac{F^2}{2},$$

$$\int \frac{(k_1 + k_2)s^2}{\cos\alpha_1 \cos\alpha_2} dG_s = 4\pi F, \qquad \int \left(\frac{f_1}{\cos\alpha_1} + \frac{f_2}{\cos\alpha_2}\right) dG_s = 2\pi \int_{\partial K} f\, d\sigma,$$

$$\int \left(\frac{f_1}{\cos\alpha_1} + \frac{f_2}{\cos\alpha_2}\right) s^3\, dG_s = 3V \int_{\partial K} f\, d\sigma,$$

where f_1, f_2 are the values of the function f, defined on ∂K, at P_1, P_2. See [395, 397, 546].

3. *An inequality of* Ambarcumjan. Let G be a line that intersects the convex body K. Let s be the length of the chord $G \cap K$ and let $d\sigma_1, d\sigma_2$ denote the area elements of ∂K at the intersection points $G \cap \partial K$. Then Ambarcumjan [7] has proved the following inequality (for any integer $n > 0$)

$$\int_{G \cap K \neq \emptyset} s^n\, dG \leqslant \frac{n}{\pi(n+2)} \int_{\partial K \times \partial K} s^{n-2}\, d\sigma_1 \wedge d\sigma_2.$$

For $n = 2$ we get $I_2 \leqslant F^2/2\pi$.

4. *Support flats to convex bodies*. Imagine a family of q-planes in E_n constructed in the following way: to each q-plane L_q tangent to the unit sphere U_{n-1}, centered at the origin O, we make correspond a translate L_q' in the L_{q+1} containing L_q and O. What conditions are necessary and sufficient for the family $\{L_q'\}$ to be the full set of support q-planes to some convex body K? We say that L_q' supports K if $L_q' \cap K$ is a nonempty subset of ∂K. The cases in which $q = 0$, $q = n - 1$ are classical [63, p. 26]. Firey [194]

ISBN 0-201-13500-0

has given the answer as a set of conditions on the distance r of L_q' from O. Calling $r/(n-q)^{1/2} = h_q$ the q-support function of K, Firey has also given formulas for the quermassintegrale of K in terms of h_q.

5. *Isoperimetric inequalities of* Chakerian. (a) Let K be a convex body in E_n and let $\Delta(\mathbf{u})$ be the breadth of K in the direction $\mathbf{u}$. Let $\mathbf{u}_1, \ldots, \mathbf{u}_p$ ($p \geqslant 1$) be fixed unit vectors that form a p-frame, not necessarily orthogonal, which we assume moving about the origin O with the kinematic density $d\mathfrak{M}_{[0]}$ (12.7). Consider the average of the product of the breadths $\Delta(\mathbf{u}_i)$, $J_p = (2O_1 \cdots O_{n-1})^{-1} \int \Delta(\mathbf{u}_1) \cdots \Delta(\mathbf{u}_p)\, d\mathfrak{M}_{[0]}$, where the integration is extended over all of the group $\mathfrak{M}_{[0]}$ (Chapter 12, Section 1). Chakerian [98] has proved the following inequalities: (i) $J_p \geqslant 2^p n W_{n-p} O_{n-1}^{-1}$ ($1 \leqslant p \leqslant n$); (ii) $J_p \geqslant \binom{p}{n} 2^p n I_{p-n} O_{n-1}^{-1}$ ($p \geqslant n$), where W_{n-p} is the $(n-p)$th quermassintegral of K and I_h is defined by $\int_K r^h\, dP$, where r is the distance from the point $P \in K$ to ∂K. For $p > 1$, equality holds in (i) and (ii) if and only if K is an n-dimensional sphere.

(b) Let K be a convex body in E_n, with volume V, surface area F, and ith quermassintegral W_i. Let K_u be the projection of K onto a hyperplane L_{n-1} perpendicular to the direction $\mathbf{u}$. Let σ_u be the $(n-1)$-dimensional volume of K_u and let λ_u be one half the mean breadth of K_u relative to L_{n-1}. Chakerian [99] has established the inequalities

$$2W_{n-1} \geqslant W_{n-2}\lambda_u^{-1} + n^{-1} O_n \lambda_u,$$

$$F \geqslant (n-1)\left[\frac{(n-1)\sigma_u}{O_{n-1}}\right]^{-1/(n-1)} V + \left[\frac{(n-1)O_n}{nO_{n-1}}\right]\sigma_u.$$

Additional inequalities can be derived by integration over the unit sphere. For $n = 2$ the inequalities are due to D. C. Benson [26].

ISBN 0-201-13500-0

CHAPTER 14

Linear Subspaces, Convex Sets, and Compact Manifolds

1. Sets of r-Planes That Intersect a Convex Set

Let K be a convex set in E_n. We want to find the measure of all r-planes L_r that intersect K. Using expression (12.38) for dL_r and taking (13.2) into account, we see that this measure is equal to $I_r(K)$ or, in terms of the quermass-integrale (13.8),

$$m(L_r; L_r \cap K \neq 0) = \int_{L_r \cap K \neq 0} dL_r = \frac{nO_{n-2}O_{n-3}\cdots O_{n-r-1}}{(n-r)O_{r-1}\cdots O_1 O_0} W_r(K). \qquad (14.1)$$

Thus we can state

The measure of all r-planes that intersect a convex set in E_n is given by (14.1) *for* $r = 1, 2, \ldots, n-1$.

For $r = 0$, the measure of points of K is equal to the volume $V(K)$. Using relation (13.45) we can also write

$$m(L_r; L_r \cap K \neq 0) = \frac{O_{n-2}\cdots O_{n-r-1}}{(n-r)O_{r-1}\cdots O_0} M_{r-1}(\partial K). \qquad (14.2)$$

If K is a convex body contained in an L_q, then in (14.1) and (14.2) the terms W_r and M_{r-1} must be replaced by the values given in Section 7 of Chapter 13.

For $n = 3$, we have the cases in which $m(L_1; L_1 \cap K \neq 0) = (\pi/2)F$, $m(L_2; L_2 \cap K \neq 0) = M_1$, which are the measures of lines and planes that meet a convex set in E_3. For a direct proof see [586, 335, 144].

ENCYCLOPEDIA OF MATHEMATICS and Its Applications, Gian-Carlo Rota (ed.).
1, Luis A. Santaló, Integral Geometry and Geometric Probability

ISBN 0-201-13500-0

Next we want to compute the integrals of the quermassintegrale $W_i^{(r)}(K \cap L_r)$ over all L_r that intersect K. To this end we shall use formula (12.52), that is,

$$dL_{i+1}^{(r)} \wedge dL_r^* = dL_{r[i+1]} \wedge dL_{i+1}, \qquad i + 1 \leqslant r. \tag{14.3}$$

Consider the integral

$$I = \int_{L_{i+1}^{(r)} \cap K \neq \emptyset} dL_{i+1}^{(r)} \wedge dL_r^*. \tag{14.4}$$

Keeping L_r fixed and integrating $dL_{i+1}^{(r)}$, we get, using (14.1),

$$I = \frac{rO_{r-2} \cdots O_{r-i-2}}{(r-i-1)O_i \cdots O_1 O_0} \int_{L_r \cap K \neq \emptyset} W_{i+1}^{(r)}(K \cap L_r)\, dL_r^*. \tag{14.5}$$

On the other hand, by (14.3) and (12.36) we have

$$I = \int_{L_{i+1} \cap K \neq \emptyset} dL_{i+1} \int_{\text{Total}} dL_{r[i+1]} = \frac{2O_{n-i-2} \cdots O_{n-r}}{O_{r-i-2} \cdots O_0} \int_{L_{i+1} \cap K \neq \emptyset} dL_{i+1} \tag{14.6}$$

where the factor 2 arises from the fact that L_r is assumed oriented. Thus, by (14.1) and (14.5), assuming L_r unoriented, we have

$$\int_{L_r \cap K \neq \emptyset} W_{i+1}^{(r)}(K \cap L_r)\, dL_r = \frac{(r-i-1)O_{n-i-2}O_{n-2} \cdots O_{n-r} n}{(n-i-1)O_{r-i-2}O_{r-2} \cdots O_1 O_0 r} W_{i+1}(K), \tag{14.7}$$

or, according to the identity

$$2\pi O_{i-2} = (i-1)O_i, \tag{14.8}$$

we can write

$$\int_{L_r \cap K \neq \emptyset} W_{i+1}^{(r)}(K \cap L_r)\, dL_r = \frac{O_{n-2} \cdots O_{n-r}}{O_{r-2} \cdots O_0} \frac{O_{n-i}}{O_{r-i}} \frac{n}{r} W_{i+1}(K). \tag{14.9}$$

Note that for $i = r - 1$, according to (13.10) we have $W_r^{(r)} = O_{r-1}/r$ and (14.9) coincides with (14.1).

In terms of the integrals of mean curvature, taking (13.45) into account, we can write (14.9) in the form

$$\int_{L_r \cap K \neq \emptyset} M_i^{(r)}(\partial(K \cap L_r))\, dL_r = \frac{O_{n-2} \cdots O_{n-r}}{O_{r-2} \cdots O_0} \frac{O_{n-i}}{O_{r-i}} M_i(\partial K). \tag{14.10}$$

ISBN 0-201-13500-0

The formulas in this section were given by Santaló [565, 572] and Hadwiger [274].

2. Geometric Probabilities

The preceding formulas give the solution of the following typical problems in geometrical probability.

1. *Let K_0, K_1 be convex sets in E_n such that $K_1 \subset K_0$. The probability that a randomly chosen r-plane L_r ($r = 1, 2, \ldots, n-1$) that meets K_0 also meets K_1 is*

$$p(L_r \cap K_1 \neq \emptyset) = \frac{W_r(K_1)}{W_r(K_0)} = \frac{M_{r-1}(\partial K_1)}{M_{r-1}(\partial K_0)}. \tag{14.11}$$

The result is an immediate consequence of (14.1) and (14.2).

2. *Let K be a convex set in E_n. Assume that $p + q \geqslant n$, and let L_p and L_q be two independent random subspaces both intersecting K. Find the probability that $L_p \cap L_q$ intersects K.*

Solution. We need the value of the integral

$$m(L_p, L_q; L_p \cap L_q \cap K \neq \emptyset) = \int_{L_p \cap L_q \cap K \neq \emptyset} dL_p \wedge dL_q. \tag{14.12}$$

Fixing L_q and integrating over all L_p that meet $L_q \cap K$, we have

$$m = \frac{O_{n-2} \cdots O_{n-p-1}}{(n-p) O_{p-1} \cdots O_0} \int_{L_q \cap K \neq \emptyset} M_{p-1}(L_q \cap \partial K)\, dL_q. \tag{14.13}$$

In order to compute the last integral we distinguish two cases:
(i) $p + q \geqslant n + 1$. According to (13.62) we have

$$M_{p-1} = \frac{\binom{q-1}{p+q-n-1}}{\binom{n-1}{p-1}} \frac{O_{p-1}}{O_{p+q-n-1}} M^{(q)}_{p+q-n-1} \tag{14.14}$$

and by (14.10)

$$\int_{L_q \cap K \neq \emptyset} M^{(q)}_{p+q-n-1}(L_q \cap \partial K)\, dL_q = \frac{O_{n-2} \cdots O_{n-q}}{O_{q-2} \cdots O_0} \frac{O_{2n-p-q+1}}{O_{n-p+1}} M_{p+q-n-1}(\partial K). \tag{14.15}$$

ISBN 0-201-13500-0

Substituting (14.14) and (14.15) in (14.13), we get the measure of all pairs L_p, L_q such that $L_p \cap L_q \cap K \neq \emptyset$. The measure of all pairs L_p, L_q that intersect K is equal to the product of measures $m(L_p; L_p \cap K \neq \emptyset)$ and $m(L_q; L_q \cap K \neq \emptyset)$ given by (14.2). By division we get the desired probability

$$p(L_p \cap L_q \cap K \neq \emptyset \text{ when } p + q > n)$$
$$= \frac{2(p-1)!(q-1)!O_{p-1}O_{q-1}O_{2n-p-q+1}}{(p+q-n-1)!(n-1)!O_{n-p+1}O_{n-q+1}O_{p+q-n-1}} \frac{M_{p+q-n-1}(\partial K)}{M_{p-1}(\partial K)M_{q-1}(\partial K)}. \tag{14.16}$$

(ii) $p + q = n$. In this case, (13.63) gives

$$M_{p-1} = \binom{n-1}{p-1}^{-1} O_{n-q-1}\sigma_q(L_q \cap K) \tag{14.17}$$

where $\sigma_q(L_q \cap K)$ indicates the q-dimensional volume of $L_q \cap K$. Moreover, using expression (12.38) for dL_q and since the integral of $\sigma_q(L_q \cap K)\, d\sigma_{n-q}$ over the projection K'_{n-q} of K onto $L_{n-q[0]}$ is the volume $V(K)$ of K, we have

$$\int_{L_q \cap K \neq \emptyset} \sigma_q(L_q \cap K)\, dL_q = V(K) \int_{G_{n-q,q}} dL_{n-q[0]} = \frac{O_{n-1} \cdots O_{n-q}}{O_{q-1} \cdots O_0} V(K). \tag{14.18}$$

Therefore in this case we have

$$m(L_p, L_q; L_p \cap L_q \cap K \neq \emptyset)$$
$$= \frac{O_{n-2} \cdots O_{n-p-1}}{(n-p)O_{p-1} \cdots O_0} \frac{O_{n-q-1}}{\binom{n-1}{p-1}} \frac{O_{n-1} \cdots O_{n-q}}{O_{q-1} \cdots O_0} V(K) \tag{14.19}$$

and the desired probability becomes

$$p(L_p \cap L_q \cap K \neq \emptyset \text{ when } p + q = n) = \frac{p!q!O_{n-1}V(K)}{(n-1)!M_{p-1}(\partial K)M_{q-1}(\partial K)}. \tag{14.20}$$

Examples. (a) For $p = 1$, $q = n - 1$, we have

$$p = \frac{O_{n-1}V(K)}{F(\partial K)M_{n-2}(\partial K)} \tag{14.21}$$

where $F(\partial K)$ denotes the surface area of ∂K.

(b) If K is a ball, (14.21) gives $p = 1/n$ and we can state that *the probability of a hyperplane and a line that intersect a ball having an intersection inside the ball is* $1/n$.

ISBN 0-201-13500-0

3. The method above gives the solution of the following general problem: Given h randomly chosen subspaces L_{r_i} ($i = 1, 2, \ldots, h$), such that $r_1 + r_2 + \cdots + r_h \geqslant (h-1)n$, that intersect a convex set K, find the probability that $L_{r_1} \cap L_{r_2} \cap \cdots \cap L_{r_h} \cap K \neq \emptyset$.

The general formula is involved, but the method above easily gives the solution for each particular case. For instance, the probability that three planes in E_3 intersecting K have their common point inside K is $\pi^4 V/M^3$.

3. Crofton's Formulas in E_n

We want to extend to E_n two classical formulas of Crofton that were given for E_2 in Chapter 4.

Chord formula. Let P_1, P_2 be two points in E_n and let G be the line determined by them. Let t_1, t_2 be the abscissas of P_1, P_2 on G. If $d\sigma_{n-1}$ denotes the volume element on the hyperplane perpendicular to G at P_1, the volume element of E_n at P_1 can be written $dP_1 = d\sigma_{n-1} \wedge dt_1$. On the other hand, the volume element at P_2 can be written $dP_2 = t^{n-1}\, du_{n-1} \wedge dt_2$ where $t = |t_2 - t_1|$ and du_{n-1} denotes the solid angle element corresponding to the direction of G. Thus we again have formula (12.23),

$$dP_1 \wedge dP_2 = t^{n-1}\, dG \wedge dt_1 \wedge dt_2. \tag{14.22}$$

Integrating over all pairs of points P_1, P_2 inside the convex body K and using the relation

$$\int_0^\sigma\int_0^\sigma |t_1 - t_2|^{n-1}\, dt_1 \wedge dt_2 = \frac{2}{n(n+1)}\sigma^{n+1} \tag{14.23}$$

where σ denotes the length of the chord intercepted by G, we have

$$\int_{G \cap K \neq \emptyset} \sigma^{n+1}\, dG = \frac{n(n+1)}{2} V^2 \tag{14.24}$$

where V is the volume of K.

This generalization of Crofton's formula for chords to E_n is due to Hadwiger ([269]; see also [337]). Since the measure of lines that intersect K is $[O_{n-2}/2(n-1)]F$ (according to (14.2)), it follows that the mean value of σ^{n+1} is given by

$$E(\sigma^{n+1}) = \frac{n(n^2-1)}{O_{n-2}} \frac{V^2}{F}.$$

ISBN 0-201-13500-0

More generally, multiplying both sides of (14.22) by a power of t and integrating as before, we get the relation

$$2I_m = m(m-1)J_{m-n-1} \tag{14.25}$$

where

$$I_m = \int_{G \cap K \neq \emptyset} \sigma^m \, dG, \qquad J_m = \int_{P_1, P_2 \in K} t^m \, dP_1 \wedge dP_2. \tag{14.26}$$

Between I_m and J_m there are inequalities similar to those given in Chapter 4 for E_2. Hadwiger [269] has proved that $F^2 \geqslant 2(n-1)(n-2)J_{-2}$, where the equality sign holds only for the ball.

From (14.25) and (14.26) it follows that the mean distance between two points of a convex set is given by

$$E(r) = \frac{2I_{n+2}}{(n+1)(n+2)V^2}.$$

For the three-dimensional ball of radius R a direct computation gives

$$I_m = \frac{2^{m+2}}{m+2} \pi^2 R^{m+2}.$$

Some known inequalities between the integrals I_m for $n = 3$ are the following

$$8I_0^3 - 9\pi^2 I_1^2 \geqslant 0, \qquad 4I_0^3 - 3\pi^4 I_4 \geqslant 0, \qquad 3^4 I_1^4 - 2^6 \pi^2 I_2^3 \geqslant 0,$$

$$3^5 I_1^5 - 5^3 \pi^4 I_3^3 \geqslant 0, \qquad 7^3 \pi^8 I_5^3 - 3^7 I_1^7 \geqslant 0,$$

in all of which the equality sign holds if and only if the convex set is a ball (see [35, 81]).

Angle formula. Next we want to extend to E_n Crofton's formula (4.26). Let L_{n-1} and L_{n-1}^* be two hyperplanes that intersect a convex set K. The total measure of such pairs of hyperplanes is

$$\int_{\substack{L_{n-1} \cap K \neq \emptyset \\ L_{n-1}^* \cap K \neq \emptyset}} dL_{n-1} \wedge dL_{n-1}^* = M_{n-2}^2(\partial K). \tag{14.27}$$

We compute the measure of the pairs of hyperplanes whose intersection $L_{n-1} \cap L_{n-1}^*$ intersects K. Using (14.13), (14.14), and (14.15) for $p = n-1$, $q = n-1$, we get (assuming $n > 2$)

$$M(L_{n-1} \cap L_{n-1}^* \cap K \neq \emptyset) = \frac{n-2}{n-1} \frac{O_{n-2}^2}{O_{n-3}} \frac{\pi}{4} M_{n-3}(\partial K). \tag{14.28}$$

ISBN 0-201-13500-0

On the other hand, in order to compute the measure of all pairs of hyperplanes that intersect K but whose intersection does not intersect K, we apply formula (12.43). Denoting by ϕ the angle between the two supporting hyperplanes of K through L_{n-2} and putting

$$\Phi_{n-1}(\phi) = \int_0^\phi \int_0^\phi |\sin^{n-1}(\phi_2 - \phi_1)| \, d\phi_1 \wedge d\phi_2, \tag{14.29}$$

we find that the desired measure is equal to $\int \Phi_{n-1}(\phi) \, dL_{n-2}$ extended over all L_{n-2} exterior to K. Since the sum of this measure and (14.28) must be equal to (14.27), we get

$$\int_{L_{n-2} \cap K = \emptyset} \Phi_{n-1}(\phi) \, dL_{n-2} = M_{n-2}^2(\partial K) - \frac{n-2}{n-1} \frac{O_{n-2}^2}{O_{n-3}} \frac{\pi}{4} M_{n-3}(\partial K). \tag{14.30}$$

This is the generalization of Crofton's formula (4.26) to E_n for $n > 2$. The case in which $n = 2$ has been considered directly in Chapter 4. The explicit value of $\Phi_{n-1}(\phi)$ is as follows.

(a) For $n - 1$ even

$$\begin{aligned}\Phi_{n-1}(\phi) = -\frac{2}{n-1} \Bigg[& \frac{1}{n-1} \sin^{n-1} \phi \\ & + \sum_{i=1}^{(n-3)/2} \frac{(n-2)\cdots(n-2i)}{(n-3)\cdots(n-1-2i)} \sin^{n-1-2i} \Phi \Bigg] \\ & + \frac{(n-2)\cdot \cdots \cdot 3 \cdot 1}{(n-1) \cdot \cdots \cdot 4 \cdot 2} \Phi^2 \end{aligned} \tag{14.31}$$

(b) For $n - 1$ odd

$$\begin{aligned}\Phi_{n-1}(\phi) = -\frac{2}{n-1} \Bigg[& \frac{1}{n-1} \sin^{n-1} \phi \\ & + \sum_{i=1}^{n/2-1} \frac{(n-2)\cdots(n-2i)}{(n-3)\cdots(n-1-2i)} \sin^{n-1-2i} \phi \\ & - \frac{(n-2)\cdot \cdots \cdot 4 \cdot 2}{(n-3) \cdot \cdots \cdot 3 \cdot 1} \phi \Bigg]. \end{aligned} \tag{14.32}$$

Formula (14.30) was given by Santaló [565]. For $n = 3$ we have

$$\int_{L_1 \cap K \neq \emptyset} (\phi^2 - \sin^2 \phi) \, dL_1 = 2M^2 - (\pi^3/2)F, \tag{14.33}$$

which was established by Herglotz [51, 306].

ISBN 0-201-13500-0

4. Some Relations between Densities of Linear Subspaces

Let O be a fixed point (origin) and let $L_{q[0]}$ be a fixed q-plane through O. Let $L_{r[0]}$ be a moving r-plane through O and assume $q + r > n$, so that $L_{q[0]} \cap L_{r[0]}$ is, in general, an $(r + q - n)$-plane through O, which we represent by $L_{r+q-n[0]}$. We want to express $dL_{r[0]}$ as a product of $dL_{r[r+q-n]}$ (density of L_r about $L_{r+q-n[0]}$) and $dL^{(q)}_{r+q-n[0]}$ (density of $L_{r+q-n[0]}$ as subspace of the fixed $L_{q[0]}$). To this end we consider the following two orthonormal moving frames:

Moving frame I

(a) $\mathbf{e}_1, \mathbf{e}_2, \ldots, \mathbf{e}_{r+q-n}$ span $L_{q[0]} \cap L_{r[0]}$.

(b) $\mathbf{e}_{r+q-n+1}, \ldots, \mathbf{e}_r$ lie on $L_{r[0]}$.

(c) $\mathbf{e}_{r+1}, \ldots, \mathbf{e}_n$ are arbitrary unit vectors that complete orthonormal frame I.

Moving frame II

(a) $\mathbf{e}_1, \mathbf{e}_2, \ldots, \mathbf{e}_{r+q-n}$ span $L_{q[0]} \cap L_{r[0]}$.

(b) $\mathbf{b}_{r+q-n+1}, \ldots, \mathbf{b}_r$ are constant unit vectors in the $(n - q)$-plane $L_{n-q[0]}$ perpendicular to $L_{q[0]}$.

(c) $\mathbf{b}_{r+1}, \ldots, \mathbf{b}_n$ are contained in $L_{q[0]}$ and such that together with $\mathbf{e}_1, \ldots, \mathbf{e}_{r+q-n}$ they form an orthonormal frame in $L_{q[0]}$.

With these notations, according to (12.27) and (12.3) we have

$$dL_{r[0]} = \bigwedge_{\alpha,i} (\mathbf{e}_{r+\alpha}, d\mathbf{e}_i) \bigwedge_{\alpha,h} (\mathbf{e}_{r+\alpha}, d\mathbf{e}_h) \tag{14.34}$$

with the following ranges of indices, which will be used throughout the rest of this section.

$$\alpha = 1, 2, \ldots, n - r; \qquad i = 1, 2, \ldots, r + q - n; \qquad h = r + q - n + 1, \ldots, r. \tag{14.35}$$

According to (12.26) we have

$$dL_{r[r+q-n]} = \bigwedge_{\alpha,h} (\mathbf{e}_{r+\alpha}, d\mathbf{e}_h), \tag{14.36}$$

$$dL^{(q)}_{r+q-n[0]} = \bigwedge_{\alpha,i} (\mathbf{b}_{r+\alpha}, d\mathbf{e}_i). \tag{14.37}$$

Put

$$\mathbf{e}_{r+\alpha} = \sum_h u_{r+\alpha,h}\mathbf{b}_h + \sum_k u_{r+\alpha,k}\mathbf{b}_k \tag{14.38}$$

where the range of h is given by (14.35) and $k = r + 1, r + 2, \ldots, n$.

ISBN 0-201-13500-0

Since $\mathbf{b}_h$ are constant vectors, we have $(\mathbf{b}_h \cdot d\mathbf{e}_i) = -(\mathbf{e}_i \cdot d\mathbf{b}_h) = 0$ and thus

$$(\mathbf{e}_{r+\alpha} \cdot d\mathbf{e}_i) = \sum_k u_{r+\alpha,k}(\mathbf{b}_k \cdot d\mathbf{e}_i). \tag{14.39}$$

From (14.34) and (14.39) we get the desired formula

$$dL_{r[0]} = \Delta^{r+q-n}\, dL_{r[r+q-n]} \wedge dL^{(q)}_{r+q-n[0]} \tag{14.40}$$

where

$$\Delta = \det(\mathbf{e}_{r+\alpha} \bullet \mathbf{b}_k). \tag{14.41}$$

Integrating (14.40) over all $L_{r[0]}$, we obtain on the left-hand side the known value (12.35) and on the right-hand side we can integrate $dL^{(q)}_{r+q-n[0]}$, applying the same formula (12.35) for $n \to q$, $r \to r+q-n$, since Δ depends only on $L_{r[r+q-n]}$. The result is

$$\int \Delta^{r+q-n}\, dL_{r[r+q-n]} = \frac{O_{n-1}O_{n-2}\cdots O_q}{O_{r-1}O_{r-2}\cdots O_{r+q-n}} \tag{14.42}$$

where the integral is extended over all $L_{r[q+r-n]}$.

We can give this formula another useful form. Recall that $dL_{r[q]} = dL^{(n-q)}_{r-q[0]}$ (Chapter 12, Section 4), which can be written

$$dL_{r[r+q-n]} = dL^{(2n-r-q)}_{n-q[0]} \tag{14.43}$$

and therefore (14.42) becomes

$$\int_{G_{n-q,n-r}} \Delta^{r+q-n}\, dL^{(2n-r-q)}_{n-q[0]} = \frac{O_{n-1}\cdots O_q}{O_{r-1}\cdots O_{r+q-n}}. \tag{14.44}$$

Making the change of notation $r+q-n = N$, $2n-r-q = \nu$, $n-q = \rho$, we have for this equation

$$\int_{G_{\rho,\nu-\rho}} \Delta^N\, dL^{(\nu)}_{\rho[0]} = \frac{O_{N+\nu-1}\cdots O_{N+\nu-\rho}}{O_{N+\rho-1}\cdots O_N}. \tag{14.45}$$

Recall the meaning of the determinant Δ: through the origin O there are ρ fixed orthonormal vectors $\mathbf{e}_1{}^0, \mathbf{e}_2{}^0, \ldots, \mathbf{e}_\rho{}^0$ and a moving $L_{\rho[0]}$ spanned by the orthonormal vectors $\mathbf{e}_1, \mathbf{e}_2, \ldots, \mathbf{e}_\rho$. Then $\Delta = \det(\mathbf{e}_i{}^0 \bullet \mathbf{e}_j)$ where $1 \leqslant i, j \leqslant \rho$.

We consider further consequences of (14.40). Consider the fixed $L_{q[0]}$ and a moving L_r, $r+q>n$. Let x be a point of the intersection $L_r \cap L_{q[0]}$ and consider moving frames I and II above with reference to point x. In order to apply (14.40) we observe that $dL_{n-r[0]} = dL_{r[0]} = dL_{r[x]}$ and $d\sigma_{n-r} = (dx \cdot \mathbf{e}_{r+1}) \wedge \cdots \wedge (dx \cdot \mathbf{e}_n)$. Hence we have

ISBN 0-201-13500-0

$$dL_r = dL_{r[x]} \bigwedge_\alpha (dx \cdot \mathbf{e}_{r+\alpha}), \qquad \alpha = 1, 2, \ldots, n - r. \tag{14.46}$$

Since the vectors $\mathbf{b}_h$ $(h = r + q - n + 1, \ldots, r)$ are perpendicular to $L_r \cap L_{q[0]}$ and $x \in L_r \cap L_{q[0]}$, we have $dx \cdot \mathbf{b}_h = 0$ and thus (14.38) gives

$$dx \cdot \mathbf{e}_{r+\alpha} = \sum_k u_{r+\alpha,k}(dx \cdot \mathbf{b}_k). \tag{14.47}$$

It follows that

$$\bigwedge_\alpha (dx \cdot \mathbf{e}_{r+\alpha}) = \Delta \bigwedge_k (dx \cdot \mathbf{b}_k), \qquad k = r + 1, \ldots, n. \tag{14.48}$$

Since $dL_{r+q-n}^{(q)} = dL_{r+q-n[0]}^{(q)} \bigwedge_k (dx \cdot \mathbf{b}_k)$, from the last relations and (14.39) it follows that

$$dL_r = \Delta^{r+q-n+1}\, dL_{r[r+q-n]} \wedge dL_{r+q-n}^{(q)}. \tag{14.49}$$

Let $F(L_r)$ be an integrable function that depends only on $L_{r+q-n}^{(q)} = L_r \cap L_{q[0]}$. We get

$$\int F(L_r)\, dL_r = \int \Delta^{r+q-n+1}\, dL_{r[r+q-n]} \int F(L_{r+q-n}^{(q)})\, dL_{r+q-n}^{(q)} \tag{14.50}$$

where the integrals are extended over all possible values of the integrands. According to (14.43), we have

$$\int \Delta^{r+q-n+1}\, dL_{r[r+q-n]} = \int \Delta^{r+q-n+1}\, dL_{n-q[0]}^{(2n-r-q)} \tag{14.51}$$

and applying (14.45) to the case in which $N = r + q - n + 1$, $\rho = n - q$, and $\nu = 2n - r - q$, we obtain

$$\int \Delta^{r+q-n+1}\, dL_{r[r+q-n]} = \frac{O_n O_{n-1} \cdots O_{q+1}}{O_r O_{r-1} \cdots O_{r+q-n+1}} \tag{14.52}$$

and finally

$$\int F(L_r)\, dL_r = \frac{O_n O_{n-1} \cdots O_{q+1}}{O_r O_{r-1} \cdots O_{r+q-n+1}} \int F(L_{r+q-n}^{(q)})\, dL_{r+q-n}^{(q)}. \tag{14.53}$$

Formulas (14.44) and (14.53) are the work of Chern [112].

As an application of (14.53), let K_q be a convex set contained in $L_{q[0]}$ and let F be the function that is equal to 1 if $L_r \cap K_q \neq \emptyset$ and to 0 otherwise. Then, the left-hand side of (14.53) is the measure of all L_r that intersect K_q. Thus, by (14.2) we have

ISBN 0-201-13500-0

$$\int_{L_r \cap K_q \neq \emptyset} dL_r = \frac{O_{n-2} \cdots O_{n-r-1}}{(n-r)O_{r-1} \cdots O_1 O_0} M_{r-1}^{(n)} \tag{14.54}$$

where $M_{r-1}^{(n)}$ denotes the $(r-1)$th integral of mean curvature of K_q as a convex body of E_n. The integral on the right-hand side in (14.53) is the measure of all $L_{r+q-n}^{(q)}$ of $L_{q[0]}$ that intersect K_q. Therefore, according to the same formula (14.2) we have

$$\int_{L_{r+q-n}^{(q)} \cap K_q \neq \emptyset} dL_{r+q-n}^{(q)} = \frac{O_{q-2} \cdots O_{n-r-1}}{(n-r)O_{r+q-n-1} \cdots O_0} M_{r+q-n-1}^{(q)} \tag{14.55}$$

where $M_{r+q-n-1}^{(q)}$ is the $(r+q-n-1)$th integral of mean curvature of K_q as a convex set of $L_{q[0]}$. Thus, (14.53) gives

$$M_{r-1}^{(n)} = \frac{O_{r+q-n} O_n O_{n-1}}{O_r O_q O_{q-1}} M_{r+q-n-1}^{(q)}. \tag{14.56}$$

This formula must coincide with (13.62). To verify this coincidence it suffices to take into account the value (1.22) of O_i and the following well-known properties of the gamma function.

$$\Gamma(z) = (z-1)!, \qquad \Gamma(z)\Gamma(z+\tfrac{1}{2}) = 2^{1-2z}\pi^{1/2}\Gamma(2z). \tag{14.57}$$

5. Linear Subspaces That Intersect a Manifold

Let $(x; \mathbf{e}_i)$ be a moving orthonormal frame and consider the r-plane L_r determined by $x, \mathbf{e}_1, \mathbf{e}_2, \ldots, \mathbf{e}_r$. The density for L_r is, by (12.18),

$$dL_r = \bigwedge_i \omega_i \bigwedge_{h,j} \omega_{hj}, \qquad i, h = r+1, \ldots, n, \quad j = 1, 2, \ldots, r. \tag{14.58}$$

The exterior product $\bigwedge \omega_i = \bigwedge (dx \cdot \mathbf{e}_i)$ $(i = r+1, \ldots, n)$ is equal to the volume element $d\sigma_{n-r}(x)$ of the $(n-r)$-plane $L_{n-r[x]}$ orthogonal to L_r at x. The exterior product $\bigwedge \omega_{hj} = \bigwedge (\mathbf{e}_j \cdot d\mathbf{e}_h)$ is the density $dL_{r[x]}$ of the r-planes about x. Hence we have (compare with (12.38)),

$$dL_r = d\sigma_{n-r}(x) \wedge dL_{r[x]}. \tag{14.59}$$

Let M^q be a compact differentiable manifold of dimension q embedded in E_n, assumed piecewise smooth. Assume $r + q \geqslant n$ and consider the set of r-planes that have common point with M^q. The intersection $L_r \cap M^q$ is in general a manifold of dimension $r + q - n$. Take the frame $(x; \mathbf{e}_i)$ such that $x \in L_r \cap M^q$ and $\mathbf{e}_1, \mathbf{e}_2, \ldots, \mathbf{e}_{r+q-n}$ are orthonormal tangent vectors to $L_r \cap M^q$. Let $\mathbf{b}_{r+1}, \mathbf{b}_{r+2}, \ldots, \mathbf{b}_n$ be a set of orthonormal vectors such that the tangent

ISBN 0-201-13500-0

space to M^q at x is spanned by $\mathbf{e}_1, \ldots, \mathbf{e}_{r+q-n}, \mathbf{b}_{r+1}, \ldots, \mathbf{b}_n$. Since we consider only r-planes that intersect M^q, we may assume that point x in (14.59) is such that

$$dx = \sum_{i=1}^{r+q-n} \lambda_i \mathbf{e}_i + \sum_{k=r+1}^{n} \beta_k \mathbf{b}_k \tag{14.60}$$

where λ_i and β_k are differential 1-forms. We have

$$\omega_{r+\alpha} = dx \cdot \mathbf{e}_{r+\alpha} = \sum_{k=r+1}^{n} \beta_k (\mathbf{b}_k \cdot \mathbf{e}_{r+\alpha}), \qquad \alpha = 1, 2, \ldots, n-r, \tag{14.61}$$

and thus

$$d\sigma_{n-r}(x) = \bigwedge_{\alpha=1}^{n-r} \omega_{r+\alpha} = \Delta \bigwedge_k \beta_k, \qquad k = r+1, \ldots, n, \tag{14.62}$$

where $\Delta = \det(\mathbf{b}_k \cdot \mathbf{e}_{r+\alpha}) = \det(\cos \phi_{k,r+\alpha})$, $\phi_{k,r+\alpha}$ being the angle between $\mathbf{b}_k$ and $\mathbf{e}_{r+\alpha}$.

If $d\sigma_{r+q-n}(x)$ denotes the volume element of $L_r \cap M^q$ and $d\sigma_q(x)$ the volume element of M^q at x, since $\bigwedge \beta_k$ is the $(n-r)$-dimensional volume element of M^q orthogonal to $L_r \cap M^q$, we have

$$\bigwedge_{k=r+1}^{n} \beta_k \wedge d\sigma_{r+q-n}(x) = d\sigma_q(x) \tag{14.63}$$

and from (14.59), (14.62), and (14.63) we obtain

$$d\sigma_{r+q-n}(x) \wedge dL_r = \Delta \, d\sigma_q(x) \wedge dL_{r[x]}. \tag{14.64}$$

Note that (a) Δ depends on the position of L_r with respect to the tangent q-plane to M^q at x, but is independent of x.

(b) If $r + q - n = 0$, formula (14.64) remains true and becomes

$$dL_r = \Delta \, d\sigma_q(x) \wedge dL_{r[x]}, \qquad r + q = n. \tag{14.65}$$

Let $\sigma(M^q)$ denote the q-dimensional volume of M^q. Integrating both sides of (14.64) over all r-planes that intersect M^q, we get

$$\int_{L_r \cap M^q \neq \emptyset} \sigma_{r+q-n}(M^q \cap L_r) \, dL_r = c\sigma_q(M^q) \tag{14.66}$$

where $c = \int \Delta \, dL_{r[x]}$ is a constant that we want to calculate. To this end we shall calculate directly the left-hand side of (14.66) for the q-dimensional unit sphere U_q in E_n. Let $L_m^{(q+1)}$, $m \leqslant q$, denote the m-planes of the $(q+1)$-plane

ISBN 0-201-13500-0

that contains U_q. Consider first the integral

$$\int \sigma_{m-1}(U_q \cap L_m^{(q+1)})\, dL_m^{(q+1)} \tag{14.67}$$

over all $L_m^{(q+1)}$ that intersect U_q. If O is the center of U_q and ρ denotes the distance from O to $L_m^{(q+1)}$, then $U_q \cap L_m^{(q+1)}$ is a sphere of dimension $m-1$ and radius $(1-\rho^2)^{1/2}$, whence $\sigma_{m-1}(U_q \cap L_m^{(q+1)}) = (1-\rho^2)^{(m-1)/2} O_{m-1}$. On the other hand, since the volume element of the $(q+1-m)$-plane perpendicular to $L_m^{(q+1)}$ through O at its intersection point with $L_m^{(q+1)}$ is equal to $\rho^{q-m}\, du_{q-m} \wedge d\rho$, we have $dL_m^{(q+1)} = \rho^{q-m}\, du_{q-m} \wedge d\rho \wedge dL_{q+1-m[0]}^{(q+1)}$. Taking into account that $\int_0^1 \rho^{q-m}(1-\rho^2)^{(m-1)/2}\, d\rho = O_{q+1}(O_{q-m}O_m)^{-1}$ and using (12.35) applied to the Grassmann manifold $G_{q+1-m,m}$, we have

$$\int_{L_m^{(q+1)} \cap U_q \neq \emptyset} \sigma_{m-1}(U_q \cap L_m^{(q+1)})\, dL_m^{(q+1)} = \frac{O_{q+1}O_q \cdots O_{m+1}O_{m-1}}{O_{q-m}O_{q-m-1} \cdots O_0}. \tag{14.68}$$

Now return to the case of a general L_m of E_n. We apply formula (14.54) to the case in which F is the volume of $U_q \cap L_m$. Replacing $q+1$ by q, since U_q is contained in a fixed $(q+1)$-plane, and using (14.68) for $m \to r+q+1-n$, we obtain

$$\int_{U_q \cap L_r \neq \emptyset} \sigma_{r+q-n}(U_q \cap L_r)\, dL_r = \frac{O_n O_{n-1} \cdots O_{r+1} O_{r+q-n}}{O_{n-r-1} \cdots O_1 O_0}.$$

Comparing with (14.66) we get the value of the constant c and substituting in (14.66) we have the final result

$$\int_{M^q \cap L_r \neq \emptyset} \sigma_{r+q-n}(M^q \cap L_r)\, dL_r = \frac{O_n \cdots O_{n-r} O_{r+q-n}}{O_r \cdots O_0 O_q} \sigma_q(M^q). \tag{14.69}$$

It is noteworthy that this formula holds without change for any space of constant curvature, that is, for euclidean and non-euclidean geometry (see [561, 565]). For $r+q=n$, (14.69) becomes

$$\int_{M^{n-r} \cap L_r \neq \emptyset} N(M^{n-r} \cap L_r)\, dL_r = \frac{O_n \cdots O_{n-r+1}}{O_r \cdots O_1} \sigma_q(M^{n-r}) \tag{14.70}$$

where $N(M^{n-r} \cap L_r)$ denotes the number of points of the intersection $M^{n-r} \cap L_r$.

Formula (14.69) includes a great number of special cases. We state the following:

ISBN 0-201-13500-0

Example 1. For the plane, $n = 2$, we have two possibilities:

(a) $r = 1$, $q = 1$. Then M^1 is a curve, $\sigma_1(M^1)$ is its length, and $\sigma_0(M^1 \cap L_1)$ is the number of intersection points of M^1 with the line L_1. Formula (14.69) reduces to (3.17).

(b) $r = 1$, $q = 2$. Then M^2 is a plane domain of area $\sigma_2(M^2)$. The integrand $\sigma_1(M^2 \cap L_1)$ is the length of the chord $M^2 \cap L_1$ and (14.69) coincides with (3.6).

Example 2. For the space $n = 3$, we have the following cases:

(a) $r = 1$, $q = 2$. L_1 are lines that intersect a fixed surface M^2 of area F and (14.69) becomes

$$\int_{L_1 \cap M^2 \neq 0} N \, dL_1 = \pi F \tag{14.71}$$

where N is the number of intersection points of L_1 and M^2.

(b) $r = 1$, $q = 3$. L_1 are lines that intersect a fixed domain D. If σ_1 denotes the length of the chord $L_1 \cap D$, we get

$$\int_{L_1 \cap D \neq 0} \sigma_1 \, dL_1 = 2\pi V, \qquad V = \text{volume of } D. \tag{14.72}$$

(c) $r = 2$, $q = 1$. L_2 are planes that intersect a curve C of length L. We get

$$\int_{L_2 \cap C \neq 0} N \, dL_2 = \pi L \tag{14.73}$$

where N is the number of intersection points of L_2 and C. This formula (14.73), which we have proved for piecewise smooth curves, holds good for any rectifiable curve.

(d) $r = 2$, $q = 2$. L_2 are planes that intersect a fixed surface M^2. Then (14.69) gives

$$\int_{L_2 \cap M^2 \neq 0} \lambda \, dL_2 = (\pi^2/2) F \tag{14.74}$$

where λ is the length of $L_2 \cap M^2$ and F the surface area of M^2.

White [721] has given integral formulas relating the length and the total curvature of curves cut from M^2 by planes passing through a fixed point to invariants of the surface.

(e) $r = 2$, $q = 3$. L_2 are planes that intersect a fixed domain D of volume V. Then (14.69) becomes

ISBN 0-201-13500-0

$$\int_{L_2 \cap D \neq 0} \sigma_2 \, dL_2 = 2\pi V \tag{14.75}$$

where σ_2 is the area of the intersection $L_2 \cap D$.

From (14.74) and (14.75), using the fact that $m(L_2; L_2 \cap K \neq 0) = M$, we get the following mean values (for convex bodies K).

$$E(\lambda) = \pi^2 F/2M, \qquad E(\sigma_2) = 2\pi V/M. \tag{14.76}$$

Example 3. Let $\omega^{(n-r)}$ be an $(n-r)$-form defined on M^{n-r}. Then (14.70) may be generalized to an integral formula of the form

$$\int_{L_r \cap M^{n-r} \neq 0} \left(\sum_i \omega^{(n-r)}(P_i) \right) dL_r = c \int_{M^{n-r}} \omega^{(n-r)}$$

where P_i are the intersection points $M^{n-r} \cap L_r$ and c is a constant. Formulas of this kind have been applied to the proof of the Stokes formula $\int_{\partial M^{n-r}} \omega = \int_{M^{n-r}} d\omega$ (by, e.g., Maak [379] and Horneffer [312, 313]).

6. Hypersurfaces and Linear Spaces

We want to extend formula (14.10) to the case of a body Q, not necessarily convex, intersected by moving r-planes. We assume that ∂Q is a hypersurface of class C^2. Let L_r be an r-plane that intersects Q and let $x \in L_r \cap \partial Q$. For $q = n - 1$, formula (14.64) becomes

$$d\sigma_{r-1}(x) \wedge dL_r = \Delta \, d\sigma_{n-1}(x) \wedge dL_{r[x]}. \tag{14.77}$$

Let $\rho_1, \ldots, \rho_{r-1}$ be the principal radii of curvature of the $(r-1)$-dimensional manifold $\partial Q \cap L_r$ at x. Multiplying both sides of (14.77) by $\{1/\rho_{h_1}, \ldots, 1/\rho_{h_i}\}$ and integrating over all values of the variables, we get, on the left-hand side, the integral $\binom{r-1}{i} \int M_i^{(r)} \, dL_r$ over all L_r such that $L_r \cap \partial Q \neq 0$. In order to calculate the integral on the right-hand side, we observe that the curvatures $1/\rho_h$ $(h = 1, 2, \ldots, r-1)$ can be expressed in terms of the principal curvatures $1/R_s$ $(s = 1, 2, \ldots, n-1)$ of ∂Q at x and the angles $\phi_{h,s}$ of the vectors $\mathbf{e}_h$ and $\mathbf{b}_s$ of Section 5 (because the classical theorems of Euler and Meusnier hold for hypersurfaces in a way analogous to that for surfaces of E_3, see [165]). In consequence of this, putting $\{1/\rho_{h_1}, \ldots, 1/\rho_{h_i}\}\Delta = F(1/R_i, \phi_{h,s})$, we have that the integral $\int F \, dL_{r[x]}$ extended over all L_r through x depends only on $R_1, R_2, \ldots, R_{n-1}$. A direct evaluation of this integral seems to be difficult. However, since we know that for the case of convex bodies it is equal to the symmetric function $\{1/R_{h_1}, \ldots, 1/R_{h_i}\}$ (up to a constant factor, formula (14.10)) and this "local" result cannot be influenced by the properties in the large

ISBN 0-201-13500-0

of ∂Q, we deduce that (14.10) holds in general. That is, we can state the formula

$$\int_{Q \cap L_r \neq \emptyset} M_i^{(r)}(\partial Q \cap L_r)\, dL_r = \frac{O_{n-2} \cdots O_{n-r} O_{n-i}}{O_{r-2} \cdots O_0 O_{r-i}} M_i(\partial Q) \qquad (14.78)$$

for any body whose boundary is of class C^2.

If Q is a convex body and we take $i = r - 1$, we have $M_{r-1}^{(r)} = O_{r-1}$ and (14.78) coincides with (14.2). For any body, not necessarily convex, using (13.38) we have

$$\int_{Q \cap L_r \neq \emptyset} \chi(Q \cap L_r)\, dL_r = \frac{O_{n-2} \cdots O_{n-r} O_{n-r+1}}{O_{r-2} \cdots O_0 O_1 O_{r-1}} M_{r-1}(\partial Q)$$

$$= \frac{O_{n-2} \cdots O_{n-r-1}}{(n-r) O_{r-1} \cdots O_0} M_{r-1}(\partial Q). \qquad (14.79)$$

If we replace $M_r(\partial Q) = nW_{r+1}(Q)$, (14.79) can be taken as the definition of $W_{r+1}(Q)$ for nonconvex bodies [274, p. 240].

7. Notes

1. *Favard measure and dimension.* If A is a subset of E_n and k is a positive integer less than n, then the k-dimensional Favard measure of A is defined by

$$m_F{}^k(A) = \frac{O_{n-k} \cdots O_1}{O_n \cdots O_{k+1}} \int_{L_{n-k} \cap A \neq \emptyset} N(A \cap L_{n-k})\, dL_{n-k}$$

where $N(A \cap L_{n-k})$ denotes the number (possibly infinite) of points of $A \cap L_{n-k}$. If s and n are integers, $n > 0$, $0 \leqslant s \leqslant n$, and $m_F{}^s(A) = 0$, then $\dim A \leqslant s - 1$.

The following formula (compare with (14.69)) can be proved.

$$m_F{}^k(A) = \frac{O_{n-k+h} \cdots O_0 O_k}{O_n \cdots O_{k-h} O_h} \int_{L_{n-k+h} \cap A \neq \emptyset} m_F{}^h(A \cap L_{n-k+h})\, dL_{n-k+h}.$$

The properties of the Favard measure, its comparison with other measures (Caratheodory, Hausdorff), and its relation to dimension have been treated in important papers by Federer [174, 175].

2. *Sets of r-planes that support a convex body.* Let ω be a set of points on the unit sphere U of all directions in E_n. Let $S(K; \omega)$ denote the $(n-1)$-dimensional area of the set of all those points in the boundary of the convex body K in E_n that lie in at least one supporting hyperplane whose outer normal falls in ω. If B is the unit ball and $\lambda \geqslant 0$, then $S(K + \lambda B; \omega)$ is a

ISBN 0-201-13500-0

polynomial in λ whose coefficients define the so-called area functions $S_{n-q-1}(K; \omega)$ for K ($q = 0, 1, \ldots, n-1$). For each $u \in U$ there is a unique hyperplane that supports $K + \lambda B$ and has outer normal u. In this hyperplane, let $C_q(u, \lambda)$ be the set of all q-planes that have points in common with $K + \lambda B$. For each set ω and for each $\eta > 0$ let $F_q(K; \omega, \eta) = \bigcup C_q(u, \lambda)$ ($u \in \omega$, $0 < \lambda < \eta$). Let $\mu_q(K; \omega, \eta)$ be the invariant measure of $F_q(K; \omega, \eta)$ as a set of q-planes in E_n. Firey [193] proves that $\lim_{\eta \to 0^+}(F(K; \omega, \eta)/\eta) = S_{n-q-1}(K; \omega)$. If $\omega = U$, then $S_{n-q-1}(K; U) = nW_{q+1}(K)$ where $W_{q+1}(K)$ is the $(q+1)$-quermassintegral of K. Consequently, $W_{q+1}(K)$, which is known to be the measure of the set of $(q+1)$-planes that meet K up to a constant factor (14.1), can also be considered as the measure of the q-planes that "support" K.

3. *An integral geometric formula for n-dimensional ellipsoids.* Let K be an n-dimensional ellipsoid centered at the origin and let $G_{r,n-r}$ denote the Grassmann manifold of r-planes through the origin in E_n. Furstenberg and Tzkoni [209a] have given the formula

$$c_{n,r} m(G_{r,n-r})(\sigma_n(K))^r = \int_{G_{r,n-r}} [\sigma_r(K \cap L_{r[0]})]^n \, dL_{r[0]}$$

where $m(G_{r,n-r})$ is the measure (12.35) of the Grassmann manifold $G_{r,n-r}$, σ_h denotes h-dimensional measure, and $c_{n,r} = [\Gamma(n/2)(n/2)]^r[\Gamma(r/2)(r/2)]^n$. This formula can be iterated to the case of flag manifolds consisting of all m-tuples of planes $L_{s_1} \subset L_{s_2} \subset \cdots \subset L_{s_m}$. For related results see those of Miles [421] and Guggenheimer [259].

4. *Sets of strips.* By a strip of breadth a in E_n we mean the part of the space between two parallel hyperplanes at a distance a from each other. The position of a strip B can be determined by the position of its midhyperplane and the density for strips is the same as for hyperplanes, $dB = dp \wedge du_{n-1}$ (12.40).

Let K be a convex set in E_n. Since the integral of mean curvature of the convex body $K_{a/2}$ parallel to K in a distance $a/2$ is $M_{n-2}(\partial K) + (O_{n-1}/2)a$ (13.55), we have

$$m(B; B \cap K \neq \emptyset) = M_{n-2}(\partial K) + (O_{n-1}/2)a. \tag{14.79a}$$

If a is less than the diameter of K we also have

$$m(B; B \supset K) = (O_{n-1}/2)a - M_{n-2}(\partial K).$$

Let K_0 be a convex set of constant breadth D_0, so that $M_{n-2}(\partial K_0) = (O_{n-1}/2)D_0$. Let K_1 be a convex set such that $K_1 \subset K_0$. Instead of assuming B moving, we can assume a fixed sequence of parallel strips B a distance D_0 apart (Fig. 5.3) and then K_0 placed at random in space, together with K_1 in its interior. Since K_0 will always meet one and only one of the strips B, and it is known that every convex set of diameter D_0 is a subset of a set of constant breadth D_0 [63, p. 130], we can state

Consider a set of parallel strips of breadth a in the space E_n at a distance D_0 apart and let a convex set K_1 of diameter $D_1 \leqslant D_0$ be placed at random in the space. The probability that K_1 will intersect one of the strips is

ISBN 0-201-13500-0

$$p = \frac{2M_{n-2}(\partial K_1) + O_{n-1}a}{O_{n-1}(D_0 + a)}.$$

In particular, if $a = 0$, we have

$$p = \frac{2M_{n-2}(\partial K_1)}{O_{n-1}D_0}$$

and if K_1 reduces to a line segment of length b,

$$p = \frac{2O_{n-2}b}{(n-1)O_{n-1}D_0}.$$

These formulas generalize to E_n the classical needle problem of Buffon (Chapter 5, Section 2). For another approach see that of Stoka [649] and Ambarcumjan [11].

5. *Mean values for convex polyhedrons.* 1. From (14.1) and (14.73) it follows that the mean number of vertices of the polygonal section of a given convex polyhedron by a random plane is $E(v) = \pi L/M$, where M is the integral of mean curvature and L is the total length of the edges of the polyhedron.

2. Let Q be a convex polyhedron and let A_i ($i = 1, \ldots, N$) be its vertices. Let ϕ_i denote the interior solid angle of Q at A_i. A line through A_i is called exterior to Q if A_i is the only point that it has in common with Q. Clearly the probability that a random line through A_i is exterior to Q is $p_i = 1 - \phi_i/2\pi$. The number of vertices of the orthogonal projection of Q onto a plane E is equal to the number of lines through the vertices A_i that are perpendicular to E and are exterior to Q. Thus, the expected value of the number of vertices of the polygonal projection of Q onto a random plane is $E(v_1) = N - (1/2\pi)\sum_1^N \phi_i$. This result generalizes to E_n (see [412]).

6. *Integral formulas for space curves.* We state some results about curves in E_3 that were obtained by [481].

Let Γ be a closed curve in E_3 that is differentiable of class C^3 and has no cusps or double points. For each line G let $\lambda(G)$ denote the linking number of G with Γ (the algebraic number of times that Γ twists around G) and for each plane E let $P_1, P_2, \ldots$ denote the points of $E \cap \Gamma$. Call $r(P_i, P_j)$ the distance from P_i to P_j and let i_{P_j} denote the intersection number of E and Γ at P_j ($+1$ or -1, according to the orientation of the frame formed by the tangent to the curve at P_j and two ordered vectors of E, assumed oriented). Then Pohl proves the integral formula

$$\pi \int \lambda^2 \, dG = -\int \left(\sum_{i,j} r(P_i, P_j) i_{P_i} i_{P_j} \right) dE \tag{14.80}$$

where the first integral is extended over all lines and the second over all planes of E_3.

The right-hand side of (14.80) may be put in a different form in the following way. To each ordered pair of points P, Q of Γ associate the unit vector $\mathbf{e} = \mathbf{e}(P, Q)$ directed from P to Q and the distance $r(P, Q)$. Let $\mathbf{t}(P)$, $\mathbf{t}(Q)$

ISBN 0-201-13500-0

denote the tangent vectors to Γ at P and Q and let σ_P and σ_Q be the angles $(t(P), \mathbf{e})$ and $(t(Q), \mathbf{e})$, respectively. Let $\tau(P, Q)$ denote the angle between the oriented planes spanned by $(\mathbf{t}(P), \mathbf{e})$ and $(\mathbf{t}(Q), \mathbf{e})$. Let ds denote the arc element and let

$$dI = -r^{-1} \cos\tau \sin\sigma_P \sin\sigma_Q \, ds(P) \wedge ds(Q) \tag{14.81}$$

for (P, Q) at which the quantities involved are defined and $dI = 0$ otherwise. Then

$$-\int \left(\sum_{i,j} r(P_i, P_j) i_{P_i} i_{P_j} \right) dE = (\pi/2) \int_{\Gamma \times \Gamma} r \, dI. \tag{14.82}$$

Another integral formula of Pohl is

$$2L = \int_{\Gamma \times \Gamma} dI \tag{14.83}$$

where L is the length of Γ.

For plane curves, (14.81) and (14.82) give

$$2\pi \int_{E_2} w^2 \, dP = \int_{\Gamma \times \Gamma} r \, dI \tag{14.84}$$

where dP is the area element in E_2 and w is the winding number of P with respect to Γ (algebraic number of times the curve winds around the point P). These formulas generalize to E_n [481].

Prove, as an exercise, that $dI = dG^*$ (12.72).

7. *Linking numbers for space curves.* Let Γ, Γ' be two closed, disjoint, oriented curves in E_3 of class C^1. We assume that Γ is the boundary of a disk D, assumed oriented compatibly with Γ. The linking number $L(\Gamma, \Gamma')$ is defined to be the intersection number of D and Γ'; hence, it is an integer. In order to give an integral formula for the linking number, we assign to each pair of points $P \in \Gamma$, $Q \in \Gamma'$ the unit vector $\mathbf{e}$ from P to Q and denote by du_2 the area element on the unit sphere U_2 corresponding to the direction of the vector $\mathbf{e}$. Then it can be shown that L is given by the Gauss integral

$$L(\Gamma, \Gamma') = (1/4\pi) \int_{\Gamma \times \Gamma'} du_2. \tag{14.85}$$

Let $\mathbf{t}_1 = \mathbf{t}_1(P)$, $\mathbf{t}_2 = \mathbf{t}_2(Q)$ denote the tangent unit vectors to Γ, Γ' at P, Q, respectively. Then an easy calculation shows that the right-hand side of (14.85) can be written

$$L(\Gamma, \Gamma') = (1/4\pi) \int_{\Gamma \times \Gamma'} r^{-2}(\mathbf{e}\mathbf{t}_1\mathbf{t}_2) \, ds_1 \wedge ds_2 \tag{14.86}$$

ISBN 0-201-13500-0

where ds_1, ds_2 denote the arc elements of Γ, Γ' at P and Q, respectively, and $(\mathbf{e}\mathbf{t}_1\mathbf{t}_2)$ is the scalar triple product of the vectors $\mathbf{e}$, $\mathbf{t}_1$, $\mathbf{t}_2$.

Calugareanu [77–79] raised the question of the significance of $L(\Gamma, \Gamma')$ if $\Gamma \equiv \Gamma'$. By considering a parallel curve to Γ in a distance ε (suitably defined) and letting $\varepsilon \to 0$, he found that if Γ is of class C^3 and has non-vanishing curvature, then

$$(1/4\pi) \int_{\Gamma \times \Gamma} du_2 + (1/2\pi) \int_{\Gamma} \tau\, ds = SL(\Gamma) \tag{14.87}$$

is an integer, where τ is the torsion of Γ. This integer $SL(\Gamma)$ is called the self-linking number of Γ.

A simplified proof of this formula of Calugareanu, together with topological questions related to it, has been given by Pohl [483]. The generalization to submanifolds of a differentiable manifold of higher dimensions has been carried out by White [720].

8. *The isoperimetric inequality of Banchoff and Pohl.* The isoperimetric inequality of Banchoff and Pohl (of Chapter 3, Section 5, 1) may be generalized to arbitrary dimension and codimension. Let M be a compact oriented manifold of dimension m in E_n. Let P_1, P_2 be two points of M and let $r = r(P_1, P_2)$ denote the chord length from P_1 to P_2. Let dP_1, dP_2 denote the volume elements on M at P_1, P_2, respectively. Then Banchoff and Pohl [20] have obtained the inequality

$$\int_{M \times M} r^{-m+1}\, dP_1 \wedge dP_2 - (1 + m)O_m c_{m,n} \int \lambda^2\, dL_{n-m-1} \geqslant 0$$

where λ is the linking number of L_{n-m-1} with M and $c_{m,n}$ is a constant $(c_{m,n} = k_{m,m+2}k_{m,m+3}\cdots k_{m,n}$ where $k_{m,j} = \pi^{-(m+1)/2}\Gamma((j+1)/2)/\Gamma((j-m)/2))$. The second integral is extended over all $(n - m - 1)$-planes in E_n. Equality holds only for one or several coincident spheres with coincident orientations, or $(n = 1)$ one or several coincident circles all traversed in the same direction, each a number of times.

The last integral $A(M) = c_{m,n} \int \lambda^2\, dL_{n-m-1}$ has the following properties: (a) $A(M)$ is just the volume bounded by M if $n = m + 1$. (b) If $M \subset E_n \subset E_N$, then $A(M)$ in the sense of submanifolds of E_N is the same as $A(M)$ in the sense of submanifolds of E_n. (c) $A(M)$ has the so-called reproductive property $\int A(M \cap L_q)\, dL_q = b_{m,n,q}A(M)$ where $b_{m,n,q}$ is a constant. For details and other similar results see [20].

9. *Mean values and curvatures.* Let M^n be a compact n-dimensional differentiable manifold (without boundary) in E_{n+N}. To each point $P \in M^n$ we attach the qth tangent fiber over P, that is, the plane spanned by the vectors $\partial/\partial x_1, \ldots, \partial/\partial x_n$; $\partial^2/\partial x_1{}^2, \partial^2/\partial x_1\, \partial x_2, \ldots, \partial^2/\partial x_n{}^2; \ldots;$ $\partial^q/\partial x_1{}^q, \ldots, \partial^q/\partial x_n{}^q$ where x_i are local coordinates in a neighborhood of P whose dimension is

$$\rho(n, q) = \sum_{i=1}^{q} \binom{n + i - 1}{i}.$$

ISBN 0-201-13500-0

Assuming $1 \leqslant r \leqslant n + N - 1$, $\rho \leqslant n + N - 1$, we define the rth total absolute curvature of order q of M^n as follows.

(a) *Case in which* $1 \leqslant r \leqslant \rho$. Let O be a fixed point of E_{n+N} and consider an $(n + N - 1)$-plane $L_{n+N-r[0]}$ through O. Let Γ_r denote the set of r-planes L_r of E_{n+N} that are contained in some of the fibers, pass through the corresponding base point P of the fiber, and are orthogonal to $L_{n+N-r[0]}$. The intersection $\Gamma_r \cap L_{n+N-r[0]}$ will be a compact variety in $L_{n+N-r[0]}$ whose dimension is $\delta = r\rho + n - r(n + N)$. Let $\mu(\Gamma_r \cap L_{n+N-r[0]})$ be the measure of this variety as subvariety of the euclidean space $L_{n+N-r[0]}$; if $\delta = 0$, then μ means the number of intersection points of Γ_r and $L_{n+N-r[0]}$; if $\delta < 0$, we put $\mu = 0$. Then we define the rth total absolute curvature of order q of $M^n \subset E_{n+N}$ as the mean value of the measures μ for all $L_{n+N-r[0]}$; that is, making use of (12.35), we have

$$K_{r,N}^{(q)}(M^n) = \frac{O_1 \cdots O_{n+N-r-1}}{O_r \cdots O_{n+N-1}} \int_{G_{n+N-r,r}} \mu(\Gamma_r \cap L_{n+N-r[0]})\, dL_{n+N-r[0]}.$$

Note that the coefficient of the right-hand side may be replaced by $O_1 \cdots O_{r-1}/O_{n+N-r} \cdots O_{n+N-1}$ and that it differs by a factor of 2 from (12.35) because we now consider unoriented $(n + N - r)$-planes.

(b) *Case in which* $\rho \leqslant r \leqslant n + N - 1$. Instead of the set of L_r that are contained in some qth tangent fiber, we consider the set of L_r that contain some of them and are orthogonal to $L_{n+N-r[0]}$. As before, we represent this set by Γ_r and the rth total absolute curvature of order q of M^n is defined by the same mean value as earlier. The dimension of $\Gamma_r \cap L_{n+N-r[0]}$ is now $\delta = r\rho + n - r(n + N)$.

We want to consider some particular cases.

1. The case in which $n = N$, $r = 1$, $q = 1$ has the following geometric interpretation. Let U_{2n-1} denote the unit $(2n - 1)$-sphere centered at O. Let $L_{n[0]}$ be the n-plane through O parallel to the tangent space T_P at $P \in M^n$. The intersection $U_{2n-1} \cap L_{n[0]}$ is an $(n - 1)$-dimensional great circle of U_{2n-1}. If we assume identified the pairs of antipodal points in U_{2n-1}, we have the $(2n - 1)$-dimensional elliptic space P_{2n-1} and the intersections $U_{2n-1} \cap L_{n[0]}$ define an n-parameter family of $(n - 1)$-planes in P_{2n-1}, say C_{n-1}. Let $v_{n-1}(y)$ be the number of $(n - 1)$-planes of C_{n-1} that contain the point $y \in P_{2n-1}$ and let $v_{2n-1}(\eta)$ denote the number of $(n - 1)$-planes of C_{n-1} that are contained in the hyperplane η of P_{2n-1}. Let $d\sigma_{2n-1}(y)$ denote the volume element in P_{2n-1} at y and let $dL_{2n-1}(\eta)$ denote the density for hyperplanes of P_{2n-1} at η (see Chapter 17, Section 3). Then the curvatures above are clearly equal to

$$K_{1,n}^{1}(M^n) = \frac{2}{O_{2n-1}} \int_{P_{2n-1}} v_{n-1}(y)\, d\sigma_{2n-1}(y),$$

$$K_{2n-1,n}^{1} = \frac{2}{O_{2n-1}} \int_{L_{2n-1} \subset P_{2n-1}} v_{2n-1}(\eta)\, dL_{2n-1}(\eta).$$

ISBN 0-201-13500-0

For $n = 2$, $N = 2$, $r = 1$, we have a congruence of lines C_1 in P_3 and, in a certain sense, the foregoing curvatures are the mean "order" and the mean "class" of the congruence C_1.

2. The case in which $r = n + N - 1$ gives rise to the curvature defined by Chern and Lashof [116].

3. *The case of curves*, $n = 1$. For curves, $n = 1$, we have the two possibilities $\rho = N$, $r = 1$ and $\rho = 1$, $r = N$. For $\rho = N$, $r = 1$ the corresponding curvature is

$$K_{1,N}^{(N)}(M^1) = \frac{1}{O_N} \int_{G_{N,1}} \nu_1 \, dL_{N[0]}$$

where ν_1 is the number of lines in E_{n+N} that are orthogonal to $L_{N[0]}$ and are contained in some Nth tangent fiber of the curve M^1. Note that $G_{N,1}$ is the unit sphere U_N and $dL_{N[0]}$ is the area element of this sphere. If $\mathbf{e}_1, \ldots, \mathbf{e}_{N+1}$ are the principal normals of M^1, it is not hard to see that the right-hand side of the last equation is equal to the length of the spherical curve $\mathbf{e}_{N+1}(s)$ (s = arc length of M^1) up to the factor $1/\pi$. That is, if κ_N is the Nth curvature of M^1 (see [165, p. 107]), we have

$$K_{1,N}^{(N)} = \frac{1}{\pi} \int_{M^1} |\kappa_N| \, ds.$$

For curves in E_3, κ_N is the torsion of the curve.

For the case in which $\rho = 1$, $r = N$ we have the curvature

$$K_{N,N}^{(1)} = \frac{1}{O_N} \int_{G_{1,N}} \nu_N \, dL_{1[0]}$$

where ν_N is the number of hyperplanes L_N of E_{N+1} that are orthogonal to $L_{1[0]}$ and contain some tangent line to M^1. The right-hand side of the last equation is equal to the length of the curve $\mathbf{e}_1(s)$ (spherical image of M^1), up to the factor $1/\pi$. Therefore if κ_1 denotes the first curvature of M^1, we have

$$K_{N,N}^{(1)}(M^1) = \frac{1}{\pi} \int_{M^1} |\kappa_1| \, ds.$$

Noting that for each direction $L_{1[0]}$ there are at least two hyperplanes orthogonal to $L_{1[0]}$ that contain a tangent line to M^1, we see that the mean value $K_{N,N}^{(1)}$ is $\geqslant 2$ and hence $\int_{M^1} |\kappa_1| \, ds \geqslant 2\pi$, which is a classical inequality of Fenchel [187]. If the curve M^1 has at least four hyperplanes orthogonal to an arbitrary direction $L_{1[0]}$ that contain a tangent line to M^1 (as happens for knotted curves in E_3), the mean value of $K_{N,N}^{(1)}$ is $\geqslant 4$ and we have the inequality $\int |\kappa_1| \, ds \geqslant 4\pi$, which is due to Fáry [168]. For details and complements see [588, 590, 591]. Related results have been obtained by Slavskii [614].

ISBN 0-201-13500-0

ISBN 0-201-13500-0

Let M^h be a compact differentiable manifold of dimension h enclosed by a sphere of radius r in E_n. The problem arises of finding inequalities connecting the area and the total absolute curvatures of M^h. For instance, for $n = 3$, $h = 2$ we have $F \leqslant (4/\pi)r^2K$ where F is the area and K the total absolute curvature (Gaussian curvature) of M^2. This inequality is due to Fáry [169]. In this connection, see [96, 97].

CHAPTER 15

The Kinematic Density in E_n

1. Formulas on Densities

In Section 1 of Chapter 12 we defined the volume element of the group of motions in E_n. Now we shall investigate in detail the group of special motions, that is, the transformation group $x' = ax + b$, where a is an orthogonal matrix and $\det a = +1$. As is usual, the kinematic density for this group will be represented by dK and is given by

$$dK = \bigwedge_i \omega_i \bigwedge_{j<h} \omega_{jh} \tag{15.1}$$

where $\omega_i = dx \cdot \mathbf{e}_i$, $\omega_{jh} = \mathbf{e}_j \cdot d\mathbf{e}_h = -\mathbf{e}_h \cdot d\mathbf{e}_j$ $(i, j, h = 1, 2, \ldots, n)$ (12.3).

Denoting by $dP = \bigwedge \omega_i$ $(i = 1, 2, \ldots, n)$ the volume element of E_n at point x and by $dK_{[x]}$ the kinematic density of the group of special rotations about x (isomorphic to $SO(n)$), we have (12.13)

$$dK = dP \wedge dK_{[x]}. \tag{15.2}$$

Note that $dK_{[x]}$ and $dM_{[x]}$ (12.7) are the same differential form. The only difference is that the total measure of $M_{[x]}$ is twice the total measure of the group of special rotations.

Let M^q be a fixed q-dimensional manifold and M^r a moving one of dimension r, both assumed piecewise smooth of class C^1, having finite volumes $\sigma_q(M^q)$ and $\sigma_r(M^r)$, respectively. Let $q + r \geqslant n$ and consider positions of M^r such that $M^q \cap M^r \neq \emptyset$. Let $x \in M^q \cap M^r$ and choose the orthonormal vectors $\mathbf{e}_1, \mathbf{e}_2, \ldots, \mathbf{e}_n$ such that $\mathbf{e}_1, \mathbf{e}_2, \ldots, \mathbf{e}_{r+q-n}$ are tangent to $M^q \cap M^r$ and $\mathbf{e}_{r+q-n+1}, \ldots, \mathbf{e}_r$ are tangent to M^r. Let $\mathbf{b}_1, \ldots, \mathbf{b}_{n-r}$ be orthonormal vectors such that $\mathbf{e}_1, \mathbf{e}_2, \ldots, \mathbf{e}_{r+q-n}, \mathbf{b}_1, \ldots, \mathbf{b}_{n-r}$ span the tangent q-plane to M^q at x.

ENCYCLOPEDIA OF MATHEMATICS and Its Applications, Gian-Carlo Rota (ed.).
1, Luis A. Santaló, Integral Geometry and Geometric Probability

ISBN 0-201-13500-0

Since $x \in M^q$, we have

$$dx = \sum_{h=1}^{r+q-n} \alpha_h \mathbf{e}_h + \sum_{j=1}^{n-r} \beta_j \mathbf{b}_j \tag{15.3}$$

where α_h and β_j are 1-forms. Thus

$$\omega_{r+h} = dx \cdot \mathbf{e}_{r+h} = \sum_{j=1} \beta_j(\mathbf{b}_j \cdot \mathbf{e}_{r+h}), \qquad h = 1, 2, \ldots, n - r, \tag{15.4}$$

and

$$\bigwedge_{h=1}^{n-r} \omega_{r+h} = \Delta \bigwedge_{j=1}^{n-r} \beta_j \tag{15.5}$$

where Δ is the $(n - r) \times (n - r)$ determinant

$$\Delta = |(\mathbf{b}_j \cdot \mathbf{e}_{r+h})| \; (j, h = 1, 2, \ldots, n - r). \tag{15.6}$$

The exterior product $\bigwedge \beta_j$ $(j = 1, 2, \ldots, n - r)$ is the $(n - r)$-dimensional volume element on M^q in the direction of the tangent $(n - r)$-plane orthogonal to $M^q \cap M^r$. Therefore, denoting by $d\sigma_{q+r-n}(x)$ the volume element of $M^q \cap M^r$ at x, we can write

$$d\sigma_{q+r-n}(x) \bigwedge_{j=1}^{n-r} \beta_j = d\sigma_q(x) \tag{15.7}$$

where $d\sigma_q(x)$ is the q-dimensional volume element of M^q at x.

On the other hand, the exterior product $\omega_1 \wedge \omega_2 \wedge \cdots \wedge \omega_r$ is equal to the volume element $d\sigma_r(x)$ of M^r at x. Thus, multiplying (15.5) by $d\sigma_{r+q-n}(x) \wedge \omega_1 \wedge \cdots \wedge \omega_r$ and taking (15.7) into account, we get

$$d\sigma_{r+q-n}(x) \bigwedge_{i=1}^{n} \omega_i = \Delta \, d\sigma_q(x) \wedge d\sigma_r(x). \tag{15.8}$$

Multiplying by $dK_{[x]} = \bigwedge \omega_{jh}$ $(j < h; j, h = 1, 2, \ldots, n)$ we get

$$d\sigma_{r+q-n}(x) \wedge dK = \Delta \, d\sigma_q(x) \wedge d\sigma_r(x) \wedge dK_{[x]}, \tag{15.9}$$

which is a very useful differential formula, similar to (14.64). For $r + q - n = 0$, we have

$$dK = \Delta \, d\sigma_q(x) \wedge d\sigma_r(x) \wedge dK_{[x]}. \tag{15.10}$$

Note that Δ depends only on the angles between the unit vectors $\mathbf{e}_{r+h}$ and $\mathbf{b}_j$ $(h, j = 1, 2, \ldots, n - r)$, not on the point x.

Another important differential formula is the following. Let L_h be the h-plane through x spanned by $\mathbf{e}_1, \ldots, \mathbf{e}_h$. We have

$$dL_h = \bigwedge_i \omega_i \bigwedge_{j,k} \omega_{jk} \qquad (j = 1, 2, \ldots, h; \; i, k = h + 1, \ldots, n). \tag{15.11}$$

ISBN 0-201-13500-0

Let dK^h denote the kinematic density in L_h, that is,

$$dK^h = \bigwedge_j \omega_j \bigwedge_{s<m} \omega_{sm}, \qquad j, s, m = 1, 2, \ldots, h, \tag{15.12}$$

and let $dK_{[x]}^{n-h}$ denote the kinematic density about x in the $(n - h)$-plane orthogonal to L_h (density for rotations about L_h, given by (12.7)). Then

$$dK_{[x]}^{n-h} = \bigwedge_{i<k} \omega_{ik}, \qquad i, k = h + 1, \ldots, n. \tag{15.13}$$

From (15.1), (15.11), (15.12), and (15.13) it follows that

$$dK = dL_h^* \wedge dK^h \wedge dK_{[x]}^{n-h}, \tag{15.14}$$

which holds good for $h = 1, 2, \ldots, n - 1$. We have used L_h^* in order to indicate that L_h must be supposed "oriented," so that the factor 2 arises when the right-hand side of (15.14) is integrated.

We leave to the reader the proof of the following analogue of (15.14). Let $L_{h[0]}$ be the h-plane through O that is parallel to the h-plane determined by $(x, \mathbf{e}_1, \ldots, \mathbf{e}_h)$ and let $L_{n-h[0]}$ be the $(n - h)$-plane orthogonal to $L_{h[0]}$ through O. Then

$$dK = dL_{h[0]}^* \wedge dK^h \wedge dK^{n-h} \tag{15.15}$$

where dK^h and dK^{n-h} denote the kinematic density on $L_{h[0]}$ and $L_{n-h[0]}$, respectively.

2. Integral of the Volume $\sigma_{r+q-n}(M^r \cap M^q)$

Integrating (15.9) over all positions of M^r we have

$$\int_{M^q \cap M^r \neq \emptyset} \sigma_{r+q-n}(M^q \cap M^r)\, dK = c\sigma_q(M^q)\sigma_r(M^r) \tag{15.16}$$

where c is a constant (independent of x and independent of the manifolds M^q, M^r) that depends on the dimensions q and r and is given by the integral

$$c = \int \Delta\, dK_{[x]}$$

taken over all positions of M^r about x. In order to calculate c, let us consider the case when M^r is a finite domain of an r-plane L_r. Expression (15.14) for dK (for $h = r$) leads to

$$\int_{M^q \cap M^r \neq \emptyset} \sigma_{r+q-n}(M^q \cap M^r)\, dL_r^* \wedge dK^r \wedge dK_{[x]}^{n-r} = c\sigma_q(M^q)\sigma_r(M^r) \tag{15.17}$$

ISBN 0-201-13500-0

where we can set $dK^r = dP^r \wedge dK^r_{[x]}$, dP^r being the volume element in L_r at x. Keeping L_r and x fixed and applying (12.10), recalling that the right-hand side must be divided by 2, we have

$$\int dK^r_{[x]} = O_{r-1} \cdots O_1, \qquad \int dK^{n-r}_{[x]} = O_{n-r-1} \cdots O_1 \tag{15.18}$$

where the integrals are taken over all positions of M^r about x. Furthermore, the integral of dP^r (by fixed L_r) gives the volume $\sigma_r(M^r)$, so that (15.17) becomes

$$O_{r-1} \cdots O_1 O_{n-r-1} \cdots O_1 \int_{M^q \cap M^r \neq \emptyset} \sigma_{r+q-n}(M^q \cap M^r)\, dL_r^* = c\sigma_q(M^q). \tag{15.19}$$

Thus, applying (14.69), we get $c = O_n \cdots O_1 O_{r+q-n}(O_r O_q)^{-1}$ and the final result is

$$\int_{M^q \cap M^r \neq \emptyset} \sigma_{r+q-n}(M^r \cap M^q)\, dK = \frac{O_n \cdots O_1 O_{r+q-n}}{O_q O_r} \sigma_q(M^q)\sigma_r(M^r). \tag{15.20}$$

For $r + q - n = 0$, $\sigma_{r+q-n}(M^q \cap M^r)$ denotes the number of points of the intersection $M^q \cap M^r$, so that (15.20) can be written

$$\int_{M^r \cap M^{n-r} \neq \emptyset} N(M^r \cap M^{n-r})\, dK = \frac{O_n \cdots O_1 O_0}{O_r O_{n-r}} \sigma_r(M^r)\sigma_{n-r}(M^{n-r}). \tag{15.21}$$

The case of several moving manifolds. Let M^q be a fixed q-dimensional manifold in E_n and consider h moving manifolds $M^{r_1}, \ldots, M^{r_h}$ of dimensions $r_1, \ldots, r_h$, respectively. Assuming $q + r_1 + \cdots + r_h - nh \geqslant 0$ and calling dK_i the kinematic density for M^{r_i}, we find by induction that

$$\int \sigma_{q+r_1+\ldots+r_h-nh}(M^q \cap M^{r_1} \cap \cdots \cap M^{r_h})\, dK_1 \wedge \cdots \wedge dK_h$$

$$= \frac{(O_n \cdots O_1)^h O_{q+r_1+\ldots+r_h-hn}}{O_q O_{r_1} \cdots O_{r_h}} \sigma_q(M^q)\sigma_{r_1}(M^{r_1}) \cdots \sigma_{r_h}(M^{r_h}) \tag{15.22}$$

where the integral is taken over all positions of M^i for which $M^q \cap M^{r_1} \cap \cdots \cap M^{r_h} \neq \emptyset$.

It is interesting to consider the case of moving hyperspheres. Let $r_1 = r_2 = \cdots = r_h = n - 1$ and let each M^{r_i} be a hypersphere Σ_i of radius R_i. We may take the origin of moving frames to be the center of the respective hypersphere. Then from $dK_i = dP_i \wedge dK_{i[x]}$, using the fact that the integral of $dK_{i[x]}$ is equal to $O_{n-1} \cdots O_1$ (12.10), we get

ISBN 0-201-13500-0

$$\int \sigma_{q-h}(M^q \cap \Sigma_1 \cap \cdots \cap \Sigma_h)\, dP_1 \wedge dP_2 \wedge \cdots \wedge dP_h = \frac{O_n{}^h O_{q-h}}{O_q} R^{(n-1)h} \sigma_q(M^q). \tag{15.23}$$

This formula holds for $h = 1, 2, \ldots, q$. For $h = q$, the integrand $\sigma_0(M^q \cap \Sigma_1 \cap \cdots \cap \Sigma_q)$ is the number of points of the intersection $M^q \cap \Sigma_1 \cap \cdots \cap \Sigma_q$, say $N_{12\ldots q}$, and (15.23) becomes

$$\int N_{12\ldots q}\, dP_1 \wedge \cdots \wedge dP_q = (2O_n{}^q/O_q) R^{q(n-1)} \sigma_q(M^q) \tag{15.24}$$

where the integral is extended over the whole space $E_n \times E_n \times \cdots \times E_n$ (q times), $N_{12\ldots q}$ being zero when $M^q \cap \Sigma_1 \cap \cdots \cap \Sigma_q = \emptyset$. Some simple special cases are the following:

(a) $n = 2$, $q = 1$. Then M^1 is a curve of the plane and (15.24) becomes $\int N_1\, dP_1 = 4LR$ where L denotes the length of M^1.

(b) $n = 3$, $q = 1$. Then M^1 is a curve of 3-space and (15.24) becomes $\int N_1\, dP_1 = 2\pi R^2 L$.

We have assumed that M^q is a piecewise smooth manifold of class C^1. Conversely, (15.24) conduces to define the q-dimensional measure of a continuum of points in E_n by the formula

$$\sigma_q(M^q) = \frac{O_q}{2O_n{}^q R^{q(n-1)}} \int N_{12\ldots q}\, dP_1 \wedge \cdots \wedge dP_q \tag{15.25}$$

provided the integral on the right-hand side exists. Integral geometry has been applied to the definition of q-dimensional measure for continua of points by Federer [174, 175], Hadwiger [274], Nöbeling [452, 453], Santaló [559], and Maak [380]. (See Federer's book [178].)

If M^r is a line segment of unit length ($r = 1$) and M^{n-1} is contained in a domain of volume V in E_n, the expected number N of intersection points $M^1 \cap M^{n-1}$, according to (15.21), is $E(N) = (O_n/\pi O_{n-1})\sigma_{n-1}(M^{n-1})/V$. This formula has been used to define the surface area of a given hypersurface by Kulle and Reich [350].

3. A Differential Formula

One of the most important results in integral geometry is the so-called fundamental formula, which we saw for the case of the plane in Chapter 7. Our purpose here is to extend this formula to n-dimensional euclidean space. The proof we shall give in the next section is due to Chern [108] and is a generalization of Blaschke's proof for $n = 3$ in [42] (see also [586]). First, however, we shall prove an interesting differential formula.

ISBN 0-201-13500-0

Let S_0 and S_1 be two piecewise smooth hypersurfaces in E_n of class C^2. We assume S_0 fixed and S_1 moving with the kinematic density dK_1. Consider a generic position of S_1 in which the intersection $S_0 \cap S_1$ is a manifold of dimension $n-2$. Let $x \in S_0 \cap S_1$ and consider the orthonormal frame $(x; \mathbf{e}_1, \ldots, \mathbf{e}_n)$ such that $\mathbf{e}_1, \mathbf{e}_2, \ldots, \mathbf{e}_{n-2}$ span the tangent $(n-2)$-plane to $S_0 \cap S_1$ and $\mathbf{e}_{n-1}$, $\mathbf{e}_n$ are, respectively, tangent and normal unit vectors to S_1 at x. Let $\mathbf{e}'_{n-1}$, $\mathbf{e}_n'$ be the unit vectors that are, respectively, tangent and normal to S_0 at x, so that $(x; \mathbf{e}_1, \ldots, \mathbf{e}_{n-2}, \mathbf{e}'_{n-1}, \mathbf{e}_n')$ is a second orthonormal frame of origin x. The kinematic density is

$$dK_1 = \bigwedge_i (dx \cdot \mathbf{e}_i) \bigwedge_{h<j} (d\mathbf{e}_h \cdot \mathbf{e}_j), \qquad i, j, h = 1, 2, \ldots, n. \tag{15.26}$$

Since the densities are always considered in absolute value, there is no question of sign. We define the kinematic density dT_1 on S_1 (i.e., the density for sets of frames $(x; \mathbf{e}_1, \ldots, \mathbf{e}_{n-1}, \mathbf{e}_n)$ such that $\mathbf{e}_n$ is the unit normal to S_1) by

$$dT_1 = \bigwedge_i (dx \cdot \mathbf{e}_i) \bigwedge_{h<j} (\mathbf{e}_h \cdot d\mathbf{e}_j), \qquad i, h, j = 1, 2, \ldots, n-1, \tag{15.27}$$

so that

$$dK_1 = dT_1 \wedge (dx \cdot \mathbf{e}_n) \wedge (d\mathbf{e}_1 \cdot \mathbf{e}_n) \wedge \cdots \wedge (d\mathbf{e}_{n-1} \cdot \mathbf{e}_n). \tag{15.28}$$

Similarly, the kinematic density on S_0 (density for frames $(x; \mathbf{e}_1, \ldots, \mathbf{e}_{n-2}, \mathbf{e}'_{n-1}, \mathbf{e}_n')$ tangent to S_0) is defined by

$$dT_0 = \bigwedge_i (dx \cdot \mathbf{e}_i) \bigwedge_{h<j} (\mathbf{e}_h \cdot d\mathbf{e}_j) \wedge (dx \cdot \mathbf{e}'_{n-1}) \wedge (d\mathbf{e}_1 \cdot \mathbf{e}'_{n-1}) \wedge \cdots \wedge (d\mathbf{e}_{n-2} \cdot \mathbf{e}'_{n-1}) \tag{15.29}$$

where the range of indices is $i, h, j = 1, 2, \ldots, n-2$.

The kinematic density on $S_0 \cap S_1$ (i.e., density for frames $(x; \mathbf{e}_1, \ldots, \mathbf{e}_{n-2})$ attached to $S_0 \cap S_1$) is

$$dT_{01} = \bigwedge_i (dx \cdot \mathbf{e}_i) \bigwedge_{h<j} (\mathbf{e}_h \cdot d\mathbf{e}_j), \qquad i, h, j = 1, 2, \ldots, n-2. \tag{15.30}$$

Thus dT_0 can be written

$$dT_0 = dT_{01} \wedge (dx \cdot \mathbf{e}'_{n-1}) \wedge (d\mathbf{e}_1 \cdot \mathbf{e}'_{n-1}) \wedge \cdots \wedge (d\mathbf{e}_{n-2} \cdot \mathbf{e}'_{n-1}). \tag{15.31}$$

Let ϕ be the angle between $\mathbf{e}_n'$ and $\mathbf{e}_n$. On the plane normal to $S_0 \cap S_1$, determined by $\mathbf{e}_n$ and $\mathbf{e}_n'$, we have

$$\mathbf{e}_n = \sin\phi\, \mathbf{e}'_{n-1} + \cos\phi\, \mathbf{e}_n', \qquad \mathbf{e}_{n-1} = \cos\phi\, \mathbf{e}'_{n-1} - \sin\phi\, \mathbf{e}_n' \tag{15.32}$$

and hence

ISBN 0-201-13500-0

$$dx \cdot \mathbf{e}_n = \sin\phi\,(dx \cdot \mathbf{e}'_{n-1}) + \cos\phi\,(dx \cdot \mathbf{e}_n')$$

$$d\mathbf{e}_i \cdot \mathbf{e}_n = \sin\phi\,(d\mathbf{e}_i \cdot \mathbf{e}'_{n-1}) + \cos\phi\,(d\mathbf{e}_i \cdot \mathbf{e}_n'), \qquad i = 1, 2, \ldots, n-2,$$

$$d\mathbf{e}_{n-1} \cdot \mathbf{e}_n = -\,d\phi - \mathbf{e}'_{n-1} \cdot d\mathbf{e}_n'. \tag{15.33}$$

Since $\mathbf{e}_n'$ is normal to S_0, we have $dx \cdot \mathbf{e}_n' = 0$ and since S_0 is a fixed hypersurface, the unit vector $\mathbf{e}_n'$ does not depend on the position of S_1, so that $d\mathbf{e}_i \cdot \mathbf{e}_n' = -\,\mathbf{e}_i \cdot d\mathbf{e}_n' = 0$ and $d\mathbf{e}'_{n-1} \cdot \mathbf{e}_n' = 0$. Hence, (15.28) and (15.33) give, up to the sign,

$$dK_1 = \sin^{n-1}\phi\, d\phi \wedge dT_1 \wedge (dx \cdot \mathbf{e}'_{n-1}) \wedge (d\mathbf{e}_1 \cdot \mathbf{e}'_{n-1}) \wedge \cdots \wedge (d\mathbf{e}_{n-2} \cdot \mathbf{e}'_{n-1}). \tag{15.34}$$

Multiplying both sides of this formula by dT_{01} and using (15.31), we get

$$dT_{01} \wedge dK_1 = \sin^{n-1}\phi\, d\phi \wedge dT_1 \wedge dT_0, \tag{15.35}$$

which is the desired formula. Note that by integrating over all locations of S_1 such that $S_0 \cap S_1 \neq \emptyset$ we obtain (15.20) for the case in which $r = q = n - 1$.

4. The Kinematic Fundamental Formula

We are now in a position to prove the kinematic fundamental formula following Chern [108]. Let D_0 and D_1 be two domains in E_n bounded by the hypersurfaces ∂D_0 and ∂D_1, which we assume to be of class C^2. Moreover, we assume that D_0 and D_1 are such that for all positions of D_1 the intersection $D_0 \cap D_1$ has a finite number of components. Suppose D_0 fixed and D_1 moving. Then, if dK_1 denotes the kinematic density for D_1 and M_i^{0}, M_i^{1} are the ith integrals of mean curvature of ∂D_0 and ∂D_1, respectively, the kinematic fundamental formula in E_n is

$$\int_{D_0 \cap D_1 \neq \emptyset} \chi(D_0 \cap D_1)\, dK_1 = O_1 \cdots O_{n-2}\Bigg[O_{n-1}\chi(D_0)V_1 + O_{n-1}\chi(D_1)V_0 + \frac{1}{n}\sum_{h=0}^{n-2}\binom{n}{h+1} M_h^{0} M_{n-h-2}^{1}\Bigg] \tag{15.36}$$

where χ denotes the Euler–Poincaré characteristic and V_0 and V_1 are the volumes of D_0 and D_1, respectively.

To prove (15.36) we use (13.38), that is,

$$\chi(D_0 \cap D_1) = (1/O_{n-1}) M_{n-1}(\partial(D_0 \cap D_1)) \tag{15.37}$$

where $M_{n-1}(\partial(D_0 \cap D_1))$ is the volume of the spherical image of $\partial(D_0 \cap D_1)$. The boundary $\partial(D_0 \cap D_1)$ consists of a finite number of hypersurfaces $\partial D_0 \cap D_1$ and $D_0 \cap \partial D_1$ that intersect in $\partial D_0 \cap \partial D_1$, the $(n-2)$-dimensional

ISBN 0-201-13500-0

edges of the intersection. We have

$$M_{n-1}(\partial(D_0 \cap D_1)) = M_{n-1}(\partial D_0 \cap D_1) + M_{n-1}(D_0 \cap \partial D_1) + M_{n-1}(\partial D_0 \cap \partial D_1). \tag{15.38}$$

The last term on the right-hand side corresponds to the spherical image of the normals belonging to the angle subtended by the outward normals to ∂D_0 and ∂D_1 at the points of $\partial D_0 \cap \partial D_1$. In order to calculate the volume of this spherical image, we use the notation of the last section and proceed as follows. Let $\mathbf{e}_n$ and $\mathbf{e}_n'$ be the unit normal vectors to ∂D_0 and ∂D_1, respectively, and denote by $\mathbf{v}$, $\mathbf{w}$ two unit vectors in the directions of the angle bisectors of $\mathbf{e}_n$, $\mathbf{e}_n'$. Let ϕ be the angle between $\mathbf{e}_n$ and $\mathbf{e}_n'$. We have

$$\mathbf{e}_n = \left(\cos\frac{\phi}{2}\right)\mathbf{v} - \left(\sin\frac{\phi}{2}\right)\mathbf{w}, \qquad \mathbf{e}_n' = \left(\cos\frac{\phi}{2}\right)\mathbf{v} + \left(\sin\frac{\phi}{2}\right)\mathbf{w} \tag{15.39}$$

and therefore

$$\mathbf{v} = [2\cos(\phi/2)]^{-1}(\mathbf{e}_n + \mathbf{e}_n'), \qquad \mathbf{w} = [2\sin(\phi/2)]^{-1}(\mathbf{e}_n' - \mathbf{e}_n). \tag{15.40}$$

Let ξ be a variable unit vector between $\mathbf{e}_n$ and $\mathbf{e}_n'$. Let α denote the angle between ξ and $\mathbf{v}$ and define the unit vector η by

$$\xi = \cos\alpha\,\mathbf{v} + \sin\alpha\,\mathbf{w}, \qquad \eta = -\sin\alpha\,\mathbf{v} + \cos\alpha\,\mathbf{w} \tag{15.41}$$

for $-\phi/2 \leqslant \alpha \leqslant \phi/2$. The volume element on the unit sphere U_{n-1} corresponding to the direction ξ is

$$\begin{aligned} du_{n-1} &= (d\xi\cdot\mathbf{e}_1) \wedge (d\xi\cdot\mathbf{e}_2) \wedge\cdots\wedge (d\xi\cdot\mathbf{e}_{n-2}) \wedge d\alpha \\ &= \bigwedge_{i=1}^{n-2} [\cos\alpha(d\mathbf{v}\cdot\mathbf{e}_i) + \sin\alpha(d\mathbf{w}\cdot\mathbf{e}_i)] \wedge d\alpha. \end{aligned} \tag{15.42}$$

By (15.40) we have

$$\cos\alpha(d\mathbf{v}\cdot\mathbf{e}_i) + \sin\alpha(d\mathbf{w}\cdot\mathbf{e}_i) = \frac{\sin(\phi/2 - \alpha)}{\sin(\phi)}(d\mathbf{e}_n\cdot\mathbf{e}_i) + \frac{\sin(\phi/2 + \alpha)}{\sin(\phi)}(d\mathbf{e}_n'\cdot\mathbf{e}_i) \tag{15.43}$$

and therefore

$$du_{n-1} = \frac{1}{\sin^{n-2}(\phi)} \bigwedge_{i=1}^{n-2} \left[\sin\left(\frac{\phi}{2} - \alpha\right)(d\mathbf{e}_n\cdot\mathbf{e}_i) + \sin\left(\frac{\phi}{2} + \alpha\right)(d\mathbf{e}_n'\cdot\mathbf{e}_i)\right] \wedge d\alpha. \tag{15.44}$$

Let us introduce the unit tangent vectors $\mathbf{v}_1, \ldots, \mathbf{v}_{n-1}$ to ∂D_1 in the principal directions and the unit tangent vectors $\mathbf{v}_1', \ldots, \mathbf{v}_{n-1}'$ to ∂D_0 in the principal

ISBN 0-201-13500-0

directions at point x. We have

$$\mathbf{e}_i = \sum_{h=1}^{n-1} c_{ih}\mathbf{v}_h = \sum_{h=1}^{n-1} c'_{ih}\mathbf{v}_h' \qquad (i = 1, 2, \ldots, n-2),$$

$$\mathbf{e}_{n-1} = \sum_{h=1}^{n-1} c_{n-1,h}\mathbf{v}_h, \qquad \mathbf{e}'_{n-1} = \sum_{h=1}^{n-1} c'_{n-1,h}\mathbf{v}_h'. \tag{15.45}$$

Since the vector sets $\{\mathbf{e}_i, \mathbf{e}'_{n-1}\}$, $\{\mathbf{v}_h\}$, $\{\mathbf{v}_h'\}$ are orthonormal systems, the matrices (c_{ih}) and (c'_{ih}) are orthogonal matrices, and therefore we have

$$\mathbf{v}_h = \sum_{i=1}^{n-1} c_{ih}\mathbf{e}_i, \qquad \mathbf{v}_h' = \sum_{i=1}^{n-2} c'_{ih}\mathbf{e}_i + c'_{n-1,h}\mathbf{e}'_{n-1}. \tag{15.46}$$

The equations of Rodrigues for hypersurfaces in E_n give $\mathbf{v}_h \cdot d\mathbf{e}_n = -\kappa_h(dx \cdot \mathbf{v}_h)$, where κ_h denotes the principal curvature of ∂D_1 in the $\mathbf{v}_h$ direction. Therefore we have $(i = 1, 2, \ldots, n-2)$

$$d\mathbf{e}_n \cdot \mathbf{e}_i = \sum_{h=1}^{n-1} c_{ih}(d\mathbf{e}_n \cdot \mathbf{v}_h) = -\sum_{h=1}^{n-1} c_{ih}\kappa_h(dx \cdot \mathbf{v}_h)$$

$$= -\sum_{h,j=1}^{n-1} c_{ih}\kappa_h c_{jh}(dx \cdot \mathbf{e}_j). \tag{15.47}$$

Similarly, for ∂D_0 we have

$$d\mathbf{e}_n' \cdot \mathbf{e}_i = \sum_{h=1}^{n-1} c'_{ih}\kappa_h' \left(\sum_{j=1}^{n-2} c'_{jh}(dx \cdot \mathbf{e}_j) + c'_{n-1,h}(dx \cdot \mathbf{e}'_{n-1}) \right). \tag{15.48}$$

In order to compute du_{n-1} by (15.44) we use that x displaces on $\partial D_0 \cap \partial D_1$ so that $dx \cdot \mathbf{e}_{n-1} = 0$, $dx \cdot \mathbf{e}'_{n-1} = 0$, and thus

$$\sin\left(\frac{\phi}{2} - \alpha\right)(d\mathbf{e}_n \cdot \mathbf{e}_i) + \sin\left(\frac{\phi}{2} + \alpha\right)(d\mathbf{e}_n' \cdot \mathbf{e}_i)$$

$$= -\sum_{j=1}^{n-2}\left[\sin\left(\frac{\phi}{2} - \alpha\right)\sum_{h=1}^{n-1} c_{ih}c_{jh}\kappa_h + \sin\left(\frac{\phi}{2} + \alpha\right)\sum_{h=1}^{n-1} c'_{ih}c'_{jh}\kappa_h'\right](dx \cdot \mathbf{e}_j). \tag{15.49}$$

Substituting in (15.44) and using the fact that the product $(dx \cdot \mathbf{e}_1)(dx \cdot \mathbf{e}_2) \cdots (dx \cdot \mathbf{e}_{n-2})$ is the volume element $d\sigma_{n-2}$ of $\partial D_0 \cap \partial D_1$, we get

$$du_{n-1} = \frac{H}{\sin^{n-2}(\phi)} d\sigma_{n-2} \wedge d\alpha \tag{15.50}$$

where H is the determinant of type $(n-2) \times (n-2)$ whose elements are

$$H_{ij} = -\sin\left(\frac{\phi}{2} - \alpha\right)\sum_{h=1}^{n-1} c_{ih}c_{jh}\kappa_h - \sin\left(\frac{\phi}{2} + \alpha\right)\sum_{h=1}^{n-1} c'_{ih}c'_{jh}\kappa_h'. \tag{15.51}$$

ISBN 0-201-13500-0

Therefore H has the form

$$H = \sum_{p=0}^{n-2} H_p \sin^{n-2-p}\left(\frac{\phi}{2} - \alpha\right) \sin^p\left(\frac{\phi}{2} + \alpha\right) \tag{15.52}$$

where

$$H_p = \sum_{i_h, j_m} A_{i_1 \ldots i_q j_1 \ldots j_p} \kappa_{i_1} \cdots \kappa_{i_q} \kappa'_{j_1} \cdots \kappa'_{j_p} \tag{15.53}$$

where $p + q = n - 2$.

The coefficients $A_{i_1 \ldots i_q j_1 \ldots j_p}$ are functions of c_{ih}, c'_{ih} and the sum is extended over all combinations $(i_1, \ldots, i_q)$ and $(j_1, \ldots, j_p)$. That is, the only terms in H that depend on ∂D_0, ∂D_1 are the products $\kappa_{i_1} \cdots \kappa_{i_q} \kappa'_{j_1} \cdots \kappa'_{j_p}$; the remaining coefficients depend only on the position of ∂D_1 about x, and they are the same for every pair of hypersurfaces ∂D_0, ∂D_1.

Since $M_{n-1}(\partial D_0 \cap \partial D_1)$ is equal to the volume of the spherical image corresponding to the $(n - 2)$-dimensional edges $\partial D_0 \cap \partial D_1$ of the hypersurface $\partial(D_0 \cap D_1)$, by (15.50) we have

$$M_{n-1}(\partial D_0 \cap \partial D_1) = \int \frac{H}{\sin^{n-2}(\phi)} \, d\sigma_{n-2} \wedge d\alpha \tag{15.54}$$

where the integral is extended over $-\phi/2 \leqslant \alpha \leqslant \phi/2$ with respect to α and over $\partial D_0 \cap \partial D_1$ with respect to the remaining variables. Using (15.35) and noting that the density (15.30) can be written $dT_{01} = d\sigma_{n-2} \wedge du_{n-3} \wedge \cdots \wedge du_1$, we have

$$\int_{\partial D_0 \cap \partial D_1 \neq \emptyset} M_{n-1}(\partial D_0 \cap \partial D_1) \, dK_1 = (O_{n-3} \cdots O_1)^{-1} \int_{\partial D_0 \cap \partial D_1 \neq \emptyset} H \sin \phi \, d\phi \wedge d\alpha \wedge dT_1 \wedge dT_2. \tag{15.55}$$

The densities dT_0 and dT_1 on ∂D_0 and ∂D_1 are $dT_0 = d\sigma^0_{n-1} \wedge du_{n-2} \wedge \cdots \wedge du_1$ and $dT_1 = d\sigma^1_{n-1} \wedge du_{n-2} \wedge \cdots \wedge du_1$, respectively. First keeping x fixed and integrating (15.55) over all positions of ∂D_1 about x and then moving x over ∂D_0, with the aid of (15.52) and (15.53) we obtain

$$\int_{\partial D_0 \cap \partial D_1 \neq \emptyset} M_{n-1}(\partial D_0 \cap \partial D_1) \, dK_1 = \sum_{p=0}^{n-2} c_p M_p^{\,0} M^1_{n-2-p} \tag{15.56}$$

where c_p are constants and $M_p^{\,0}$, M^1_{n-2-p} are the integrals of mean curvature of ∂D_0 and ∂D_1, respectively.

It remains to calculate the integrals of the first and second terms on the right-hand side in (15.38). To this end, let P be a point of $\partial D_0 \cap D_1$ and let

ISBN 0-201-13500-0

$du_{n-1}(P)$ denote the volume element of the spherical image of P. Fixing D_1 and then letting P vary on $\partial D_0 \cap D_1$, we get

$$\int_{P\in\partial D_0\cap D_1} du_{n-1}(P) \wedge dK_1 = \int_{D_0\cap D_1\neq\emptyset} M_{n-1}(\partial D_0 \cap D_1)\, dK_1 \qquad (15.57)$$

and fixing P and then rotating D_1 about this point and letting it vary over D_1 and ∂D_0, we have

$$\int_{P\in\partial D_0\cap D_1} du_{n-1} \wedge dK_1 = O_{n-1}\cdots O_1 V_1 M_{n-1}^0 \qquad (15.58)$$

where $M_{n-1}^0 = M_{n-1}(\partial D_0)$. Thus, from (15.57) and (15.58) we have

$$\int_{D_0\cap D_1\neq\emptyset} M_{n-1}(\partial D_0 \cap D_1)\, dK_1 = O_{n-1}\cdots O_1 V_1 M_{n-1}^0 \qquad (15.59)$$

where V_1 denotes the volume of D_1.

Similarly, by the invariance of the kinematic measure under inversion of the motion,

$$\int_{D_0\cap D_1\neq\emptyset} M_{n-1}(D_0 \cap \partial D_1)\, dK_1 = O_{n-1}\cdots O_1 V_0 M_{n-1}^1 \qquad (15.60)$$

where V_0 denotes the volume of D_0 and we have put $M_{n-1}^1 = M_{n-1}(\partial D_1)$. From (15.37), (15.38), (15.56), (15.59), and (15.60) it follows that

$$\int_{D_0\cap D_1\neq\emptyset} \chi(D_0 \cap D_1)\, dK_1 = O_{n-1}\cdots O_1(V_1\chi_0 + V_0\chi_1) + \sum_{p=0}^{n-2} c_p M_p{}^0 M_{n-2-p}^1. \qquad (15.61)$$

It remains to determine the constants c_p. This can be done by taking D_0 to be a convex body and D_1 a hypersphere of radius ρ. Then $\chi(D_1 \cap D_0) = 1$, $\chi_0 = \chi_1 = 1$, and (15.61) gives

$$V_{1\rho}O_{n-1}\cdots O_1 = O_{n-1}\cdots O_1(V_1 + V_0) + \sum_{p=0}^{n-2} c_p M_p{}^0 M_{n-2-p}^1 \qquad (15.62)$$

where $V_{1\rho}$ denotes the volume of the parallel body of D_0 in the distance ρ. Comparing with (13.44) we get

$$c_p = \frac{O_{n-2}\cdots O_1}{r+1}\binom{n-1}{r}$$

and the proof of (15.36) is complete.

ISBN 0-201-13500-0

5. Fundamental Formula for Convex Sets

1. If D_0 and D_1 are convex bodies, we have $\chi(D_0) = \chi(D_1) = \chi(D_0 \cap D_1) = 1$ and the fundamental formula becomes

$$\int_{D_0 \cap D_1 \neq \emptyset} dK_1 = O_{n-2} \cdots O_1 \left[O_{n-1}(V_0 + V_1) + \frac{1}{n} \sum_{h=0}^{n-2} \binom{n}{h+1} M_h^{\,0} M_{n-2-h}^1 \right]. \tag{15.63}$$

Introducing the quermassintegrale W_h, we can write

$$\int_{D_0 \cap D_1 \neq \emptyset} dK_1 = O_{n-2} \cdots O_1 \sum_{h=0}^{n} n \binom{n}{h} W_h^{\,0} W_{n-h}^1. \tag{15.64}$$

Though we have proved (15.63) for bodies D_0, D_1 whose boundaries were assumed of class C^2, since any convex body can be approximated by smooth convex bodies (actually, by convex bodies whose boundaries are analytic manifolds [63]) and the quermassintegrale are continuous functionals, it follows that (15.64) holds for any pair of convex sets in E_n.

From the point of view of the theory of geometric probability, formula (15.64) may be interpreted as follows: If D, D_0 are convex bodies such that $D \subset D_0$, the probability that a random convex body D_1 intersecting D_0 also intersects D is

$$p = \frac{\sum_{h=0}^{n} \binom{n}{h} W_h W_{n-h}^1}{\sum_{h=0}^{n} \binom{n}{h} W_h^{\,0} W_{n-h}^1} \tag{15.65}$$

where W_i, $W_i^{\,0}$, $W_i^{\,1}$ are the quermassintegrale of D, D_0, and D_1, respectively.

2. If D_0 is composed of m congruent nonintersecting convex bodies K_0 and D_1 is a convex body K_1, then $\chi(D_0 \cap D_1) = N$, the number of convex bodies K_0 that are met by K_1, and the fundamental formula becomes

$$\int N\, dK_1 = O_{n-2} \cdots O_1 m \left[O_{n-1}(V_0 + V_1) + \frac{1}{n} \sum_{h=0}^{n-2} \binom{n}{h+1} M_h^{\,0} M_{n-2-h}^1 \right]. \tag{15.66}$$

6. Mean Values for the Integrals of Mean Curvature

We will pursue the situation studied in the preceding sections. We want to evaluate the integral

ISBN 0-201-13500-0

$$\int_{D_0 \cap D_1 \neq \emptyset} M_{q-1}(\partial(D_0 \cap D_1))\, dK_1, \qquad q = 1, 2, \ldots, n-1. \tag{15.67}$$

Consider first the case in which D_0 and D_1 are convex bodies. The fundamental kinematic formula is (15.63). If D_0 is a q-dimensional convex set $D_0{}^q$ contained in a q-plane L_q, formula (15.63) is still valid, but the integrals of mean curvature $M_h{}^0$ have the limit values given in Section 7 of Chapter 13. Therefore we have

$$\int_{D_0{}^q \cap D_1 \neq \emptyset} dK_1 = O_{n-2} \cdots O_1 \left[O_{n-1} V_1 + \frac{\binom{n}{n-q}}{n\binom{n-1}{q}} O_{n-q-1} \sigma_q(D_0{}^q) M^1_{q-1} \right.$$
$$\left. + \frac{1}{n} \sum_{h=n-q}^{n-2} \binom{n}{h+1} \frac{\binom{q-1}{h+q-n}}{\binom{n-1}{h}} \frac{O_h}{O_{h+q-n}} M^{0(q)}_{h+q-n} M^1_{n-2-h} \right] \tag{15.68}$$

where $\sigma_q(D_0{}^q)$ denotes the q-dimensional volume of $D_0{}^q$ and $M^{0(q)}_{h+q-n}$ denotes the $(h + q - n)$th integral of mean curvature of the boundary of $D_0{}^q$ as a convex body of L_q. In order to calculate (15.67) we use the following device. Let D_0, D_1 be two convex bodies. Let D_0 be fixed and D_1 moving with kinematic density dK_1. Let L_q be a moving q-plane and consider the integral

$$I = \int_{D_0 \cap D_1 \cap L_q \neq \emptyset} dK_1 \wedge dL_q. \tag{15.69}$$

First keeping D_1 fixed, according to (14.2) we have

$$I = \frac{O_{n-2} \cdots O_{n-q-1}}{(n-q) O_{q-1} \cdots O_0} \int_{D_0 \cap D_1 \neq \emptyset} M_{q-1}(\partial(D_0 \cap D_1))\, dK_1, \tag{15.70}$$

and then keeping L_q fixed and integrating dK_1, by (15.68) we have

$$I = O_{n-2} \cdots O_1 \int \left[O_{n-1} V_1 + \frac{\binom{n}{n-q}}{n\binom{n-1}{q}} O_{n-q-1} \sigma_q(D_0 \cap L_q) M^1_{q-1} \right.$$

ISBN 0-201-13500-0

$$+\frac{1}{n}\sum_{h=n-q}^{n-2}\binom{n}{h+1}\frac{\binom{q-1}{h+q-n}}{\binom{n-1}{h}}\frac{O_h}{O_{h+q-n}}M_{h+q-n}(\partial(D_0\cap L_q))M^1_{n-2-h}\Bigg]dL_q. \tag{15.71}$$

The last integrals can be evaluated by (14.69) and (14.10) and the obtained value, when compared with (15.70), gives the final result

$$\int_{D_0\cap D_1\neq\emptyset} M_{q-1}(\partial(D_0\cap D_1))\,dK_1$$

$$= O_{n-2}\cdots O_1\Bigg[O_{n-1}(V_1M^0_{q-1}+V_0M^1_{q-1})$$

$$+\frac{(n-q)O_{q-1}}{O_{n-q-1}}\sum_{h=n-q}^{n-2}\frac{\binom{q-1}{q+h-n}O_{2n-h-q}O_h}{(h+1)O_{n-h}O_{h+q-n}}M^1_{n-2-h}M^0_{h+q-n}\Bigg], \tag{15.72}$$

which holds for $q = 1, 2, \ldots, n-1$. For $q = n$ it must be replaced by (15.36). Division of (15.72) by (15.63) gives the expected value of $M_{q-1}(\partial(D_0\cap D_1))$.

Formula (15.72) can be generalized to m convex bodies $D_1, D_2, \ldots, D_m$ intersecting a fixed convex body D_0. The result is the expected value of $M_{q-1}(\partial(D_0\cap\cdots\cap D_m))$ (see [657]). Applications to geometric probabilities have been made by Stoka [648].

Generalization to nonconvex domains. Formula (15.72) holds for bounded domains D_0, D_1, not necessarily convex, provided the mean curvatures $M_h(\partial D_0)$, $M_h(\partial D_1)$, and $M_h(\partial(D_0\cap D_1))$ exist in the sense of Section 6 of Chapter 13. In order to show this, consider the identity

$$M_{q-1}(\partial(D_0\cap D_1)) = M_{q-1}(\partial D_0\cap D_1) + M_{q-1}(D_0\cap\partial D_1) + M_{q-1}(\partial D_0\cap\partial D_1). \tag{15.73}$$

An argument similar to that used in proving (15.59) and (15.60) gives

$$\int M_{q-1}(D_0\cap\partial D_1)\,dK_1 = O_{n-1}\cdots O_1M^1_{q-1}V_0,$$

$$\int M_{q-1}(\partial D_0\cap D_1)\,dK_1 = O_{n-1}\cdots O_1M^0_{q-1}V_1. \tag{15.74}$$

ISBN 0-201-13500-0

For the integral of the last term of (15.73) the same method that yields (15.56) gives an expression of the form

$$\int M_{q-1}(\partial D_0 \cap \partial D_1)\, dK_1 = \sum c_{nqh} M^1_{n-2-h} M^0_{h+q-n}$$

$$(h = n - q, n - q + 1, \ldots, n - 2) \tag{15.75}$$

whose coefficients c_{nqh} do not depend on the shape of D_0, D_1 (they are "local" coefficients). Consequently, they must be the same as in (15.72) and thus formula (15.72) holds for any pair of domains D_0, D_1 provided that the integrals of mean curvature $M_h{}^0$, $M_h{}^1$, and $M_h(\partial(D_0 \cap D_1))$ exist ($h = n - q$, $n - q + 1, \ldots, n - 2$).

7. Fundamental Formula for Cylinders

Let O be a fixed point in E_n and let $L_{n-p[0]}$ be an $(n - p)$-plane through O. Let D_{n-p} be a bounded domain in $L_{n-p[0]}$ whose boundary $\partial D_{n-p} \subset L_{n-p[0]}$ is smooth of class C^2. For each point $x \in D_{n-p}$ we consider the p-plane L_p orthogonal to $L_{n-p[0]}$. The set of all such L_p constitutes a cylinder Z_p. The p-planes L_p are the generators of Z_p and D_{n-p} is a normal cross section. The density for congruent cylinders in E_n is

$$dZ_p = dL^*_{n-p[0]} \wedge dK^{n-p} \tag{15.76}$$

where dK^{n-p} denotes the kinematic density on $L_{n-p[0]}$ and the asterisk means that $L_{n-p[0]}$ must be considered as oriented. From (15.76) and (15.14) it follows that the kinematic density in E_n, which we will now denote by dK^n, can be written

$$dK^n = dZ_p \wedge dK^p \tag{15.77}$$

where dK^p is the kinematic density on L_p.

Since dK^n can also be written $dK^n = dL_p{}^* \wedge dK_{[p]} \wedge dK^p$, we have

$$dZ_p = dL_p{}^* \wedge dK_{[p]} \tag{15.78}$$

where $dK_{[p]}$ denotes the density for rotations about L_p (Chapter 12, Section 1). A cylinder Z_p is determined, up to a motion, by its cross section D_{n-p} and thus we may define the integrals of mean curvature of Z_p as those of ∂D_{n-p} considered as an $(n - p - 1)$-dimensional manifold in L_{n-p}. That is, we shall define

$$M_i(Z_p) = M_i^{n-p}(\partial D_{n-p}), \qquad i = 0, 1, \ldots, n - p - 1, \tag{15.79}$$

and $M_i(Z_p) = 0$ for $i = n - p, n - p + 1, \ldots, n$. We represent by $V(Z_p)$ the $(n - p)$-dimensional volume of D_{n-p}.

ISBN 0-201-13500-0

We want to extend the fundamental formula (15.36) to the case in which D_1 is a cylinder Z_p. In order to do this, we first consider the bounded body of all p-dimensional balls $U_p(\rho)$ of radius ρ whose centers are the points of the cross section D_{n-p} and are contained in the corresponding generator L_p orthogonal to $L_{n-p[0]}$. Let $Z_p(\rho)$ denote this body (so that $Z_p = Z_p(\infty)$). To find the integrals of mean curvature of $Z_p(\rho)$, consider its parallel body in the distance ε and then let $\varepsilon \to 0$. We easily obtain

$$M_i(Z_p(\rho)) = h_i \qquad \text{for} \quad i > n-p-1,$$

$$M_i(Z_p(\rho)) = \frac{\binom{n-p-1}{i}}{\binom{n-1}{i}} M_i(Z_p) \frac{O_{p-1}}{p} \rho^p + h_i \qquad \text{for} \quad 0 \leqslant i \leqslant n-p-1, \tag{15.80}$$

where h_i is the part of the integral of mean curvature corresponding to the locus of the boundaries of the balls $U_p(\rho)$, so that $h_i/\rho^p \to 0$ as $\rho \to \infty$. Note that $(O_{p-1}/p)\rho^p$ is the volume of the p-dimensional ball of radius ρ. We write the fundamental formula (15.36) for D_0 and $Z_p(\rho)$, using expression (15.77) for dK_1. On the left, when the boundaries of the balls $U_p(\rho)$ do not intersect D_0, the integral of dK^p yields $O_1 \cdots O_{p-1}(O_{p-1}/p)\rho^p$ (asymptotically for large ρ). On the other hand, the measure of $Z_p(\rho)$ such that the boundaries of $U_p(\rho)$ intersect D_0 is bounded above by an expression of the form $\rho^{p-1} \times$ constant. Therefore, dividing both sides of the fundamental formula by $O_1 \cdots O_{p-1}(O_{p-1}/p)\rho^p$ and letting $\rho \to \infty$, taking (15.80) into account, we get the *fundamental kinematic formula for cylinders*

$$\begin{aligned} &\int_{D_0 \cap Z_p \neq \emptyset} \chi(D_0 \cap Z_p)\, dZ_p \\ &= O_{n-1} \cdots O_p \chi(D_0) V(Z_p) \\ &\quad + O_{n-2} \cdots O_p \sum_{\nu=p-1}^{n-2} \frac{1}{n-p} \binom{n-p}{\nu-p+1} M_\nu(D_0) M_{n-\nu-2}(Z_p). \end{aligned} \tag{15.81}$$

Particular cases. 1. When D_{n-p} reduces to a single point, then Z_p reduces to an "oriented" L_p. In this case, using (15.78) and (12.14), formula (15.81) reduces to (14.79).

2. When D_0 reduces to a point P, we have

$$\int_{P \in Z_p} dZ_p = O_{n-1} \cdots O_p V(Z_p). \tag{15.82}$$

ISBN 0-201-13500-0

3. If Z_p is a convex cylinder and D_0 is a convex set, (15.81) becomes

$$\int_{D_0 \cap Z_p \neq \emptyset} dZ_p = O_{n-1} \cdots O_p V(Z_p)$$

$$+ \frac{O_{n-2} \cdots O_p}{n-p} \sum_{\nu=p-1}^{n-2} \binom{n-p}{\nu-p+1} M_\nu(D_0) M_{n-\nu-2}(Z_p) \quad (15.83)$$

or, in terms of the quermassintegrale W_i,

$$\int_{D_0 \cap Z_p \neq \emptyset} dZ_p = nO_{n-2} \cdots O_p \sum_{p-1}^{n-1} \binom{n-p}{\nu-p+1} W_{\nu+1}(D_0) W_{n-1-\nu}(Z_p) \quad (15.84)$$

where we have put $W_{n-1-\nu}(Z_p) = W_{n-1-\nu}^{n-p}(\partial D_{n-p})$.

This result may be stated as follows: If D_0 and D_1 are convex sets such that $D_0 \subset D_1$, the probability that a random cylinder Z_p intersecting D_1 also intersects D_0 is $m(Z_p; Z_p \cap D_0 \neq \emptyset)/m(Z_p; Z_p \cap D_1 \neq \emptyset)$ where the numerator and denominator are given by (15.84) applied to D_0 and D_1, respectively.

4. In euclidean space E_3 we have the cases in which $p = 1, 2$. If $p = 1$, the cross section of Z_1 is a plane domain D_2; let f_Z be the area, u_Z the perimeter, and c_Z the total curvature of this domain (Chapter 7, Section 3). Then we have $M_0(Z_1) = u_Z$, $M_1(Z_1) = c_Z = 2\pi\chi(Z_1)$, $V(Z_1) = f_Z$, and (15.81) becomes

$$\int_{D_0 \cap Z_1 \neq \emptyset} \chi(D_0 \cap Z_1)\, dZ_1 = 8\pi^2 \chi(D_0) f_Z + \pi F c_Z + 2\pi M_1 u_Z \quad (15.85)$$

where F and M_1 are the area and mean curvature integral of ∂D_0, respectively.

If $p = 2$, then Z_2 is a strip or a set of parallel strips in E_3. Assuming that Z_2 is a strip of breadth a, we have $M_1(Z_2) = 0$, $M_0(Z_2) = 2$, $V(Z_2) = a$, and (15.81) becomes (compare with (14.79a)),

$$\int_{D_0 \cap Z_2 \neq \emptyset} \chi(D_0 \cap Z_2)\, dZ_2 = 4\pi a \chi(D_0) + 2M_1(\partial D_0). \quad (15.86)$$

8. Some Mean Values

1. Let $K \equiv K_i$ ($i = 1, 2, \ldots, m$) be m fixed congruent convex sets. Let n be the number of K_i intersected by a moving convex cylinder Z_p. Applying the fundamental formula (15.81) for D_0 as the union of all sets K_i, we have

$$\int n\, dZ_p = nmO_{n-2} \cdots O_p \sum_{\nu=p-1}^{n-1} \binom{n-p}{\nu-p+1} W_{\nu+1}(K) W_{n-1-\nu}(Z_p). \quad (15.87)$$

ISBN 0-201-13500-0

Hence we have *if m congruent convex bodies $K_i \equiv K$ are interior to a convex body K_0, the expected number of them intersected by a random convex cylinder intersecting K_0 is*

$$E(n) = \frac{m \sum_{\nu=p-1}^{n-1} \binom{n-p}{\nu-p+1} W_{\nu+1}(K) W_{n-1-\nu}(Z_p)}{\sum_{\nu=p-1}^{n-1} \binom{n-p}{\nu-p+1} W_{\nu+1}(K_0) W_{n-1-\nu}(Z_p)}. \tag{15.88}$$

2. Now consider m congruent cylinders $Z_p{}^i$ $(i = 1, 2, \ldots, m)$ that meet at random a fixed convex set K_0. We wish to find the mean value of the fraction of volume of K_0 that is covered exactly by r cylinders $(r \leqslant m)$. To this end, consider the integral $\int dP \wedge dZ_p{}^1 \wedge \cdots \wedge dZ_p{}^m$ taken over all cylinders $Z_p{}^i$ that intersect K_0 and over all points $P \in K_0$ that are covered exactly by r cylinders. Fixing either first the cylinders and integrating dP or first P and integrating $dZ_p{}^i$ and equating the results, we obtain

$$\int V_r \, dZ_p{}^1 \wedge \cdots \wedge dZ_p{}^m$$

$$= \binom{m}{r} (O_{n-2} \cdots O_p)^m (O_{n-1} V(Z_p))^r$$

$$\cdot \left[n \sum_{\nu=p-1}^{n-1} \binom{n-p}{\nu-p+1} W_{\nu+1}(K_0) W_{n-1-\nu}(Z_p) - O_{n-1} V(Z_p) \right]^{m-r} V(K_0). \tag{15.89a}$$

As a consequence of (15.84) and (15.89a) we have

Consider m random congruent cylinders intersecting a fixed convex set K_0. The mean value of the fraction of volume of K_0 covered exactly by $r \leqslant m$ cylinders is

$$E(V_r) = \binom{m}{r} \frac{O_{n-1}^r}{n^m} \frac{\left[n \sum_{p-1}^{n-1} \binom{n-p}{\nu-p+1} W_{\nu+1}(K_0) W_{n-1-\nu}(Z_p) - O_{n-1} V(Z_p) \right]^{m-r}}{\left[\sum_{p-1}^{n-1} \binom{n-p}{\nu-p+1} W_{\nu+1}(K_0) W_{n-i-\nu}(Z_p) \right]^m}. \tag{15.89b}$$

This formula holds for $r = 0, 1, \ldots, m$.

Exercise. Assuming that $m \to \infty$ in such a way that the total volume of the cross sections remains constant, that is, $mV(Z_p) = S$ (constant), show that, as $m \to \infty$,

ISBN 0-201-13500-0

$$E(V_r) \to \frac{V(K_0)}{r!}\left[\frac{(n-p)\,O_{n-1}S}{nO_{n-p-1}W_p(K_0)}\right]^r \exp\left(-\frac{(n-p)O_{n-1}S}{nO_{n-p-1}W_p(K_0)}\right).$$

9. Lattices in E_n

The extension to E_n of the results of Chapter 8 is straightforward. Assume E_n divided by a lattice of fundamental regions $\alpha_i = T_i\alpha_0$ where T_i $(i = 0, 1, \ldots)$ are the elements of a discrete subgroup of the group of motions in E_n that leaves invariant the lattice. We assume that each point of E_n belongs to one and only one α_i. Let D_0 be a set of points contained in α_0 and consider the set T_iD_0 $(i = 0, 1, 2, \ldots)$. Applying (15.36) and using the same method as in Chapter 8 for the plane, we obtain

$$\int_{P\in\alpha_0} \sum_i \chi(T_iD_0 \cap D)\,dK = O_{n-1}\cdots O_1(V\chi_0 + V_0\chi)$$

$$+\, O_{n-2}\cdots O_1 \frac{1}{n}\sum_{h=0}^{n-2}\binom{n}{h+1} M_h{}^0 M_{n-2-h} \qquad (15.90)$$

where the integral on the left-hand side is taken over all locations of the moving set D for which the origin P of the moving frame is contained in α_0.

We shall consider two consequences of (15.90).

1. Assume E_n partitioned by the lattice of cubes of edges parallel to the coordinate axis and length a. Assume that α_0 is the cube $0 \leqslant x_i < 1$ $(i = 1, 2, \ldots, n)$ and that D_0 is the closure of α_0. Let D be a topological ball, so that $\chi = \chi(D) = 1$, $\chi_0 = \chi(D_0) = 1$, and $\sum_i \chi(T_iD_0 \cap D)$ is equal to the number of pieces, say v, into which D is divided by the lattice of cubes. Hence, applying (15.90) and using the values (13.48) of the mean curvature integrals of a cube of edge a, we have

$$E(v) = 1 + \frac{V}{a^n} + \frac{1}{O_{n-1}}\sum_{h=0}^{n-2}\binom{n}{h+1}\frac{O_h}{(1+h)a^{h+1}} M_{n-2-h}. \qquad (15.91)$$

Since the number of fundamental regions that have common point with D, say N, is always equal to or less than v, we have $E(N) \leqslant E(v)$ and hence we can state

Every body D that is a topological ball can be covered by cubes of edge a so that their number is not greater than the quantity (15.91).

By considering the n-dimensional balls of radius $r = (\sqrt{n}/2)a$ circumscribed to the cubes α_i we can also state

Every body D that is a topological ball can be covered by n-dimensional balls of radius r so that their number is not greater than the quantity

ISBN 0-201-13500-0

$$1 + \frac{V}{2^n r^n} n^{n/2} + \frac{1}{O_{n-1}} \sum_{h=0}^{n-2} \binom{n}{h+1} \frac{O_h n^{(h+1)/2}}{(h+1)2^{h+1} r^{h+1}} M_{n-2-h}.$$

For $n = 2$ these inequalities are due to Hadwiger [263] and their extension to E_n was obtained by Santaló [543]. See also the work of Hadwiger [278] and Trandafir [677a].

2. Assume that D_0 is a convex body and D is a line segment of length s. Moreover, assume that D cannot meet more than one of the bodies $T_i D_0$. Then, using (15.90) and the values (13.49) for the mean curvature integrals of a line segment of length s, we have the following result.

Consider a lattice of convex bodies formed by D_0 and its translates $T_i D_0$ $(i = 0, 1, \ldots)$. Given a randomly chosen line segment of length s, such that it can intersect at most one of the bodies $T_i D_0$, the probability that it intersects one of them is

$$p = \frac{1}{|\alpha_0|}\left(V_0 + \frac{O_{n-2}}{(n-1)O_{n-1}} s F_0\right)$$

where $|\alpha_0|$ denotes the volume of α_0 and V_0, F_0 are the volume and surface area, respectively, of D_0.

10. Notes and Exercise

1. *Problems on random sets.* Let A be a right parallelepiped in E_n consisting of the points $x(x_1, \ldots, x_n)$ such that $0 \leqslant x_i \leqslant a_i$ $(i = 1, \ldots, n)$. The volume of A is $V_A = a_1 a_2 \cdots a_n$. Let $Y_1, \ldots, Y_m$ be m congruent right parallelepipeds whose edges have the lengths $b_1, \ldots, b_n$, so that their volumes are all $V_Y = b_1 \cdots b_n$. Let P_i be the center of Y_i and let dP_i be the volume element of E_n at P_i. Assume that the Y_i have all their edges parallel to the coordinate axis and that they are moving by translations (not by rotations). The measure of a set of translates of Y_i is the volume covered by their centers P_i. Put $X = (Y_i \cup Y_2 \cup \cdots \cup Y_m) \cup A$ and consider the integral $I = \int dP \wedge dP_1 \wedge \cdots \wedge dP_m$ taken over the set of points $P_1, \ldots, P_m$ such that $P \in A$, $Y_i \cap A \neq 0$, and $P \notin Y_i$ $(i = 1, \ldots, m)$. In order to calculate I, note that $Y_i \cap A \neq 0$ is satisfied if P_i belongs to the right parallelepiped that has the same center as A and has edges $a_i + b_i$. Note also that $\int dP_i = V_Y$ (the integral taken over $P \in Y_i$). Keeping P fixed first, we have $I = (V_0 - V_Y)^m V_A$ where we have put $V_0 = (a_1 + b_1) \cdots (a_n + b_n)$. On the other hand, integrating P while keeping the parallelepipeds V_Y fixed, we obtain $I = \int (V_A - X) \cdot dP_1 \wedge \cdots \wedge dP_m$. Hence we have $\int X \, dP_1 \wedge \cdots \wedge dP_m = V_A[V_0^m - (V_0 - V_Y)^m]$. Dividing by V_0^m, we get the expected value of the volume X:

$$E(X) = V_A\left[1 - \left(1 - \frac{V_Y}{V_0}\right)^m\right]. \tag{15.92}$$

ISBN 0-201-13500-0

The expression for the higher-order moments turns out to be more complicated. For these and related problems see the articles by Robbins [512], Votaw [705], and Bronowski and Neyman [71]. For $n = 2$ we can consider the case where the random rectangles also have random orientation [552]. Instead of parallelepipeds Y_i we can consider random congruent spheres (see [512, 213, 335]). For a general approach see [2]. If the random sets are not uniformly distributed, the results are more involved (see [432, 616, 617]).

2. *A general kinematic formula of Chern and Federer.* Let M be a compact Riemann manifold of dimension k. Suppose M embedded in E_n and suppose that T_ρ is the set of all points a distance ρ from M. Then for ρ sufficiently small, Weyl [716] proved that the volume of T_ρ is given by the formula

$$V(T_\rho) = O_{m-1} \sum_e \frac{(e-1)(e-3)\cdots 1}{(m+e)(m+e-2)\cdots m} \mu_e(M)\rho^{m+e} \tag{15.93}$$

(e even, $0 \leqslant e \leqslant k$, $m = n - k$) where $\mu_e(M)$ are integral invariants of M that can be computed from the components of the curvature tensor of M. In particular, $\mu_0(M)$ is the total volume of M and if k is even,

$$\mu_k(M) = \frac{(2\pi)^{k/2}}{(k-1)(k-3)\cdots 1} \chi(M) \tag{15.94}$$

where $\chi(M)$ is the Euler–Poincaré characteristic of M.

Let M^p and M^q be two compact submanifolds (without boundary) of dimensions p and q in E_n. Let M^p be fixed and M^q moving with kinematic density dK. Then

$$\int_{M^p \cap M^q \neq \emptyset} \mu_e(M^p \cap M^q)\, dK = \sum_{\substack{0 \leqslant i \leqslant e \\ i \text{ even}}} c_i \mu_i(M^p)\mu_{e-i}(M^q) \tag{15.95a}$$

where c_i are constants, rather involved, depending on n, p, q, e.

If dL_q means the density for q-planes in E_n, we have

$$\int_{M^p \cap L^q \neq \emptyset} \mu_e(M^p \cap L_q)\, dL_q = \frac{O_n \cdots O_{n-q} O_{p+q-n+1} O_{p+q-n} O_{p+1-e}}{O_q \cdots O_1 O_{p+1} O_p O_{p+q-n+1-e}} \mu_e(M^p). \tag{15.95b}$$

For $e = 0$, (15.95a) coincides with (15.20); for $p = n - 1$, (15.95b) coincides with (14.78). General formulas (15.95a) and (15.95b) were given by Chern [112] and are similar to those given earlier by Federer [177]. Flaherty [200] has extended the curvature measures μ_k to bounded Borel sets is E_n (see also the book by Sulanke and Wintgen [664]). Nijenhuis [451] pointed out that there exists a normalization of the μ's and a kinematic density, so that the curvature polynomials defined by $\mu(X, \lambda) = \sum_e \mu_e(X)\lambda^e$ satisfy $\int \mu(M^p \cap M^q)\, dK = \mu(M^p, \lambda)\mu(M^q, \lambda)(\text{mod } \lambda^{p+q-n+1})$.

Chen [103] has given some kinematic formulas that involve extrinsic invariants of M^p, M^q, and $M^p \cap M^q$. For instance, let $\tau(C)$ denote the total square curvature of a curve C in E_3 and for a surface M^2 with mean curvature

ISBN 0-201-13500-0

H and Gaussian curvature K let $H^*(M^2) = \int H^2\, d\sigma$, $K^*(M^2) = \int K\, d\sigma$, $\sigma(M^2)$ = surface area of M^2. Chen proves that

$$\int \tau(M_0{}^2 \cap M^2)\, dK = 2\pi^3(3H^*(M_0{}^2) - K^*(M_0{}^2))\sigma_0(M^2) + 2\pi^3(3H^*(M^2) - K^*(M^2))\sigma_0(M_0{}^2). \quad (15.96)$$

For surfaces in E_3 see also [687].

General formula (15.95a) can be extended to moving cylinders. Let M^h be an orientable, compact, differentiable manifold (without boundary) contained in an $(n-m)$-plane of E_n $(h + m < n)$. Through each point of M^h we consider the m-plane L_m perpendicular to L_{n-m}. The set of all these L_m is a cylinder $Z_{h,m}$ of dimension $h + m$, whose generators are the m-planes L_m and cross section M^h. The density for $Z_{h,m}$ is $dZ_{h,m} = dL_{n-m[0]} \wedge dK^{n-m}$, where dK^{n-m} is the kinematic density on $L_{n-m[0]}$. The kinematic formula (15.95a) becomes

$$\int_{M^p \cap Z_{h,m} \neq \emptyset} \mu_e(M^p \cap Z_{h,m})\, dZ_{h,m} = \sum_{e-h \leqslant i \leqslant e} \frac{c_i}{O_1 \cdots O_{m-1}} \frac{\binom{h}{e-i}}{\binom{h+m}{e-i}} \mu_i(M^p)\mu_{e-i}(M^h) \quad (15.97)$$

(e even, $0 \leqslant e \leqslant p + h + m - n$, $i \geqslant 0$, i even).

3. *Extensions of Poincaré's formulas.* Formulas such as (15.95a) concerning the integral of the intersection of a fixed manifold M^p with a moving manifold M^q are usually known in integral geometry as Poincaré's formulas. Formulas (7.11) and (15.20) are particular cases. This kind of formula has been generalized to homogeneous spaces by Kurita [351], Federer [177], and Brothers [73]. The following are examples of the type of formula given by Brothers.

Let M be an n-dimensional Riemann manifold with a transitive group of isometries $\mathfrak{G}$. Let M^p, M^q be proper p-, q-dimensional submanifolds of class C^1 of M, and let $A \subset M^p$, $B \subset M^q$ be Borel sets. Assume that $\mathfrak{G}$ acts transitively on the set of tangent spaces of M^p and of M^q, respectively. Let dg be the left-invariant measure on $\mathfrak{G}$ and suppose $p + q \geqslant n$. Then there exists a constant α depending only on M^p, M^q, and $\mathfrak{G}$ such that

$$\int_{\mathfrak{G}} m(A \cap gB)\, dg = \alpha m(A) \int_B \Delta\, d\sigma_B \quad (15.98)$$

where $m(A \cap gB)$ is the $(p + q - n)$-dimensional measure (Hausdorff measure) of $A \cap gB$, $d\sigma_B$ is the element of q-dimensional measure on M^q, and Δ is a positive function that depends on M and $\mathfrak{G}$. The constant α is $\neq 0$ if and only if for some $g \in \mathfrak{G}$ there exists a point $a \in M^p \cap gM^q$ for which the union of the tangent space $T_a(M^p)$ with $T_a(gM^q)$ spans $T_a(M)$.

ISBN 0-201-13500-0

Formula (15.98) may be generalized in the following way. Let $\mathscr{E}$ be a set of closed q-dimensional submanifolds of M such that $\mathfrak{G}$ acts transitively on $\mathscr{E}$ and if $E \in \mathscr{E}$, then $\mathfrak{G} \cap \{g; gE = E\}$ is transitive on $\mathscr{E}$. Also assume that $\mathscr{E}$ has a $\mathfrak{G}$-invariant density dE. Then we have

$$\int_{\mathscr{E}} m(A \cap E)\, dE = \beta\alpha m(A) \tag{15.99}$$

where $m(A \cap E)$ is the $(p + q - n)$-dimensional measure of $A \cap E$ and β is a positive constant depending on the density dE (which, in general, is not unique).

4. *Fields of convex sets and fields of cylinders.* Consider a countable set of congruent convex sets $K = K_1 = K_2 = \cdots$ in E_n, not necessarily disjoint and such that if $m(r)$ denotes the number of convex sets intersecting an n-ball S_r whose center is the point P and that has radius r, the asymptotic relation

$$m(r) = \rho\kappa_n r^n + o(r^n), \qquad \kappa^n = \pi^{n/2}(\Gamma(1 + n/2))^{-1} \tag{15.100}$$

holds for $r \to \infty$; that is, the limit

$$\lim_{r\to\infty} \frac{m(r)}{\kappa_n r^n} = \rho \tag{15.101}$$

exists and is independent of the center P. We will say that we have in E_n a field of convex sets K of density ρ. Assuming that a convex set K_0 is placed at random in E_n, what is the probability that it meets exactly s sets of the field?

To solve the problem we proceed as follows. Suppose that we place the convex set K_0 at random inside the ball S_r. Then, the probability that a random convex set K intersecting S_r also intersects K_0 is given by (15.65), which can be written, denoting by W_h^1 the quermassintegrale of S^r (13.46),

$$p = \frac{B_0}{B_1}, \quad B_i = \kappa_n^{-1} \sum_{h=0}^{n} \binom{n}{h} W_h^i W_{n-h} \quad (i = 0, 1). \tag{15.102}$$

If we consider m random convex sets congruent to K that intersect S_r, the probability that exactly s of them meet K_0 will be

$$p_s(r) = \binom{m}{s} p^s(1 - p)^{m-s}.$$

Allowing both $m, r \to \infty$ in such a way that relation (15.101) holds, we find, by a simple calculation, that

$$p_s = \lim_{r\to\infty} p_s(r) = \frac{(\rho B_0)^s}{s!} \exp(-\rho B_0). \tag{15.103}$$

That is, the number of convex sets K of a field of density ρ that are intersected by a convex set K_0 placed at random in E_n (a "probe") is a Poisson random variable with parameter ρB_0 (see [228]).

ISBN 0-201-13500-0

Now consider a countable set of congruent convex cylinders Z_p, not necessarily disjoint, such that if $m(r)$ denotes the number of cylinders intersecting a ball S_r of radius r, the asymptotic relation

$$\lim_{r\to\infty} \frac{m(r)}{\kappa_n r^{n-p}} = \rho$$

holds, independently of the center of S_r. We shall say that we have in E_n a field of cylinders Z_p of density ρ. By considerations similar to those above, if a probe convex set K_0 is placed at random in E_n, the number of cylinders that are intersected by it is a Poisson random variable with parameter $\rho B_0{}^*$, where

$$B_0{}^* = \kappa_{n-p}^{-1} \sum_{i=p-1}^{n-1} \binom{n-p}{i-p+1} W_{i+1}(K_0) W_{n-1-i}(Z_p).$$

(See Exercise 1.)

5. *Agglomerations of convex sets.* Imagine a fixed convex set K_0 in E_3 and n convex sets $K_1, K_2, \ldots, K_n$ that intersect K_0. Let $K_{01\ldots n} = K_0 \cap K_1 \cap \cdots \cap K_n$ and call $V_{01\ldots n}$ the volume, $F_{01\ldots n}$ the surface area, and $M_{01\ldots n}$ the integral of mean curvature of $K_{01\ldots n}$. We can prove that

$$\int V_{01\ldots n}\, dK_1 \wedge dK_2 \wedge \cdots \wedge dK_n = (8\pi^2)^n V_0 V_1 \cdots V_n,$$

$$\int F_{01\ldots n}\, dK_1 \wedge dK_2 \wedge \cdots \wedge dK_n = (8\pi^2)^n \sum_i V_0 \cdots V_{i-1} F_i V_{i+1} \cdots V_n,$$

$$\int M_{01\ldots n}\, dK_1 \wedge dK_2 \wedge \cdots \wedge dK_n$$

$$= \tfrac{1}{2}\pi^4 (8\pi^2)^{n-1} \sum_{i,j} V_0 \cdots V_{i-1} F_i V_{i+1} \cdots V_{j-1} F_j V_{j+1} \cdots V_n$$

$$+ (8\pi^2)^n \sum_i V_0 \cdots V_{i-1} M_i V_{i+1} \cdots V_n,$$

$$\int dK_1 \wedge dK_2 \wedge \cdots \wedge dK_n$$

$$= (8\pi^2)^n \sum_i V_0 \cdots V_{i-1} V_{i+1} \cdots V_n$$

$$+ 2\pi (8\pi^2)^{n-1} \sum_{i,j} V_0 \cdots V_{i-1} F_i V_{i+1} \cdots V_{j-1} M_j V_{j+1} \cdots V_n$$

$$+ \pi^5 (8\pi^2)^{n-2} \sum_{i,j,k} V_0 \cdots V_{i-1} F_i V_{i+1} \cdots V_{j-1} F_j V_{j+1} \cdots V_{k-1} F_k V_{k+1} \cdots V_n$$

where the integrals are taken over the locations of K_i such that $K_0 \cap K_1 \cap \cdots \cap K_n \neq \emptyset$. Note that the least integral divided by the product of measures $m(K_i;\, K_i \cap K_0 \neq \emptyset)$ gives the probability that the common intersection of

ISBN 0-201-13500-0

the n convex sets K_i hitting K_0 and K_0 itself is not empty. See [524] and, for the generalization to E_n, [657]; see also [190, 191].

Let P_i be n random points inside a convex set K. What is the probability that they may be enclosed by a ball of radius r? Assume that each point is the center of a ball of radius r. The required probability is equal to the probability that all such balls have a nonvoid intersection. Let K_{-r} denote the interior parallel set of K in the distance r and let V_{-r}, F_{-r}, and M_{-r} be its volume, area, and integral of mean curvature, respectively. Applying the last result, we get the following value for the required probability.

$$p(r) = (4\pi/3)^{n-1}r^{3n-3}V^{-n}[(4\pi/3)r^3 + n(M_{-r}r^2 + F_{-r}r + V_{-r})$$
$$+ 3n(n-1)((\pi^2/32)F_{-r}r + V_{-r}) + (3\pi^2/32)n(n-1)(n-2)V_{-r}].$$

Hadwiger and Streit [281] extended this result to the case of cylinders or strips intersecting a fixed convex set K_0 and then used it to solve the problem of finding the probability that n random lines (or random planes) that intersect a convex set K_0 form an "almost bundle," that is, there exists a ball of radius r that has common point with all the lines (or planes). Note that the problem requires only that such a ball exist, without prescribing its position within K_0.

6. *Cylinders in* E_3. Let Z be a right cylinder in E_3 whose cross section is a plane domain of area f and perimeter u. The density for Z, according to (15.78), can be written $dZ = dG^* \wedge d\phi$ where G^* is an oriented line parallel to the generators of the cylinder and ϕ denotes a rotation about G^*. We quote the following results.

(a) If K is a fixed convex body of area F and integral of mean curvature M, fundamental formula (15.85) becomes, assuming $C_Z = 2\pi$,

$$\int_{K\cap Z\neq\emptyset} dZ = 2\pi(\pi F + 4\pi f + Mu).$$

(b) If Γ is a rectifiable curve of length L and n denotes the number of intersection points with ∂Z, then

$$\int_{\Gamma\cap\partial Z\neq\emptyset} n\,dZ = 4\pi^2 uL.$$

(c) If Σ is a surface of area F and λ denotes the length of the curve $\Sigma \cap \partial Z$,

$$\int_{\Sigma\cap\partial Z\neq\emptyset} \lambda\,dZ = 2\pi^3 uF.$$

(d) Assume a fixed plane E divided into fundamental domains of area α, each domain containing a convex set K_E of area f and perimeter u. Assume that each K_E is the cross section of a right cylinder Z, so that they define a lattice of convex cylinders in E_3 (Fig. 15.1). Let K be a convex set of area F and integral of mean curvature M that cannot meet more than one cylinder

ISBN 0-201-13500-0

of the lattice. Then, the probability that K, placed at random in space, meets a cylinder of the lattice is

$$p = (\pi F + 4\pi f + Mu)/4\pi\alpha.$$

In particular, if K is a line segment of length b, we have $F = 0$, $M = \pi b$, and the probability becomes $p = (4f + bu)/4\alpha$. If the cylinders reduce to lines ($f = u = 0$), we have $p = F/4\alpha$. These are new generalizations of Buffon's problem. For other related results see [51].

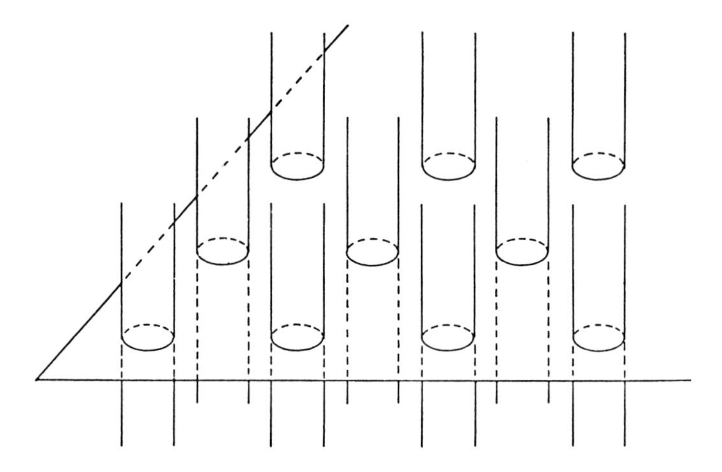

Figure 15.1.

EXERCISE 1. Apply the last result in Note 4 to the case of fields of lines or planes of density ρ, considered as cylinders whose cross sections reduce to a single point.

ISBN 0-201-13500-0

CHAPTER 16

Geometric and Statistical Applications; Stereology

1. Size Distribution of Particles Derived from the Size Distribution of Their Sections

Let us consider convex particles distributed at random in E_3. The determination of the size distribution of these particles from the size distribution of their sections with random figures of known shape (e.g., a convex body, a cylinder, a plane, a strip, or a line) is one of the basic problems of so-called stereology, which is an interdisciplinary field relating such seemingly disparate disciplines as biology, mineralogy, metallurgy, and geometry. Elias [166] has proposed the following definition: Stereology deals with a body of methods for the exploration of three-dimensional space when only two-dimensional sections through solid bodies or their projections are available. The chief methods of stereology are closely related to integral geometry, as we shall show in this chapter with some typical examples. As a main reference see [166]; for a discussion of the fundamental equations, see [226, 227].

We first state some formulas from the preceding chapter for the case in which $n = 3$. Let K be a fixed convex body in E_3 and K_1 a moving one with kinematic density dK_1. The kinematic fundamental formula (15.63) becomes

$$\int_{K \cap K_1 \neq \emptyset} dK_1 = 8\pi^2(V + V_1) + 2\pi(FM_1 + F_1M) \tag{16.1}$$

where V, F, M denote the volume, surface area, and integral of mean curvature, respectively.

ENCYCLOPEDIA OF MATHEMATICS and Its Applications, Gian-Carlo Rota (ed.).
1, Luis A. Santaló, Integral Geometry and Geometric Probability

ISBN 0-201-13500-0

Formula (15.20) for $n = 3$, $q = r = 3$ and formula (15.72) for $n = 3$, $q = 1$ and $n = 3$, $q = 2$ give

$$\int_{K \cap K_1 \neq \emptyset} V_{01}\, dK_1 = 8\pi^2 V V_1, \qquad \int_{K \cap K_1 \neq \emptyset} F_{01}\, dK_1 = 8\pi^2(F V_1 + V F_1),$$

$$\int_{K \cap K_1 \neq \emptyset} M_{01}\, dK_1 = 8\pi^2(V_1 M + M_1 V) + (\pi^4/2) F F_1 \tag{16.2}$$

where V_{01} is the volume of $K \cap K_1$ and F_{01}, M_{01} are the surface area and the mean curvature integral, respectively, of $\partial(K \cap K_1)$.

If the moving figure is a convex cylinder Z_1 and u, f denote the perimeter and the area of its normal cross section, the preceding formulas become

$$\int_{K \cap Z_1 \neq \emptyset} dZ_1 = 2\pi(\pi F + 4\pi f + Mu), \tag{16.3}$$

$$\int_{K \cap Z_1 \neq \emptyset} V_{01}\, dZ_1 = 8\pi^2 V f, \qquad \int_{K \cap Z_1 \neq \emptyset} F_{01}\, dZ_1 = 8\pi^2(F f + V u), \tag{16.4}$$

$$\int_{K \cap Z_1 \neq \emptyset} M_{01}\, dZ_1 = 2\pi^2(4Mf + 4\pi V + \tfrac{1}{4}\pi^2 F u) \tag{16.5}$$

where (16.3) is a particular case of (15.83) and the remaining formulas follow easily from (16.2) by reasoning similar to that in Section 7 of Chapter 15.

If the moving figure is a strip B consisting of two parallel planes in distance Δ, we have

$$\int_{K \cap B \neq \emptyset} dB = M + 2\pi\Delta, \qquad \int_{K \cap B \neq \emptyset} V_{01}\, dB = 2\pi V \Delta, \tag{16.6}$$

$$\int_{K \cap B \neq \emptyset} F_{01}\, dB = 2\pi(2V + F\Delta), \qquad \int_{K \cap B \neq \emptyset} M_{01}\, dB = (\pi^3/4)F + 2\pi M \Delta \tag{16.7}$$

where the density dB for strips is equal to the density for the midplane of the strip.

We next assume congruent convex particles K distributed at random in E_3 and consider the problem of determining the particle quantities ρ (number of particles per unit volume), V (volume), F (surface area), M (integral of mean curvature) from the quantities V_1, F_1, M_1 of a moving convex body K_1 (the probe) and the total number of particles N^*, total volume V^*, total surface area F^*, and total integral of mean curvature M^* of the intersection of K_1 with the particles, averaged over all positions of K_1.

ISBN 0-201-13500-0

Assuming the particles inside a ball of radius R and letting $R \to \infty$, using (16.1) and (16.2), we have $V^* = \rho V V_1$, $F^* = (F_1 V + V_1 F)\rho$, and thus

$$V = \frac{V^*}{\rho V_1}, \qquad F = \frac{F^* V_1 - V^* F_1}{\rho V_1^2}. \tag{16.8}$$

Analogously we get

$$M = \frac{1}{\rho V_1^3}\left[V_1^2 M^* - \frac{\pi^2}{16} F_1 V_1 F^* + \left(\frac{\pi^2}{16} F_1^2 - M_1 V_1\right) V^*\right] \tag{16.9}$$

and using (15.66) for $n = 3$, we obtain

$$\rho = \frac{N^*}{V_1} - \frac{F_1 M^*}{4\pi V_1^2} + \left(\frac{\pi}{32} F_1^2 - \frac{V_1 M_1}{2\pi}\right)\frac{F^*}{2V_1^3}$$
$$+ \left(\frac{V_1 F_1 M_1}{2\pi} - V_1^2 - \frac{\pi}{64} F_1^3\right)\frac{V^*}{V_1^4}. \tag{16.10}$$

These formulas permit an estimation of ρ, V, F, M from N^*, V^*, F^*, M^*, using the known quantities V_1, F_1, M_1 of the probe K_1.

If the moving figure is a cylinder Z_1, from (16.3)–(16.5) we easily get

$$V^* = \rho V, \qquad F^* = \rho F + \rho V(u/f), \qquad M^* = \rho M + \rho\pi(V/f) + \rho\pi^2 Fu/16f,$$
$$N^* = \rho(F/4f + Mu/4\pi f + 1) \tag{16.11}$$

where N^*, V^*, F^*, M^* are, respectively, the number of particles, volume, surface area, and mean curvature integral of the intersection of Z_1 with the particles (averaged over all positions of Z_1) per unit volume of Z_1.

From (16.11) we deduce

$$V = \frac{V^*}{\rho}, \qquad F = \frac{F^*f - V^*u}{\rho f}, \qquad M = \frac{M^*}{\rho} - \frac{\pi V^*}{\rho f} - \frac{\pi^2 u}{16\rho f^2}(F^*f - V^*u),$$
$$\rho = N^* - \frac{uM^*}{4\pi f} + \left(\frac{\pi u^2}{16f} - 1\right)\frac{F^*}{4f} + \left(1 - \frac{\pi u^2}{32f}\right)\frac{uV^*}{2f^2}. \tag{16.12}$$

If the moving figure is a strip B of breadth Δ, from (16.6) and (16.7) it is easy to deduce that

$$V^* = \rho V\Delta, \qquad F^* = \rho(F\Delta + 2V), \qquad M^* = \rho\left(M\Delta + \frac{\pi^2}{8}F\right),$$
$$N^* = \rho\left(\Delta + \frac{M}{2\pi}\right) \tag{16.13}$$

and hence

ISBN 0-201-13500-0

$$V = \frac{V^*}{\rho \Delta}, \qquad F = \frac{F^* \Delta - 2V^*}{\rho \Delta^2}, \qquad M = \frac{8M^* \Delta^2 - \pi^2 F^* \Delta + 2\pi^2 V^*}{8\rho \Delta^3},$$

$$\rho = \frac{N^*}{\Delta} - \frac{8M^* \Delta^2 - \pi^2 F^* \Delta + 2\pi^2 V^*}{16\pi \Delta^4} \tag{16.14}$$

where N^*, V^*, F^*, M^* refer to the hits per unit area of the strip.

If the particles are not congruent, then V, F, M denote the averaged quantities over all particles. We can derive the values of ρ and M by using

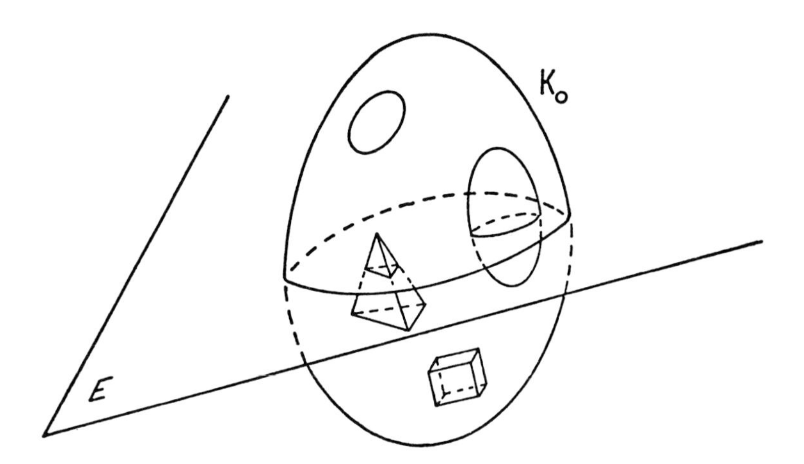

Figure 16.1.

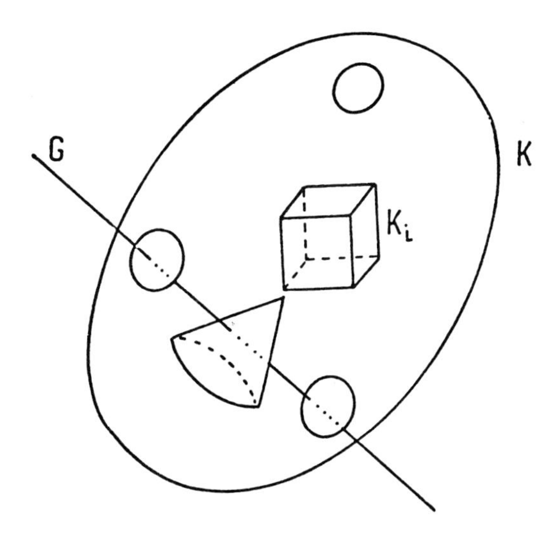

Figure 16.2.

ISBN 0-201-13500-0

measurements of N^* in at least two strips of different breadth. For details and complements see the paper of G. Bach in Elias' book [166]. Usually the moving figure is a random plane E (strip with $\Delta = 0$) (Fig. 16.1) or a random line G (cylinder with $f = 0$, $u = 0$) (Fig. 16.2). We are going to consider these cases separately.

2. Intersection with Random Planes

Consider a convex body Q that contains a certain number of convex nonoverlapping particles distributed at random. Suppose that all particles are similar to a convex body K and let λ be the ratio of similitude: we shall denote by K_λ the convex body similar to K with ratio λ, so that $K_1 = K$. Let $H(\lambda)\, d\lambda$ be the number of particles per unit volume in Q whose ratio lies in the range λ, $\lambda + d\lambda$.

We intersect Q by a random plane E and let $h(\sigma)\, d\sigma$ be the number of sections per unit area in $Q \cap E$ that have area between σ and $\sigma + d\sigma$. The problem that we shall now consider is that of finding $H(\lambda)$ from $h(\sigma)$, a problem that has applications in several fields (see [335, p. 86] and references therein).

Let $\phi(\sigma)$ be the probability distribution of the area σ of $E \cap K$, so that $\phi(\sigma)\, d\sigma$ is the probability of a plane chosen at random having an intersection of area in the range σ, $\sigma + d\sigma$. Denoting by σ_m the maximal value of σ, we have

$$\int_0^{\sigma_m} \phi(\sigma)\, d\sigma = 1, \qquad \int_0^{\sigma_m} \sigma\phi(\sigma)\, d\sigma = 2\pi V/M, \tag{16.15}$$

where the second equality is equivalent to the second formula (14.76). If $\phi(\sigma, \lambda)$ denotes the probability distribution of the areas of $E \cap K_\lambda$, so that $\phi(\sigma, 1) = \phi(\sigma)$, we have $\phi(\sigma, \lambda)\, d(\lambda^2\sigma) = \phi(\sigma/\lambda^2)\, d\sigma$, or

$$\phi(\sigma, \lambda) = \frac{1}{\lambda^2}\, \phi\left(\frac{\sigma}{\lambda^2}\right). \tag{16.16}$$

The function $\phi(\sigma)$ is in general not simple. It should be interesting to have it for simple bodies, such as a simplex or a cube. We will give it for the unit sphere. In this case, putting $x^2 = 1 - r^2$, we find that the probability of a plane section chosen at random having a radius in the range r, $r + dr$ is $|dx| = r(1 - r^2)^{-1/2}\, dr$. Since $\sigma = \pi r^2$, $d\sigma = 2\pi r\, dr$, the probability of having an area in the range σ, $\sigma + d\sigma$ is $2[\pi(\pi - \sigma)]^{-1/2}\, d\sigma$ and thus

$$\phi(\sigma) = 1/2\sqrt{\pi}\,(\pi - \sigma)^{1/2}. \tag{16.17}$$

ISBN 0-201-13500-0

For the sphere of radius λ, according to (16.16), we have

$$\phi(\sigma, \lambda) = 1/2\sqrt{\pi}\,\lambda(\pi\lambda^2 - \sigma)^{1/2} \tag{16.18}$$

or, since $\sigma_m = \pi\lambda^2$,

$$\phi(\sigma, \lambda) = 1/2\sqrt{\pi}\,\lambda(\sigma_m - \sigma)^{1/2}. \tag{16.19}$$

Let M_λ and M_Q denote the integrals of mean curvature of K_λ and Q, respectively. The total number of particles whose ratio λ lies in the range $\lambda, \lambda + d\lambda$ is $VH(\lambda)\,d\lambda$, where V denotes the volume of Q, and the mean value of the number of particles that are intersected by a random plane E that cuts Q is $(M_\lambda/M_Q)VH(\lambda)\,d\lambda$. Multiplying by the probability distribution $\phi(\sigma, \lambda)$, we obtain the mean value of plane sections $E \cap K_\lambda$ with λ in the range $\lambda, \lambda + d\lambda$ and σ in the range $\sigma, \sigma + d\sigma$. By integration over λ, from $\lambda = (\sigma/\sigma_m)^{1/2}$ to $\lambda = \infty$, we get the mean value of intersections $K_\lambda \cap E$ whose area lies between σ and $\sigma + d\sigma$, namely

$$d\sigma \int_0^\infty (M_\lambda/M_Q)VH(\lambda)\phi(\sigma, \lambda)\,d\lambda. \tag{16.20}$$

Since the number of intersected particles per unit area in $Q \cap E$ whose area lies between σ and $\sigma + d\sigma$ is $h(\sigma)\,d\sigma$ and the mean area of the intersections $Q \cap E$ is $2\pi V/M_Q$, it follows that mean value (16.20) is equal to $(2\pi V/M_Q)h(\sigma)\,d\sigma$. Taking (16.16) into account, we obtain the equation

$$\int_{(\sigma/\sigma_m)^{1/2}}^\infty \lambda^{-1}\phi(\sigma/\lambda^2)H(\lambda)\,d\lambda = (2\pi/M)h(\sigma) \tag{16.21}$$

where we have applied the relation $M_\lambda = \lambda M_1$ and have put $M = M_1$. Equation (16.21) is the integral equation that relates $H(\lambda)$ and $h(\sigma)$. Using measurements of the intersection of Q by random planes, we can estimate $h(\sigma)$ and then calculate $H(\lambda)$ from (16.21). This integral equation contains the function $\phi(\sigma)$, which depends on the shape of the particles K and is in general difficult to calculate. We will consider some particular cases.

(a) *Spherical particles.* If K are spheres, we can take $\lambda = r$ (radius of the spheres). Then we have $M = 4\pi$, $\sigma_m = \pi$, and using (16.17), equation (16.21) becomes

$$\int_{(\sigma/\pi)^{1/2}}^\infty \frac{H(\lambda)\,d\lambda}{(\pi\lambda^2 - \sigma)^{1/2}} = \sqrt{\pi}\,h(\sigma). \tag{16.22}$$

ISBN 0-201-13500-0

Putting $\pi\lambda^2 = s$ and $H_1(s) = H((s/\pi)^{1/2})/\sqrt{s}$, $h_1(\sigma) = 2\pi h(\sigma)$, we find that (16.20) becomes

$$\int_\sigma^\infty \frac{H_1(s)\,ds}{(s-\sigma)^{1/2}} = h_1(\sigma), \tag{16.23}$$

which is an integral equation of Abel's type. The solution is (see, e.g., [125, p. 158]),

$$H_1(s) = -\frac{1}{\pi}\int_s^\infty \frac{h_1'(\sigma)\,d\sigma}{(\sigma - s)^{1/2}} \tag{16.24}$$

where the prime denotes derivative. In terms of $H(\lambda)$ and $h(\sigma)$ this solution can be written

$$H(\lambda) = -\,2\pi^{1/2}\lambda\int_{\pi\lambda^2}^\infty \frac{h'(\sigma)}{(\sigma - \pi\lambda^2)^{1/2}}\,d\sigma. \tag{16.25}$$

If we introduce the distribution function $g(r)$ of the radii ($g(r)\,dr = h(\sigma)\,d\sigma$ = number of intersected spheres per unit area in $Q \cap E$ whose intersections have radii in the range r, $r + dr$), we have $\sigma = \pi r^2$, $h(\sigma) = g(r)/2\pi r$, $h'(\sigma) = (g(r)/r)'/4\pi^2 r$, and (16.25) becomes

$$H(\lambda) = -\frac{\lambda}{\pi}\int_\lambda^\infty \left(\frac{g(r)}{r}\right)' \frac{dr}{(r^2 - \lambda^2)^{1/2}}, \tag{16.26}$$

which is a formula due to Wicksell [724].

(b) *Nearly spherical particles.* An analogous formula holds for particles whose function $\phi(\sigma)$ may be assumed of the form

$$\phi(\sigma) = a(\sigma_m - \sigma)^{-\mu} \tag{16.27}$$

for $0 \leqslant \sigma \leqslant \sigma_m$ and $\phi(\sigma) = 0$ for $\sigma > \sigma_m$, where a and μ are constants. The sphere corresponds to $\mu = \frac{1}{2}$, $a = 1/2\sqrt{\pi}$. From (16.15) we have

$$\frac{a}{1-\mu}\,\sigma_m{}^{1-\mu} = 1, \qquad \frac{a}{(1-\mu)(2-\mu)}\,\sigma_m{}^{2-\mu} = \frac{2\pi V}{M}, \tag{16.28}$$

from which we get

$$a = p\sigma_m{}^{-p}, \qquad \mu = 1 - p, \qquad p = (M\sigma_m/2\pi V) - 1. \tag{16.29}$$

ISBN 0-201-13500-0

Since for any convex body we have the known inequalities $\sigma_m \geqslant F/4$ and $MF \geqslant 12\pi V$, it follows that $p \geqslant MF(8\pi V)^{-1} - 1 \geqslant \frac{1}{2}$ and hence $\mu \leqslant \frac{1}{2}$. From (16.27) and (16.21) we have

$$aM \int_{(\sigma\sigma_m)^{1/2}}^{\infty} \frac{\lambda^{2\mu-1} H(\lambda)}{(\sigma_m \lambda^2 - \sigma)^{\mu}} \, d\lambda = 2\pi h(\sigma). \tag{16.30}$$

This is a generalized Abel's type equation whose solution is (see [125, p. 159])

$$H(\lambda) = -\frac{4\sigma_m \lambda^{2(1-\mu)} \sin \mu\pi}{aM} \int_{\sigma_m \lambda^2}^{\infty} \frac{h'(\sigma)}{(\sigma - \sigma_m \lambda^2)^{1-\mu}} \, d\sigma. \tag{16.31}$$

(c) *Particles of the same shape and size.* If all particles embedded in Q are congruent and there are N of them per unit volume, we can consider $H(\lambda)$ as N times a Dirac delta function, more precisely, N times the translated function $\delta(\lambda)$ such that $\delta(\lambda) = 0$ if $\lambda \neq 1$ and $\int \delta(\lambda)\, d\lambda = 1$ when the region of integration includes the point $\lambda = 1$. Then, for an arbitrary function $g(\lambda)$ that is continuous at $\lambda = 1$, the equation $\int g(\lambda)\, \delta(\lambda)\, d\lambda = g(1)$ is valid, where again the integration includes the point $\lambda = 1$. Then (16.21) gives $N = 2\pi h(\sigma)/M\phi(\sigma)$. The case of ellipsoidal particles has been considered by De Hoff [142, a, b]

3. Intersection with Random Lines

With the same notation as above, we shall now consider the intersections of Q by random lines G. Let $h(\sigma)\, d\sigma$ be the number of intersected particles per unit length of $Q \cap G$ whose chords have their length in the range $\sigma, \sigma + d\sigma$. The problem is to relate $H(\lambda)$ and $h(\sigma)$.

Since the mean length of the chords that G determines in Q is $4V_Q/F_Q$, where V_Q is the volume of Q and F_Q is the surface area of Q, the mean number of particles intersecting G in chords whose length lies in the range $\sigma, \sigma + d\sigma$ is $(4V_Q/F_Q)h(\sigma)\, d\sigma$. On the other hand, the mean number of particles intersecting G and having a similitude ratio in the range $\lambda, \lambda + d\lambda$ is $(F_\lambda/F_Q)VH(\lambda)\, d\lambda$ where F_λ denotes the surface area of K_λ. Let K_1 be a particle that corresponds to $\lambda = 1$ and let $\phi(\sigma)\, d\sigma$ be the probability that the length of the chord $G \cap K_1$ lies between σ and $\sigma + d\sigma$. The same probability for a particle K_λ is $\phi(\sigma, \lambda) = \lambda^{-1}\phi(\sigma/\lambda)\, d\sigma$ and the number of chords whose lengths lie in the range $\sigma, \sigma + d\sigma$ becomes

$$d\sigma \int_{\sigma/\sigma_m}^{\infty} (F_\lambda/F_Q) VH(\lambda)\phi(\sigma, \lambda)\, d\lambda \tag{16.32}$$

ISBN 0-201-13500-0

where σ_m denotes the greatest chord cut from G by K_1. By equating this value with $(4V_Q/F_Q)h(\sigma)\,d\sigma$ and using that $F_\lambda = \lambda^2 F$, where $F = F_1$ is the area of K_1, we get the integral equation

$$\int_{\sigma/\sigma_m}^{\infty} \lambda\phi(\sigma/\lambda)H(\lambda)\,d\lambda = (4/F)h(\sigma). \tag{16.33}$$

The solution of this equation depends on $\phi(\sigma)$. For spherical particles we have $\phi(\sigma)\,d\sigma = (\sigma/2)\,d\sigma$, $\sigma_m = 2$, $F = 4\pi$, and (16.33) becomes

$$\int_{\sigma/2}^{\infty} H(\lambda)\,d\lambda = (2/\sigma\pi)h(\sigma), \tag{16.34}$$

which has the immediate solution

$$H(\lambda) = -(1/\pi)(h(2\lambda)/\lambda)'. \tag{16.35}$$

For the case in which all particles have the same shape and size, assuming that there are N particles per unit volume, we can consider $H(\lambda)$ to be N times a Dirac delta function and (16.33) gives $N = 4h(\sigma)/F\phi(\sigma)$.

4. Notes

1. *Further results.* For details and bibliography relating to the topics in this chapter, see the pertinent works by Duffin [154], Kendall and Moran [335], Moran [430], De Hoff and Rhines [142], Bodzioni [54, 55], Matheron [400], and Santaló [573]. See also a number of related investigations in [166]. Nicholson [450] considers in detail the case of cylindrical particles. Watson [708] warns on the bad statistical properties of some of the often-used estimators. Miles [416] has generalized many stereological results to higher-dimensional spaces. See also [321, 322].

Marriot [391] considers rods of fixed length whose centers are uniformly distributed in a planar domain. If the rods make a random angle ϕ with some fixed direction, where ϕ has density function $f(\phi) = 2(1 + k\cos 2\phi)/\pi$ $(-1 \leqslant k \leqslant 1)$, k an unspecified parameter, we can derive the probability that the rod is intersected by a system of parallel lines making an angle α with the reference direction. The number of intersections of the rods with two systems of parallel lines with different α enables us to estimate the total rod length per unit area and the parameter k.

Sidak [609] considers an aggregate of spheres and a slice section of breadth Δ through the aggregate. Let Y be the maximum diameter of that portion of the sphere falling in the section and let X be the diameter of the original sphere. If two sections of different thickness Δ_1, Δ_2 are taken, then $E(X)$ can be expressed in terms of $E_1(Y)$, $E_2(Y)$, Δ_1, and Δ_2.

ISBN 0-201-13500-0

Tallis [667] considers planar sectioning of aggregates of ellipsoids whose principal axes have fixed directions. The case of spherical particles whose diameters follow a normal distribution is considered by Giger and Riedwyl [229]. Computational methods for some stereological problems are reported by Jakeman and Anderssen [321, 322].

Duffin, Meussner, and Rhin [154a] (see also [154]) consider the problem of determining the true size distribution of spherical particles from the observation of linear or planar samples. They define the following distribution functions: (a) the sphere distribution function $G_3(s)$ is the average number of spheres per cubic centimeter having diameter greater than s; (b) the circle distribution function $G_2(s)$ is the average number of circles per square centimeter having diameter greater than s (the circles are the intersection of a plane with the spherical particles); (c) the segment distribution $G_1(s)$ is the average number of segments per centimeter having length greater than s (the segments are the intersection of a line with the spherical particles). Then, the cumulative distribution functions are related by the following Stieltjes integrals:

$$\text{(i)}\quad G_2(s) = -\int_s^\infty (u^2-s^2)^{1/2}\,dG_3(u), \quad G_3(s) = -\frac{2}{\pi}\int_s^\infty (u^2-s^2)^{-1/2}\,dG_2(u),$$

$$\text{(ii)}\quad G_1(s) = -\left(\frac{\pi}{4}\right)\int_s^\infty (u^2-s^2)\,dG_3(u), \quad G_3(s) = -\frac{2}{\pi s}\frac{dG_1}{ds},$$

$$\text{(iii)}\quad G_1(s) = -\int_s^\infty (u^2-s^2)^{1/2}\,dG_2(u), \quad G_2(s) = -\frac{2}{\pi}\int_s^\infty (u^2-s^2)^{-1/2}\,dG_1(u).$$

A distribution function is termed Gaussian if it has the form $G(s) = A\exp(-ks^2)$ where A and k are constants. Then the following theorem holds:

If any two of the distribution functions G_1, G_2, G_3 are proportional, then they are all Gaussian. If any one of them is Gaussian, all are Gaussian.

2. *Moments of the distributions of sizes of particles.* Instead of the functions $H(\lambda)$ and $g(r)$ in (16.26), it is sometimes best to introduce the probability density function of the diameters D of the spherical particles embedded in an opaque medium, say $G(D)$, and the probability density of the diameters of the circles of intersection with a random plane, say $\phi(D)$. If v denotes the mean number of spheres in a unit volume and D_0 is the mean diameter of a sphere chosen at random, we have $F(D/2) = 2vG(D)$, $g(D/2) = 2vD_0\phi(D)$. Thus, (16.22) and (16.26) become

$$\phi(D) = \frac{D}{D_0}\int_D^\infty \frac{G(t)}{(t^2-D^2)^{1/2}}\,dt, \qquad H(D) = -\frac{2DD_0}{\pi}\int_r^\infty (t^2-D^2)^{-1/2}\left(\frac{\phi(t)}{t}\right)'\,dt.$$

ISBN 0-201-13500-0

Following Kendall and Moran [335], we can express the moments of one distribution in terms of those of the other. Putting

$$M_h = \int_0^\infty t^h G(t)\, dt, \qquad m_h = \int_0^\infty t^h \phi(t)\, dt,$$

we obtain the result

$$m_h = J_{h+1} D_0^{-1} M_{h+1} \qquad \text{for} \quad h = -1, 0, 1, 2, \ldots$$

where

$$J_h = \int_0^{\pi/2} \sin^h \theta \, d\theta = \begin{cases} \dfrac{2 \cdot 4 \cdot \cdots \cdot (h-1)}{1 \cdot 3 \cdot \cdots \cdot h} & \text{if } h \text{ is odd,} \\[2ex] \dfrac{1 \cdot 3 \cdot \cdots \cdot (h-1)}{2 \cdot 4 \cdot \cdots \cdot h} \dfrac{\pi}{2} & \text{if } h \text{ is even.} \end{cases}$$

For D_0 we have the value

$$D_0 = \left(\frac{\pi}{2}\right) \left[\int_0^\infty \frac{\phi(t)}{t}\, dt \right]^{-1}.$$

If v_0 denotes the mean number of circles per unit area, we have $v_0 = vD_0$ and thus v can be estimated by

$$v = (2v_0/\pi) \int_0^\infty (\phi(t)/t)\, dt.$$

We can also introduce in (16.35) the probability density function $G(D)$ and the probability density function $s(\sigma)$ of the lengths of the chords. Then we have $h(\sigma) = s(\sigma) v \pi D_2 / 4$, where D_2 is the mean value of D^2 and (16.35) becomes $G(D) = -(D_2/2)(s(D)/D)'$ [335], [309c].

3. *Nearly spherical particles*. When the particles embedded in a medium are not spherical, the results of this chapter are less satisfactory. Wicksell [724] considers the case of ellipsoidal particles and further references can be seen in [335].

A formula analogous to (16.35) may be applied for particles that do not depart very far from sphericity in the sense that the probability density $\phi(\sigma)$ may be assumed of the form $\phi(\sigma) = (a\sigma)^\mu$, where a, μ are constants determined by the conditions

$$\int_0^\infty \phi(\sigma)\, d\sigma = 1, \qquad \int_0^\infty \sigma \phi(\sigma)\, d\sigma = 4V/F$$

ISBN 0-201-13500-0

where V is the volume and F the surface area of the particle corresponding to $\lambda = 1$. We get

$$\mu = \frac{\sigma_m F - 8V}{4V - \sigma_m F}, \qquad a = [(\mu + 1)\sigma_m{}^{-(\mu+1)}]^{1/\mu}.$$

Then (16.33) takes the form

$$\int_{\sigma/\sigma_m}^{\infty} \lambda^{1-\mu} H(\lambda)\, d\lambda = \frac{4}{(a\sigma_m)^\mu F} h(\sigma)$$

and thus

$$H(\lambda) = -\frac{4\lambda^{\mu-1}}{(a\sigma_m)^\mu F}\left(\frac{h(\sigma_m\lambda)}{\lambda^\mu}\right),$$

which generalizes (16.35).

4. *Size distribution of cubic particles.* If the number of particles per unit volume is N_V, then the number of particle intersections N_A per unit area of a sectioning plane is given by $N_A = N_V M/2\pi$ ((16.13) for $\Delta = 0$). Where the particles are congruent cubes randomly oriented, we have $M = 3\pi a$, where a is the cube edge, and thus $N_A = (3/2)N_V a$. The possible sections observed have three, four, five, or six sides, with four sides comprising, on the average, 48.7% of the total. Since some sections will be nearly square, a can be estimated closely and the equations above used to determine N_V.

Such a dispersion of cubic particles may also be analyzed by lineal traverses. Then, the number of particle intersections N_L per unit length of the sectioning line is $N_L = (f/4)N_V$, where f is the area of the particles. For cubes, we have $f = 6a^2$, $N_L = (3/2)N_V a^2$. Combining these equations, we have $N_V = 2N_A{}^2/3N_L$ and $a = N_L/N_A$ (see [442]).

5. *Sectioning by random spheres.* Consider the case of spherical particles whose centers form a Poisson field with a mean of λ centers per unit volume. We may intersect by a random sphere S_R of radius R greater than the radii of the particles and then study how the size distribution of the particles can be estimated from the distribution of the sizes of their intersection circles with S_R. If a sphere S_r of radius r does intersect S_R, under the assumption that $R > r$, the probability that the radius ρ of the intersection circle lies in the interval $\rho, \rho + d\rho$ is

$$\frac{3(R^2 + 3r^2 - 4\rho^2)\rho}{r(3R^2 + r^2)(r^2 - \rho^2)^{1/2}}\, d\rho. \tag{16.36}$$

Let $F(r)\, dr$ be the probability of a random particle's having a radius between r and $r + dr$. The expected number of particles whose radii lie in the range $r, r + dr$ and that intersect an arbitrary sphere of radius R in a unit area of the latter is $(2/3)\lambda(3r + r^3/R^2)F(r)\, dr$; hence, the probability density of the distribution of radii of spheres that intersect S_R is

ISBN 0-201-13500-0

$$f(r) = \frac{(r + r^3/3R^2)F(r)}{\int_0^\infty (r + r^3/3R^2)F(r)\, dr}.$$

Hence, writing r_1, r_3 for the moments of order 1, 3 and using (16.36), we get that the probability distribution of the observed radii, that is, of the radii of the intersection circles of the particles with S_R, is

$$\phi(\rho) = \frac{3\rho}{r_1 + r_3/3R^2} \int_\rho^R \frac{(R^2 + 3r^2 - 4\rho^2)(1 + r^2/3R^2)F(r)}{(3R^2 + r^2)(r^2 - \rho^2)^{1/2}}\, dr.$$

This is the integral equation that relates $\phi(\rho)$ with $F(r)$. The radius ρ is related to the spherical radius α of the intersection $S_R \cap S_r$, considered as a circle of S_R, by $\rho = R \sin \alpha$.

Consider now the space divided at random into convex polyhedra by a Poisson field of infinite random planes. If the resulting density of polyhedra is λ per unit volume, the density of the resulting convex areas on a random intersecting plane is $(9\pi/16)^{1/3}\lambda^{2/3}$ [430]. The density of the resulting convex spherical areas on a random intersecting sphere of radius R such that $(4/3)\pi R\lambda > 1$ is $\lambda_s = (9\pi/16)^{1/3}\lambda^{2/3} - ((6\pi^2)^{1/3}/16R)\lambda^{1/3}$. This would give a means of estimating λ from cross sections by a random sphere of radius R.

6. *Pattern analysis.* Integral geometry is related to some problems in pattern recognition, that is, the study of how to restore pure geometric objects when we can observe only deformed or partial versions of them. Typical problems are those of restoring a pattern in a certain n-dimensional domain A from a sample in the euclidean space that contains A, or of recognizing a set of pure images given a set of deformed images. See [15, 16, 237] and references therein; see also [199, 457], [206a].

7. *Division of a body by random planes.* Let D be a domain in E_3 topologically equivalent to a ball, but not necessarily convex. Let V, F, M denote, respectively, the volume, surface area, and integral of mean curvature of D. Let D_0 be the convex hull of D. Consider n independently random planes $E_1, E_2, \ldots, E_n$ intersecting D. Then we have the following integral formulas (from Santaló [544, 553]):

(i) If N_i denotes the number of intersection points of three planes that belong to D, then

$$\int N_i\, dE_1 \wedge dE_2 \wedge \cdots \wedge dE_n = \pi^4 \binom{n}{3} M_0{}^{n-3} V \tag{16.37}$$

where M_0 denotes the mean curvature integral of ∂D_0 and the integral is taken over all n-tuples of planes meeting D.

(ii) If N_s denotes the number of points in which the intersection line of two planes meets ∂D, then

$$\int N_s\, dE_1 \wedge \cdots \wedge dE_n = (\pi^3/2) \binom{n}{2} M_0{}^{n-2} F. \tag{16.38}$$

ISBN 0-201-13500-0

(iii) If R denotes the number of regions into which the planes divide D, then

$$\int R \, dE_1 \wedge \cdots \wedge dE_n = \pi^4 \binom{n}{3} M_0{}^{n-3} V + \frac{\pi^3}{4} \binom{n}{2} M_0{}^{n-2} F + n M_0{}^{n-1} M + M_0{}^n. \tag{16.39}$$

From these formulas we deduce the following mean values.

$$E(N_i) = \pi^4 \binom{n}{3} \frac{V}{M_0{}^3}, \qquad E(N_s) = \left(\frac{\pi^3}{2}\right) \binom{n}{2} \frac{F}{M_0{}^2},$$

$$E(R) = \pi^4 \binom{n}{3} \frac{V}{M_0{}^3} + \left(\frac{\pi^3}{4}\right) \binom{n}{2} \frac{F}{M_0{}^2} + n \frac{M}{M_0} + 1. \tag{16.40}$$

Consider a convex body K in E_3. Consider the regions into which K is divided by n planes $E_1, \ldots, E_n$ intersecting K. Noting that each interior vertex belongs to eight regions and each vertex on ∂K belongs to four regions, we have that the mean number of vertices of each region is $(8N_i + 4N_s)/R$. Hence, if the variable N_R denotes the number of vertices of each region, we can define the following quotient of mean values.

$$E^*(N_R) = \frac{E(8N_i + 4N_s)}{E(R)} = \frac{32\binom{n}{3}\pi^4 V + 8\binom{n}{2}\pi^3 MF}{4\binom{n}{3}\pi^4 V + \binom{n}{2}\pi^3 MF + 4(n+1)M^3}. \tag{16.41}$$

Similarly, as the mean number of edges of each region we can take $E^*(A_R) = E(4A_i + 2A_s)/E(R)$, where A_i is the number of interior edges and A_s the number of edges on ∂K. Since $6N_i + N_s = 2A_i$, $4N_s = 2A_s$, it follows that

$$E^*(A_r) = \frac{12E(N_i) + 6E(N_s)}{E(R)} = \frac{48\binom{n}{3}\pi^4 V + 12\binom{n}{2}\pi^3 FM}{4\binom{n}{3}\pi^4 V + \binom{n}{2}\pi^3 MF + 4(n+1)M^3}. \tag{16.42}$$

The mean number of faces, say C_R, of each region follows from the last mean values and the Euler relation $N_R - A_R + C_R = 2$. The result is

$$E^*(C_R) = \frac{24\binom{n}{3}\pi^4 V + 6\binom{n}{2}\pi^3 FM + 8(n+1)M^3}{4\binom{n}{3}\pi^4 V + \binom{n}{2}\pi^3 FM + 4(n+1)M^3}. \tag{16.43}$$

To calculate the mean area of each region, note that if we call σ_i the area of $E_i \cap K$, we have

ISBN 0-201-13500-0

$$\int \sum_{1}^{n} \sigma_i \, dE_1 \wedge \cdots \wedge dE_n = 2\pi n V M^{n-1}, \qquad E\left(\sum_{1}^{n} \sigma_i\right) = 2\pi n (V/M), \quad (16.44)$$

and since the interior faces belong to two regions and the faces on ∂K have their total area equal to F, the mean area of each region can be defined as the quotient

$$E^*(S_R) = \frac{1}{E(R)}\left(\frac{4\pi n V}{M} + F\right) \qquad (16.45)$$

and for the mean volume of each region we have

$$E^*(V_R) = V/E(R) \qquad (16.46)$$

where $E(R)$ is given by (16.40).

8. *Averages for polyhedrons formed by random planes.* We know that if K_0 is a line segment of length s contained in the convex set K, the probability that a random plane intersecting K also intersects K_0 is $\pi s/M$. Hence, if there are n random planes intersecting K, the probability that exactly m of them intersect K_0 is

$$p_m = \binom{n}{m} (\pi s/M)^m (1 - \pi s/M)^{n-m}.$$

Assuming that K expands to the whole space and the number of random planes grows in such a way that

$$\frac{n}{M} \to \frac{\lambda}{\pi}, \qquad \lambda \text{ a positive constant,} \qquad (16.47)$$

it follows easily that p_m tends to the Poisson law

$$p_m{}^* = \lim p_m = ((\lambda s)^m/m!)e^{-\lambda s}. \qquad (16.48)$$

Such a distribution of planes in space is called a *Poisson field* or a *homogeneous Poisson system* of parameter λ. The mean value of m is $E(m) = \lambda s$, so that λ is equal to the mean number of planes crossing any line segment of unit length. Noting that for any convex body K that expands to the whole space we have $F/V \to 0$, $M/F \to 0$, $M/V \to 0$, independently of its shape, from (16.41)–(16.46) we deduce the following mean values E^* for the polyhedrons into which the space is partitioned by planes of a homogeneous Poisson system of parameter λ:

$$E^*(N_R) \to 8, \quad E^*(A_R) \to 12, \quad E^*(C_R) \to 6, \quad E^*(S_R) \to \frac{24}{\pi\lambda^2}, \quad E^*(V_R) \to \frac{6}{\pi\lambda^3}.$$

For the true mean values E of the quantities N_R, S_R, etc. (suitably defined) and their second moments, Miles [418] found the following values.

$$E(N_R) = 8, \qquad E(S_R) = 24/\pi\lambda^2, \qquad E(V_R) = 6/\pi\lambda^3, \qquad E(C_R) = 6,$$

ISBN 0-201-13500-0

$$E(N_R^2) = (13\pi^2 + 96)/3, \qquad E(N_R S_R) = 28\pi/\lambda^2, \qquad E(S_R^2) = 240/\lambda^4,$$

$$E(N_R V_R) = 8\pi/\lambda^3, \qquad E(S_R V_R) = 96/\lambda^5, \qquad E(V_R^2) = 48/\lambda^6,$$

$$E(V_R^3) = 1344\pi/\lambda^9, \qquad E(C_R V_R) = \frac{4(\pi^2 + 3)}{\pi\lambda^3}, \qquad E(C_R S_R) = \frac{2(7\pi^2 + 24)}{\pi\lambda^2}.$$

Furthermore, if L_R denotes the total edge length and M_R the mean curvature integral of a polyhedron, Miles showed that

$$E(L_R) = 12/\lambda, \qquad E(M_R) = 3/2\lambda, \qquad E(N_R L_R) = 2(5\pi^2 + 12)/\lambda,$$

$$E(L_R^2) = (24\pi^2 + 24)/\lambda^2, \qquad E(N_R M_R) = (13\pi^2 + 48)/12\lambda$$

$$E(L_R M_R) = (5\pi^2 + 12)/2\lambda^2, \quad E(M_R^2) = (13\pi^2 + 48)/48\lambda^2,$$

$$E(C_R L_R) = \frac{(5\pi^2 + 36)}{\lambda},$$

$$E(L_R S_R) = 72\pi/\lambda^3, \quad E(M_R S_R) = (7\pi^2 + 12)/\pi\lambda^3, \quad E(C_R M_R) = \frac{13\pi^2 + 120}{12\lambda},$$

$$E(L_R V_R) = 24\pi/\lambda^4, \quad E(M_R V_R) = 2(\pi^2 + 3)/\pi\lambda^4, \quad E(C_R^2) = \frac{13\pi^2 + 336}{12}.$$

If the random planes that divide the space into polyhedrons are assumed perpendicular to three orthogonal axes and are uniformly distributed on each of these three directions, then it is easy to show that $E(V_R^2) = 288/\pi^2\lambda^6$.

Interesting and promising results on random sets, with applications to random polyhedral regions defined by Poisson fields of hyperplanes in E_n ($n = 2, 3, \ldots$), are to be found in [399–401]; see also [333].

9. *Random division of space into cells.* The random tessellations described in Section 4 of Chapter 2 generalize easily to E_n. We will consider the case in which $n = 3$ for simplicity. The following models are of interest [405, 230]. Let P_i ($i = 1, 2, \ldots$) be a countable set of points uniformly distributed in E_3 with a density of ρ points per unit volume. Next, subdivide the space into regions or cells by the following rule: the cell C_i contains all points in space closer to P_i than to any P_j ($j \neq i$). Then C_i is a convex polyhedron, because it is the intersection of several half spaces. P_i is called the center of C_i. Figure 16.3 shows an example for the plane. The model is of interest in mineralogy. The centers P_i represent the location of the original nucleus or seed crystals from which C_i grow. We assume that the sets for all crystals start growing at the same instant and that they grow at the same rate in all directions, staying fixed in space without pushing apart as they grow into contact. Call S_R the surface area of a cell, L_R the total length of the edges of a cell; N_R, A_R, C_R the number of vertices, number of edges, and number of faces of a cell; and ν the number of crystal sections per unit area cut by a plane. Then Meijering [405] derived the following mean values.

$$E(S_R) = 5.821\rho^{-2/3}, \qquad E(L_R) = 17.50\rho^{-1/3}, \qquad E(N_R) = 27.07,$$

$$E(A_R) = 40.61, \qquad E(C_R) = 15.54, \qquad E(\nu) = 1.458\rho^{2/3}.$$

ISBN 0-201-13500-0

The mean number of sides of a face is $3E(N_R)/E(C_R) = 5.23$. Comparing with the values for polyhedrons obtained by random planes in the last section, we note that $E(N_R)$, $E(A_R)$, $E(C_R)$ differ by a factor of 3. It should be interesting to compare this factor for dimensions $n > 3$.

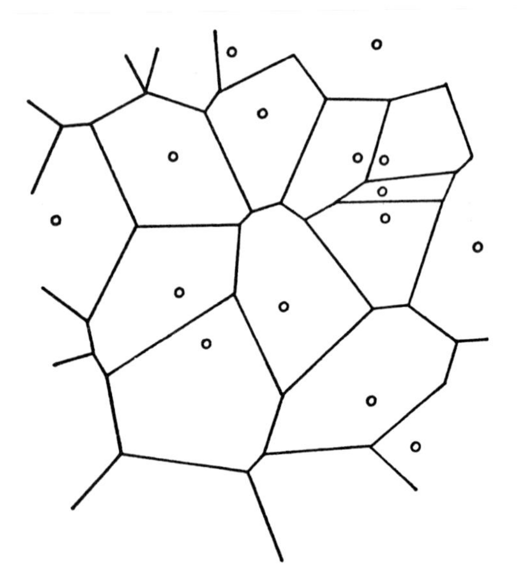

Figure 16.3.

Johnson and Mehl [325] considered more complicated random tessellations, assuming that seeds appear at a constant rate of α seeds per second per unit volume, starting at an initial time $t = 0$. In this model the cells do not have plane faces and they need not be convex. With the notation used above, Meijering obtained the following mean values for the Johnson–Mehl model:

$$E(S_R) = 5.143\rho^{-2/3}, \qquad E(L_R) = 14.71\rho^{-1/3}, \qquad E(N_R) = 22.56,$$

$$E(A_R) > 33.84, \qquad E(C_R) > 13.28, \qquad E(v) = 1.225\rho^{2/3},$$

where $\rho = 0.8960(\alpha/v)^{3/4}$ and v denotes the growth rate of the cells. Other mean values and variances for these models have been computed by Gilbert [230]. For instance, the volume variance of a cell for the first model is $0.180\rho^{-2}$ and for the Johnson–Mehl model it is $1.136\rho^{-2}$. These values are much less than the values corresponding to the division of the space by random planes, as was to be expected. For details, complements, and further references, see the important paper of Miles [418].

ISBN 0-201-13500-0

PART IV

Integral Geometry in Spaces of Constant Curvature

CHAPTER 17

Noneuclidean Integral Geometry

1. The n-Dimensional Noneuclidean Space

Let P_n be the real n-dimensional projective space. Let $x_0, x_1, \ldots, x_n$ be the homogeneous coordinates of a point x and consider the quadric

$$\Phi \equiv x_0^2 + x_1^2 + \cdots + x_{n-1}^2 + \varepsilon x_n^2 = 0 \tag{17.1}$$

where ε can take the value $+1$ or -1.

We assume the homogeneous coordinates of the points x that do not belong to Φ normalized so that

$$\Phi(x) = 1/\varepsilon K \tag{17.2}$$

where K is a *positive* constant. For $\varepsilon = -1$ the points such that $\Phi(x) > 0$ have complex coordinates and are called *ideal* or *imaginary* points.

The real n-dimensional noneuclidean space (or n.e. space) is the set of all points of P_n if $\varepsilon = +1$, and the set of all points of P_n such that $\Phi(x) < 0$ if $\varepsilon = -1$, with the group of motions and the metric we shall define below. The r-planes of P_n are also the r-planes of the n.e. space ($r = 0, 1, \ldots, n-1$). The quadric Φ is called the fundamental or absolute quadric; for $\varepsilon = +1$,

ENCYCLOPEDIA OF MATHEMATICS and Its Applications, Gian-Carlo Rota (ed.).
1, Luis A. Santaló, Integral Geometry and Geometric Probability

ISBN 0-201-13500-0

Φ has no real points and the space is called the *elliptic* n.e. space, and for $\varepsilon = -1$, Φ is a real quadric and the set of points interior to Φ (i.e., the points such that $\Phi(x) < 0$) constitute the *hyperbolic* n.e. space. In both cases εK is called the *curvature* of the space. The elliptic space can be identified with the n-dimensional hemisphere with the points of its boundary antipodally identified. For details see, for instance, the book by Busemann and Kelly [76].

The inner product of two points $a(a_0, a_1, \ldots, a_n)$ and $b(b_0, b_1, \ldots, b_n)$ is defined by

$$\langle a, b \rangle = \langle b, a \rangle = \sum_{i=0}^{n-1} a_i b_i + \varepsilon a_n b_n, \qquad \langle a, a \rangle = 1/\varepsilon K. \tag{17.3}$$

In matrix notation, quadric (17.1) can be written

$$\Phi(x) \equiv x^t Q x = 0, \qquad Q = \begin{pmatrix} 1 & 0 & \cdots & 0 \\ 0 & 1 & \cdots & 0 \\ \cdot & \cdot & \cdots & \cdot \\ 0 & 0 & \cdots & \varepsilon \end{pmatrix} \tag{17.4}$$

where x is now an $n \times 1$ matrix whose elements are the homogeneous coordinates of point x and x^t is its transpose. By means of Q the inner product can be written $\langle a, b \rangle = a^t Q b = b^t Q a$.

The collineations of P_n that carry Φ into itself are called noneuclidean motions (n.e. motions) and they form a group that we will represent by $\mathfrak{M}^*$. A necessary and sufficient condition that a collineation $x' = Ax$ be an n.e. motion is that $\Phi(x') = x^t A^t Q A x = x^t Q x = \Phi(x)$ for all points x, and hence $A^t Q A = Q$.

Note that the inner product is invariant under n.e. motions, namely, $\langle a', b' \rangle = a^t A^t Q A b = a^t Q b = \langle a, b \rangle$. Two points such that $\langle a, b \rangle = 0$ are called *conjugate* points. In the hyperbolic space ($\varepsilon = -1$) the conjugate points of a real point are ideal points.

Let $a^0, a^1, \ldots, a^n$ be $n + 1$ points that are vertices of a self-conjugate simplex, that is, such that $\langle a^i, a^j \rangle = \delta_{ij}/\varepsilon K$, where δ_{ij} is the Kronecker symbol. Assume that a^0 is a real point and consider the matrix

$$A = ((\varepsilon K)^{1/2} a^1, (\varepsilon K)^{1/2} a^2, \ldots, (\varepsilon K)^{1/2} a^n, K^{1/2} a^0) \tag{17.5}$$

whose columns are the coordinates of the points $(\varepsilon K)^{1/2} a^1, \ldots, K^{1/2} a^0$. We have $A^t Q A = Q$ and thus $x' = Ax$ is an n.e. motion. According to the method of moving frames, the Maurer–Cartan forms for the n.e. motions group $\mathfrak{M}^*$ are defined by

$$d((\varepsilon K)^{1/2} a^i) = \sum_{j=1}^{n} \omega_{ji} (\varepsilon K)^{1/2} a^j + \omega_{0i} K^{1/2} a^0 \tag{17.6}$$

ISBN 0-201-13500-0

for $i = 1, 2, \ldots, n$ and

$$d(K^{1/2}a^0) = \sum_{j=1}^{n} \omega_{j0}(\varepsilon K)^{1/2} a^j. \tag{17.7}$$

Multiplying (17.6) by a^j and by a^0 we get

$$\omega_{ji} = \varepsilon K \langle a^j, da^i \rangle = -\omega_{ij} \qquad (i, j = 1, 2, \ldots, n),$$

$$\omega_{0i} = \sqrt{\varepsilon}\, \varepsilon K \langle a^0, da^i \rangle = -\sqrt{\varepsilon}\, \varepsilon K \langle a^i, da^0 \rangle. \tag{17.8}$$

Multiplying (17.7) by a^j we have $\omega_{j0} = \sqrt{\varepsilon}\, K \langle a^j, da^0 \rangle = -\sqrt{\varepsilon}\, K \langle a^0, da^j \rangle$ and thus we have the relations

$$\omega_{j0} + \varepsilon \omega_{0j} = 0, \qquad \omega_{0j} + \varepsilon \omega_{j0} = 0. \tag{17.9}$$

Differentiating (17.6), we get the structure equations

$$d\omega_{ji} = \sum_{h=0}^{n} \omega_{ih} \wedge \omega_{hj}, \qquad d\omega_{0i} = \varepsilon \sum_{h=1}^{n} \omega_{ih} \wedge \omega_{h0}. \tag{17.10}$$

We shall give the definition of the n.e. distance between two points a, b. The distance s between a and b is defined by

$$\langle a, b \rangle = \frac{\cos(\varepsilon K)^{1/2} s}{\varepsilon K}. \tag{17.11}$$

If a and b are real points, the distance s is a real number. In the case in which $\varepsilon = -1$ we can put

$$\cos \sqrt{-K}\, s = \cosh \sqrt{K}\, s. \tag{17.12}$$

The distance between two conjugate points satisfies the condition $\cos(\varepsilon K)^{1/2} s_1 = 0$ and thus $s_1 = \pi/2(\varepsilon K)^{1/2}$, which is real for $\varepsilon = +1$ and complex for $\varepsilon = -1$. Since s_1 and $-s_1$ correspond to the same point (the conjugate point of a^0), it follows that the lines of the elliptic space of curvature K are closed lines of length $\pi/\sqrt{K}$.

Let $a = a^0$ and assume that b lies in the line joining the pair of conjugate points a^0, a^i. Then $b = \alpha a^0 + \beta a^i$, $\langle b, a^0 \rangle = \alpha(\varepsilon K)^{-1}$ and (17.11) gives $\alpha = \cos(\varepsilon K)^{1/2} s$. On the other hand, the relation $\langle b, b \rangle = (\varepsilon K)^{-1}$ gives $\alpha^2 + \beta^2 = 1$. Therefore, denoting by s_i the distance from b to a^0, we have

$$b = \cos((\varepsilon K)^{1/2} s_i) a^0 + \sin((\varepsilon K)^{1/2} s_i) a^i. \tag{17.13}$$

Assuming a^0, a^i fixed and b moving on the line $a^0 a^i$, we have

$$db = (\varepsilon K)^{1/2}(-\sin((\varepsilon K)^{1/2} s_i) a^0 + \cos((\varepsilon K)^{1/2} s_i) a^i)\, ds_i. \tag{17.14}$$

ISBN 0-201-13500-0

In particular, for $s_i = 0$, we have $da^0 = (\varepsilon K)^{1/2} a^i\, ds_i$. From this relation and (17.9) we obtain

$$ds_i = K^{-1/2}\omega_{i0} = -\varepsilon K^{-1/2}\omega_{0i}. \tag{17.15}$$

The volume element at point a^0 is the product $ds_1 \wedge ds_2 \wedge \cdots \wedge ds_n$; that is,

$$dv = K^{-n/2}\omega_{10} \wedge \omega_{20} \wedge \cdots \wedge \omega_{n0} = (-\varepsilon)^n K^{-n/2}\omega_{01} \wedge \cdots \wedge \omega_{0n}. \tag{17.16}$$

The angle between two lines through point a^0 that lie in the plane $a^0 a^i a^j$ is defined by

$$\cos\phi = \varepsilon K\langle a', b'\rangle \tag{17.17}$$

where a', b' are any pair of conjugate points to a^0 that lie respectively in the given lines. In particular, if $a' = a^i$, we have $\cos\phi = \varepsilon K\langle a^i, b'\rangle$. Assuming b' moving on the line $a^i a^j$ (i.e., assuming that the line $a^0 b'$ rotates about a^0 in the plane $a^0 a^i a^j$), we get $-\sin\phi\, d\phi = \varepsilon K\langle a^i, db'\rangle$. If $b' = a^j$, we have $\phi = \pi/2$ and the angle element on the plane $a^0 a^i a^j$ about a^0, corresponding to the direction $a^0 a^j$ in the sense from a^j to a^i, takes the form

$$d\phi_{ij} = \varepsilon K\langle a^i, da^j\rangle = \omega_{ij}. \tag{17.18}$$

Therefore the solid angle element corresponding to the direction $a^0 a^j$ through a^0 is given by

$$du_{n-1}(a^0 a^j) = \omega_{1j} \wedge \omega_{2j} \wedge \cdots \wedge \omega_{j-i,j} \wedge \omega_{j+1,j} \wedge \cdots \wedge \omega_{nj}. \tag{17.19}$$

2. The Gauss-Bonnet Formula for Noneuclidean Spaces

We need the so-called Gauss–Bonnet formula for compact, orientable hypersurfaces of the n-dimensional n.e. space. This formula is a particular case of the generalized Gauss–Bonnet formula of Allendoerfer and Weil [5] and Chern [106]. For the special case of hypersurfaces of the n.e. space, the formula was given independently by Herglotz [307].

Let Q be a connected domain in the n.e. space that is bounded by a compact hypersurface ∂Q of class C^3. At each point a^0 of ∂Q we attach the self-conjugate simplex $a^0 a^1 \cdots a^n$ such that the points $a^0, a^1, \ldots, a^{n-1}$ define the tangent hyperplane and $a^0 a^n$ is the normal of ∂Q at a^0. Assuming that $a^0 a^i$ $(i = 1, 2, \ldots, n-1)$ are the principal directions at a^0, the principal radii of curvature R_i are defined by the Rodrigues formulas $\omega_{in} = d\phi_{in} = -ds_i/R_i$ and the ith integral of mean curvature is defined by

$$M_i = \frac{1}{\binom{n-1}{i}} \int_{\partial Q} \left\{\frac{1}{R_{h_1}} \cdots \frac{1}{R_{h_i}}\right\} df \tag{17.20}$$

ISBN 0-201-13500-0

where { } denotes the elementary symmetric function of degree i of the curvatures $1/R_h$ and df denotes the area element of ∂Q. Then the Gauss–Bonnet formula for the boundary ∂Q of a domain Q states that

$$c_{n-1}M_{n-1} + c_{n-3}M_{n-3} + \cdots + c_1 M_1 + (\varepsilon K)^{n/2} V = \tfrac{1}{2} O_n \chi(Q) \quad (17.21)$$

for n even, and

$$c_{n-1}M_{n-1} + c_{n-3}M_{n-3} + \cdots + c_2 M_2 + c_0 F = \tfrac{1}{2} O_n \chi(Q) \quad (17.22)$$

for n odd. In these formulas M_i are the mean curvature integrals and

$$c_h = \binom{n-1}{h} \frac{O_n}{O_h O_{n-1-h}} (\varepsilon K)^{(n-1-h)/2}. \quad (17.23)$$

If n is odd, we can use the equality

$$\chi(Q) = \tfrac{1}{2}\chi(\partial Q), \qquad n \text{ odd}. \quad (17.24)$$

If ∂Q is only piecewise of class C^2, the Gauss–Bonnet formula holds in all cases where the boundary of the parallel set Q_ε of Q in the distance ε is of class C^2. Then the integrals M_i are defined to be the mean curvature integrals of ∂Q_ε as $\varepsilon \to 0$. Federer [177, 178] extended the Gauss–Bonnet formula to sets with positive reach (the reach of a subset A of the n.e. space is the largest ε such that if the distance from point x to A is smaller than ε, then A contains a unique point nearest to x) and Hadwiger [274, 276] to sets of the convex ring (sets representable as the finite union of convex sets). All formulas in this chapter hold for these kinds of sets.

Examples. For $n = 2$ (17.21) becomes

$$\int_{\partial Q} \kappa \, ds + \varepsilon K F = 2\pi \chi(Q)$$

where κ is the curvature (geodesic curvature) of ∂Q and F is the area of Q. On the unit sphere we have $\varepsilon = +1$, $K = 1$ and on the hyperbolic plane $\varepsilon = -1$, $K = 1$.

For $n = 3$, (17.22) becomes

$$M_2 + \varepsilon K F = 4\pi \chi(Q). \quad (17.25)$$

Dual formulas in elliptic space. Consider the elliptic case $\varepsilon = +1$. For simplicity, without loss of generality, we can take $K = 1$. Then the hypersurface parallel to ∂Q in a distance $\pi/2$ is called the hypersurface "dual" or "polar" of ∂Q. This hypersurface is the boundary of a domain Q^P (which does not contain Q). It can be proved that the mean curvature integrals of

ISBN 0-201-13500-0

∂Q and those of its polar ∂Q^P satisfy the equations

$$M_i(\partial Q^P) = M_{n-1-i}(\partial Q), \qquad i = 0, 1, \ldots, n-1 \tag{17.26}$$

(for a proof see [558, 580]). Writing (17.21) for Q^P and using that $\chi(Q) = \chi(Q^P)$, we get (putting $M_0 = F$, $M_0{}^P = F^P$)

$$c_{n-1}F + c_{n-3}M_2 + \cdots + c_1 M_{n-2} + V^P = \tfrac{1}{2}O_n\chi(Q) \qquad (n \text{ even}) \tag{17.27a}$$

and

$$c_{n-1}F^P + c_{n-3}M_{n-3} + \cdots + c_1 M_1 + V = \tfrac{1}{2}O_n\chi(Q) \qquad (n \text{ even}). \tag{17.27b}$$

Similarly, from (17.22) it follows that

$$c_{n-1}F + c_{n-3}M_2 + \cdots + c_2 M_{n-3} + c_0 F^P = \tfrac{1}{4}O_n\chi(\partial Q) \qquad (n \text{ odd}) \tag{17.28}$$

where we have applied that $\chi(Q) = \frac{1}{2}\chi(\partial Q)$.

For n odd the volume V^P of Q^P can be evaluated as the difference between twice the volume of the elliptic space (which is equal to O_n) and the volume of the body parallel to Q in the distance $\pi/2$. The result is [4]

$$c_{n-2}M_{n-2} + \cdots + c_1 M_1 + V + V^P = (1 - \tfrac{1}{2}\chi(Q))O_n \qquad (n \text{ odd}). \tag{17.29}$$

Example 1. $n = 2$. Calling L the length of the closed curve ∂Q and F the area of Q, we have

$$L + F^P = 2\pi\chi(Q), \qquad L^P + F = 2\pi\chi(Q). \tag{17.30}$$

Example 2. For $n = 3$, we have

$$F + F^P = 4\pi\chi(Q), \qquad M_1 + V + V^P = 2\pi^2 - \pi^2\chi(Q). \tag{17.31}$$

3. Kinematic Density and Density for r-Planes

According to the general theory, the kinematic density of the group $\mathfrak{M}^*$ of n.e. motions is equal to the exterior product of the 1-forms of a set of Maurer–Cartan forms. Since the kinematic density is defined up to a constant factor, using (17.16) and (17.19) we can take the normalized form

$$dK = \left(\frac{1}{K}\right)^{n/2} \bigwedge_h \omega_h \bigwedge_{i<j} \omega_{ij} = dv \wedge du_{n-1} \wedge \cdots \wedge du_1; \qquad i, j, h = 0, 1, \ldots, n, \tag{17.32}$$

where dv denotes the volume element corresponding to the point a^0 and du_h is the area element of the h-dimensional unit sphere in the linear space defined by $a^0, a^1, \ldots, a^{h+1}$ ($h = 1, 2, \ldots, n-1$). Note that (17.32) has the same

ISBN 0-201-13500-0

form as in the euclidean case (15.1). Similarly, as in the euclidean case, the kinematic density about a fixed q-plane L_q can be written

$$dK_{[q]} = du_{n-q-1} \wedge \cdots \wedge du_1 \qquad (q = 1, 2, \ldots, n-2). \qquad (17.33)$$

Consider now the r-plane L_r defined by the set of conjugate points $a^0, a^1, \ldots, a^r$. The n.e. motions that leave L_r invariant are characterized (according to (17.6) and (17.7)) by

$$\omega_{0i} = 0 \qquad \text{for} \qquad i = r+1, \ldots, n;$$

$$\omega_{jh} = 0 \qquad \text{for} \qquad j = r+1, \ldots, n, \quad h = 1, \ldots, r. \qquad (17.34)$$

Therefore, the density for r-planes can be taken as

$$dL_r = \left(\frac{1}{K}\right)^{(n-r)/2} \bigwedge \omega_{0i} \bigwedge \omega_{jh} \qquad (17.35)$$

where the indices of the exterior products run over the ranges

$$i, j = r+1, r+2, \ldots, n, \qquad h = 1, 2, \ldots, r. \qquad (17.36)$$

The constant factor $(1/K)^{(n-r)/2}$ is not essential, but it is useful in order to simplify some formulas in the sequel. The density for r-planes that contain a fixed q-plane $(q < r)$ is given by the same formula (12.26) as in the euclidean

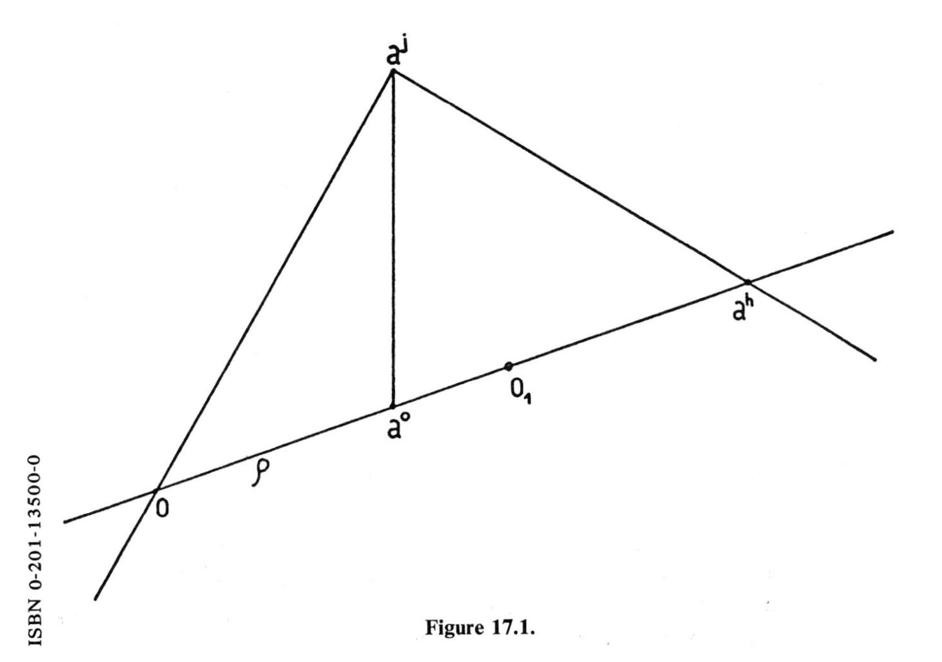

Figure 17.1.

ISBN 0-201-13500-0

case. The duality (12.28) and the total measure of r-planes through a fixed q-plane (12.36) also hold without change.

It is useful to express the density dL_r in terms of the distance ρ from L_r to a fixed point O, the density $dL_{n-r[0]}$ of the $(n - r)$-plane perpendicular to L_r through O, and the volume element $d\sigma_{n-r}$ of $L_{n-r[0]}$ at the intersection point $L_r \cap L_{n-r[0]}$. To do this we proceed as follows.

Let ρ be the distance from O to the point a^0 (Fig. 17.1). If a^h is the conjugate point of a^0 on the line Oa^0 and a^j is a conjugate point of a^0 and a^h, by (17.8) we have

$$\omega_{jh} = \varepsilon K\langle a^j, da^h\rangle = - \varepsilon K\langle a^h, da^j\rangle. \tag{17.37}$$

By (17.18) we know that ω_{jh} denotes an elementary rotation about a^0 in the plane $a^0a^ja^h$ corresponding to the direction a^0a^h. If O_1 is the conjugate point of O on the line Oa^0, using (17.13) and the fact that the distance a^ha^0 is equal to $\pi/2(\varepsilon K)^{1/2}$, we have $a^h = \cos(\varepsilon K)^{1/2}(\rho + \pi/2(\varepsilon K)^{1/2})O + \sin(\varepsilon K)^{1/2} \cdot (\rho + \pi/2(\varepsilon K)^{1/2})O_1 = -\sin((\varepsilon K)^{1/2}\rho)O + \cos((\varepsilon K)^{1/2}\rho)O_1$. Therefore, between the elementary rotation ω_{jh} about a^0 and the elementary rotation $\omega_{jh}^* = \varepsilon K\langle a^j, dO_1\rangle = - \varepsilon K\langle O_1, da^j\rangle$ about O, using that $\langle a^j, O\rangle = \langle a^j, O_1\rangle = 0$ and that $dO = 0$, we have the equation

$$\omega_{jh} = \cos((\varepsilon K)^{1/2}\rho)\omega_{jh}^*. \tag{17.38}$$

Let a^0 be the foot of the perpendicular drawn from O to L_r, the r-plane defined by $a^0, a^1, \ldots, a^r$. Take, moreover, a^{r+1} on the line Oa^0. The elementary rotations ω_{jh}^* $(j = 1, 2, \ldots, r;\ h = r + 2, \ldots, n)$ are contained in planes orthogonal to Oa^0 and thus they have the same value at the point a^0 as at the point O; that is, we have

$$\omega_{jh}^* = \omega_{jh} \qquad \text{for} \quad j = 1, 2, \ldots, r, \quad h = r + 2, \ldots, n; \tag{17.39}$$

while according to (17.38) we have

$$\omega_{j,r+1} = \omega_{j,r+1}^* \cos((\varepsilon K)^{1/2}\rho), \qquad j = 1, 2, \ldots, r. \tag{17.40}$$

Substituting (17.39) and (17.40) into (17.35) and using that the product $(1/K)^{(n-r)/2} \bigwedge \omega_{0i}$ $(i = r + 1, \ldots, n)$ is the volume element $d\sigma_{n-r}$ of $L_{n-r[0]}$ at point a^0, we get

$$dL_r = \cos^r((\varepsilon K)^{1/2}\rho)\, d\sigma_{n-r} \wedge dL_{n-r[0]} \qquad (r = 1, 2, \ldots, n - 1) \tag{17.41}$$

where $dL_{n-r[0]} = \bigwedge \omega_{jh}^*$ $(j = r + 1, \ldots, n;\ h = 1, 2, \ldots, r)$ is the density for $(n - r)$-planes through O. This formula (17.41), which generalizes (12.38) of euclidean space, is the desired expression for dL_r. Another expression for dL_r in terms of matrices has been given by Müller [437].

ISBN 0-201-13500-0

Let us consider the case in which $r = 1$. In it, $dL_{n-1[0]} = dL_{1[0]}$ is the solid angle element of vertex O corresponding to the normal of the $(n-1)$-plane perpendicular to L_1 through O. In order to replace this solid angle element $dL_{1[0]}$ by the analogous element $dL_{1[a^0]}$ of vertex a^0, since L_1 is defined by a^0 and a^1, using (17.19), (17.39), and (17.40), we have

$$dL_1 = d\sigma_{n-1} \wedge du_{n-1} \tag{17.42}$$

where du_{n-1} is the solid angle element corresponding to the direction of L_1 at point a^0 and $d\sigma_{n-1}$ is the volume element of the $(n-1)$-plane perpendicular to L_1 through a^0. If ds denotes the arc element on L_1 at a^0, we have $d\sigma_{n-1} \wedge ds = dv$, volume element of the space at a^0, and (17.42) gives

$$dL_1 \wedge ds = dv \wedge du_{n-1}. \tag{17.43}$$

Consider, for instance, a domain Q of the n.e. space and integrate both sides of (17.43) over all points $a^0 \in Q$. The right-hand member gives $\frac{1}{2} VO_{n-1}$ (where the factor $\frac{1}{2}$ arises from the fact that the lines L_1 are considered unoriented) and on the left-hand side the integral of ds gives the length λ of the chord $Q \cap L_1$. The result is the integral formula

$$\int_{L_1 \cap Q \neq 0} \lambda \, dL_1 = \tfrac{1}{2} O_{n-1} V. \tag{17.44}$$

Note that (17.43) and (17.44) are independent of the curvature of the space. Actually, they hold for any Riemann space.

Formula (17.41) fails for $r = 0$. In this case, dL_0 is the density for points and the formula that corresponds to (17.41) is the expression of the volume element in polar coordinates. In order to find it, note that with the same notation as above we have $a^0 = \cos((\varepsilon K)^{1/2}\rho)O + \sin((\varepsilon K)^{1/2}\rho)O_1$ and therefore (since O is fixed and thus $\langle O, da^i \rangle = -\langle a^i, dO \rangle = 0$) we have

$$\omega_{i0} = -(\varepsilon K)^{1/2} \langle a^0, da^i \rangle = -(\varepsilon K)^{1/2} \sin((\varepsilon K)^{1/2}\rho) \langle O_1, da^i \rangle. \tag{17.45}$$

On the other hand, considering the self-conjugate simplex $O, O_1, a^2, \ldots, a^n$, the solid angle element of vertex O corresponding to the direction Oa^0, according to (17.19), is $du_{n-1} = \omega'_{21} \wedge \omega'_{31} \wedge \cdots \wedge \omega'_{n1}$, where $\omega'_{i1} = \varepsilon K \langle a^i, dO_1 \rangle = -\varepsilon K \langle O_1, da^i \rangle = \sqrt{\varepsilon}\, \omega_{i0} / \sin(\varepsilon K)^{1/2} \rho \; (i = 2, 3, \ldots, n)$. Therefore, applying (17.16) we have

$$dv = (\sqrt{\varepsilon}/(\varepsilon K)^{n/2}) \sin^{n-1}((\varepsilon K)^{1/2}\rho)\, \omega_{10} \wedge du_{n-1}.$$

According to (17.15) we have $d\rho = ds_1 = (1/\sqrt{K})\omega_{10}$ and hence

$$dL_0 = dv = \frac{1}{(\varepsilon K)^{(n-1)/2}} \sin^{n-1}((\varepsilon K)^{1/2}\rho)\, d\rho \wedge du_{n-1}. \tag{17.46}$$

ISBN 0-201-13500-0

This expression for the volume element of the n.e. space in polar coordinates immediately gives the volume of the n.e. sphere of radius ρ, namely

$$V_n(\rho) = \frac{O_{n-1}}{(\varepsilon K)^{(n-1)/2}} \int_0^\rho \sin^{n-1}((\varepsilon K)^{1/2}\rho)\, d\rho, \tag{17.47}$$

and the area $A_n(\rho) = dV_n(\rho)/d\rho$ of the n.e. sphere of radius ρ,

$$A_n(\rho) = \frac{O_{n-1}}{(\varepsilon K)^{(n-1)/2}} \sin^{n-1}((\varepsilon K)^{1/2}\rho). \tag{17.48}$$

For completeness we are going to compute the mean curvature integrals $M_r(\rho)$ of the n.e. sphere of radius ρ. We need only recall that if R_i is the principal radius of curvature of a hypersurface at the point P corresponding to the line of curvature C_i, and ρ_i denotes the distance from P to the contact point of the normal to the hypersurface at P with the envelope of the normals along C_i, we have the relation (see, e.g., [165, p. 214]),

$$R_i = (\varepsilon K)^{-1/2} \tan((\varepsilon K)^{1/2}\rho_i). \tag{17.49}$$

Thus, for an n.e. sphere of radius ρ, according to definition (17.20), we have

$$M_r(\rho) = \frac{O_{n-1}}{(\varepsilon K)^{(n-1-r)/2}} \sin^{n-r-1}((\varepsilon K)^{1/2}\rho) \cos^r((\varepsilon K)^{1/2}\rho). \tag{17.50}$$

Measure of the r-planes that meet an n.e. sphere. As an application of (17.41) we want to compute the measure of the r-planes that intersect a fixed n.e. sphere Σ_ρ of radius ρ. We can take the center of Σ_ρ as a fixed point O. Then we have

$$d\sigma_{n-r} = A_{n-r-1}(\rho)\, d\rho = (\varepsilon K)^{-(n-r-1)/2} \sin^{n-r-1}((\varepsilon K)^{1/2}\rho) O_{n-r-1}\, d\rho; \tag{17.51}$$

hence, applying (17.48) and (12.35), which holds without change for n.e. spaces, we obtain

$$\int_{L_r \cap \Sigma_\rho \neq 0} dL_r = \frac{O_{n-1} \cdots O_r}{(\varepsilon K)^{(n-r-1)/2} O_{n-r-2} \cdots O_1 O_0} \cdot \int_0^\rho \cos^r((\varepsilon K)^{1/2}\rho) \sin^{n-r-1}((\varepsilon K)^{1/2}\rho)\, d\rho. \tag{17.52}$$

For the elliptic case, $\varepsilon = +1$, the space is finite and the measure of all L_r is also finite. Its value follows from (17.52) for $\rho = \pi/2(\varepsilon K)^{1/2}$. Since

ISBN 0-201-13500-0

$$\int_0^{\pi/2} \cos^r(\alpha)\sin^{n-r-1}(\alpha)\,d\alpha = \frac{O_n}{O_r O_{n-r-1}}, \tag{17.53a}$$

it follows that

$$\int_{\text{Total}} dL_r = \frac{O_n O_{n-1}\cdots O_{r+1}}{K^{(n-r)/2} O_{n-r-1}\cdots O_0} \qquad (\varepsilon = +1). \tag{17.53b}$$

Note. It is sometimes advantageous to write (17.52) and (17.53b) in the equivalent forms

$$\int_{L_r \cap \Sigma_\rho \neq \emptyset} dL_r = \frac{O_{n-1}\cdots O_{n-r-1}}{(\varepsilon K)^{(n-r-1)/2} O_{r-1}\cdots O_0} \int_0^\rho \cos^r((\varepsilon K)^{1/2}\rho)\sin^{n-r-1}((\varepsilon K)^{1/2}\rho)\,d\rho \tag{17.52*}$$

$$\int_{\text{Total}} dL_r = \frac{O_n O_{n-1}\cdots O_{n-r}}{K^{(n-r)/2} O_r\cdots O_0}. \tag{17.53b*}$$

For instance, for $r = n - 1$, (17.52) is not directly applicable, but (17.52)* gives

$$\int_{L_{r-1} \cap \Sigma_\rho \neq \emptyset} dL_{n-1} = O_{n-1} \int_0^\rho \cos^{n-1}((\varepsilon K)^{1/2}\rho)\,d\rho. \tag{17.54}$$

4. Sets of r-Planes That Meet a Fixed Body

Since dL_r and $dL_{r[0]}$ have the same form, up to a constant factor, in the euclidean as in the n.e. case, it follows that formula (14.77) holds good in both cases. Then since the theorems on local differential geometry that are used in proving (14.78) are also valid in n.e. differential geometry, it follows that (14.78) is also valid for any n.e. space of curvature εK. In like manner, (14.69) holds without change in n.e. geometry. In particular, for $q = n$ we have

$$\int_{Q \cap L_r \neq \emptyset} \sigma_r(Q \cap L_r)\,dL_r = \frac{O_{n-1}\cdots O_{n-r}}{O_{r-1}\cdots O_0}\sigma_n(Q) \tag{17.55}$$

where $\sigma_n(Q)$ denotes the volume of the domain Q and $\sigma_r(Q \cap L_r)$ is the r-dimensional volume of $Q \cap L_r$. For $r = 1$ we again get (17.44).

Consider $Q \cap L_r$ as a body of L_r and let $M_i^{(r)}$ $(i = 0, 1, \ldots, r-1)$ be the mean curvature integrals of $\partial(Q \cap L_r)$. The generalized Gauss–Bonnet formula gives, for $r = 2r'$ (r even)

ISBN 0-201-13500-0

$$\tfrac{1}{2}O_n\chi(Q \cap L_r) = (\varepsilon K)^{r'}\sigma_r(Q \cap L_r) + \sum_{i=1}^{r'} \binom{r-1}{2i-1} \frac{O_r}{O_{2i-1}O_{r-2i}} (\varepsilon K)^{r'-i} M_{2i-1}^{(r)} \quad (17.56)$$

and for $r = 2r' + 1$ (r odd),

$$\tfrac{1}{2}O_n\chi(Q \cap L_r) = \sum_{i=0}^{r'} \binom{r-1}{2i} \frac{O_r}{O_{2i}O_{r-2i-1}} (\varepsilon K)^{r'-i} M_{2i}^{(r)}. \quad (17.57)$$

Multiplying both sides of these equations by dL_r and integrating over all L_r that intersect Q, we have, for $r = 2r'$,

$$\int_{Q \cap L_r \neq 0} \chi(Q \cap L_r)\, dL_r = \frac{O_{n-2} \cdots O_{n-r}}{O_r \cdots O_1} \Bigg[(\varepsilon K)^{r'} O_{n-1}\sigma(Q) + \sum_{i=1}^{r'} \binom{r-1}{2i-1} \frac{O_r O_{r-1} O_{n-2i+1}}{O_{2i-1}O_{r-2i}O_{r-2i+1}} (\varepsilon K)^{r'-i} M_{2i-1} \Bigg] \quad (17.58)$$

and for $r = 2r' + 1$,

$$\int_{Q \cap L_r \neq 0} \chi(Q \cap L_r)\, dL_r = \frac{O_{n-2} \cdots O_{n-r}}{O_{r-1} \cdots O_1} \sum_{i=0}^{r'} \binom{r-1}{2i} \frac{O_{r-1}O_{n-2i}}{O_{2i}O_{r-2i-1}O_{r-2i}} (\varepsilon K)^{r'-i} M_{2i}. \quad (17.59)$$

For $r = 0$, dL_0 is the volume element and $\chi(Q \cap L_0) = 1$, so that integral (17.59) is the volume of Q.

Note that for $r = 1$ ($r' = 0$) (17.59) is not directly applicable. However, using the identity

$$\frac{O_{n-2} \cdots O_{n-r}}{O_{r-1} \cdots O_1} = \frac{O_{n-2} \cdots O_r}{O_{n-r-1} \cdots O_1}, \quad (17.60)$$

we get

$$\int_{Q \cap L_1 \neq 0} \chi(Q \cap L_1)\, dL_1 = (O_n/4\pi)F. \quad (17.61)$$

If Q is a convex set, then $\chi(Q \cap L_r) = 1$ and we can state

The measure of all L_r intersecting a convex set Q in the n.e. space of constant curvature εK is given by the right-hand side of (17.58) *or* (17.59), *according to the parity of r.*

ISBN 0-201-13500-0

For instance, for $n = 3$, assuming Q convex, we have

$$\int_{Q \cap L_1 \neq 0} dL_1 = (\pi/2)F, \qquad \int_{Q \cap L_2 \neq 0} dL_2 = M_1 + \varepsilon K V. \tag{17.62}$$

For the elliptic integral geometry, see the work of Blaschke [43] and T. J. Wu [731, 732, 732a].

5. Notes

1. *An inequality for tetrahedrons in n.e. geometry.* Let T be a tetrahedron in three-dimensional n.e. space of curvature εK. The mean curvature integral M_1 of T can be computed by the same method as in the euclidean case, and we get the same result (13.58). Therefore we have

$$\int_{L_2 \cap T \neq 0} dL_2 = \tfrac{1}{2}\pi \sum_{i=1}^{6} a_i - \tfrac{1}{2} \sum_{i=1}^{6} \alpha_i a_i + \varepsilon K V, \qquad \int_{L_2 \cap T \neq 0} N\, dL_2 = \pi \sum_{i=1}^{6} a_i$$

where a_i are the lengths of the edges, α_i are the corresponding dihedral angles, V is the volume, and N denotes the number of edges that are intersected by L_2. Therefore either $N = 3$ or $N = 4$, except for locations of L_2 that belong to a set of measure zero.

Calling m_i the measure of planes that intersect i edges, from the last formulas we deduce

$$m_3 = \pi \sum_{i=1}^{6} a_i - 2 \sum_{i=1}^{6} \alpha_i a_i + 4\varepsilon K V,$$

$$m_4 = \tfrac{3}{2} \sum_{i=1}^{6} \alpha_i a_i - \tfrac{1}{2}\pi \sum_{i=1}^{6} a_i - 3\varepsilon K V.$$

Since these measures are nonnegative, we have the inequalities

$$\pi \sum_{i=1}^{6} a_i + 4\varepsilon K V \geqslant 2 \sum_{i=1}^{6} \alpha_i a_i, \qquad 3 \sum_{i=1}^{6} \alpha_i a_i \geqslant \pi \sum_{i=1}^{6} a_i + 6\varepsilon K V.$$

These inequalities generalize to n.e. space those given by Pólya and Szego [489] for Euclidean space. They may be of some interest because, as is well known, V cannot be expressed in terms of elementary functions of a_i and α_i. The volume of a tetrahedron in n.e. geometry has been studied by several authors (including Coxeter [128] and Bohm [57–59]).

2. *An integral formula.* Let Q be a connected domain in the elliptic n-dimensional space and assume that the boundary ∂Q is of class C^3. For each point P of the elliptic space we draw the normals PA_i $(i = 1, 2, \ldots)$ to ∂Q and let C_{ih} $(h = 1, 2, \ldots, n-1)$ be the $n-1$ lines of curvature of ∂Q at A_i. Set $\varepsilon_{ih} = -1$ if PA_i is a relative maximum of the distances from P to C_{ih} and $\varepsilon_{ih} = +1$ if PA_i is a relative minimum of these distances; set $\varepsilon_{ih} = 0$

ISBN 0-201-13500-0

otherwise. Putting $v_i = \varepsilon_{i1}\varepsilon_{i2}\cdots\varepsilon_{i,n-1}$ and $N = \sum v_i$ where the sum is over all normals from P to Q, we find that the following integral formulas hold

$$\int N\,dP = O_n\chi(Q) \quad (n \text{ odd}), \qquad \int N\,dP = O_n\chi(Q) - 2V^P \quad (n \text{ even})$$

where dP is the volume element and the integrals are extended over the whole elliptic space. For $n = 2, 3$ these formulas are the work of Blaschke [45] (see [570]).

3. *Geometric probability in n.e. spaces.* From the results of this chapter, we can easily obtain the solution of typical problems on geometric probability in n.e. spaces. We state some of them, leaving the proofs to the reader.

1. *If K_1 is a convex set contained in a convex set K_0 in n-dimensional n.e. space of curvature εK, the probability that a random secant of K_0 also intersects K_1 is F_1/F_0, where F_1 and F_0 are the surface areas of ∂K_1 and ∂K_0, respectively.*

2. *The mean length of a secant of a convex body K in n.e. n-dimensional space is $E(\sigma) = 2\pi O_{n-1}V/O_nF$.*

3. *If K_1 is a convex set interior to a convex set K_0 in n.e. space of three dimensions, the probability that a random plane intersecting K_0 also intersects K_1 is*

$$p = \frac{M_1 + \varepsilon K V_1}{M_0 + \varepsilon K V_0}$$

where M_i and V_i are, respectively, the mean curvature integrals and the volumes of K_i $(i = 0, 1)$.

4. *Let K be a convex body in three-dimensional n.e. space and let L_1 and L_2 be a line and a plane that meet K independently and at random. Then the probability that they meet inside K is*

$$p = \frac{4\pi V}{(M + \varepsilon K V)F}.$$

5. *Let K be a convex body in three-dimensional n.e. space and let L_2, L_2^* be two independently random planes, both intersecting K. The probability that their line of intersection intersects K is*

$$p = \frac{\pi^3 F}{4(M + \varepsilon K V)^2}.$$

4. *Geometric probability on the sphere.* We state some problems concerning the theory of geometric probabilities on the two-dimensional sphere U. They are easily solved using the results of this chapter, taking into account that the geometry on U is locally the geometry on the elliptic plane, the lines being the great circles on the sphere.

1. *The probability that two random great circles intersect inside a convex set K of area F is $p = F/2\pi$. If the great circles are assumed secant to K, then the probability that they intersect inside K is $p = 2\pi F/L^2$.*

ISBN 0-201-13500-0

By duality we have

2. *The probability that the circle defined by two random points on the unit sphere does not intersect a fixed convex set K of perimeter L is* $p = (2\pi - F)/L$.

3. *The probability that the arc of great circle* $P_1P_2 \leqslant \pi$ *joining two random points* P_1, P_2 *cuts a fixed convex set K is* $p = (L + 2F)/8\pi$.

4. *The mean distance between two random points on the unit sphere is* $\pi/2$.

5. *Let* P_1, P_2, P_3 *be three points chosen at random on the unit sphere. The mean perimeter of the triangle* $P_1P_2P_3$ *is* $3\pi/2$ *and the mean area is* $\pi/2$.

6. *Let* H_1, H_2 *be the two hemispheres defined by a given great circle* G_0 *on the unit sphere. The mean distance between two points* P_1, P_2 *chosen at random on* H_1 *and* H_2, *respectively, is* $E(\theta) = \pi - 4/\pi$.

Consider the hemisphere H of unit radius. Because of the lack of symmetry, mean value problems on the hemisphere are in general more involved than those on the whole sphere. We consider some examples:

7. *The mean distance between two points on the unit hemisphere is* $4/\pi$.

8. *The mean perimeter of a triangle defined by three random points* P_1, P_2, P_3 *on the hemisphere is* $12/\pi$, *and the mean area is* $(12/\pi) - \pi$. *If the triangle is defined by three random great circles, the mean perimeter is* $E^*(L) = 3\pi - 12/\pi$ *and the mean area* $E^*(F) = 2\pi - 12/\pi$.

9. *The probability that four random points on the unit hemisphere form a convex spherical quadrilateral is* $p = 3 - 24/\pi^2 = 0.569\ldots$ *(Sylvester's problem on the hemisphere).*

Second-order moments. Let $f_H(L)$ denote the probability density function of the perimeter L of the triangle $P_1P_2P_3$ generated by three independent random points P_1, P_2, P_3 on a hemisphere H of the unit sphere U and let $f(L)$ be the probability density function of L when P_1, P_2, P_3 are chosen randomly on the unit sphere U. Then, given at random three points P_1, P_2, P_3 and a great circle G, we have $p(P_1, P_2, P_3 \in H)\cdot f_H(L) = f(L)\cdot p(G \cap P_1P_2P_3 = \emptyset)$, where $p(P_1, P_2, P_3 \in H)$, the probability that P_1, P_2, P_3 all lie in some single hemisphere of those into which G partitions U is equal to $\frac{1}{4}$, and $p(G \cap P_1P_2P_3 = \emptyset)$, the probability that the great circle G does not intersect the triangle $P_1P_2P_3$ is $1 - L/2\pi$. Hence we have

$$f_H(L) = 4f(L) - (2/\pi)Lf(L).$$

Multiplying both sides by L and integrating from 0 to 2π, we get

$$E_H(L) = 4E(L) - (2/\pi)E(L^2)$$

and since $E_H(L) = 12/\pi$ and $E(L) = 3\pi/2$, we get $E(L^2) = 3\pi^2 - 6$. By duality, if E^* denotes the mean value when the spherical triangles are generated by three random great circles, we have $E^*(F) = E((2\pi - L)^2) = \pi^2 - 6$.

These and other second-order moments of certain quantities of random spherical figures have been given by Miles [413]. In this paper, Miles gives a complete survey of geometrical probability on the sphere, including the tessellation determined by n uniform random great circles and the first and second moments of the area, perimeter, and number of angles of the polygons of the tessellation.

ISBN 0-201-13500-0

5. *Integral formulas with tangents.* Let K be a convex set on the unit two-dimensional sphere U. Let t_1, t_2 be the lengths of the arcs of great circles tangent to ∂K through the point P measured from P to the contact

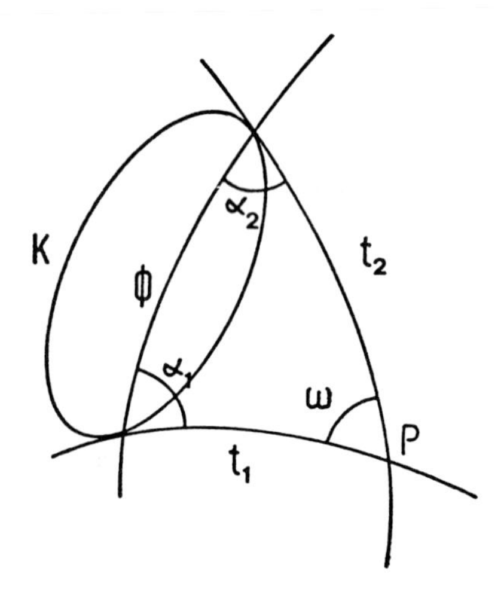

Figure 17.2.

point ($t_i \leqslant \pi$) and let ω be the angle between these tangent great circles (Fig. 17.2). Then the integral formula

$$\int_{P \notin K} \frac{\sin \omega}{\sin t_1 \sin t_2} dP = \tfrac{1}{2}(2\pi - F)^2$$

holds, where the points situated at the extremities of a diameter are considered as a single point. By duality we get

$$\int_{G \cap K \neq \emptyset} \frac{\sin \phi}{\sin \alpha_1 \sin \alpha_2} dG = \tfrac{1}{2}L^2$$

where ϕ is the length of the chord $G \cap K$ and α_1, α_2 are the angles that G makes with the great circles tangent to ∂K at the intersection points $G \cap K$ [539].

6. *Density for horocycles.* When the center of a circle in a hyperbolic plane moves off to infinity, still requiring the circle to pass through a fixed point, then the circle becomes a circle of infinite radius, which is called a horocycle. Assuming a system of polar coordinates (r, α) with origin at O, the density for horocycles is $dH_+ = e^r \, dr \wedge d\alpha$ if the horocycle has its convexity toward the origin O and $dH_- = e^{-r} \, dr \wedge d\alpha$ if the convexity of the horocycle is turned away from the origin. We shall write dH for the density of horocycles, with the convention of taking dH_+ or dH_-, according to the case. It is easy to see that two points A, B determine two horocycles joining them. A set

ISBN 0-201-13500-0

of points K is said to be h-convex (horocycle convex) if for each pair of points $A, B \in K$ the entire horocycle segments joining A and B are contained in K. Then, if σ denotes the length of the chord $H \cap K$, we can prove that

$$\int_{H \cap K \neq 0} \sigma \, dH = 2\pi F, \qquad \int_{H \cap K \neq 0} \sigma^3 \, dH = 6F^2$$

where F is the area of K.

Let K be an h-convex set with smooth boundary. Let $h(\theta)$ be the distance from a fixed point $O \in K$ to the horocycle that is tangent to K and is perpendicular to the line emanating from O and making the angle θ with a reference direction through O, K being on the convex side of H. The function $h(\theta)$ is called the support function with respect to horocycles that have K on the convex side. Then the length of ∂K is given by

$$L = \tfrac{1}{2} \int_0^{2\pi} (e^h - e^{-h} + e^{-h} h'^2) \, d\theta. \tag{17.63}$$

If h^* is the support function for K constructed as was h but with K on the concave side of the horocycle, (17.63) becomes

$$L = \tfrac{1}{2} \int_0^{2\pi} (e^{-h*} + e^{-h*} - e^{-h*} h^{*\prime 2}) \, d\theta. \tag{17.64}$$

If $h - h^* = B$ is a constant, K is said to be of constant breadth with respect to horocycles, and then

$$L = \tfrac{1}{2}(1 - e^{-B}) \int_0^{2\pi} e^h \, d\theta. \tag{17.65}$$

Formulas (17.63), (17.64), and (17.65) are the work of J. P. Fillmore [189]. Any convex set of constant breadth with respect to horocycles is also of constant breadth with respect to its tangent lines and then the perimeter L and area F are related to the constant breadth B by the formula $L = (2\pi + F) \tanh(B/2)$ [548, 589]. For applications of integral geometry to the theory of curves in three-dimensional hyperbolic space, see the work of D. Jusupov [326].

ISBN 0-201-13500-0

CHAPTER 18

Crofton's Formulas and the Kinematic Fundamental Formula in Noneuclidean Spaces

1. Crofton's Formulas

We want to generalize formula (14.24) to convex bodies of the n.e. space of curvature εK. Let P_1, P_2 be two points and let L_1 denote the line P_1P_2. Let t_1 and t_2 denote the coordinates of P_1, P_2 on L_1. Assuming that P_1 is the origin of a polar coordinate system, the volume element at P_2, according to (17.46), may be written

$$dP_2 = \frac{\sin^{n-1}((\varepsilon K)^{1/2}|t_2 - t_1|)}{(\varepsilon K)^{(n-1)/2}}\, dt_2 \wedge du_{n-1}. \tag{18.1}$$

Exterior multiplication by the volume element dP_1, using (17.43) (where dv is now denoted by dP_2), gives [283]

$$dP_1 \wedge dP_2 = \frac{\sin^{n-1}((\varepsilon K)^{1/2}|t_2 - t_1|)}{(\varepsilon K)^{(n-1)/2}}\, dt_1 \wedge dt_2 \wedge dL_1. \tag{18.2}$$

Let Q be a convex body and integrate (18.2) over all pairs of points $P_1, P_2 \in Q$. On the right-hand side we have the integral

$$\Phi_{n-1}(\sigma, \varepsilon K) = \int_0^\sigma\!\!\int_0^\sigma \frac{\sin^{n-1}((\varepsilon K)^{1/2}|t_2 - t_1|)}{(\varepsilon K)^{(n-1)/2}}\, dt_2 \wedge dt_1, \tag{18.3}$$

which has the value

ENCYCLOPEDIA OF MATHEMATICS and Its Applications, Gian-Carlo Rota (ed.).
1, Luis A. Santaló, Integral Geometry and Geometric Probability

ISBN 0-201-13500-0

$$\Phi_{n-1}(\sigma, \varepsilon K) = -\frac{2}{(n-1)(\varepsilon K)^{(n+1)/2}}\left[\frac{1}{n-1}\sin^{n-1}((\varepsilon K)^{1/2}\sigma)\right.$$

$$\left. + \sum_{i=1}^{(n-3)/2} \frac{(n-2)\cdots(n-2i)}{(n-3)\cdots(n-1-2i)^2}\sin^{n-1-2i}((\varepsilon K)^{1/2}\sigma)\right]$$

$$+ \frac{(n-2)\cdots 1}{(n-1)\cdots 2}(\varepsilon K)^{-(n-1)/2}\sigma^2 \tag{18.4}$$

for n odd, and

$$\Phi_{n-1}(\sigma, \varepsilon K) = -\frac{2}{(n-1)(\varepsilon K)^{(n+1)/2}}\left[\frac{1}{n-1}\sin^{n-1}((\varepsilon K)^{1/2}\sigma)\right.$$

$$+ \sum_{i=1}^{(n-2)/2} \frac{(n-2)\cdots(n-2i)}{(n-3)\cdots(n-1-2i)}\sin^{n-1-2i}((\varepsilon K)^{1/2}\sigma)$$

$$\left. - \frac{(n-2)\cdots 2}{(n-3)\cdots 1}(\varepsilon K)^{1/2}\sigma\right] \tag{18.5}$$

for n even and $n > 2$. For $n = 2$ we have

$$\Phi_1(\sigma, \varepsilon K) = 2((\varepsilon K)^{1/2}\sigma - \sin((\varepsilon K)^{1/2}\sigma))(\varepsilon K)^{-3/2}. \tag{18.6}$$

Therefore, integrating both sides of (18.2), we have

$$\int_{L_1 \cap Q \neq \emptyset} \Phi_{n-1}(\sigma, \varepsilon K)\, dL_1 = V^2, \tag{18.7}$$

which is the desired generalization. For instance, for $n = 2$, we have

$$\frac{1}{\varepsilon K}\int_{L_1 \cap Q \neq \emptyset}\left(\sigma - \frac{\sin((\varepsilon K)^{1/2}\sigma)}{(\varepsilon K)^{1/2}}\right) dL_1 = \frac{1}{2}F^2 \tag{18.8}$$

and for $n = 3$

$$\frac{1}{\varepsilon K}\int_{L_1 \cap Q \neq \emptyset}\left(\sigma^2 - \frac{1}{\varepsilon K}\sin^2((\varepsilon K)^{1/2}\sigma)\right) dL_1 = 2V^2. \tag{18.9}$$

From (18.8) with the help of (17.44) it follows that, for $\varepsilon = +1$ (elliptic plane) we have

$$\int_{Q \cap L_1 \neq \emptyset} \sigma\, dL_1 = \pi F, \qquad \int_{Q \cap L_1 \neq \emptyset} \sin\sigma\, dL_1 = \pi F - \tfrac{1}{2}F^2 \tag{18.10}$$

and for $\varepsilon = -1$ (hyperbolic plane)

ISBN 0-201-13500-0

$$\int_{Q \cap L_1 \neq 0} \sigma \, dL_1 = \pi F, \qquad \int_{Q \cap L_1 \neq 0} \sinh \sigma \, dL_1 = \pi F + \tfrac{1}{2} F^2. \tag{18.11}$$

Prove, as an exercise, that passing to the limit as $K \to 0$, (18.8) and (18.9) become (4.9) and (14.24) (for $n = 3$), respectively.

2. Dual Formulas in Elliptic Space

In the elliptic space, $\varepsilon = 1$, the so-called principle of duality holds, which affirms that every theorem remains true when we interchange "point" and "hyperplane" and make the consequent alterations in wording (e.g., "contains" is interchanged with "is contained in," "dimension r" is interchanged with "dimension $n - r - 1$"). As we saw in Section 2 of Chapter 17, to each L_r there corresponds by duality an L^P_{n-r-1}, called the dual of L_r. We want to translate by duality some of the formulas of the preceding section.

Consider first formula (17.44). Assume Q convex. The dual of a line L_1 is an L^P_{n-2}, and to the length λ of the chord $Q \cap L_1$ there corresponds the angle $\pi - \phi^P$, where ϕ^P is the angle between the tangent hyperplanes to the convex body Q^P, dual of Q, through L^P_{n-2}. Thus, (17.44) gives

$$\int_{Q^P \cap L^P_{n-2} = 0} (\pi - \phi^P) \, dL^P_{n-2} = \tfrac{1}{2} O_{n-1} V. \tag{18.12}$$

The measure of all L^P_{n-2} that do not meet Q^P is equal to the measure of all L_1 that meet Q; hence, by (17.61) it is equal to $(O_n/4\pi)F$. Consequently (18.12) yields

$$\int_{Q^P \cap L^P_{n-2} = 0} \phi^P \, dL^P_{n-2} = (O_n/4)F - (O_{n-1}/2)V \tag{18.13}$$

or, interchanging Q and Q^P, we have

$$\int_{Q \cap L_{n-2} \neq 0} \phi \, dL_{n-2} = (O_n/4)F^P - (O_{n-1}/2)V^P \tag{18.14}$$

where F^P and V^P are the area and volume of Q^P, whose values can be deduced from the formulas in Section 2 of Chapter 17.

For instance, for $n = 2$, (18.14) becomes

$$\int_{P \notin Q} \phi \, dP = \pi(L - F) \tag{18.15}$$

ISBN 0-201-13500-0

where ϕ is the angle between the tangents to Q from the exterior point Q. For $n = 3$ we get

$$\int_{Q \cap L_1 = 0} \phi \, dL_1 = 2\pi(M_1 + V) - \tfrac{1}{2}\pi^2 F \tag{18.16}$$

where we have applied (17.31).

We want to translate by duality formula (18.7). Note that for n even, according to (18.5) we have ($n > 2$)

$$\Phi_{n-1}(\pi - \phi, 1) = \Phi_{n-1}(\phi, 1) + \frac{2(n-2)\cdots 2}{(n-1)\cdots 1}\pi - \frac{4(n-2)\cdots 2}{(n-1)\cdots 1}\phi. \tag{18.17}$$

Therefore, using (18.12) and (18.7), we get

$$\int_{Q \cap L_{n-2} = 0} \Phi_{n-1}(\phi, 1)\, dL_{n-2} = (V^P)^2 + \frac{(n-2)\cdots 2}{2(n-1)\cdots 1}(O_n F^P - 4O_{n-1} V^P) \tag{18.18}$$

where ϕ is the angle between the tangent hyperplanes to Q through L_{n-2} and $\Phi_{n-1}(\phi, 1)$ is given by (18.5).

Similarly, for n odd we have

$$\int_{Q \cap L_{n-2} = 0} \Phi_{n-1}(\phi, 1)\, dL_{n-2} = (V^P)^2 + \frac{(n-2)\cdots 1}{(n-1)\cdots 2}\frac{\pi}{4}(O_n F^P - 4O_{n-1} V^P). \tag{18.19}$$

Examples. For $n = 2$, (18.6) yields $\Phi_1(\pi - \phi, 1) = \Phi_1(\phi, 1) + 2(\pi - 2\phi)$ and (18.18) becomes

$$\int_{P \notin Q} (\phi - \sin\phi)\, dP = \tfrac{1}{2}L^2 - \pi F \tag{18.20}$$

which is the classical result of Crofton (4.24) extended to the elliptic plane. From (18.20) and (18.14) it follows that

$$\int_{P \notin Q} \sin\phi \, dP = \pi L - \tfrac{1}{2}L^2. \tag{18.21}$$

For $n = 3$, using (17.31), we obtain from (18.19)

$$\frac{1}{2}\int_{L_1 \cap Q = 0} (\phi^2 - \sin^2\phi)\, dL_1 = (M_1 + V)^2 - \frac{\pi^3}{4}F, \tag{18.22}$$

which is the generalization to elliptic three-dimensional space of the formula of Herglotz (14.33).

ISBN 0-201-13500-0

3. The Kinematic Fundamental Formula in Noneuclidean Spaces

Formula (15.72) of the euclidean n-dimensional space is also valid for domains with sufficiently smooth boundary of the n-dimensional n.e. space of curvature εK. In fact, all the steps of the proof are based on formulas of the local differential geometry of hypersurfaces, which have the same form for euclidean and n.e. spaces (namely, the Meusnier, Euler, and Rodrigues formulas). Formula (15.72) is valid for $q = 1, 2, \ldots, n-1$. For $q = n$ it must be replaced by (15.36), taking (15.37) into account. That is, as a complement of (15.72) we have (for $q = n$)

$$\int_{D_0 \cap D_1 \neq \emptyset} M_{n-1}(\partial(D_0 \cap D_1)\, dK_1 = O_{n-1} \cdots O_1 \left[M^0_{n-1} V_1 + M^1_{n-1} V_0 + \frac{1}{n} \sum_{h=0}^{n-2} \binom{n}{h+1} M_h{}^0 M^1_{n-2-h} \right], \tag{18.23}$$

which is valid for any pair of domains D_0, D_1 with sufficiently smooth boundaries of the n.e. space of curvature εK.

With these preliminaries, it is easy to obtain the fundamental kinematic formula for n.e. spaces. We consider separately the cases of n even and n odd.

1. *n even*. We apply the Gauss–Bonnet formula (17.21) to the intersection $D_0 \cap D_1$ and integrate over all positions of D_1. The integral of the volume $V(D_0 \cap D_1)$ is immediate.

$$\int_{D_0 \cap D_1 \neq \emptyset} V(D_0 \cap D_1)\, dK_1 = O_{n-1} \cdots O_1 V_0 V_1. \tag{18.24}$$

Then, (15.72) and (18.23), after some rearrangements, give

$$\int_{D_0 \cap D_1 \neq \emptyset} \chi(D_0 \cap D_1)\, dK_1$$

$$= -\frac{2O_{n-1} \cdots O_1}{O_n} (\varepsilon K)^{n/2} V_0 V_1$$

$$+ O_{n-1} \cdots O_1 (V_1 \chi_0 + V_0 \chi_1) + O_{n-2} \cdots O_1 \frac{1}{n} \sum_{h=0}^{n-2} \binom{n}{h+1} M_h{}^0 M^1_{n-2-h}$$

$$+ O_{n-2} \cdots O_1 \left\{ \sum_{i=0}^{n/2-2} \binom{n-1}{2i+1} \frac{n-2i-2}{O_{n-2i-3}} \frac{2}{O_{n-2i-2}} (\varepsilon K)^{(n-2i-2)/2} \right.$$

ISBN 0-201-13500-0

$$\sum_{h=n-2i-2}^{n-2} \frac{\binom{2i+1}{n-h-1} O_{2n-h-2i-2}}{(h+1)O_{n-h}} \frac{O_h}{O_{2i+h-n+2}} M^1_{n-2-h} M^0_{h+2i+2-n}\Bigg\}. \tag{18.25}$$

2. *n odd*. Applying (17.22) to $D_0 \cap D_1$ and integrating both sides over all positions of D_1, using (15.72), after some rearrangements, we get

$$\int_{D_0 \cap D_1 \neq 0} \chi(D_0 \cap D_1)\, dK_1$$

$$= O_{n-1} \cdots O_1 (V_1 \chi_0 + V_0 \chi_1) + O_{n-2} \cdots O_1 \frac{1}{n} \sum_{h=0}^{n-2} \binom{n}{h+1} M_h^0 M^1_{n-2-h}$$

$$+ O_{n-2} \cdots O_1 \Bigg\{ \sum_{i=0}^{(n-3)/2} \binom{n-1}{2i} \frac{2}{O_{n-2i-1}} \frac{n-2i-1}{O_{n-2i-2}} (\varepsilon K)^{(n-1-2i)/2}$$

$$\sum_{h=n-2i-1}^{n-2} \binom{2i}{n-h-1} \frac{O_h}{O_{2i+h-n+1}} \frac{O_{2n-h-2i-1}}{(h+1)O_{n-h}} M^1_{n-2-h} M^0_{h+2i+1-n}\Bigg\}. \tag{18.26}$$

For $n = 2$ formula (18.25) must be written as

$$\int_{D_0 \cap D_1 \neq 0} \chi(D_0 \cap D_1)\, dK_1 = -(\varepsilon K) F_0 F_1 + 2\pi (F_1 \chi_0 + F_0 \chi_1) + L_0 L_1 \tag{18.27}$$

and for $n = 3$

$$\int_{D_0 \cap D_1 \neq 0} \chi(D_0 \cap D_1)\, dK_1 = 8\pi^2 (V_1 \chi_0 + V_0 \chi_1) + 2\pi (F_0 M_1 + F_1 M_0). \tag{18.28}$$

Note that according to (18.25) and (18.26), $n = 3$ is the only case such that the kinematic fundamental formula is independent of the curvature of the space. For the "dual" fundamental formula in the elliptic case, see [732].

4. Steiner's Formula in Noneuclidean Spaces

Consider the special case when D_1 is a ball of radius ρ and D_0 is a convex set Q. Then $\chi(Q \cap D_1) = 1$ if $Q \cap D_1 \neq 0$ and $\chi(Q \cap D_1) = 0$ otherwise. Using expression (17.32) for dK_1 and choosing the center of the ball D_1 as the origin of the moving frame, for each position of D_1 we can integrate $du_{n-1} \wedge \cdots \wedge du_1$ and the kinematic fundamental formula gives the volume

ISBN 0-201-13500-0

of the convex body Q_ρ parallel to Q in the distance ρ (i.e., the generalized Steiner formula for n.e. spaces). The resulting formula is rather complicated, so we will write it only for the particular cases in which $n = 2, 3$. For a different approach, see that of Allendoerfer [4].

(a) $n = 2$. With the natural change of notation $V_0 \to F_0$, $V_1 \to F_1$, $M_0{}^0 \to L_0$, $M_0{}^1 \to L_1$ where L_0, F_0 are the perimeter and area of Q and L_1, F_1 the perimeter and area of the disk of radius ρ, that is,

$$L_1 = (2\pi/(\varepsilon K)^{1/2}) \sin((\varepsilon K)^{1/2}\rho), \qquad F_1 = (2\pi/\varepsilon K)[1 - \cos((\varepsilon K)^{1/2}\rho)], \tag{18.29}$$

we get the following expression for the area of Q_ρ [690]

$$F_\rho = F_0 \cos((\varepsilon K)^{1/2}\rho) + \frac{L_0}{(\varepsilon K)^{1/2}} \sin((\varepsilon K)^{1/2}\rho) + \frac{2\pi}{\varepsilon K}[1 - \cos((\varepsilon K)^{1/2}\rho)]. \tag{18.30}$$

The perimeter of Q_ρ can be calculated by the formula $L_\rho = dF_\rho/d\rho$.

(b) $n = 3$. For the ball of radius ρ, by (17.47), (17.48), and (17.50), we have

$$V_1 = \frac{2\pi}{(\varepsilon K)^{3/2}}[(\varepsilon K)^{1/2}\rho - \sin((\varepsilon K)^{1/2}\rho)\cos((\varepsilon K)^{1/2}\rho)],$$

$$F_1 = M_0{}^1 = \frac{4\pi}{\varepsilon K}\sin^2((\varepsilon K)^{1/2}\rho),$$

$$M_1 = M_1{}^1 = \frac{4\pi}{(\varepsilon K)^{1/2}}\sin((\varepsilon K)^{1/2}\rho)\cos((\varepsilon K)^{1/2}\rho), \qquad \chi_1 = 1. \tag{18.31}$$

Substituting into (18.26) and dividing by $8\pi^2$, we get

$$V_\rho = V_0 + \frac{F_0}{(\varepsilon K)^{1/2}}\sin((\varepsilon K)^{1/2}\rho)\cos((\varepsilon K)^{1/2}\rho) + M_1\frac{\sin^2((\varepsilon K)^{1/2}\rho)}{\varepsilon K}$$
$$+ \frac{2\pi}{(\varepsilon K)^{3/2}}[(\varepsilon K)^{1/2}\rho - \sin((\varepsilon K)^{1/2}\rho)\cos((\varepsilon K)^{1/2}\rho)]. \tag{18.32}$$

The surface area F_ρ of Q_ρ can be calculated by the formula $F_\rho = dV_\rho/d\rho$.

5. An Integral Formula for Convex Bodies in Elliptic Space

Consider the elliptic n-dimensional space ($\varepsilon = 1$, $K = 1$). We know that the lines L_1 are closed and have the length π. Let L_1 be such a line. We can consider L_1 as a degenerate body for which

ISBN 0-201-13500-0

$$\chi_1 = 0, \quad V_1 = 0, \quad M_0 = M_1 = \cdots = M_{n-3} = M_{n-1} = 0,$$

$$M_{n-2} = [\pi/(n-1)]O_{n-2} \tag{18.33}$$

and then apply fundamental formula (18.25) or (18.26) to the case in which $D_1 = L_1$ and $D_0 = Q_0$, a fixed convex body. We have $\chi(Q_0 \cap L_1) = 1$. Let P be a point of L_1 and take the kinematic density (17.32) with P as origin of the moving frame. The integral on the left-hand side is equal to

$$O_{n-1} \cdots O_1 V_0 + 2O_{n-2} \cdots O_1 \int_{P \notin Q_0} \Omega \, dP \tag{18.34}$$

where Ω is the solid angle under which Q_0 is seen from the point P (exterior to Q_0). The factor 2 before the second summand arises from the fact that each point P is the common vertex of two angles Ω that subtend Q_0 (because L_1 is closed). The integral on the left-hand side, in view of (18.33), reduces to $O_{n-2} \cdots O_1 O_{n-2}(\pi/(n-1))M_0{}^1$. Therefore, denoting by $F_0 = M_0{}^1$ the area of ∂Q_0, we have the formula

$$\int_{P \notin Q_0} \Omega \, dP = \frac{\pi}{2(n-1)} O_{n-2} F_0 - \frac{1}{2} O_{n-1} V_0. \tag{18.35}$$

For $n = 2$ we again have (18.15).

6. Notes

1. *Integral formulas relating to the intersection of manifolds in n.e. space.* Formula (15.20), which was proved for euclidean space, is valid without change for noneuclidean space. That is, we may state

Let M^q be a fixed q-dimensional compact manifold in n-dimensional noneuclidean space and let M^r be an r-dimensional compact manifold, moving with kinematic density dK. Assume $r + q - n \geqslant 0$ and let $\sigma_{r+q-n}(M^q \cap M^r)$ denote the $(r + q - n)$-dimensional volume of the intersection $M^q \cap M^r$. Then we have

$$\int_{M^q \cap M^r \neq 0} \sigma_{r+q-n}(M^q \cap M^r) \, dK = \frac{O_n O_{n-1} \cdots O_1 O_{r+q-n}}{O_q O_r} \sigma_q(M^q) \sigma_r(M^r) \tag{18.36}$$

where $\sigma_q(M^q)$ and $\sigma_r(M^r)$ are the volumes of M^q and M^r, respectively.

If $r + q - n = 0$, then $\sigma_{r+q-n}(M^q \cap M^r)$ denotes the number of intersection points of M^q and M^r. For instance, if Γ_0, Γ_1 are two curves of the n.e. plane, we have

$$\int_{\Gamma_0 \cap \Gamma_1 \neq 0} n \, dK_1 = 4L_0 L_1 \tag{18.37}$$

ISBN 0-201-13500-0

where L_0, L_1 are the lengths of the curves and n is the number of intersections. This shows that Poincaré's formula (7.11) is independent of the curvature of the space.

2. *The isoperimetric inequality.* Let D_0, D_1 be two congruent domains in the n.e. plane, of area F and bounded by a single curve of length L. Assuming D_0 fixed and D_1 moving with kinematic density dK_1, we have $\chi(D_0 \cap D_1) = \nu =$ the number of pieces of the intersection $D_0 \cap D_1$, and using that $\chi(D_0) = \chi(D_1) = 1$, we can write the fundamental formula (18.27) as

$$\int_{D_0 \cap D_1 \neq 0} \nu \, dK_1 = 4\pi F + L^2 - (\varepsilon K)F^2. \tag{18.38}$$

Since $\nu \leqslant n/2$, from (18.37) and (18.38) it follows that

$$L^2 + \varepsilon K F^2 - 4\pi F \geqslant 0, \tag{18.39}$$

which is the isoperimetric inequality for the n.e. plane of curvature εK. Proceeding as in Section 5 of Chapter 7, we can prove the following stronger inequalities (for convex domains).

(a) For the elliptic plane ($\varepsilon = 1$, $K = 1$), putting $\Delta = L^2 + F^2 - 4\pi F$, we have

$$\Delta \geqslant 4\pi^2 \sin^2\left(\frac{r_M - r_m}{2}\right), \qquad \Delta \geqslant 4\pi^2 \tan^2\left(\frac{r_M - r_m}{2}\right) \quad [62],$$

$$\Delta \geqslant [F \cot(r_m/2) - L]^2, \qquad \Delta \geqslant [L - F \cot(r_M/2)]^2.$$

(b) For the hyperbolic plane ($\varepsilon = -1$, $K = 1$), putting $\Delta = L^2 - F^2 - 4\pi F$, we have

$$\Delta \geqslant [L - F \coth(r_m/2)]^2, \qquad \Delta \geqslant [F \coth(r_M/2) - L]^2.$$

In both cases r_M is the radius of the smallest circle that contains the convex domain D and r_m is the radius of the greatest circle that is contained in D [539, 541]. For isoperimetric inequalities in the elliptic plane, see [341]; for the hyperbolic plane, see [458].

3. *The theorem of Hadwiger for the n.e. plane.* The same argument that produced Hadwiger's theorem in Section 4 of Chapter 7 now gives

(a) For the elliptic plane ($\varepsilon = 1$, $K = 1$): D_0 and D_1 being two domains bounded by simple closed piecewise smooth curves, the inequalities

$$L_0 L_1 - F_1(4\pi - F_0) \geqslant [L_0^2 L_1^2 - F_0 F_1(4\pi - F_0)(4\pi - F_1)]^{1/2}$$

and

$$F_0(4\pi - F_1) - L_0 L_1 \geqslant [L_0^2 L_1^2 - F_0 F_1(4\pi - F_0)(4\pi - F_1)]^{1/2}$$

are two sufficient (but not necessary) conditions in order that D_1 can be contained in D_0.

ISBN 0-201-13500-0

(b) For the hyperbolic plane ($\varepsilon = -1$, $K = 1$) the analogous conditions are

$$L_0 L_1 - F_1(4\pi + F_0) \geqslant [L_0^2 L_1^2 - F_0 F_1 (4\pi + F_0)(4\pi + F_1)]^{1/2},$$

$$F_0(4\pi + F_1) - L_0 L_1 \geqslant [L_0^2 L_1^2 - F_0 F_1 (4\pi + F_0)(4\pi + F_1)]^{1/2}.$$

In particular, if D_1 is a circle of radius r on the unit sphere, we have $L_1 = 2\pi \sin r$, $F_1 = 2\pi(1 - \cos r)$ and recalling that the elliptic plane is locally the unit sphere, we can state

In order that a domain D_0 on the unit sphere, bounded by a simple closed curve, contain a circular cap of angular radius r, it is sufficient that one of the inequalities

$$\tan\left(\frac{r}{2}\right) < \frac{L_0 - \Delta_0^{1/2}}{4\pi - F_0}, \qquad \cot\left(\frac{r}{2}\right) > \frac{L_0 + \Delta_0^{1/2}}{F_0}$$

hold, where $\Delta_0 = L_0^2 - 4F_0 + F_0^2$. In like manner, either of the conditions

$$\cot\left(\frac{R}{2}\right) < \frac{L_0 - \Delta_0^{1/2}}{F_0}, \qquad \tan\left(\frac{R}{2}\right) > \frac{L_0 + \Delta_0^{1/2}}{4\pi - F_0}$$

is a sufficient condition for D_0 to be contained in a cap of angular radius R.

Another result of the same kind is the following: let D be a domain, not necessarily convex, of area F and perimeter L, bounded by a single simple closed curve contained in a hemisphere of the unit two-dimensional sphere. Then there exists a spherical cap of angular radius ρ contained in D such that $\tan \rho \geqslant F/L$. The inequality is the best possible in the sense that it cannot be replaced by $\tan \rho \geqslant c(F/L)$ with a constant $c > 1$. If D is convex, then the angular radius of the largest cap in D satisfies the inequality $\rho \geqslant F/4$. This inequality is due to D. J. White [719] and can be extended to convex domains of the n-dimensional unit sphere. Its form is then $F \leqslant (\rho/\pi)O_n$, where O_n is the area of the n-dimensional unit sphere. This inequality is the best possible, as we see by taking D to be the intersection of two closed hemispheres.

4. *Some covering problems.* The geometry of elliptic space is locally the geometry of the sphere. The fundamental formula of Section 3 for $\varepsilon = 1$, $K = 1$ gives the kinematic fundamental formula of the n-dimensional unit sphere. We will apply it to some covering problems on the unit two-dimensional sphere.

Let K_0 be a fixed convex domain, of area F_0 and perimeter L_0, on the two-dimensional unit sphere U. Let $K_1, K_2, \ldots, K_n$ be n congruent convex domains that are moving on U and have area F and perimeter L; let dK_i denote the corresponding kinematic density. Let $F_{01\ldots n}$ denote the area and $L_{01\ldots n}$ the perimeter of the intersection $K_0 \cap K_1 \cap \cdots \cap K_n$. Then in a way analogous to that in Section 7 of Chapter 6 we can prove that

$$\int F_{01\ldots n}\, dK_1 \wedge dK_2 \wedge \cdots \wedge dK_n = (2\pi)^n F^n F_0, \tag{18.40}$$

ISBN 0-201-13500-0

$$\int L_{01\ldots n}\, dK_1 \wedge dK_2 \wedge \cdots \wedge dK_n = (2\pi)^n F^n L_0 + n(2\pi)^n F^{n-1} F_0 L, \quad (18.41)$$

where the integrals are extended over all locations of K_i $(i = 1, 2, \ldots, n)$.

Hence we have the following mean values.

$$E(F_{01\ldots n}) = F^n F_0 (4\pi)^{-n}, \qquad E(L_{01\ldots n}) = (F^n L_0 + nF^{n-1} F_0 L)(4\pi)^{-n}. \quad (18.42)$$

Furthermore, by successive application of (18.27) for $\varepsilon = K = 1$, we get

$$I_{12\cdots n} = \int dK_1 \wedge dK_2 \wedge \cdots \wedge dK_n = (2\pi)^n (F^n + nF_0 F^{n-1})$$
$$+ (2\pi)^{n-1}\left[nF^{n-1} L_0 L + \binom{n}{2} F_0 L^2 F^{n-2} - nF_0 F^n\right] \quad (18.43)$$

where the integration is extended over all locations of $K_1, \ldots, K_n$ such that $K_0 \cap K_1 \ldots \cap K_n \neq \emptyset$.

This result may be stated as follows.

Given a fixed convex set K_0 on the two-dimensional unit sphere, the probability that n random congruent convex sets $K_1, K_2, \ldots, K_n$ have a nonvoid intersection interior to K_0 is

$$p = (8\pi^2)^{-n} I_{12\ldots n} \quad (18.44)$$

where $I_{12\ldots n}$ is given by (18.43).

Assume $n + 1$ points chosen at random on the unit sphere. We want the probability that they may be covered by a circular cap of radius r. If we consider each point as the center of a spherical cap of radius r, then the desired probability is equal to the probability that n circular caps of radius r have a nonvoid intersection with a fixed cap of radius r. The probability is given by (18.44) with the values $L_0 = L = 2\pi \sin r$, $F_0 = F = 2\pi(1 - \cos r)$.

In particular, for $r = \pi/2$ we have

The probability that $n + 1$ random points on the unit two-dimensional sphere belong to the same hemisphere is

$$p^* = \frac{1}{2^n}\left[(n + 1) + \frac{n(n-1)}{2}\right]. \quad (18.45)$$

By duality we have that the probability that $n + 1$ random hemispheres will have a nonvoid intersection is given by the same formula (18.45). If the $n + 1$ hemispheres have a common point P, then the opposite point P^* will not belong to any hemisphere and therefore the $n + 1$ random hemispheres will not cover the unit sphere U. Hence, the probability of covering the whole unit sphere U by $n + 1$ random hemispheres is $1 - p^*$.

Wendel [714] has proved the more general result that the probability that n random points on the surface of the unit sphere in euclidean m-dimensional space belong to the same hemisphere is

$$p_{n,m} = 2^{-(n-1)} \sum_{h=0}^{m-1} \binom{n-1}{h} \quad (18.46)$$

ISBN 0-201-13500-0

for $n > m$ and $p_{n,m} = 1$ for $n \leqslant m$. The probability of covering the whole $(m - 1)$-sphere by n random hemispheres is $1 - p_{n,m}$.

More difficult and not yet solved is the problem of finding the probability of covering the unit sphere U with N circular caps of angular radius $r < \pi/2$ given at random on U. An approximate solution is given by Moran and Fazekas de St. Groth [431], and some bounds for the probability have been given by Gilbert [231].

Miles [412] has considered the following general problem. Consider n arbitrary subsets of a set X. Suppose that the point $x \in X$ lies in $H(x)$ of these subsets and define $H_* = \min H(x)$, $H^* = \max H(x)$ for $x \in X$. That is, the least and most covered regions of X are, respectively, H_*-covered and H^*-covered, with $0 \leqslant H_* \leqslant H^* \leqslant n$. Each subset is taken to be the uniform random image of a fixed subset of X; that is, they are n different positions of a fixed subset of X, the n positions assumed to be given independently. Find the asymptotic values, as $n \to \infty$, of the probabilities $p(H_* \leqslant m)$ and $p(H^* \geqslant n - m)$. Assume that X is the surface U of the unit sphere in E_3 and let Y be a spherical polygon, that is, a subset of U bounded by a simple closed curve consisting of great circle arcs. Write F and L for the area and perimeter of Y. Let $Y_1, \ldots, Y_n$ be a set of independent random positions of Y. Then Miles proves that as $n \to \infty$

$$p\left(\bigcup_{i=1}^{n} Y_i \neq X\right) \sim n(n-1)L^2(4\pi - F)^{n-2}(4\pi)^{-n},$$

$$p\left(\bigcap_{i=1}^{n} Y_i \neq 0\right) \sim n(n-1)L^2F^{n-2}(4\pi)^{-n}.$$

Both $p(H_* \leqslant m)$ and

$$p(H_* = m) \sim \binom{n}{m+2}(m+1)(m+2)L^2F^m(4\pi - F)^{n-m-2}(4\pi)^{-n}.$$

Both $p(H^* \geqslant n - m)$ and

$$p(H^* = n - m) \sim \binom{n}{m+2}(m+1)(m+2)L^2F^{n-m-2}(4\pi - F)^m(4\pi)^{-n}.$$

The last results generalize to the case in which $Y \subset X$ is bounded by a simple closed curve with bounded (spherical) curvature. They also generalize to the m-dimensional unit sphere [412].

5. *Tessellations on the sphere.* A set of n great circles in general position on the unit two-dimensional sphere U partitions U into $\alpha_n = n(n-1) + 2$ convex spherical polygons. This aggregate of polygons is called a tessellation on the sphere and it has $v_n = n(n-1)$ vertices and $e_n = 2n(n-1)$ edges. If the n great circles are given independently at random, Miles [413] has considered the problem of finding the moments E_n^* of the area F, perimeter L, and number of vertices N of the resulting polygons (faces of the tessellation). The first-order moments are immediate:

$$E_n^* = 4\pi/\alpha_n, \qquad E_n^*(L) = 4\pi n/\alpha_n, \qquad E_n^*(N) = 4n(n-1)/\alpha_n \quad (18.47)$$

ISBN 0-201-13500-0

for all $n \geqslant 2$. The second-order moments are

$$E_n^*(F^2) = 8\pi^2\gamma_n/\alpha_n \quad (n \geqslant 1), \qquad E_n^*(L^2) = 2\pi^2(4 + \beta_{n,2})/\alpha_n \quad (n \geqslant 2),$$

$$E_n^*(N^2) = (12n(n-1) + \beta_{n,4}/2)/\alpha_n \quad (n \geqslant 4),$$

$$E_n^*(LN) = \pi(8n + \beta_{n,3})/\alpha_n \quad (n \geqslant 3),$$

$$E_n^*(NF) = 2\pi\beta_{n,2}/\alpha_n \quad (n \geqslant 2), \qquad E_n^*(LF) = 4\pi^2\beta_{n,1}/\alpha_n \quad (n \geqslant 1),$$

where

$$\alpha_n = n(n-1) + 2, \qquad \beta_{n,i} = n(n-1)\cdots(n-i+1)\gamma_{n-i},$$

$$\gamma_j = 1 - \frac{j(j-1)}{\pi^2} + \frac{j(j-1)(j-2)(j-3)}{\pi^4} - \cdots$$

$$+ \begin{cases} (-1)^{j/2}2(j!)\pi^{-j}, & j \text{ even}, \\ (-1)^{(j-1)/2}(j!)\pi^{-j}, & j \text{ odd}. \end{cases}$$

For instance, when $n = 3$ (triangles given by three random great circles),

$$E_3^*(F^2) = \pi^2 - 6, \quad E_3^*(L^2) = (5/2)\pi^2, \quad E_3^*(FL) = (3/2)\pi^2 - 6.$$

If the triangles are generated by three random points instead of random great circles, by duality we have

$$E_3(F^2) = \pi^2/2, \qquad E_3(L^2) = 3\pi^2 - 6, \qquad E_3(LF) = E_3^*(LF).$$

These and similar results are the work of Miles [413].

6. *Random polyhedral convex cones.* If N hyperplanes in E_n are so placed that every n but no $n + 1$ have a common point, the number of regions into which they decompose the space is

$$f(N, n) = \binom{N}{0} + \binom{N}{1} + \cdots + \binom{N}{n}. \tag{18.48}$$

If the given N hyperplanes go through the origin O and are in "general position," then they decompose the space into

$$C(N, n) = 2\left[\binom{N-1}{0} + \binom{N-1}{1} + \cdots + \binom{N-1}{n-1}\right] \tag{18.49}$$

polyhedral convex cones. A set of N vectors in E_n is said to be in "general position" if every n-element subset is linearly independent, and a set of N hyperplanes through the origin is said to be in general position if the corresponding set of normal vectors is in general position.

Formulas (18.48) and (18.49) are due to Schläfli [597]. Using (18.49), Cover and Efron [126] have established the following results:

(a) Let W be a random polyhedral convex cone spanned by N random half lines through the origin. Then, the expected number of r-faces of W,

ISBN 0-201-13500-0

conditioned on W being a proper cone (i.e., given that the N half lines all lie in some single half space), is given by

$$E(R_r(W)) = 2^r \binom{N}{r} \frac{C(N-r, n-r)}{C(N, n)} \tag{18.50}$$

and

$$\lim_{N\to\infty} E(R_r(W)) = 2^r \binom{n-1}{r}. \tag{18.51}$$

(b) Let W^* be the random polyhedral convex cone resulting from the intersection of N random half spaces in E_n through the origin. Then the expected number of r-faces $R_r(W^*)$ of W^*, conditioned on $W^* \neq 0$, is given by $(r = 1, 2, \ldots, n-1)$

$$E(R_r(W^*)) = 2^{n-r} \binom{N}{n-r} \frac{C(N-n+r, r)}{C(N, n)} \tag{18.52}$$

and we have

$$\lim_{N\to\infty} E(R_r(W^*)) = 2^{n-r} \binom{n-1}{n-r}. \tag{18.53}$$

If the random half lines of (a) or the random half spaces of (b) are intersected by the unit $(n-1)$-dimensional sphere U_{n-1}, theorems (a) and (b) give theorems on the expected values of random polyhedral regions on U_{n-1}. In particular, theorem (a) affirms that for N random points in U_{n-1}, given that they all lie in some single hemisphere, the mean number of vertices of their convex hull (taken with great $(n-2)$-dimensional spheres), is

$$E(R_1(W)) = 2NC(N-1, n-1)/C(N, n)$$

and hence

$$\lim_{N\to\infty} E(R_1(W)) = 2(n-1). \tag{18.54}$$

Cover and Efron [126] note that for N points chosen at random on the unit sphere in E_3, so that they all lie in some single hemisphere, the mean number of extreme points of their convex hull does not grow without bound as N increases, but rather approaches the limit 4. In the case of the plane, however, this limit goes to infinity (Chapter 2, Section 5, Note 3). On the other hand, from (18.54) we deduce, for N great circles chosen at random on the unit two-dimensional sphere, that they decompose the sphere into regions having an expected number of sides 4 as N goes to infinity (as follows also from the last formula in (18.47)). This agrees with the case of polygons formed by random lines in the plane (4.50) and of polygons formed by random lines in the hyperbolic plane [592].

ISBN 0-201-13500-0

CHAPTER 19

Integral Geometry in Foliated Spaces; Trends in Integral Geometry

1. Foliated Spaces

In the preceding chapters we have considered integral geometry from the point of view of homogeneous spaces, that is, spaces with a transitive transformation group $\mathfrak{G}$. The measure for sets of geometric objects embedded in such space has been defined by the condition of being invariant under $\mathfrak{G}$, a condition that determines the measure up to a constant factor. Integral geometry in Riemann spaces of nonconstant curvature cannot be based on the same principle because in general such spaces do not admit a transitive transformation group that preserves the metric nor a transitive transformation group that maps geodesics into geodesics. Thus, the density for sets of geodesics cannot be defined by the property of being invariant under a certain group of transformations.

Nevertheless, it is possible to start from another point of view and to define a measure for sets of geodesics and sets of points which, though it is not invariant under any group, has some invariance properties that make it a measure of geometric interest. The method is based on the theory of foliated spaces and was developed by Hermann [308] and Vidal Abascal [696–699]. We want to give a résumé of the method and apply it to the integral geometry of Riemann manifolds.

Let X be a differentiable manifold of dimension $m + n$. Let F be a field of tangent linear subspaces F_x of dimension m, that is, a map $F: x \to F_x \subset T_x$ (T_x = tangent space to X at the point x). Assume that F is completely integrable. That means that about every point $x \in X$ there is a local coordinate

ENCYCLOPEDIA OF MATHEMATICS and Its Applications, Gian-Carlo Rota (ed.).
1, Luis A. Santaló, Integral Geometry and Geometric Probability

ISBN 0-201-13500-0

system $(x_1, x_2, \ldots, x_{n+m})$ such that the system $x_{m+1} = \text{constant}, \ldots, x_{m+n} = \text{constant}$, represents locally differentiable submanifolds L of X whose tangent spaces are the subspaces F_x. Then F is called a *foliation* on X and the submanifolds L are called the *leaves* of the foliation (of dimension m). The set of leaves of the foliation F on X is represented by X/F. The global study of this set is in general not easy (see the work of Palais [466]). For our purposes we shall assume that X/F is a differentiable manifold of dimension n and that there exists a map $\theta: U \to X/F$ of every open subset U of X into X/F such that it is constant along the leaves; that is, for all points $x \in U$ that belong to the leaf, $\theta(x)$ is the same point of X/F.

A differential form ω on X is said to be an invariant form of the foliation F providing that (a) it is invariant under changes of local coordinates on X; (b) it is invariant by displacements on the leaves of the foliation. This last property means that ω reduces to a form on X/F or, more precisely, that for any map $\theta: U \to X/F$ there exists a differential form ω_θ on X/F such that $\theta^*(\omega_\theta) = \omega$ (recall Chapter 9, Section 3).

Every nonzero invariant form ω of the foliation gives rise, by integration of ω_θ over X/F, to a measure for sets of leaves. We are going to apply these definitions to the case of geodesics of an n-dimensional Riemann manifold.

2. Sets of Geodesics in a Riemann Manifold

Let M be a Riemann manifold of dimension n endowed with the fundamental quadratic form (in a local coordinate system)

$$ds^2 = \sum_{i,j=1}^{n} g_{ij}\, dx_i\, dx_j. \tag{19.1}$$

Putting

$$\Gamma = \left(\sum_{i,j=1}^{n} g_{ij}\dot{x}_i\dot{x}_j \right)^{1/2}, \qquad p_i = \frac{\partial \Gamma}{\partial \dot{x}_i}, \qquad i = 1, 2, \ldots, n, \tag{19.2}$$

we have

$$\sum_{i=1}^{n} p_i\dot{x}_i = \Gamma, \qquad \sum_{i,j=1}^{n} g^{ij} p_i p_j = 1 \tag{19.3}$$

where g^{ij} are the elements of the inverse matrix $(g_{ij})^{-1}$ and $\dot{x}_i$ is the derivative with respect to a parameter t. Since g_{ij} is a covariant tensor and $\dot{x}_i$ is a tangent vector, it follows from (19.2) that Γ is a scalar invariant and equations (19.3) say that p_i is a unit covector of M.

Take as manifold X the bundle of all unit covectors of M. A local coordinate system for X is $(x_1, \ldots, x_n, p_1, \ldots, p_n)$ where the coordinates p_i satisfy the

ISBN 0-201-13500-0

second of conditions (19.3). Hence, the dimension of X is $2n - 1$. The geodesic curves of M are, by definition, the integral curves of the Euler equations

$$\frac{d}{dt}\left(\frac{\partial \Gamma}{\partial \dot{x}_i}\right) - \frac{\partial \Gamma}{\partial x_i} = 0, \qquad i = 1, 2, \ldots, n. \tag{19.4}$$

Any integral curve of these second-order differential equations is defined by a point $(x_1, \ldots, x_n)$ and a direction $(p_1, \ldots, p_n)$; that is, through each point of X goes one and only one integral curve. It follows that the geodesics of M define a foliation F_G on X. The geodesics G are the leaves of the foliation and, since $\dim G = 1$, it follows that $\dim X/F_G = 2n - 2$.

Consider the 2-form on X

$$dG^1 = \sum_{i=1}^{n} dp_i \wedge dx_i. \tag{19.5}$$

We want to prove that dG^1 is an invariant form of the foliation F_G. To this end we have to prove the following invariance properties.

1. *Invariance with respect to a change of coordinates.* Let $x_1', \ldots, x_n'$ be a second coordinate system, given by $x_i = x_i(x_1', \ldots, x_n')$ $(i = 1, 2, \ldots, n)$ where the functions $x_i(x_1', \ldots, x_n')$ and the inverse $x_i'(x_1, \ldots, x_n)$ are differentiable functions with a nonvanishing Jacobian. Since p_h is a covector, we have

$$p_h' = \sum_{i=1}^{n} \frac{\partial x_i}{\partial x_h'} p_i, \qquad dp_h' = \sum_{i=1}^{n} \frac{\partial x_i}{\partial x_h'} dp_i + \sum_{i,k=1}^{n} \frac{\partial^2 x_i}{\partial x_h' \partial x_k'} p_i dx_k' \tag{19.6}$$

and consequently

$$\sum_{h=1}^{n} dp_h' \wedge dx_h' = \sum_{h,i,j=1}^{n} \frac{\partial x_h'}{\partial x_j} \frac{\partial x_i}{\partial x_h'} dp_i \wedge dx_j$$

$$+ \sum_{i,k,h=1}^{n} \frac{\partial^2 x_i}{\partial x_h' \partial x_k'} p_i dx_k' \wedge dx_h'. \tag{19.7}$$

Using the fact that $dx_k' \wedge dx_h' = - dx_h' \wedge dx_k'$ and that

$$\sum_{h=1}^{n} \frac{\partial x_h'}{\partial x_j} \frac{\partial x_i}{\partial x_h'} = \frac{\partial x_i}{\partial x_j} = \delta_j^{\,i},$$

we get

$$\sum_{i=1}^{n} dp_i' \wedge dx_i' = \sum_{i=1}^{n} dp_i \wedge dx_i, \tag{19.8}$$

which proves the first invariance property.

ISBN 0-201-13500-0

2. *Invariance by displacements on the leaves.* We must prove that the integral of dG^1 over any two-dimensional set of geodesics is invariant under displacements on the geodesics. The following proof is due to Firey.

Let S be a family of geodesics depending on two parameters. The equation of the geodesics will have the form $x_i = x_i(\alpha, \beta; t)$ $(i = 1, 2, \ldots, n)$ where α, β define the geodesic among the elements of S, and t is the parameter on the geodesic. According to (19.2), we have $p_i(\alpha, \beta; t) = \partial \Gamma(x_i(\alpha, \beta; t), \dot{x}_i(\alpha, \beta; t))/\partial \dot{x}_i$. Let R be the set of values (α, β) corresponding to the geodesics of the family S. We want to prove that

$$m(S) = \int_S dG^1 = \int_S \sum_{i=1}^{n} dp_i \wedge dx_i = \int_R \sum_{i=1}^{n} \frac{\partial(p_i, x_i)}{\partial(\alpha, \beta)} d\alpha \wedge d\beta \tag{19.9}$$

is independent of t for all R.

By Stokes' theorem we have

$$m(S) = -\int_C \sum_{i=1}^{n} p_i(\sigma; t) \frac{\partial x_i(\sigma; t)}{\partial \sigma} d\sigma \tag{19.10}$$

where $C\colon \alpha = \alpha(\sigma)$, $\beta = \beta(\sigma)$ is the boundary of R, and we have set $p_i(\sigma; t) = p_i(\alpha(\sigma), \beta(\sigma); t)$, $x_i(\sigma; t) = x_i(\alpha(\sigma), \beta(\sigma); t)$. Then

$$\frac{dm(S)}{dt} = -\int_C \sum_{i=1}^{n} \left(\frac{\partial p_i}{\partial t} \frac{\partial x_i}{\partial \sigma} + p_i \frac{\partial^2 x_i}{\partial \sigma\, \partial t} \right) d\sigma.$$

Making use of (19.2) and (19.4), we obtain

$$\frac{dm(S)}{dt} = -\int_C \sum_{i=1}^{n} \left(\frac{\partial \Gamma}{\partial x_i} \frac{\partial x_i}{\partial \sigma} + \frac{\partial \Gamma}{\partial \dot{x}_i} \frac{\partial \dot{x}_i}{\partial \sigma} \right) d\sigma = -\int_C d\Gamma = 0. \tag{19.11}$$

Since this holds for any set S of geodesics, dG^1 is independent of t.

Note that we have taken dG^1 with its sign in order that Stokes' theorem could be applied. If we take the absolute value $|dG^1|$, as is usual in integral geometry, the independence with respect to t is still true. To prove this we decompose R into R^+ and R^- such that the sum $\sum_1^n \partial(p_i, x_i)/\partial(\alpha, \beta)$ is $\geqslant 0$ over R^+ and < 0 over R^- at t, and suppose that these sets are bounded by a finite number of simple closed curves that are piecewise smooth. Then we have $\int_R |dG^1| = \int_{R^+} dG^1 - \int_{R^-} dG^1$. The last two measures are invariant separately under a change in the choice of t and so this is true for the left-hand member.

Thus we have proved invariance properties 1 and 2. Since dG^1 is an invariant form of the foliation F_G, the exterior powers

ISBN 0-201-13500-0

$$dG^h = (dG^1)^h = \sum_{(i_1,\ldots,i_h)} dp_{i_1} \wedge dx_{i_1} \wedge \cdots \wedge dp_{i_h} \wedge dx_{i_h} \tag{19.12}$$

are also invariant forms ($h = 1, 2, \ldots, n-1$). They define a measure for $2h$-dimensional sets of geodesics. The most interesting cases, which we will study separately, are $h = 1$ and $h = n - 1$.

3. Measure of Two-Dimensional Sets of Geodesics

Let M be a Riemann manifold of dimension $n > 2$. Consider a two-dimensional family of geodesics on it, such that there is a transversal surface B that cuts each geodesic in one and only one point. Such a set is called a *congruence* of geodesics. If we choose a rectangular coordinate system so that the equations of B are (locally) $x_3 = 0, x_4 = 0, \ldots, x_n = 0$, then

$$ds^2 = \sum_{i=1}^{n} g_{ii}\, dx_i^2, \qquad p_i = g_{ii} \frac{dx_i}{ds}. \tag{19.13}$$

The angle α_i of the geodesic G with the x_i curve is given by $\cos \alpha_i = (g_{ii})^{1/2}(dx_i/ds)$ and therefore we have

$$p_i = (g_{ii})^{1/2} \cos \alpha_i,$$

$$dp_i = -(g_{ii})^{1/2} \sin \alpha_i \, d\alpha_i + \sum_{h=1}^{n} \frac{\partial (g_{ii})^{1/2}}{\partial x_h} \cos \alpha_i \, dx_h. \tag{19.14}$$

Making use of the second invariance property, we may determine G by its intersection point P with B, that is, by the point $(x_1, x_2, 0, \ldots, 0)$. Then we get

$$\begin{aligned} dG^1 &= dp_1 \wedge dx_1 + dp_2 \wedge dx_2 \\ &= -(g_{11})^{1/2} \sin \alpha_1 \, d\alpha_1 \wedge dx_1 - (g_{22})^{1/2} \sin \alpha_2 \, d\alpha_2 \wedge dx_2 \\ &\quad + \left(\frac{\partial (g_{22})^{1/2}}{\partial x_1} \cos \alpha_2 - \frac{\partial (g_{11})^{1/2}}{\partial x_2} \cos \alpha_1 \right) dx_1 \wedge dx_2. \end{aligned} \tag{19.15}$$

Let us consider a pencil of geodesics whose intersection with B is a domain R bounded by a closed piecewise smooth curve ∂R. Contrary to the general convention of taking the densities in absolute value, we define as measure of this pencil of geodesics the integral of dG^1 with the corresponding sign, so that we can apply the Stokes formula; we get

$$m_1(G) = \int_R dG^1 = \int_{\partial R} (g_{11})^{1/2} \cos \alpha_1 \, dx_1 + (g_{22})^{1/2} \cos \alpha_2 \, dx_2. \tag{19.16}$$

If ϕ denotes the angle of G with the tangent to ∂R at its intersection point, we have

ISBN 0-201-13500-0

$$\cos\phi = (g_{11})^{1/2}\cos\alpha_1\frac{dx_1}{ds} + (g_{22})^{1/2}\cos\alpha_2\frac{dx_2}{ds} \tag{19.17}$$

where we have applied that the unit tangent vector to G has the components $(\cos\alpha_i)/(g_{ii})^{1/2}$ and the unit tangent vector to ∂R has the components dx_i/ds. Therefore we have

$$m_1(G) = \int_R dG^1 = \int_{\partial R} \cos\phi\, ds. \tag{19.18}$$

This integral, which was discovered by E. Cartan [85], has interesting properties. Note that $\cos\phi\, ds$ is the projection of ds onto the corresponding geodesic. Hence, if we consider a tube made up of the geodesic lines going through a closed contour and AEA' is an orthogonal trajectory of the geodesics around the tube, then $m_1(G)$ is equal to the arc AA' (Fig. 19.1).

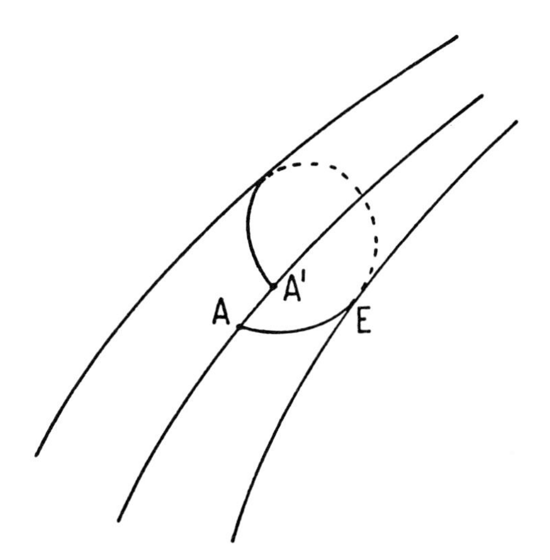

Figure 19.1.

Consider the case in which $n = 3$. Then, if all geodesics of the congruence are orthogonal to a surface B, the congruence is said to be a *normal congruence*. In this case $A \equiv A'$ and $m_1(G) = 0$. Conversely, if $m_1(G) = 0$ for any set of geodesics of a given congruence, then all normal trajectories from a point A will generate a surface B to which the geodesics are normal (locally about A). Therefore, the congruence is a normal congruence and we have *a necessary and sufficient condition that a congruence of geodesics of a three-dimensional Riemann space be a normal congruence is that* $m_1(G) = 0$ *for any set of geodesics; that is, that* $dG^1 = 0$.

ISBN 0-201-13500-0

Another property of the measure $m_1(G)$ is its behavior under refraction by a surface B. Indeed, if a geodesic passes through a surface B according to the refraction law of physics $\sin i = n \sin i'$ (where n is the refraction index, Fig. 19.2), since $i = \pi/2 - \phi$ and $i' = \pi/2 - \phi'$, we have $\cos \phi = n \cos \phi'$ and therefore $m_1(G) = nm_1(G')$, where G' denotes the refracted geodesic. Hence, *when a congruence of geodesics is refracted by a medium of index of refraction n, the measure $m_1(G)$ goes into $nm_1(G')$. In particular, the property of being a normal congruence is conserved by refraction.*

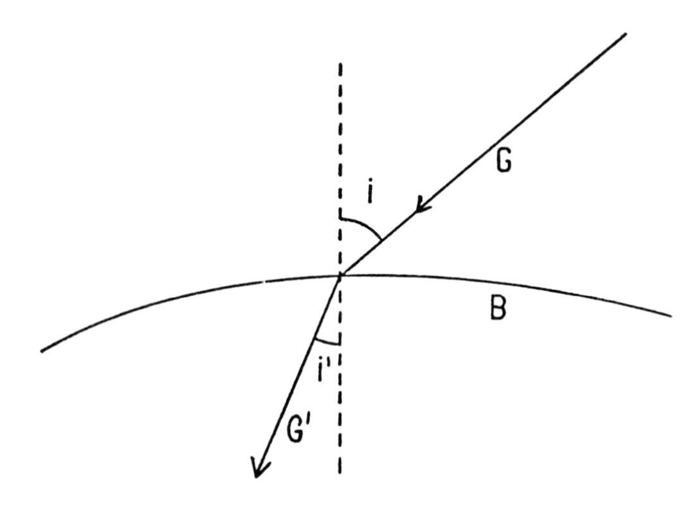

Figure 19.2.

Case of the lines of E_3. If the Riemann manifold M is the euclidean space E_3, then the geodesics are the straight lines and the density dG^1 coincides with the density dG^* (12.72). Indeed, if the transversal surface B is a plane E, then $ds^2 = dx^2 + dy^2$, $g_{11} = g_{22} = 1$, and (19.15) reduces to $dG^1 = -\sin \alpha_1 \, d\alpha_1 \wedge dx - \sin \alpha_2 \, d\alpha_2 \wedge dy$. All properties given here for congruences of geodesics obviously hold for congruences of lines in E_3. In particular, $dG^* = 0$ is a necessary and sufficient condition that a congruence of lines be a congruence of normal lines to a surface.

4. Measure of $(2n - 2)$-Dimensional Sets of Geodesics

Sets of geodesics depending on $2h$ parameters can be measured by means of the density dG^h (19.12). We have considered the case in which $h = 1$ and we now want to consider the case in which $h = n - 1$. The intermediate cases $h = 2, 3, \ldots, n - 2$ are little known.

ISBN 0-201-13500-0

For $h = n - 1$ the density becomes

$$dG^{n-1} = \sum_{i=1}^{n} dp_1 \wedge dx_1 \wedge \cdots \wedge dp_{i-1} \wedge dx_{i-1} \wedge dp_{i+1} \wedge dx_{i+1} \wedge \cdots \wedge dp_n \wedge dx_n. \tag{19.19}$$

In order to give a geometric interpretation to this density, consider a hypersurface S_{n-1} that intersects all the geodesics of the given $(2n - 2)$-dimensional set. Let G be a geodesic and let P be its intersection point with S_{n-1}. In a neighborhood of P we may assume that S_{n-1} is defined by the equation $x_n = 0$ and that the coordinate system is orthogonal, so that equations (19.13) and (19.14) hold good. In order to define G, we may choose the point P so that we have $x_n = 0$, $dx_n = 0$ and hence (19.19) becomes

$$dG^{n-1} = dp_1 \wedge dx_1 \wedge \cdots \wedge dp_{n-1} \wedge dx_{n-1} \tag{19.20}$$

or, according to (19.14), up to the sign,

$$dG^{n-1} = (g_{11} \cdots g_{n-1,n-1})^{1/2} \sin \alpha_1 \cdots \sin \alpha_{n-1}\, dx_1 \wedge \cdots \wedge dx_{n-1} \wedge d\alpha_1 \wedge \cdots \wedge d\alpha_{n-1}. \tag{19.21}$$

If $d\sigma$ denotes the $(n - 1)$-dimensional area element on S_{n-1}, we have $d\sigma = (g_{11} \cdots g_{n-1,n-1})^{1/2}\, dx_1 \wedge \cdots \wedge dx_{n-1}$ and the area element on the unit $(n - 1)$-dimensional sphere of center P corresponding to the direction of the line tangent to G at P has the value

$$du_{n-1} = \frac{\sin \alpha_1 \cdots \sin \alpha_{n-1}}{\cos \alpha_n}\, d\alpha_1 \wedge \cdots \wedge d\alpha_{n-1}.$$

Hence, (19.21) can be written (in absolute value)

$$dG^{n-1} = |\cos \alpha_n|\, d\sigma \wedge du_{n-1} \tag{19.22}$$

where α_n is the angle between the geodesic and the normal to S_{n-1} at intersection point P.

This expression (19.22) immediately gives a very general integral formula. Let $f(\sigma, \alpha_n)$ be an integrable function defined on S_{n-1}, depending on the point $P(\sigma)$ and on the direction α_n at P. Multiplying both sides of (19.22) by f and performing the integration over S_{n-1} and half of the $(n - 1)$-dimensional unit sphere (in order to consider unoriented geodesics), on the left-hand side each geodesic appears as common factor of the sum $\sum f(\sigma_i, \alpha_{n,i})$ of the values of $f(\sigma, \alpha_n)$ at the N intersection points of G with S_{n-1}. Hence we have

$$\int \sum_{i=1}^{N} f(\sigma_i, \alpha_{n,i})\, dG^{n-1} = \int_{S_{n-1}} \int_{U_{n-1}/2} f(\alpha, \alpha_n) |\cos \alpha_n|\, d\sigma \wedge du_{n-1}. \tag{19.23}$$

ISBN 0-201-13500-0

For instance, if $f = 1$, the integral of $|\cos \alpha_n| \, du_{n-1}$ is half of the projection of the unit $(n-1)$-dimensional sphere onto a diametral plane and we get

$$\int N \, dG^{n-1} = [O_{n-2}/(n-1)]F \tag{19.24}$$

where N is the number of intersection points of G with S_{n-1} and F is the surface area of S_{n-1}.

5. Sets of Geodesic Segments

Let t denote the arc length on the geodesic G. From (19.22) it follows that

$$dG^{n-1} \wedge dt = |\cos \alpha_n| \, d\sigma \wedge du_{n-1} \wedge dt. \tag{19.25}$$

The product $|\cos \alpha_n| \, dt$ equals the projection of the arc element dt on the normal to the hypersurface S_{n-1} at P. Consequently, $|\cos \alpha_n| \, dt \wedge d\sigma$ represents the volume element dP of the given Riemann manifold at P. Hence, (19.25) can be written

$$dG^{n-1} \wedge dt = dP \wedge du_{n-1}. \tag{19.26}$$

An oriented geodesic segment S can be determined either by G and t (where G is the geodesic that contains S and t is the coordinate of the origin of S on G) or by P, u_{n-1} (where P is the origin of the segment and u_{n-1} denotes the point on the unit $(n-1)$-dimensional sphere giving the direction of S). Either of the two equivalent forms in (19.26) can be taken as the density for geodesic segments. Consider, for example, the measure of the set of "oriented" segments S^* that have their origin inside a fixed domain D. The integral on the left-hand side of (19.26) gives $2\lambda \, dG^{n-1}$ where λ denotes the length of the arc of G that lies in D (the factor 2 appears because dG^{n-1} is the density for unoriented geodesics). The integral on the right-hand side in (19.26) is equal to $O_{n-1}V$ where V is the volume of D. Thus we have the integral formula

$$\int_{D \cap G \neq \emptyset} \lambda \, dG = \tfrac{1}{2} O_{n-1} V, \tag{19.27}$$

which generalizes (17.44) to Riemann manifolds. Other applications are given in [567].

6. Integral Geometry on Complex Spaces

Integral geometry on complex spaces has not been sufficiently developed and probably deserves further study. We shall give a résumé of the densities

ISBN 0-201-13500-0

and some formulas relating to the n-dimensional complex projective space $P_n(C)$ and the unitary group $\mathfrak{U}$ acting on it.

Let z_i $(i = 0, 1, \ldots, n)$ be the homogeneous coordinates of a point $z \in P_n(C)$, so that $z = (z_0, \ldots, z_n)$ and $\lambda z = (\lambda z_0, \ldots, \lambda z_n)$ define the same point $(\lambda \neq 0)$. Let $\bar{z}_i$ denote the complex conjugate of z_i. We define the hermitian inner product

$$(z, \bar{y}) = \sum_{i=0}^{n} z_i \bar{y}_i \tag{19.28}$$

and assume the homogeneous coordinates z_i normalized so that

$$(z, \bar{z}) = \sum_{i=0}^{n} z_i \bar{z}_i = 1. \tag{19.29}$$

This condition determines the coordinates z_i up to a factor of the form $\exp(i\alpha)$ $(i = \sqrt{-1}$, α real). We consider the group $\mathfrak{U}(n+1)$, called the *unitary group*, of all linear transformations $z' = Az$ that leave invariant the form (19.29). Then the $(n+1) \times (n+1)$ complex matrix A satisfies

$$A\bar{A}^t = E, \qquad A^{-1} = \bar{A}^t, \qquad \bar{A}^t A = E \tag{19.30}$$

where E is the $(n+1) \times (n+1)$ unit matrix. These relations show that $\mathfrak{U}(n+1)$ depends on $(n+1)^2$ real parameters. Because z and $z \exp(i\alpha)$ denote the same point, matrix A and matrix $A \exp(i\alpha)$ define the same linear transformation $z' = Az$; hence A can be normalized so that $\det A$ is a real number and then equations (19.30) give $\det A = 1$.

If $\mathfrak{R} \in \mathfrak{U}(n+1)$ denotes the group of matrices $\exp(i\alpha)E$, then the factor group $\mathfrak{U}(n+1)/\mathfrak{R}$ is called the hermitian elliptic group $\mathfrak{H}(n+1)$ and it defines in P_n the so-called hermitian elliptic geometry (see [89]). The elements A of $\mathfrak{H}(n+1)$ satisfy (19.30) and $\det A = 1$; hence the dimension of $\mathfrak{H}(n+1)$ is $n(n+2)$. It is easy to show that $\mathfrak{U}(n+1)$ and $\mathfrak{H}(n+1)$ are compact groups.

The Maurer–Cartan forms of $\mathfrak{U}(n+1)$ are given by the matrix

$$\Omega = A^{-1}\, dA = \bar{A}^t\, dA \qquad \text{such that} \quad \Omega + \bar{\Omega}^t = 0. \tag{19.31}$$

A system of Maurer–Cartan forms is then

$$\omega_{jk} = \sum_{h=0}^{n} \bar{a}_{hj}\, da_{hk} = (\bar{a}_j, da_k), \qquad \omega_{jk} + \bar{\omega}_{kj} = 0 \tag{19.32}$$

where a_{hk} are the elements of the matrix A.

The kinematic density of $\mathfrak{U}(n+1)$ is equal to the exterior product of all independent 1-forms ω_{jk}, $\bar{\omega}_{jk}$, that is, up to a constant factor,

$$d\mathfrak{U} = \bigwedge (\omega_{jk} \wedge \bar{\omega}_{jk}) \bigwedge \omega_{hh}, \qquad j < k,\ 0 \leqslant j, k, h \leqslant n. \tag{19.33}$$

ISBN 0-201-13500-0

Since $\mathfrak{U}(n+1)$ is compact and hence unimodular, this density is left and right invariant (Chapter 9, Section 7). The structure equations are

$$d\omega_{jk} = -\sum_{i=0}^{n} \omega_{ji} \wedge \omega_{ik}. \tag{19.34}$$

The group $\mathfrak{H}(n+1)$ has the same invariant forms (19.32) and the same structure equations (19.34). The only difference is that now the relation $\omega_{00} + \omega_{11} + \cdots + \omega_{nn} = 0$ holds, as follows by differentiating the relation $\det A = 1$. Hence the kinematic density for $\mathfrak{H}(n+1)$ is the exterior product of all the ω_{ij} except one of the forms, ω_{ii}. Calling ω^{ii} the exterior product (19.33) from which the ω_{ii} form has been omitted, we can take for $d\mathfrak{H}(n+1)$ the symmetric expression [47]

$$d\mathfrak{H}(n+1) = \omega^{00} + \omega^{11} + \cdots + \omega^{nn}. \tag{19.35}$$

Densities for linear subspaces and normal chains. We want to give the invariant densities with respect to $\mathfrak{H}(n+1)$ for linear subspaces and normal chains in $P_n(C)$. They are the same, up to a constant factor, as the invariant densities under $\mathfrak{U}(n+1)$.

Let L_r^0 be a fixed r-plane of $P_n(C)$ and let $\mathfrak{h}_r$ denote the subgroup of $\mathfrak{H}(n+1)$ that leaves L_r^0 invariant. The invariant density for r-planes is the invariant volume element of the homogeneous space $\mathfrak{H}(n+1)/\mathfrak{h}_r$. Since $\mathfrak{h}_r$ is a closed subgroup of the compact group $\mathfrak{H}(n+1)$, it is also compact and $\mathfrak{H}(n+1)/\mathfrak{h}_r$ has an invariant volume element. In order to find it, we will follow the general method of Chapter 10. Let a_k denote the point whose coordinates are the columns of matrix A. Conditions (19.30) give $(a_j, \bar{a}_k) = \delta_{jk}$ and from (19.32) it follows that $da_k = \sum_0^n \omega_{jk} a_j$. It follows that, assuming L_r^0 defined by the points $a_0, a_1, \ldots, a_r$, we have $\omega_{jk} = 0$ for $0 \leqslant k \leqslant r$ and $r+1 \leqslant j \leqslant n$. Since ω_{jk} are complex forms, from $\omega_{jk} = 0$ it follows that $\bar{\omega}_{jk} = 0$; hence, the density for r-planes invariant under $\mathfrak{H}(n+1)$ is

$$dL_r = \bigwedge (\omega_{jk} \wedge \bar{\omega}_{jk}) \qquad (0 \leqslant k \leqslant r,\ r+1 \leqslant j \leqslant n) \tag{19.36}$$

which is defined up to a constant factor. Note that dL_r is a differential form of degree $2(r+1)(n-r)$, as expected, since the r-planes in $P_n(C)$ depend on $2(r+1)(n-r)$ real parameters.

For $r = 0$ we get the density for points, that is, the volume element of $P_n(C)$ with respect to the hermitian (elliptic) geometry, which coincides with the volume element deduced from the so-called hermitian metric

$$ds^2 = (dz, d\bar{z}) - (z, d\bar{z})(\bar{z}, dz) \tag{19.37}$$

where z is normalized according to (19.29). Indeed, applying (19.37) to the

ISBN 0-201-13500-0

point a_0, we have $ds^2 = (da_0, d\bar{a}_0) - (a_0, d\bar{a}_0)(\bar{a}_0, da_0) = \sum_1^n \omega_{j0}\bar{\omega}_{j0}$ and the volume element is $\bigwedge \omega_{j0} \wedge \bar{\omega}_{j0}$ $(j = 1, 2, \ldots, n)$, which coincides with dL_0.

In the complex space $P_n(C)$ there are, besides the linear subspaces, the so-called *normal chains*. A normal chain K_n of dimension n is the set of points that may be expressed parametrically by

$$z = \sum_{i=0}^{n} \lambda_i a_i \tag{19.38}$$

where λ_i are real parameters such that $\sum_0^n \lambda_i^2 = 1$. A normal chain depends on $(n + 1)(n + 2)/2$ real parameters. In order to find the density for normal chains, we put $\omega_{rs} = \alpha_{rs} + i\beta_{rs}$ where α_{rs} and β_{rs} are real forms that satisfy, according to (19.32), the relations $\alpha_{rs} + \alpha_{sr} = 0$, $\beta_{rs} - \beta_{sr} = 0$. We have

$$dz = \sum_{h=0}^{n} \lambda_h \, da_h = \sum_{j,h=0}^{n} \lambda_h \alpha_{jh} a_j + i \sum_{j,h=0}^{n} \lambda_h \beta_{jh} a_j \tag{19.39}$$

and therefore, in order that K_n remain fixed, we have $\beta_{jh} = 0$ for $j, h = 0, 1, \ldots, n$, that is, the density for normal chains with respect to the group $\mathfrak{H}(n + 1)$ is the exterior product of all the β_{jh}, with one of the β_{ii} forms omitted because he relation $\omega_{00} + \cdots + \omega_{nn} = 0$ implies that $\beta_{11} + \cdots + \beta_{nn} = 0$. Denoting by β^{ii} the exterior product of all the β_{jh} with β_{ii} omitted, the density for normal chains becomes [47]

$$dK_n = \beta^{11} + \beta^{22} + \cdots + \beta^{nn}. \tag{19.40}$$

The density for normal chains K_r $(r < n)$ is equal to the exterior product of the density for K_r, as a normal chain of the subspace L_r that contains K_r, times the density dL_r.

Integral formulas. We shall present, without proof, some results of integral geometry on hermitian spaces. (For details, see Santaló's article [564]). Each of the groups $\mathfrak{U}(n + 1)$ and $\mathfrak{H}(n + 1)$ has a finite volume; according to (19.33) and (19.35), these volumes are

$$m(\mathfrak{U}(n + 1)) = i^{(n+1)(n+2)/2} \prod_{h=1}^{n+1} \frac{(2\pi i)^h}{(h - 1)!} \tag{19.41}$$

and

$$m(\mathfrak{H}(n + 1)) = i^{n(n+3)/2} \prod_{h=2}^{n+1} \frac{(2\pi i)^h}{(h - 1)!}, \tag{19.42}$$

respectively, where the powers of i have been added as a factor in order to get a real-valued measure.

ISBN 0-201-13500-0

The measure of all r-planes of the hermitian n-dimensional space is also finite and has the value

$$m(L_r) = \frac{(2\pi)^{(n-r)(r+1)}1!2!\cdots r!}{n!(n-1)!\cdots(n-r)!}. \tag{19.43}$$

For $r = 0$ we get the volume of the n-dimensional hermitian (elliptic) space,

$$m(L_0) = \frac{(2\pi)^n}{n!}. \tag{19.44}$$

Let C_h be an analytic manifold of complex dimension h contained in $P_n(C)$, that is, a manifold piecewise defined by a set of $n + 1$ analytic functions $z_i = z_i(t_1,\ldots, t_h)$ $(i = 0, 1,\ldots, n)$ of r complex variables $t_1,\ldots, t_r$ in a domain D. Assuming z_i normalized according to (19.29), let us consider the following differential form of degree $2h$:

$$\Omega_h = \sum_{(i_1,\ldots,i_h)} dz_{i_1} \wedge d\bar{z}_{i_1} \wedge\cdots\wedge dz_{i_h} \wedge d\bar{z}_{i_h} \tag{19.45}$$

where the sum is over all combinations of $i_1, i_2,\ldots, i_h = 0, 1,\ldots, n$. This form is invariant under $\mathfrak{U}(n + 1)$ (actually it is the only differential form of degree $2h$ that is invariant under $\mathfrak{U}(n + 1)$; see [86]) and the integral of Ω_h over an h-plane is

$$\int_{L_h} \Omega_h = \frac{(2\pi i)^h}{h!}. \tag{19.46}$$

For an analytic manifold C_h of complex dimension h, we define

$$J_h(C_h) = \frac{h!}{(2\pi i)^h}\int_{C_h}\Omega_h, \tag{19.47}$$

which has the property of being equal to the order of C_h if it is an algebraic manifold.

Assume that C_h is fixed and let L_r be an r-plane moving with density dL_r. Then the following integral formula holds (assuming $h + r - n \geqslant 0$).

$$\int_{C_h\cap L_r\neq 0} J_{h+r-n}(C_h \cap L_r)\, dL_r = m(L_r)J_h(C_h) \tag{19.48}$$

where $m(L_r)$ is given by (19.43). In particular, if $r + h - n = 0$, $J_0(C_h \cap L_r)$ denotes the number of intersection points of C_h and L_r.

For algebraic manifolds (compact manifolds) $J_0(C_h \cap L_{n-h})$ is a constant, equal to the order $J_h(C_h)$ of the manifold. For noncompact manifolds the difference between $J_h(C_h)$ and the number of points of intersection of a

ISBN 0-201-13500-0

generic $(n - h)$-plane and C_h, each counted with its proper multiplicity, can be expressed as an integral over the boundary of C_h [364, 109].

If C_r is another analytic manifold of dimension r such that $r + h - n \geqslant 0$ and uC_r denotes the transform of C_r by $u \in \mathfrak{U}(n+1)$, we have

$$\int_{\mathfrak{U}(n+1)} J_{r+h-n}(C_h \cap uC_r)\, d\mathfrak{U}(n+1) = m(\mathfrak{U}(n+1))J_h(C_h)J_r(C_r). \quad (19.49)$$

The same formula holds for $\mathfrak{H}(n+1)$ by replacing $m(\mathfrak{U}(n+1))$ by $m(\mathfrak{H}(n+1))$. For the proof of (19.48) and (19.49) see [564]. A generalization and a different proof using techniques of currents has been given by Schiffman [596].

Formulas (19.48) and (19.49) relate to the invariants (19.47). If instead of $J_h(C_h)$ we consider the volume of C_h induced by the hermitian metric (19.37), the resulting formulas are not so simple. They give rise to interesting inequalities, related to those of Wirtinger [728], which are not yet completely known. For $n = 2$ see the work of Rohde [516] and Varga [688] (see also that of De Rahm [145]).

For algebraic manifolds $J_{r+h-n}(C_h \cap uC_r)$ is independent of u and (19.49) is Bezout's theorem. For noncompact manifolds (19.49) may be considered an average Bezout theorem. The idea of extending Bezout's theorem to noncompact manifolds has given rise to important publications. We set $C_h(\rho) = C_h \cap S(\rho)$, where $S(\rho) = \{z; |z| \leqslant \rho\}$ and $N(C_h, \rho) = \int_0^\rho J_h(C_h(t))(dt/t)$. The Bezout problem is then to find estimates of $N(C_h \cap C_r, \rho)$ in terms of ρ, $N(C_h, \rho)$, and $N(C_r, \rho)$. See the articles by Carlson [82], Stoll [655, 656], Griffiths [239], and Carlson and Griffiths [83], in several points of which integral geometry plays an important role.

Integral geometry has been applied to the geometry of submanifolds in a complex projective space by Chern [110] (see also Pohl [482]). For applications to complex analytic mappings in several complex variables, see [114].

Holomorphic curves. The tools of complex integral geometry have been successfully applied to the theory of meromorphic or holomorphic curves in complex projective space in the sense of H. Weyl [717], H. Weyl and J. Weyl [718], and L. Ahlfors [1]. Consider a holomorphic curve $C_1 : y = y(t)$ in the complex projective space $P_n(C)$ defined by $n + 1$ holomorphic functions $y^i = y^i(t)$ $(i = 0, 1, \ldots, n)$ of the complex variable t ranging on a given Riemann surface. We can associate to C_1 the manifolds C_r $(r = 1, 2, \ldots, n)$ generated by the $(r-1)$ osculating linear spaces of C_1. Let Y_r $(r = 0, 1, \ldots, n)$ denote the multivector $y \wedge y' \wedge y'' \wedge \cdots \wedge y^{(r-1)}$, that is, the multivector whose components are the determinants of order r of the $r \times (n+1)$ matrix $(y^{k(i)})$ where i denotes derivatives $(i = 0, 1, \ldots, r)$ and $k = 0, 1, \ldots, n$. Then the invariants $J_r(C_r)$ can be written

ISBN 0-201-13500-0

$$J_r(C_r) = \frac{1}{2\pi i} \int_{C_1} \frac{|Y_{r-1}|^2 |Y_{r+1}|^2}{|Y_r|^4} dt \wedge d\bar{t} \tag{19.50}$$

where $|Y|^2$ denotes the scalar product $Y \cdot \bar{Y}$. For instance, for $r = 1$, we have

$$J_1(C_1) = \frac{1}{2\pi i} \int_{C_1} \frac{|y \wedge y'|^2}{|y|^4} dt \wedge d\bar{t} \tag{19.51}$$

where $y \wedge y'$ denotes the bivector whose components are $y^i y'^h - y^h y'^i$.

For a plane algebraic curve C_1: $y^0 = y^0(t)$, $y^1 = y^1(t)$, $y^2 = y^2(t)$, $J_2(C_1)$ is the *class* of C_1 (up to the sign)

$$J_2(C_1) = \frac{1}{2\pi i} \int_{C_1} \frac{|y|^2 |yy'y''|^2}{|y \wedge y'|^2} dt \wedge d\bar{t} \tag{19.52}$$

where $|yy'y''|$ denotes the absolute value of the determinant whose elements are the components of y, y', y''.

The classical Plücker formulas for algebraic curves are linear relations among the invariants J_h, namely, $s + J_{r-1} - 2J_r + J_{r+1} = -\chi$ where s is the so-called stationary index (depending on the critical points of C_1) and χ is the Euler–Poincaré characteristic of the Riemann surface on which C_1 is defined. Integral formulas (19.48), (19.49) give a geometric interpretation of the classes in terms of the average number of points of intersection of the loci of osculating h-planes and random $(n - h)$-planes or random $(n - h)$-dimensional analytic manifolds. On holomorphic curves, see the earlier work of Weyl [717] and Ahlfors [1] and the more recent contributions of Hung-Hsi Wu [730], Chern [115], and Griffiths [240].

7. Symplectic Integral Geometry

Let $Z = (z_{hk})$ be a symmetric $n \times n$ matrix with complex elements and let $\bar{Z} = (\bar{z}_{hk})$ denote the conjugate complex matrix of Z. Put $Z = X + iY$, where $X = \frac{1}{2}(Z + \bar{Z})$ and $Y = (1/2i)(Z - \bar{Z})$, and let H be the domain defined by the inequality $Y > 0$. The real dimension of H is $n(n + 1)$. The *homogeneous symplectic group* $\mathfrak{S}_0$ is the group of all real $2n \times 2n$ matrices

$$M = \begin{pmatrix} A & B \\ C & D \end{pmatrix} \tag{19.53}$$

where A, B, C, D are $n \times n$ real matrices such that

$$M^t J M = J, \qquad J = \begin{pmatrix} 0 & E \\ -E & 0 \end{pmatrix} \tag{19.54}$$

ISBN 0-201-13500-0

where E is the unit $n \times n$ matrix and 0 denotes the null matrix. Since $M^t J M$ is skew-symmetric, it follows that the dimension of $\mathfrak{S}_0$ is $2n^2 + n$.

It is easily proved that the transformations

$$Z' = (AZ + B)(CZ + D)^{-1} \tag{19.55}$$

map the domain H onto itself. They form the *symplectic group* $\mathfrak{S}$, which acts transitively on H and is obtained from $\mathfrak{S}_0$ by identifying M and $-M$. The group $\mathfrak{S}$ defines on H the so-called symplectic geometry. The quadratic differential form

$$ds^2 = \mathrm{Tr}(Y^{-1}\, dZ\, Y^{-1}\, d\bar{Z}) = \mathrm{Tr}(Y^{-1}\, dX\, Y^{-1}\, dX + Y^{-1}\, dY\, Y^{-1}\, dY) \tag{19.56}$$

is invariant under $\mathfrak{S}$ and defines a Riemannian metric on H. The volume element defined by this metric is the density for points in symplectic geometry. Up to a constant factor this density can be written

$$dP = [\mathrm{Tr}(Y^{-1}\, dZ \wedge Y^{-1}\, d\bar{Z})]^{n(n+1)/2} \tag{19.57}$$

where the power is to be understood as an exterior product.

With reference to geodesics, it can be proved that the symplectic metric (19.56) exactly defines one geodesic connecting two arbitrary points of H and that all geodesics are symplectic images of the curves represented by $Z = i \exp(sG)$, where s is the arc length and G is the diagonal matrix $(\delta_{jk} g_k)$ in which $0 < g_1 \leqslant g_2 \leqslant \cdots \leqslant g_n$ are arbitrary constants such that $\sum g_i^2 = 1$. This proves that the set of all geodesics of H splits into an $(n-1)$-dimensional bundle of symplectic equivalent classes, each class determined by the real numbers $(g_1, g_2, \ldots, g_n)$ satisfying the condition $\sum g_i^2 = 1$. The geodesics of each class and the geodesics that go through a fixed point have a density. These densities and applications to a kind of Crofton's formula for sets of geodesics that intersect a fixed hypersurface have been studied by Legrady [358].

It seems that symplectic integral geometry deserves much further study. For the necessary concepts in symplectic geometry see the fundamental paper of Siegel [610].

8. The Integral Geometry of Gelfand

The term integral geometry has been used in a sense different from that of the present book by Gelfand, Helgason, and others. They consider the following general problem.

Let X be a differentiable manifold in which certain submanifolds $M(u) = M(u_1, u_2, \ldots, u_k)$, depending analytically on the parameters $u_1, \ldots, u_k$, are

ISBN 0-201-13500-0

given. Given a function $f(x)$ on X, we associate its integrals over each $M(u)$:

$$\hat{f}(u) = \int_{M(u)} f(x)\, d\sigma(u) \tag{19.58}$$

where $d\sigma$ is a suitable differential form on $M(u)$. This gives a new function $\hat{f}(u)$ defined on the set of submanifolds. The problem is to invert the "Radon transform" $f \to \hat{f}$, that is, to determine $f(x)$ in terms of $\hat{f}(u)$. For a systematic exposition of the results on this subject obtained before 1966, see the book of Gelfand, Graev, and Vilenkin [224].

The classical case is when X is R^n and the manifolds $M(u)$ the hyperplanes in R^n. The problem was solved by Radon in 1917 [493] for $n = 2$ and $n = 3$ and generalized to any n by John [324].

John's result may be stated as follows. With the hyperplane $(u, x) = u_1 x_1 + \cdots + u_n x_n = p$, determined by the unit normal vector u and the real number p, we shall associate the differential form

$$d\sigma = (-1)^{h-1} u_h^{-1}\, dx_1 \wedge \cdots \wedge dx_{h-1} \wedge dx_{h+1} \wedge \cdots \wedge dx_n$$

(oriented area element on the hyperplane), which is independent of h. Then, for any $f(x)$, infinitely differentiable and rapidly decreasing (criteria that can be weakened), we consider the Radon transform

$$\hat{f}(u, p) = \int_{(u,x)=p} f(x)\, d\sigma. \tag{19.59}$$

The formula expressing $f(x)$ in terms of $\hat{f}(u, p)$ depends on whether the space has odd or even dimensions. If the dimension n is odd, the inverse Radon transform is

$$f(x) = \frac{(-1)^{(n-1)/2}}{2(2\pi)^{n-1}} \int_{\Sigma} \hat{f}_p^{(n-1)}(u, (u, x))\, d\sigma_1 \tag{19.60}$$

where $d\sigma_1 = \sum_1^n (-1)^{h-1} u_h\, du_1 \wedge \cdots \wedge du_{h-1} \wedge du_{h+1} \wedge \cdots \wedge du_n$ is the surface element on the unit sphere S^{n-1}, and $\hat{f}_p^{(n-1)}$ denotes the $(n-1)$th derivative of $\hat{f}(u, p)$ with respect to p. The integral is over any hypersurface Σ that encloses the origin in u-space.

If n is even, the inverse Radon transform is

$$f(x) = \frac{(-1)^{n/2}(n-1)!}{(2\pi)^n} \int_{\Sigma} \left(\int_{-\infty}^{\infty} \hat{f}(u, p)[p - (u, x)]^{-n}\, dp \right) d\sigma_1 \tag{19.61}$$

where the integral over p is understood in terms of its regularization. For details see [224], which includes a table of Radon transforms of some functions.

ISBN 0-201-13500-0

The general question of determining a function f on a space from the knowledge of the integrals of f over certain subsets of the space was initiated by Funk [210], who proved that a function f on the two-dimensional sphere symmetric with respect to the center can be determined by means of the integrals of f over the great circles (for an application to convex bodies, see [50, p. 154]). Radon [493] treated the problem of determining a function on the noneuclidean plane from the integrals of the function over all geodesics. For related results see those of Blaschke [39] and J. W. Green [235].

Radon's and John's formulas (19.60), (19.61) were generalized to hyperbolic spaces by Helgason [297], who associated with every $f(x)$ of bounded support its integrals over all possible hyperplanes (geodesic submanifolds). In a different direction, Gelfand, Graev, and Vilenkin [224] treated the same problem, associating with every such function its integrals over horospheres, since the intrinsic geometry of the horospheres in hyperbolic space is euclidean and thus they form one of the analogues of hyperplanes of a euclidean space. The result was extended by Gelfand and Graev [222] to noncompact symmetric spaces with a complex isometry group, and Helgason [299, 301] extended Radon's formula to all noncompact symmetric spaces.

Helgason [300, 301] noted that formula (19.60), when (19.59) is taken into account, contains two integrations, dual to each other: first, we integrate over the set of points in a given hyperplane through x, then this integral is integrated over the set of hyperplanes through x. Guided by this duality, Helgason [301] adopted the following very general setup.

Let G be a locally compact topological group. Let H_X, H_Ξ be two closed subgroups of G and consider the left coset spaces $X = G/H_X$, $\Xi = G/H_\Xi$. Assume that (i) G, H_X, H_Ξ, $H_X \cap H_\Xi$ are unimodular; (ii) if $h_X \in H_X$, $h_X H_\Xi \subset H_\Xi H_X$, then $h_X \in H_\Xi$; if $h_\Xi \in H_\Xi$, $h_\Xi H_X \subset H_X H_\Xi$, then $h_\Xi \in H_X$.

Two elements in X and Ξ, respectively, are called incident if as cosets in G they have a point in common. Then, for $x \in X$ put $\check{x} = \{\xi \in \Xi;\, x, \xi \text{ incident}\}$ and for $\xi \in \Xi$, put $\hat{\xi} = \{x \in X;\, x, \xi \text{ incident}\}$. These sets can be identified with coset spaces of certain subgroups of G and they have invariant densities, say $d\mu$ and dm, respectively. Then, if f and ϕ are suitable restricted functions on X and Ξ, respectively, we define

$$\hat{f}(\xi) = \int_{\hat{\xi}} f(x)\, dm(x), \qquad \check{\phi}(x) = \int_{\check{x}} \phi(\xi)\, d\mu(\xi) \tag{19.62}$$

and Helgason [301] has stated the following general problems:

(A) Relate function spaces on X and Ξ by means of integral transforms $f \to \hat{f}$ and $\phi \to \check{\phi}$;

(B) Relate directly the functions f and $(\hat{f})^\vee$ on X, ϕ and $(\check{\phi})^\wedge$ on Ξ.

ISBN 0-201-13500-0

Helgason [301] proved that for suitable normalized invariant measures dx and $d\xi$ on X and Ξ, respectively, the formula

$$\int_X f(x)\check{\phi}(x)\,dx = \int_\Xi \hat{f}(\xi)\phi(\xi)\,d\xi \tag{19.63}$$

holds for all continuous f and ϕ of compact support.

As an example, let G be the group of all rigid motions of R^n. Let H_X be the subgroup of G leaving a given h-plane L_h fixed and let H_Ξ be the subgroup of G that leaves an r-plane L_r containing L_h invariant ($h < r$, $L_h \subset L_r$). Then $X = G/H_X$ is the set of h-planes and $\Xi = G/H_\Xi$ is the set of r-planes in R^n. The density $d\mu(\xi) = dL_{r[h]}$ is the invariant density for r-planes through L_h and $dm(x) = dL_h^{(r)}$ is the invariant density for h-planes in L_r. Formula (19.63) is then a consequence of (12.52).

Formulas (19.60) and (19.61) show that for R^n there is a remarkable contrast between the inversion formulas in cases in which n is odd and those in which n is even. In the first case, (19.59) is inverted in terms of a differential operator; in the second case, by an integral operator. This contrast persists for the case of a symmetric space, but then the decisive feature is not the parity of the dimension, but whether the group of isometries has all Cartan subgroups conjugate or not [301].

Applications of the Radon transform on R^n to constant coefficient differential equations in R^n are given by John [324a] and Borovikov [65] (cf. also Rhee's article [509]). Applications of the Radon transform on symmetric spaces to the solving of differential equations on these spaces are given by Helgason [299, 303, 305].

Several examples and applications to group representations are given by Gelfand and Graev [222], Gelfand, Graev, and Vilenkin [224], and Helgason [300, 304].

Further work on the Radon transform on symmetric spaces has been done by Helgason ([302]; compact symmetric spaces of rank one and Grassmannians), Petrov and Sibasov ([472, 607]; Grassmannians), and Morimoto [433] and Kelly [331] (on lower-dimensional horocycles). A general group-theoretic approach is described in [301] and a related differential form generalization by Gelfand, Graev, and Shapiro [223], Kelly [331], Morimoto [433], and Petrov [472].

Some uniqueness theorems for various integral geometric problems have been given by Romanov [517]. For instance, Romanov considers the problem of determining a function from its integrals over a family of ellipsoids of revolution with one focus fixed and the other running over a point set of a fixed hyperplane passing through the fixed focus. The problem has applications to the study of the earth's internal structure from seismological data.

ISBN 0-201-13500-0

9. Notes

1. *Two formulas of* Hermann [308]. (a) Let X be a differentiable manifold of dimension $m + n$ with a foliation F. Let n be the dimension of the leaves and let X/F be the set of leaves. Suppose that Y is an $(m + p)$-dimensional manifold and $\alpha: Y \to X$ a differentiable map. Let $\theta^{(p)}$ be a p-form on Y and let $\omega^{(m)}$ be an invariant m-form of the foliation on X. Suppose that for each $L \in X/F$, except perhaps for a set of $\omega^{(m)}$-measure zero, $\alpha^{-1}(L)$ is a p-dimensional submanifold of Y. Define the function N almost everywhere on X/F as

$$N(L) = \int_{\alpha^{-1}(L)} \theta^{(p)}. \tag{19.64a}$$

Then

$$\int_Y \alpha^*(\omega^{(m)}) \wedge \theta^{(p)} = \int_{X/F} N\omega^{(m)}. \tag{19.64b}$$

As an application of this formula, suppose that $q: X \to Z$ is a mapping of X onto a manifold Z and $S \subset Z$ is a submanifold of Z with $\dim S = \dim Z - \dim F$. Let $Y = q^{-1}(S)$ and $\alpha: Y \to X$ be the inclusion map. Then, in formula (19.64a), $N(L)$, for $L \in X/F$, becomes the number of times the projection of L in Z intersects S and the right-hand side of (19.64a) is the measure of all leaves of F that intersect S, counted according to multiplicity.

(b) Let $\mathfrak{G}$ be a connected Lie group and $\mathfrak{L}$ a closed subgroup. Let $\mathfrak{G}/\mathfrak{L}$ be the space of left cosets of $\mathfrak{L}$ and $p: \mathfrak{G} \to \mathfrak{G}/\mathfrak{L}$ the projection $g \to g\mathfrak{L}$. Let K, K_0 be submanifolds of $\mathfrak{G}/\mathfrak{L}$ with $\dim K + \dim K_0 = \dim \mathfrak{G}/\mathfrak{L}$. Let gK denote the transform of K by $g \in \mathfrak{G}$ and let $N(g)$ be the number of points of $gK \cap K_0$. Suppose that there are cross sections $\theta: K \to \mathfrak{G}$ and $\psi: K_0 \to \mathfrak{G}$ and define the map $\alpha: K_0 \times \mathfrak{L} \times K \to \mathfrak{G}$ as

$$\alpha(y, s, x) = \psi(y)s\theta(x)^{-1}$$

for $y \in K_0$, $s \in \mathfrak{L}$, $x \in K$. Then

$$\int_{\mathfrak{G}} N\,dg = \int_{K_0 \times \mathfrak{L} \times K} \alpha^*(dg)$$

where dg denotes the left-invariant measure on $\mathfrak{G}$. The evaluation of $\alpha^*(dg)$ depends on the structure of $\mathfrak{G}$ as a Lie group.

2. *Sets of geodesics of dimension* $2h$. Consider the formula $dG^h = (dG^1)^h$ (19.12). Assume two h-dimensional differentiable submanifolds of the Riemann manifold M, say M_1 and M_2, such that any pair of points $P_1 \in M_1$, $P_2 \in M_2$ determines one and only one geodesic G. Call $d\sigma_i$ the volume element of M_i at P_i, r the length of the geodesic arc P_1P_2, α_i the angle that G makes with the tangent space T_i of M_i at P_i and θ the angle between the geodesic $(h + 1)$-dimensional manifolds containing T_i and G $(i = 1, 2)$. Then, up to

ISBN 0-201-13500-0

a constant factor, we have $dG^h = r^{-h} \cos\theta \sin\alpha_1 \sin\alpha_2\, d\sigma_1 \wedge d\sigma_2$. For lines in E_n, see Pohl [481].

3. *Vector integral geometry*. Let H_r denote the rth elementary symmetric function of the $n - 1$ principal curvatures of the boundary ∂K of a convex body K of E_n. The vectorial quermassintegrale of K are defined by

$$q_i(K) = \left(n\binom{n-1}{i-1}\right)^{-1} \int_{\partial K} x H_{i-1}\, d\sigma \quad (i = 1, 2, \ldots, n); \quad q_0(K) = \int_K x\, dv$$

where dv is the volume element of E_n, $d\sigma$ is the area element of ∂K, and x denotes the radius vector from the origin to the point. From these, we deduce the so-called curvature centroids $p_i = q_i/W_i$ $(i = 0, 1, \ldots, n)$, where W_i are the ordinary quermassintegrale (13.8) and (13.45). In particular, p_n is the Steiner point $s(K)$ of K, which has important properties (for references and an account of the history, see [247, 248]). Hadwiger and Schneider [280] extended the definitions of q_i, p_i to sets D of the convex ring, that is, sets representable as finite unions of convex bodies, and proved that many integral formulas of ordinary integral geometry have their analogues by replacing the scalar quermassintegrale W_i with q_i. For instance, if dK_1 is the kinematic density of the domain D_1 and D_0 is a fixed domain, D_1 and D_0 being elements of the convex ring, we have $\int q_h(D_0 \cap D_1)\, dK_1 = \sum c_{nhi} W_{h-i}(D_0) q_i(D_1)$ for certain constants c_{nhi}. This is the analogue of (15.72) in vector integral geometry. Details and extensions can be seen in the work of Hadwiger and Schneider [280], Schneider [599, 600], Hadwiger and Meier [279], and H. R. Müller [438, 439].

4. *Integral geometry on surfaces*. Let $ds^2 = e\, du^2 + g\, dv^2$ be the first fundamental form of a surface Σ in terms of local rectangular coordinates. If ϕ denotes the angle between the geodesic G and the u-curve at the point (u, v), the density (19.15) for geodesics takes the form

$$\begin{aligned} dG = &-\sqrt{e} \sin\phi\, d\phi \wedge du + \sqrt{g} \cos\phi\, d\phi \wedge dv \\ &+ [(\partial\sqrt{g}/\partial u) \sin\phi - (\partial\sqrt{e}/\partial v) \cos\phi]\, du \wedge dv. \end{aligned} \tag{19.65}$$

In particular, if (u, v) is a geodesic coordinate system (i.e., the curves v = constant are geodesics and u is the arc length parameter along each of these geodesics), we have $e = 1$ and (19.65) becomes

$$dG = (\partial\sqrt{g}/\partial u) \sin\phi\, du \wedge dv - \sin\phi\, d\phi \wedge du + \sqrt{g} \cos\phi\, d\phi \wedge dv. \tag{19.66}$$

If each geodesic of the set is perpendicular to a u-curve, according to the second invariance property (Section 2) we can choose on each geodesic the point at which $\phi = \pi/2$ and the density dG takes the simple form

$$dG = (\partial\sqrt{g}/\partial u)\, du \wedge dv. \tag{19.67}$$

For geodesic polar coordinates, u denotes the distance from the origin O to the geodesic G and v is the angle that the geodesic perpendicular to G from O makes with a fixed direction at O.

ISBN 0-201-13500-0

Let Γ be a curve of length L on the surface Σ. Formula (19.24) is then

$$\int_{G \cap \Gamma \neq 0} N \, dG = 2L \tag{19.68}$$

and for a geodesic segment S^* of length L_0, denoting by N the number of intersection points $S^* \cap \Gamma$, we have

$$\int_{\Gamma \cap S^* \neq 0} N \, dS^* = 4LL_0 \tag{19.69}$$

where $dS^* = dP \wedge d\theta = dG^* \wedge ds$ is the kinematic density on Σ (dP is the area element at the origin P of the segment S^* and θ is the angle of S^* with a reference direction at P (19.26)).

Let us give an application of these formulas. Let Σ be a convex surface in E_3 all of whose geodesics are closed curves. Then it can easily be shown that all the geodesics have the same length, say L, and that each pair of geodesics intersects at $N \geqslant 2$ points. If G_1 is a fixed geodesic and G is a moving one, considered as a geodesic segment of length L, according to (16.69) we have

$$4L^2 = \int N \, dP \wedge d\theta \geqslant 2 \int dP \wedge d\theta = 4\pi F \tag{19.70}$$

where F is the surface area of Σ. If $1/R_0{}^2$ is the greatest value of the Gaussian curvature K of Σ, we have $F = \int K^{-1} \, du_2 \geqslant 4\pi R_0{}^2$ and from (19.70) it follows that $L \geqslant 2\pi R_0$, which is an inequality due to Zoll [735].

Let λ be the length of the greatest geodesic segment that cannot intersect a geodesic of Σ in more than one point. Then, from (19.69) it follows that

$$4\lambda L = \int N \, dP \wedge d\theta \leqslant \int dP \wedge d\theta = 2\pi F \tag{19.71}$$

and since $\lambda \leqslant L/2$, we have $4\lambda^2 \leqslant \pi F$. If $f(P, \theta)$ denotes the distance from the point P to its first conjugate point P' along the geodesic determined by the angle θ at P and call $a = \inf f(P, \theta)$ for $P \in \Sigma$ and $0 \leqslant \theta \leqslant 2\pi$, then $a \leqslant \lambda$ and the last inequality gives $4a^2 \leqslant \pi F$. For compact orientable surfaces of Euler–Poincaré characteristic χ, this inequality (which is due to Berger: see L. W. Green [236]) generalizes to $2a^2\chi \leqslant \pi F$.

For the integral geometry on surfaces and on Finsler spaces, see [39, 95, 463, 555, 621, 646]; for other points of view, see [253, 254].

ISBN 0-201-13500-0

APPENDIX

Differential Forms and Exterior Calculus

1. Differential Forms and Exterior Product

The object of this Appendix is to give a rather informal approach to differential forms, mainly as they relate to the applications that have been made in the text. The so-called exterior differential forms, or simply differential forms, were used by Poincaré in his "new methods" on celestial mechanics and by E. Cartan in his work on continuous groups and Pfaffian systems at the end of the last century. A systematic treatment of these forms was presented by E. Cartan [87] in his theory of integral invariants (1922). Later on, mainly through the 1946 work on Lie groups of Chevalley [118], differential forms were incorporated into the domains of algebra and since that time they have been widely and successfully used in differential geometry and mathematical physics (see the works of Sternberg [628, 629] and H. Cartan [91]). In this Appendix, however, since our purpose is only to compile some definitions and some results that have been used in the text, we will follow a more classical and intuitive exposition, giving more emphasis to the practical handling of differential forms and their geometric meaning than to their rigorous and precise formulation. This geometric approach can be seen in the work of E. Cartan [90] and Flanders [201, 202].

A differential form of degree 1, or 1-form (also called a Pfaffian form), in R^n is the expression

$$\omega^{(1)} = a_1\, dx_1 + a_2\, dx_2 + \cdots + a_n\, dx_n, \tag{1}$$

which occurs under the integral sign in line integrals. That means that (1) can be integrated over oriented curves, the result being independent of the parametrization of the curve.

The coefficients a_i are real-valued functions of the variables $x_1, x_2, \ldots, x_n$, which for simplicity are assumed of class C^∞. By a change of coordinates $x_i \to x_i'$ the 1-form (1) may be written

ENCYCLOPEDIA OF MATHEMATICS and Its Applications, Gian-Carlo Rota (ed.).
1, Luis A. Santaló, Integral Geometry and Geometric Probability

ISBN 0-201-13500-0

$$\omega^{(1)} = \sum_{i=1}^{n} a_i \frac{\partial x_i}{\partial x_h'} dx_h' = \sum_{h=1}^{n} a_h' \, dx_h' \tag{2}$$

and thus the transition law relating the coefficients a_i to the a_i' is given by the equation

$$a_h' = \sum_{i=1}^{n} \frac{\partial x_i}{\partial x_h'} a_i, \tag{3}$$

which is known to define the transformation law for covectors in R^n.

An example of a 1-form is the differential of a function f,

$$df = \sum_{i=1}^{n} \frac{\partial f}{\partial x_i} dx_i. \tag{4}$$

Not all 1-forms, however, can be obtained in this manner. By a well-known theorem of calculus, the necessary and sufficient conditions for a differential form (1) to be the differential of a function is that $\partial a_i/\partial x_h = \partial a_h/\partial x_i$.

The product of a 1-form $\omega^{(1)}$ and a scalar λ is defined by $\lambda\omega^{(1)} = \sum_i \lambda a_i \, dx_i$ and the sum and difference of 1-forms $\omega^{(1)} = \sum_i a_i \, dx_i$, $\phi^{(i)} = \sum_i b_i \, dx_i$ is defined by $\omega^{(1)} \pm \phi^{(1)} = \sum_i (a_i \pm b_i) \, dx_i$. With these operations the 1-forms constitute a module on the ring of differentiable functions on R^n. The 1-forms $dx_1, dx_2, \ldots, dx_n$ are a basis for the module. Any set of independent 1-forms, that is, any set of 1-forms $\omega_h^{(1)} = \sum_i a_{hi} \, dx_i$ $(h = 1, 2, \ldots, n)$ such that $\det|a_{hi}| \neq 0$, also constitute a basis of the module.

To define the differential forms of degree two, or 2-forms, we consider first the case of the plane R^2. The expressions under the integral sign in double integrals on R^2 contain the product $dx_1 \, dx_2$, which according to the change of variables formula, by the change $x_i \to x_i'$ becomes

$$dx_1 \, dx_2 = \left(\frac{\partial x_1}{\partial x_1'} \frac{\partial x_2}{\partial x_2'} - \frac{\partial x_1}{\partial x_2'} \frac{\partial x_2}{\partial x_1'} \right) dx_1' \, dx_2'. \tag{5}$$

This shows that the "product" $dx_1 \, dx_2$ that appears under the double integral sign is not the ordinary product of the differentials $dx_i = (\partial x_i/\partial x_1') dx_1' + (\partial x_i/\partial x_2') \, dx_2'$ $(i = 1, 2)$ but another kind of product, which is called the *exterior* product of dx_1 and dx_2 and is denoted by $\bigwedge$ (usually read "wedge product"). The fundamental property of the exterior product is that it is skew-symmetric, or alternating; that is,

$$dx_1 \wedge dx_2 = - dx_2 \wedge dx_1, \qquad dx_i \wedge dx_i = 0. \tag{6}$$

Using this skew symmetry, by the change of coordinates $x_i \to x_i'$ we have

$$\begin{aligned} dx_1 \wedge dx_2 &= \left(\frac{\partial x_1}{\partial x_1'} dx_1' + \frac{\partial x_1}{\partial x_2'} dx_2' \right) \wedge \left(\frac{\partial x_2}{\partial x_1'} dx_1' + \frac{\partial x_2}{\partial x_2'} dx_2' \right) \\ &= \left(\frac{\partial x_1}{\partial x_1'} \frac{\partial x_2}{\partial x_2'} - \frac{\partial x_1}{\partial x_2'} \frac{\partial x_2}{\partial x_1'} \right) dx_1' \wedge dx_2', \end{aligned}$$

which agrees with the classical rule (5).

ISBN 0-201-13500-0

If x_1, x_2 denote rectangular coordinates, then the exterior product $dx_1 \wedge dx_2$ denotes the element of the signed area on the plane. Thus, considering an oriented parallelogram whose sides are the infinitesimal vectors of components (dx_1, dx_2) and $(\delta x_1, \delta x_2)$, respectively, we have

$$dx_1 \wedge dx_2 = \begin{vmatrix} dx_1 & dx_2 \\ \delta x_1 & \delta x_2 \end{vmatrix} = dx_1\, \delta x_2 - dx_2\, \delta x_1. \tag{7}$$

For instance, taking two displacements, respectively parallel to the coordinate axes, we have $(dx_1 = dx_1, dx_2 = 0)$, $(\delta x_1 = 0, \delta x_2 = dx_2)$ and the area element takes the usual form $dx_1\, dx_2$. Taking the first displacement in the direction of the radius vector of a polar coordinate system (ρ, θ) and the second in the perpendicular direction, we have $dx_1 = \cos\theta\, d\rho$, $dx_2 = \sin\theta\, d\rho$, $\delta x_1 = -\rho \sin\theta\, d\theta$, $\delta x_2 = \rho\cos\theta\, d\theta$ and (7) gives $dx_1 \wedge dx_2 = \rho\, d\rho \wedge d\theta$, which is the area element in polar coordinates.

From the representation (7) of the exterior product, which could be taken as a definition in any affine coordinate system, not necessarily rectangular, the following distributive laws follow easily.

$$dx_1 \wedge (dx_2 + dx_3) = dx_1 \wedge dx_2 + dx_1 \wedge dx_3, \quad dx_1 \wedge (x_3\, dx_2) = x_3\, dx_1 \wedge dx_2. \tag{8}$$

After these preliminaries we can define the 2-forms in R^n as the expressions that occur under the integral sign in double integrals, that is, the expressions

$$\omega^{(2)} = \sum_{i,j=1}^{n} a_{ij}\, dx_i \wedge dx_j, \tag{9}$$

which can be integrated over oriented two-dimensional manifolds, the result being independent of the parametrization of the manifold.

The functions a_{ij} are assumed of class C^∞. If we again call a_{ij} the difference $\frac{1}{2}(a_{ij} - a_{ji})$, it is clear that it is always possible to assume that the coefficients of the 2-forms are skew-symmetric, that is, $a_{ij} = -a_{ji}$. The condition that the 2-form (9) remains invariant under the change of coordinates $x_i \to x_i'$ conduces to the transformation law

$$a'_{ij} = \sum_{h,k=1}^{n} \frac{\partial x_h}{\partial x_i'} \frac{\partial x_k}{\partial x_j'} a_{hk}, \tag{10}$$

which proves that a_{ij} are components of a skew-symmetric covariant double tensor.

A way of generating 2-forms is the exterior product of two 1-forms $\omega^{(1)} = \sum_i a_i\, dx_i$, $\phi^{(1)} = \sum_i b_i\, dx_i$ defined by

$$\omega^{(1)} \wedge \phi^{(1)} = \sum_{i,j=1}^{n} a_i b_j\, dx_i \wedge dx_j = \sum_{i<j} \begin{vmatrix} a_i & a_j \\ b_i & b_j \end{vmatrix} dx_i \wedge dx_j, \tag{11}$$

which according to (7) can be written

ISBN 0-201-13500-0

$$\omega^{(1)} \wedge \phi^{(1)} = \sum_{i,j=1}^{n} a_i b_j (dx_1\, \delta x_j - \delta x_i\, dx_j) \tag{12}$$

or, putting $\omega^{(1)}(d) = \sum_i a_i\, dx_i$, $\omega^{(1)}(\delta) = \sum_i a_i\, \delta x_i$ and introducing the analogous expressions for $\phi^{(1)}$,

$$\omega^{(1)} \wedge \phi^{(1)} = \omega^{(1)}(d)\phi^{(1)}(\delta) - \omega^{(1)}(\delta)\phi^{(1)}(d), \tag{13}$$

which is the expression of the exterior product in terms of ordinary products of differential forms.

Obviously not all 2-forms are the product of two 1-forms. When this is the case, the 2-form is said to be *decomposable*. For instance, if $n > 3$, and $\omega_1, \omega_2, \omega_3, \omega_4$ are independent 1-forms in R^n, the 2-form $(\omega_1 \wedge \omega_2) + (\omega_3 \wedge \omega_4)$ is not decomposable.

The definition of r-forms (differential forms of degree r) follows the same lines as above. They are the expressions

$$\omega^{(r)} = \sum_{i_1, \ldots, i_r} a_{i_1 \ldots i_r}\, dx_{i_1} \wedge \cdots \wedge dx_{i_r}, \tag{14}$$

which occur under the integral sign in r-fold integrals over r-dimensional manifolds of R^n. If $r > n$, then the only r-form is the null form.

According to the skew symmetry of the wedge product, we can always assume that the coefficients $a_{i_1 \ldots i_r}$ are skew symmetric. They are the components of a covariant skew-symmetric tensor of order r.

The exterior product of the r-form (14) and the q-form

$$\omega^{(q)} = \sum b_{j_1 \ldots j_q}\, dx_{j_1} \wedge \cdots \wedge dx_{j_q}$$

is the $(r + q)$-form defined by

$$\omega^{(r)} \wedge \omega^{(q)} = \sum_{i,j} a_{i_1 \ldots i_r} b_{j_1 \ldots j_q}\, dx_{i_1} \wedge \cdots \wedge dx_{i_r} \wedge dx_{j_1} \wedge \cdots \wedge dx_{j_q}. \tag{15}$$

Note that by r interchanges we may bring dx_{j_1} to the left past $dx_{i_1}, \ldots, dx_{i_r}$ and similarly, r interchanges bring each $dx_{j_2}, \ldots, dx_{j_q}$ in turn past $dx_{i_1}, \ldots, dx_{i_r}$. Thus $\omega^{(q)} \wedge \omega^{(r)}$ is obtained from $\omega^{(r)} \wedge \omega^{(q)}$ by rq interchanges. Hence, the exterior product, which is associative and distributive with respect to addition, has the property

$$\omega^{(r)} \wedge \omega^{(q)} = (-1)^{rq} \omega^{(q)} \wedge \omega^{(r)}. \tag{16}$$

In particular, the product of 1-forms does not change if we perform an even permutation of the factors and is changed into its negative if we perform an odd permutation of the factors.

It is clear that under conventional addition and scalar multiplication the r-forms constitute a module on the ring of differentiable functions on R^n and that a basis for this module is the set of r-forms $dx_{i_1} \wedge \cdots \wedge dx_{i_r}$ with $i_1 < i_2 < \cdots < i_r$. Hence, the dimension of the module of r-forms in R^n is $\binom{n}{r}$. If $\omega_1, \omega_2, \ldots, \omega_r$ is a set of independent 1-forms, the set of r-forms

ISBN 0-201-13500-0

$$\omega_{i_1} \wedge \omega_{i_2} \wedge \cdots \wedge \omega_{i_r} \qquad (i_1 < i_2 < \cdots < i_r)$$

is also a basis for the module of r-forms.

2. Two Applications of the Exterior Product

1. If $\omega_i = \sum_h a_{ih}\, dx_h$ $(i = 1, 2, \ldots, r)$ are r 1-forms in R^n, they may be considered as linear forms in the indeterminates dx_i and then from linear algebra we know that they are linearly independent if and only if the rank of the matrix (a_{ih}) is r. On the other hand, as a generalization of (11), we have

$$\omega_1 \wedge \cdots \wedge \omega_r = \sum_{i_1 < \cdots < i_r} \begin{vmatrix} a_{1i_1} & \cdots & a_{1i_r} \\ \vdots & & \vdots \\ a_{ri_1} & \cdots & a_{ri_r} \end{vmatrix} dx_{i_1} \wedge \cdots \wedge dx_{i_r}. \tag{17}$$

Hence we have

The 1-forms $\omega_1, \omega_2, \ldots, \omega_r$ are linearly independent if and only if $\omega_1 \wedge \omega_2 \wedge \cdots \wedge \omega_r \neq 0$.

Note that for $r = n$ we have

$$\omega_1 \wedge \cdots \wedge \omega_n = \det|a_{ih}|\, dx_1 \wedge dx_2 \wedge \cdots \wedge dx_n. \tag{18}$$

2. Let $\omega_1, \ldots, \omega_r$ be 1-forms in R^n that are linearly independent and assume that there exist other 1-forms $\phi_1, \ldots, \phi_r$ such that

$$\omega_1 \wedge \phi_1 + \omega_2 \wedge \phi_2 + \cdots + \omega_r \wedge \phi_r = 0. \tag{19}$$

Then

$$\phi_i = \sum_{h=1}^{r} A_{ih}\omega_h, \qquad A_{ih} = A_{hi} \qquad (i = 1, 2, \ldots, r). \tag{20}$$

This is a useful property known as E. Cartan's lemma. For a proof choose $\omega_{r+1}, \ldots, \omega_n$ so that $\omega_1, \ldots, \omega_r, \omega_{r+1}, \ldots, \omega_n$ is a basis of the 1-forms in R^n. Then we can write

$$\phi_i = \sum_{h=1}^{r} A_{ih}\omega_h + \sum_{h=r+1}^{n} B_{ih}\omega_h$$

and substituting this expression into (19), we have

$$\sum_{1 \leq i < j \leq r} (A_{ij} - A_{ji})\omega_i \wedge \omega_j + \sum_{i=1}^{r} \sum_{h=r+1}^{n} B_{ih}\omega_i \wedge \omega_h = 0.$$

Since the set of exterior products $\omega_i \wedge \omega_j$ is a basis for the 2-forms in R^n, we deduce $A_{ij} = A_{ji}$, $B_{ih} = 0$, and this proves (20).

ISBN 0-201-13500-0

3. Exterior Differentiation

The *exterior differential* or *exterior derivative* of an r-form

$$\omega^{(r)} = \sum_{i_1,\ldots,i_r} a_{i_1\ldots i_r}\, dx_{i_1} \wedge \cdots \wedge dx_{i_r} \tag{21}$$

is defined as the $(r+1)$-form

$$\begin{aligned} d\omega^{(r)} &= \sum_{i_1,\cdots,i_r} da_{i_1\ldots i_r} \wedge dx_{i_1} \wedge \cdots \wedge dx_{i_r} \\ &= \sum_{h,i_1,\cdots,i_r} \frac{\partial a_{i_1\ldots i_r}}{\partial x_h} dx_h \wedge dx_{i_1} \wedge \cdots \wedge dx_{i_r} \end{aligned} \tag{22}$$

where the sums are extended over all permutations of the indices.

For instance, if $\omega = \sum_i a_i\, dx_i$, we have

$$d\omega = \sum_{i=1}^{n} da_i \wedge dx_i = \sum_{i,h=1}^{n} \frac{\partial a_i}{\partial h_h} dx_h \wedge dx_i = \sum_{i<h} \left(\frac{\partial a_i}{\partial x_h} - \frac{\partial a_h}{\partial x_i}\right) dx_h \wedge dx_i. \tag{23}$$

Since $a_{i_1\ldots i_r}$ may be assumed components of a skew-symmetric covariant tensor of order r, the coefficients of $d\omega^{(r)}$ are also the components of a skew-symmetric covariant tensor of order $r+1$ (the so-called rotor, or curl, of $a_{i_1\ldots i_r}$), so that definition (22) is independent of the coordinate system. To complete the definition, we agree that the exterior differential of a function f (considered as a differential form of degree 0) coincides with its ordinary differential (4).

The exterior differential has the following properties:

(a) If $\omega^{(r)}$ and $\omega^{(s)}$ are forms of degrees r, s, respectively, then

$$d(\omega^{(r)} \wedge \omega^{(s)}) = d\omega^{(r)} \wedge \omega^{(s)} + (-1)^r \omega^{(r)} \wedge d\omega^{(s)}. \tag{24}$$

The proof is an easy consequence of definition (22).

(b) For any ω,

$$d(d\omega) = 0. \tag{25}$$

Proof. Assuming $\omega = \omega^{(r)}$, we have

$$\begin{aligned} d(d\omega) &= \sum_{h,k,i_1,\cdots,i_r} \frac{\partial^2 a_{i_1\ldots i_r}}{\partial x_h\, \partial x_k} dx_k \wedge dx_h \wedge dx_{i_1} \wedge \cdots \wedge dx_{i_r} \\ &= \sum_{h<k,i_1,\cdots,i_r} \left(\frac{\partial^2 a_{i_1\ldots i_r}}{\partial x_h\, \partial x_k} - \frac{\partial^2 a_{i_1\ldots i_r}}{\partial x_k\, \partial x_h}\right) dx_k \wedge dx_h \wedge dx_{i_1} \wedge \cdots \wedge dx_{i_r} \\ &= 0, \end{aligned}$$

proving (25).

We can prove that the exterior differential operator d is uniquely determined by the condition of linearity $d(\omega + \phi) = d\omega + d\phi$ and properties (24), (25), and (4). This proves again that the exterior differential does not depend on the coordinate system (see, e.g., [628]).

ISBN 0-201-13500-0

A differential form ω is called *closed* if $d\omega = 0$. It is called *exact* if $\omega = d\phi$ for some differential form ϕ.

According to (25), every exact differential form is closed. The converse is true only locally. For instance, the form $\omega = (x_1^2 + x_2^2)^{-1}(x_1\,dx_2 - x_2\,dx_1)$ in $R^2 - \{0\}$ satisfies $d\omega = 0$ but $\omega = d\phi$ only locally. In general we have the following result, which is called Poincaré's lemma:

Let ω be an r-form $(1 \leqslant r \leqslant n)$ defined on a star-shaped open set of R^n with $d\omega = 0$. Then there exists an $(r-1)$-form ϕ such that $\omega = d\phi$.

For a proof see Sternberg [628] or H. Cartan [91].

Recall that a set U is called star-shaped if there is a point $x \in U$ such that for every $y \in U$ the line segment joining x and y is contained in U. The star-shapedness is only a sufficient condition on U that every closed form be exact. A necessary and sufficient condition has been given by Whitney [723].

4. Stokes' Formula

Let ω be an $(r-1)$-form of compact support on an oriented r-dimensional manifold V_r in R^n and let ∂V_r denote the boundary of V_r considered as an oriented $(r-1)$-dimensional manifold with the orientation induced by that of V_r. Then

$$\int_{\partial V_r} \omega = \int_{V_r} d\omega. \tag{26}$$

This is the so-called Stokes formula. For a proof see, for instance, [628, 201, 723]. Note that for $r = 1$ it is just the fundamental theorem of the calculus.

Particular cases. 1. The case in which $r = 2$, $n = 2$. In this case V_2 is a closed region in R^2 bounded by a smooth simple closed curve ∂V_2. Let $\omega = a_1\,dx_1 + a_2\,dx_2$. Then formula (26) becomes

$$\int_{\partial V_2} a_1\,dx_1 + a_2\,dx_2 = \int_{V_2} \left(\frac{\partial a_2}{\partial x_1} - \frac{\partial a_1}{\partial x_2}\right) dx_1 \wedge dx_2, \tag{27}$$

which is the classical Green theorem.

2. The case in which $r = 2$, $n = 3$. In this case V_2 is a smooth oriented surface in R^3 whose boundary ∂V_2 is a smooth simple closed curve directed in accordance with the given orientation in V_2. Let $\omega = a_1\,dx_1 + a_2\,dx_2 + a_3\,dx_3$. Then

$$\int_{\partial V_2} a_1\,dx_1 + a_2\,dx_2 + a_3\,dx_3$$

$$= \int_{V_2} \left(\frac{\partial a_3}{\partial x_2} - \frac{\partial a_2}{\partial x_3}\right) dx_2 \wedge dx_3 + \left(\frac{\partial a_1}{\partial x_3} - \frac{\partial a_3}{\partial x_1}\right) dx_3 \wedge dx_1 + \left(\frac{\partial a_2}{\partial x_1} - \frac{\partial a_1}{\partial x_2}\right) dx_1 \wedge dx_2,$$

which is often called Ostrogradsky's theorem, or the curl theorem.

ISBN 0-201-13500-0

3. The case in which $r = 3$, $n = 3$. Let V_3 be a closed region in R^3 bounded by a smooth surface ∂V_3. Let $\omega = a_1\, dx_2 \wedge dx_3 + a_2\, dx_3 \wedge dx_1 + a_3\, dx_1 \wedge dx_2$. Then

$$\int_{\partial V_3} a_1\, dx_2 \wedge dx_3 + a_2\, dx_3 \wedge dx_1 + a_3\, dx_1 \wedge dx_2$$

$$= \int_{V_3} \left(\frac{\partial a_1}{\partial x_1} + \frac{\partial a_2}{\partial x_2} + \frac{\partial a_3}{\partial x_3}\right) dx_1 \wedge dx_2 \wedge dx_3,$$

which is the so-called Gauss theorem.

4. Integration by parts. If V_r is a compact manifold without boundary ($\partial V_r = \theta$) and $\omega = \omega^{(p)} \wedge \omega^{(q)}$ $(p + q = r - 1)$, then from (24) and (26) we have

$$\int_{V_r} d\omega^{(p)} \wedge \omega^{(q)} = (-1)^{p+1} \int_{V_r} \omega^{(p)} \wedge d\omega^{(q)},$$

which is a generalization of integration by parts.

5. Comparison with Vector Calculus in Euclidean Three-Dimensional Space

Consider euclidean three-dimensional space and the standard orthonormal frame $\mathbf{e}_1, \mathbf{e}_2, \mathbf{e}_3$ such that $\mathbf{e}_i \cdot \mathbf{e}_j = \delta_{ij}$. Each 1-form $\omega^{(1)} = a_1\, dx_1 + a_2\, dx_2 + a_3\, dx_3$ defines a field of vectors $\mathbf{A} = a_1 \mathbf{e}_1 + a_2 \mathbf{e}_2 + a_3 \mathbf{e}_3$ and each 2-form $\omega^{(2)} = a_{12}\, dx_1 \wedge dx_2 + a_{23}\, dx_2 \wedge dx_3 + a_{31}\, dx_3 \wedge dx_1$, with $a_{ij} = -a_{ji}$, defines a field of skew-symmetric covariant tensors of second order a_{ij}, or a field of axial or relative vectors $\mathfrak{A} = a_{23}\mathbf{e}_1 + a_{31}\mathbf{e}_2 + a_{12}\mathbf{e}_3$. Remember that axial vectors or relative vectors are geometric objects that transform like vectors except that they change direction when the "handedness" of the original frame is changed. For instance, the vector product $\mathbf{A} \times \mathbf{B}$ of two vectors is an axial vector. Analogously, a 3-form $\omega^{(3)} = a\, dx_1 \wedge dx_2 \wedge dx_3$ defines a relative invariant $\mathfrak{a} = a$.

Consider the differential forms

$$\omega^{(1)} = a_1\, dx_1 + a_2\, dx_2 + a_3\, dx_3, \qquad \phi^{(1)} = b_1\, dx_1 + b_2\, dx_2 + b_3\, dx_3,$$

$$\omega^{(2)} = c_1\, dx_2 \wedge dx_3 + c_2\, dx_3 \wedge dx_1 + c_3\, dx_1 \wedge dx_2, \quad \omega^{(3)} = a\, dx_1 \wedge dx_2 \wedge dx_3,$$

and let ψ denote the operator that associates to each differential form the corresponding vector element, that is,

$$\psi(\omega^{(1)}) = \mathbf{A}, \qquad \psi(\omega^{(2)}) = \mathfrak{C}, \qquad \psi(\omega^{(3)}) = \mathfrak{a},$$

$$\psi(\omega^{(1)} \wedge \phi^{(1)}) = \mathbf{A} \times \mathbf{B}, \qquad \psi(\omega^{(1)} \wedge \omega^{(2)}) = \mathbf{A} \cdot \mathfrak{C}.$$

ISBN 0-201-13500-0

According to the definitions

$$\operatorname{grad} f = \frac{\partial f}{\partial x_1}\mathbf{e}_1 + \frac{\partial f}{\partial x_2}\mathbf{e}_2 + \frac{\partial f}{\partial x_3}\mathbf{e}_3,$$

$$\operatorname{curl} \mathbf{A} = \left(\frac{\partial a_3}{\partial x_2} - \frac{\partial a_2}{\partial x_3}\right)\mathbf{e}_1 + \left(\frac{\partial a_1}{\partial x_3} - \frac{\partial a_3}{\partial x_1}\right)\mathbf{e}_2 + \left(\frac{\partial a_2}{\partial x_1} - \frac{\partial a_1}{\partial x_2}\right)\mathbf{e}_3,$$

$$\operatorname{div} \mathbf{A} = \frac{\partial a_1}{\partial x_1} + \frac{\partial a_2}{\partial x_2} + \frac{\partial a_3}{\partial x_3},$$

we have

$$\psi(df) = \operatorname{grad} f, \qquad \psi(d\omega^{(1)}) = \operatorname{curl} \mathbf{A}, \qquad \psi(d\omega^{(2)}) = \operatorname{div} \mathfrak{C}.$$

Applying these relations, we obtain the following correspondence between identities of exterior calculus and identities of vector calculus:

$$d(df) = 0 \to \operatorname{curl} \operatorname{grad} f = 0,$$

$$d(d\omega^{(1)}) = 0 \to \operatorname{div} \operatorname{curl} \mathbf{A} = 0,$$

$$d(fg) = df \cdot g + f\,dg \to \operatorname{grad}(fg) = g \operatorname{grad} f + f \operatorname{grad} g,$$

$$d(f\omega^{(1)}) = df \wedge \omega^{(1)} + f\,d\omega^{(1)} \to \operatorname{curl}(f\mathbf{A}) = (\operatorname{grad} f) \times \mathbf{A} + f \operatorname{curl} \mathbf{A},$$

$$d(f\omega^{(2)}) = df \wedge \omega^{(2)} + f\,d\omega^{(2)} \to \operatorname{div}(f\mathfrak{C}) = \operatorname{grad} f \cdot \mathfrak{C} + f \operatorname{div} \mathfrak{C},$$

$$d(\omega^{(1)} \wedge \phi^{(1)}) = d\omega^{(1)} \wedge \phi^{(1)} - \omega^{(1)} \wedge d\phi^{(1)}$$

$$\to \operatorname{div}(\mathbf{A} \times \mathbf{B}) = (\operatorname{curl} \mathbf{A}) \cdot \mathbf{B} - \mathbf{A} \cdot \operatorname{curl} \mathbf{B}.$$

Poincaré's lemma translates into the following theorems of vector calculus.

(a) For $\omega = \omega^{(1)}$ we have *the condition* curl $\mathbf{A} = 0$ *is the necessary and sufficient condition for the existence of a function* f *such that* $\mathbf{A} = \operatorname{grad} f$ (*assuming that the region for which* curl $\mathbf{A} = 0$ *is star shaped*).

(b) For $\omega = \omega^{(2)}$ we have *if* $\mathfrak{A}$ *is a field of axial vectors, then the condition* div $\mathfrak{A} = 0$ *is the necessary and sufficient condition for the existence of a vector field* $\mathbf{B}$ *such that* $\mathfrak{A} = \operatorname{curl} \mathbf{B}$ (*assuming that the region for which* div $\mathfrak{A} = 0$ *is star shaped*).

Thus, we see that the exterior calculus can serve to represent and to unify the basic operations of classical vector calculus in euclidean three-dimensional space.

6. Differential Forms over Manifolds

We have seen how differential forms transform under the change of coordinates $x_i \to x_i'$. It suffices to replace each coefficient $a_{i_1 \ldots i_r}(x)$ by the composite function $a_{i_1 \ldots i_r}(x(x'))$ and replace each dx_i by the differential $dx_i = \sum_h (\partial x_i / \partial x_h')\, dx_h'$, then use the exterior product of these 1-forms. Because of

ISBN 0-201-13500-0

this simple transformation rule under mappings, the differential forms work on spaces that are defined by means of local coordinate systems, for instance, on differentiable manifolds.

On such a manifold M, local charts (U_i, h_i) exist such that the open sets U_i cover the whole manifold and the h_i are one-to-one maps of each U_i onto open sets of R^n. Differential forms over U_i are then trivial transforms of differential forms over the corresponding open sets of R^n. Moreover, it is possible to consider a form ω as existing over the whole manifold if its representations over U_i and U_j in the intersection $U_i \cap U_j$ are related by the change of coordinates $h_i \circ h_j^{-1}$.

The integration of differential forms on manifolds has the conventional meaning if the domain of integration is contained in some U_i. Otherwise, it is necessary to use the so-called "partition of unity" (see, e.g., the work of Sternberg [628] and Fleming [204]). With this device, all exterior calculus, including Stokes' formula, translates without change from R^n to any differentiable manifold M.

We conclude this Appendix with the definition of the rank of a differential form and the statement, without proof, of some theorems related to the problem of finding a normal form for linear differential forms.

An r-form ω over an n-dimensional manifold is called of rank p $(r \leqslant p \leqslant n)$ if there exist p and no less than p 1-forms ϕ_h such that

$$\omega = \sum \phi_{i_1} \wedge \cdots \wedge \phi_{i_r} \qquad (1 \leqslant i_1 < \cdots < i_r \leqslant p).$$

If $r = p$, then $\omega = \phi_1 \wedge \cdots \wedge \phi_r$ and the form is said to be decomposable.

We mention the following results:

1. The rank of a 2-form $\omega^{(2)}$ is an even integer $p = 2s$ where s is the greatest integer such that $\omega^{(2)} \wedge \omega^{(2)} \wedge \cdots \wedge \omega^{(2)} \neq 0$ (s factors).

2. If $\omega^{(2)}$ is a closed 2-form of rank $p = 2s$ on an n-dimensional manifold M, then about every point of M we can introduce coordinates $x_1, x_2, \ldots, x_s, y_1, \ldots, y_s$ such that in terms of these coordinates $\omega^{(2)} = dx_1 \wedge dy_1 + dx_2 \wedge dy_2 + \cdots + dx_s \wedge dy_s$.

3. Let ω be a 1-form such that $d\omega$ has rank p on M. If $\omega \neq 0$ and $\omega \wedge d\omega \wedge \cdots \wedge d\omega = 0$ (with s factors $d\omega$), then about every point of M we can find coordinates $x_1, x_2, \ldots, x_s, y_1, \ldots, y_s$ such that $\omega = x_1\, dy_1 + \cdots + x_s\, dy_s$. If $\omega \wedge d\omega \wedge \cdots \wedge d\omega \neq 0$ (with s factors $d\omega$) everywhere, about every point of M we can introduce coordinates $x_1, \ldots, x_s, x_{s+1}, y_1, \ldots, y_s$ so that $\omega = x_1\, dy_1 + \cdots + x_s\, dy_s + dx_{s+1}$. This property is known as Darboux's theorem [628].

ISBN 0-201-13500-0

Bibliography and References

This section lists books and articles on topics closely related to the content of the book. An extensive list of related publications may also be found in each of the following references: [51, 63, 152, 271, 274, 335, 367, 392, 428, 429, 647].

1. L. Ahlfors, "The theory of meromorphic curves," *Acta Soc. Sci. Fenn.* **A3** (1941), 1–31.
2. G. Ailam, "Moments of coverage and coverage spaces," *J. Appl. Probability* **3** (1966), 550–555.
3. H. A. Alikoski, "Über das Sylvestersche Vierpunktproblem," *Ann. Acad. Sci. Fenn.* **51** (1939), no. 7, 1–10.
4. C. B. Allendoerfer, "Steiner's formulae on a general S^n," *Bull. Amer. Math. Soc.* **54** (1948), 128–135.
5. C. B. Allendoerfer and A. Weil, "The Gauss-Bonnet theorem for Riemannian polyhedra," *Trans. Amer. Math. Soc.* **53** (1943), 101–129.
6. R. V. Ambarcumjan, "Intersections of complex curves and random straight lines," *Soviet Math. Dokl.* **10** (1969), 865–868. (Translated from *Dokl. Akad. Nauk SSSR* **187** (1969), no. 3, 487–489.)
7. R. V. Ambarcumjan, "Invariant imbedding in the theory of random lines," *Izv. Akad. Nauk Armjan. SSR* **5** (1970), 167–206.
8. R. V. Ambarcumjan, "On random plane mosaics," *Soviet Math. Dokl.* **12** (Pt. 2) (1971), 1349–1353.
9. R. V. Ambarcumjan, "Convex polygons and random tessellations," in *Stochastic Geometry* (E. F. Harding and D. G. Kendall, eds.), pp. 176–191. Wiley, New York, 1974.
10. R. V. Ambarcumjan, "Probability distributions in the geometry of clusters," *Studia Sci. Math. Hungar.* **6** (1971), 235–241.
11. R. V. Ambarcumjan, "On the solution of the Buffon–Sylvester problem in R^3," *Dokl. Akad. Nauk SSSR* **210** (1973), 1257–1260.
12. P. Armitage, "An overlap problem arising in particle counting," *Biometrika* **36** (1949), 257–266.
13. V. I. Arnold and A. Avez, *Problèmes ergodiques de la mécanique classique*. Gauthier-Villars, Paris, 1967.
14. L. Auslander, *Differential Geometry*. Harper & Row, New York, 1966.
15. F. Azorin, "Reconstruccion de patrones por muestreo sistemático espacial," *Estadist. Espanola* **57** (1971), 5–43.
16. F. Azorin, "Comparacion de supuestos y disenos en la estimacion de patrones por muestreo sistemático pluridimensional," *Trabajos Estadist.* **21** (1970), 3–23.
17. M. Balanzat, "Généralisation de quelques formules de géométrie intégrale," *C. R. Acad. Sci. Paris* **210** (1940), 596–598.
18. M. Balanzat, "Sur quelques formules de la géométrie intégrale des ensembles dans un espace à n dimensions, *Portugal. Math.* **3** (1940), 87–94.
19. R. P. Bambah, "Polar reciprocal convex bodies," *Proc. Cambridge Phil. Soc.* **51** (1954), 377–378.
20. T. F. Banchoff and W. F. Pohl, "On a generalization of the isoperimetric inequality,"

ISBN 0-201-13500-0

J. Diff. Geometry **6** (1971), 175–192.

21. E. Barbier, "Note sur le problème de l'aiguille et le jeu du joint couvert," *J. Math. Pures Appl.* (2) **5** (1860), 273–286.
22. A. E. Bates and M. E. Pillow, "Mean free path of sound in auditorium," *Proc. Phys. Soc.* **59** (1957), 535–541.
23. M. Baticle, "Le problème des répartitions," *C. R. Acad. Sci. Paris* **201** (1935), 862–864.
24. G. Baxter, "A combinatorial lemma for convex numbers," *Ann. Math. Statist.* **32** (1961), 901–904.
25. E. A. Bender, "Area–perimeter relations for two-dimensional lattices," *Amer. Math. Monthly* **69** (1962), 742–744.
26. D. C. Benson, "Sharpened form of the plane isoperimetric inequality," *Amer. Math. Monthly* **77** (1970), 29–34.
27. R. V. Benson, *Euclidean Geometry and Convexity*. McGraw-Hill, New York, 1966.
28. J. D. Bernal, "A geometrical approach to the structure of liquids," *Nature* **183** (1959), 141–147; **185** (1960), 68–70.
29. J. D. Bernal, J. Masson, and K. F. Knight, "Radial distribution of the random close packing of spheres," *Nature* **189** (1962), 956–958.
30. L. Berwald and O. Varga, [1] Über die Schiebung im Raum, *Math. Z.* **42** (1937), 710–736.

30a. L. Bieberbach, "Über eine Extremaleigenschaft des Kreises," *Jber. Deutsch. Math. Verein* **24** (1915), 247–250.

31. G. D. Birkhoff, *Dynamical Systems*. Amer. Math. Soc., Providence, R.I., 1927.
32. R. L. Bishop and S. I. Goldberg, *Tensor Analysis on Manifolds*. Macmillan, New York, 1968.

32a. B. E. Blaisdell and H. Solomon, "On random sequential packing in the plane and a conjecture of Palasti," *J. Appl. Probability* **7** (1970), 667–698.

33. W. Blaschke, *Vorlesungen über Differentialgeometrie*, Vol 2 (*Affine Differentialgeometrie*). Springer, Berlin, 1923.
34. W. Blaschke, "Einige Bemerkungen über Kurven und Flächen konstanter Breite," *Ber. Verh. Sächs. Akad. Wiss. Leipzig. Math. Natur. Kl.* **67** (1915), 290–297.
35. W. Blaschke, "Eine isoperimetrische Eigenschaft des Kreises," *Math. Z.* **1** (1918), 52–57.
36. W. Blaschke, "Integralgeometrie 1: Ermittlung der Dichten für lineare Unterräume im E_n," *Actualités Sci. Indust.* **252** (1935).
37. W. Blaschke, "Integralgeometrie 2: Zu Ergebnissen von M. W. Crofton," *Bull. Math. Soc. Roumaine Sci.* **37** (1935), 3–11.
38. W. Blaschke, "Integralgeometrie 10: Eine isoperimetrische Eigenschaft der Kugel," *Bull. Math. Soc. Roumaine Sci.* **37** (2) (1936), 3–7.
39. W. Blaschke, "Integralgeometrie 11: Zur Variationsrechnung," *Abh. Math. Sem. Univ. Hamburg* **11** (1936), 359–366.
40. W. Blaschke, "Integralgeometrie 12: Vollkommene optische Instrumente," *Abh. Math. Sem. Univ. Hamburg* **11** (1936), 409–412.
41. W. Blaschke, "Integralgeometrie 13: Zur Kinematik," *Math. Z.* **41** (1936), 465–478.
42. W. Blaschke, "Integralgeometrie 17: Über Kinematik," *Deltion* (Athens) **1936**, 3–14.
43. W. Blaschke, "Integralgeometrie 20: Zur elliptischen Geometrie," *Math. Z.* **41** (1936), 785–786.
44. W. Blaschke, "Integralgeometrie 21: Über Schiebungen," *Math. Z.* **42** (1937), 1–12.
45. W. Blaschke, "Integralgeometrie 22: Geschlossene Kurven und Flächen in der

ISBN 0-201-13500-0

elliptischen Geometrie," *Abh. Math. Sem. Univ. Hamburg* **12** (1937), 111–113.
46. W. Blaschke, "Sulla proprietà isoperimetrica del cerchio," *Rend. Sem. Mat. Univ. Roma IV* **1** (1937), 233–234.
47. W. Blaschke, "Densita negli spazi di Hermite," *Atti Accad. Naz. Lincei Rend. Cl. Sci. Fis. Mat. Natur. Ser. VI* **29** (1939), 105–108.
48. W. Blaschke, "Über Integrale in der Kinematik," *Arch. Math. (Basel)* **1** (1948), 18–22.
49. W. Blaschke, "Zur Integralgeometrie," *Rend. Circ. Mat. Palermo II* **1** (1952), 108–110.
50. W. Blaschke, *Kreis und Kugel*, 2nd ed. W. de Gruyter, Berlin, 1956.
51. W. Blaschke, *Vorlesungen über Integralgeometrie*, 3rd ed. Deutsch. Verlag Wiss., Berlin, 1955.
52. W. Blaschke and O. Varga, "Integralgeometrie 9: Über Mittelwerte an Eikörpern," *Mathematica (Cluj)* **12** (1936), 65–80.
53. H. F. Blichfeldt, "A new principle in the geometry of numbers with some applications," *Trans. Amer. Math. Soc.* **15** (1914), 227–235.
54. J. Bodziony, "On the probability of application of integral geometry methods in certain problems of liberation of mineral grains," *Bull. Acad. Polon. Sci.* Sèr. Sc. Techniques, **13** (1965), 459–467.
55. J. Bodziony, "On certain indices characterizing the geometric structure of rocks," *Bull. Acad. Polon. Sci.* Sèr. Sc. Techniques, **13** (1965), 469–475.
56. J. Bodziony, "On the liberation of mineral grains," *Bull. Acad. Polon. Sci.* Sèr. Sc. Techniques, **13** (1965), 513–518.
57. J. Bohm, "Untersuchung des Simplex-Inhaltes im Raum konstanter Krümmung beliebiger Dimensionen," *J. Reine Angew. Math. (Crelle)* **202** (1959), 16–51.
58. J. Bohm, "Über Inhaltsmessung im Raum konstanter Krümmung," *Wiss. Z. Friedrich-Schiller-Univ. Jena/Thüringen* **10** (1960–1961), 29–32.
59. J. Bohm, "Inhaltsmessung im R_5 konstanter Krümmung," *Arch. Math. (Basel)* **11** (1960), 298–309.
60. J. Bokowski, H. Hadwiger, and J. M. Wills, "Eine Ungleichung zwischen Volumen, Oberfläche und Gitterpunktzahl konvexer Körper im n-dimensionalen euklidischen Raum," *Math. Z.* **127** (1972), 363–364.
61. G. Bol, "Zur kinematischen Ordnung ebener Jordan-Kurven," *Abh. Math. Sem. Univ. Hamburg* **11** (1936), 394–408.
62. T. Bonnesen, *Les problèmes des isopérimètres et des isépiphanes.* Gauthier-Villars, Paris, 1920.
63. T. Bonnesen and W. Fenchel, *Theorie der konvexen Körper.* Ergeb. Math., Springer, Berlin, 1934.
64. E. Borel, *Principes et formules classiques du calcul des probabilités.* Gauthier-Villars, Paris, 1925.
65. V. A. Borovikov, "Fundamental solutions of partial differential equationt with constant coefficients," *Trans. Moscow Math. Soc.* **8** (1959), 199–257.
66. O. Bottema, "Eine obere Grenze für das isoperimetrische Defizit ebener Kurven," *Nederl. Akad. Wetensch. Proc.* **A36** (1933), 442–446.
67. J. L. Boursin, "Sur quelques problèmes de géométrie aléatoire," *Ann. Fac. Sci. Univ. Toulouse* (4), **28** (1965), 9–99.
68. N. Bourbaki, *General Topology*, Pt. I. Addison-Wesley, Reading, Mass., 1966.
69. N. Bourbaki, *Éléments de Mathématiques*, Livre VI (*Integration*), Chap. 7. Hermann, Paris, 1963.
70. L. G. Briarty, "Stereology methods for quantitative light and electron microscopy,"

ISBN 0-201-13500-0

Sci. Progr. **62** (1975), 1–32.
71. J. Bronowski and J. Neyman, "The variance of the measure of a two-dimensional random set," *Ann. Math. Statist.* **16** (1945), 330–341.
72. B. Brosowski, "Über die konvexe Hülle von zufälligen Pfaden," *Z. Angew. Math. Mech.* **43** (1963), Supplement, T. 40, T. 41.
73. J. E. Brothers, "Integral geometry in homogeneous spaces," *Trans. Amer. Math. Soc.* **124** (1966), 480–517.
74. H. Busemann, *Convex Surfaces.* Wiley (Interscience), New York, 1958.
75. H. Busemann, "Volume in terms of concurrent cross-sections," *Pacific J. Math.* **3** (1953), 1–12.
76. H. Busemann and P. J. Kelly, *Projective Geometry and Projective Metrics.* Academic Press, New York, 1953.
76a. J. W. Cahn, "The generation and characterization of shape," *Advances in Appl. Probability*, Spec. Suppl. (*Proc. Symp. Statist. Probabilistic Problems in Metallurgy, Seattle, 1971*) (1972), 221–242.
77. G. Calugareanu, "L'intégrale de Gauss et l'analyse des noeuds tridimensionnels," *Rev. Roumaine Math. Pures Appl.* **4** (1959), 5–20.
78. G. Calugareanu, "Sur les classes d'isotopie des noeuds tridimensionnels et leurs invariants," *Czechoslovak Math. J.* **11** (1961), 588–624.
79. G. Calugareanu, "Sur les enlacements tridimensionnels des courbes," *Comm. Acad. R. P. Roumaine* **11** (1961), 829–832.
80. I. P. Caregradskii, "A certain problem of search by networks," *Teor. Verojatnost. i Primenen.* **15** (1970), 326–330. (English transl: *Theor. Probability Appl.* **15** (1970), 315–319.)
81. T. Carleman, "Über eine isoperimetrische Aufgabe und ihre physikalischen Anwendungen," *Math. Z.* **3** (1916), 1–7.
82. J. A. Carlson, "A remark on the transcendental Bezout problem," in *Value Distribution Theory* (R. O. Kujala and A. L. Willer III, *eds.*), Pt. A, pp. 133–143. Dekker, New York, 1974.
83. J. A. Carlson and P. A. Griffiths, "The order function for entire holomorphic mappings," in *Value Distribution Theory* (R. O. Kujala and A. L. Willer III, *eds.*), Pt. A, pp. 225–248. Dekker, New York, 1974.
84. H. Carnal, "Die konvexe Hülle von n rotationssymmetrisch verteilten Punkten," *Z. Wahrscheinlichkeits.* **15** (1970), 168–176.
85. E. Cartan, "Le principe de dualité et certaines intégrales multiples de l'espace tangentiel et de l'espace reglé," *Bull. Soc. Math. France* **24** (1896), 140–177. (*Oeuvres Complètes*, Pt. II, Vol. 1, pp. 265–302.) Gauthier-Villars, Paris, 1952.
86. E. Cartan, "Sur les invariants intégraux de certains espaces homogènes clos et les propriétés topologiques de ces espaces," *Ann. Soc. Polon. Math.* **8** (1929), 181–225. (*Oeuvres Complètes*, Pt. II, Vol. 1, pp. 265–302.) Gauthier-Villars, Paris, 1952.
87. E. Cartan, *Leçons sur les invariants intégraux.* Hermann, Paris, 1922.
88. E. Cartan, *La théorie des groupes finis et continus et la géométrie différentielle traités par la méthode du repère mobile.* Gauthier-Villars, Paris, 1937.
89. E. Cartan, *Leçons sur la Géométrie Projective Complexe.* Gauthier-Villars, Paris, 1931.
90. E. Cartan, *Les systèmes différentiels extérieurs et leurs applications géométriques, Actualités Sci. Indust.* **994** (1945).
91. H. Cartan, *Formes différentielles.* Hermann, Paris, 1967.
92. J. W. S. Cassels, *An Introduction to the Geometry of Numbers.* Springer, Berlin, 1959.

ISBN 0-201-13500-0

93. A. Cauchy, "Mémoire sur la rectification des courbes et la quadrature des surfaces courbes," *Mem. Acad. Sci. Paris* **22** (1850), 3–15.
94. C. B. S. Cavallin, "Question 6571," *Educational Times* **34** (1881), 23.
95. G. D. Chakerian, "Integral geometry in the Minkowski plane," *Duke Math. J.* **29** (1962), 375–381.
96. G. D. Chakerian, "An inequality for closed space curves," *Pacific J. Math.* **12** (1962), 53–57.
97. G. D. Chakerian, "On some geometric inequalities," *Proc. Amer. Math. Soc.* **15** (1964), 886–888.
98. G. D. Chakerian, "Isoperimetric inequalities for the mean width of a convex body," *Geometria Dedicata* **1** (1973), 356–362.
99. G. D. Chakerian, "Higher-dimensional analogues of an isoperimetric inequality of Benson," *Math. Nachr.* **48** (1971), 33–41.
100. G. D. Chakerian and M. S. Klamkin, "Minimal covers for closed curves," *Math. Mag.* **46** (1973), 55–61.
101. G. D. Chakerian and S. K. Stein, "Bisected chords of a convex body," *Arch. Math. (Basel)* **17** (1966), 561–565.
102. N. G. Chebotarev, "The determination of volume in Lie groups," *Zap. Mat. Otdel. Fiz.-Mat. Fak. Kharkov Mat. Obsc.* **14** (1937), 3–20.
103. C.-S. Chen, "On the kinematic formula of square of mean curvature," *Indiana Univ. Math. J.* **22** (1972–1973), 1163–1169.
104. S. S. Chern, "Sur les invariants intégraux en géométrie," *Sci. Repts. Natl. Tsing-Hua Univ.* **A4** (1940), 85–95.
105. S. S. Chern, "On integral geometry in Klein spaces," *Ann. of Math.* (2) **43** (1942), 178–189.
106. S. S. Chern, "A simple intrinsic proof of the Gauss–Bonnet theorem for closed riemannian manifolds," *Ann. of Math.* (2) **45** (1944), 747–752.
107. S. S. Chern, "On the curvature integral of a riemannian manifold," *Ann. of Math.* (2) **46** (1945), 674–684.
108. S. S. Chern, "On the kinematic formula in the euclidean space of n dimensions," *Amer. J. Math.* **74** (1952), 227–236.
109. S. S. Chern, "Differential geometry and integral geometry," *Proc. Internat. Congr. Math., Edinburgh,* **1958** pp. 441–449. Cambridge Univ. Press, New York, 1960.
110. S. S. Chern, "Geometry of submanifolds in a complex projective space," *Symp. Internat. Top. Alg., México,* **1958** pp. 87–96.
111. S. S. Chern, "Holomorphic mappings of complex manifolds," *Enseignement Math.* **7** (1961), 179–187.
112. S. S. Chern, "On the kinematic formula in integral geometry," *J. Math. and Mech.* **16** (1966), 101–118.
113. S. S. Chern, "Curves and surfaces in euclidean space," in *Studies in Global Geometry and Analysis*, pp. 17–56. The Mathematical Association of America, 1967.
114. S. S. Chern, "The integrated form of the first main theorem for complex analytic mappings in several complex variables," *Ann. of Math.* (2) **71** (1960), 536–551.
115. S. S. Chern, "Holomorphic curves in the plane," in *Differential Geometry in Honour of K. Yano*, pp. 13–94. Kinokuniya, Tokyo, 1972.
116. S. S. Chern and R. K. Lashof, "On the total curvature of immersed manifolds, I," *Amer. J. Math.* **79** (1957), 306–318; II, *Michigan Math. J.* **5** (1958), 5–12.
117. S. S. Chern and C. T. Yien, "Sulla formula principale cinematica dello spazio ad n dimensioni," *Boll. Un. Mat. Ital.* (2) **2** (1940), 434–437.
118. C. Chevalley, *Theory of Lie Groups*. Princeton Univ. Press, Princeton, N.J., 1946.

ISBN 0-201-13500-0

119. R. Coleman, "Random paths through convex bodies," *J. Appl. Probability* **6** (1969), 430–441.
120. R. Coleman, "Sampling procedures for the lengths of random straight lines," *Biometrika* **59** (1972), 415–426.
121. R. Coleman, "The distance from a given point to the nearest end of one member of a random process of linear segments," in *Stochastic Geometry* (E. F. Harding and D. G. Kendall, eds.), pp. 192–201. Wiley, New York, 1974.
122. P. J. Cooke, Sequential Coverage in Geometrical Probability, Ph.D. dissertation, Dept. of Statist., Stanford Univ., 1971.
123. P. J. Cooke, "Bounds for coverage probabilities with applications to sequential coverage problems," *J. Appl. Probability* **11** (1974), 281–293.
124. H. Corte, "On the distribution of mean density in paper, I, II," *Papier* **23** (1969), 381–393; **24** (1970), 261–271.
125. R. Courant and D. Hilbert, *Methods of Mathematical Physics*, Vol. 1. Wiley (Interscience), New York, 1953.
126. T. Cover and B. Efron, "Geometrical probability and random points in a hypersphere," *Ann. Math. Statist.* **38** (1967), 213–220.
127. H. S. M. Coxeter, *Introduction to Geometry*. Wiley, New York, 1961.
128. H. S. M. Coxeter, *Non-Euclidean Geometry*, 3rd ed. Univ. Toronto Press, Toronto, 1957.
129. H. S. M. Coxeter, "Regular compound tessellations of the hyperbolic plane," *Proc. Roy. Acad.* **A278** (1964), 147–167.
130. I. K. Crain, "Monte Carlo simulation of random Voronoi polygons; preliminary results," *Search* **3** (1972), 220–221.
131. A. Creanga-Samboan, "Random sets, random coverings," *Stud. Cerc. Mat.* **23** (1971), 845–851.
132. M. W. Crofton, "On the theory of local probability," *Phil. Trans. Roy. Soc. London* **158** (1868), 181–199.
133. M. W. Crofton, "Probability," in *Encyclopaedia Britannica*, 9th ed., Vol. 19 (1885), pp. 768–788.
134. E. Czuber, *Geometrische Wahrscheinlichkeiten und Mittelwerte*. Teubner, Leipzig, 1884. (French transl. Hermann, Paris, 1902.)
135. T. Dalenius, J. Hajek, and S. Zubrizcki, "On plane sampling and related geometrical problems," *Proc. 4th Berkeley Symp. Math. Statist. and Probability*, 1961 Vol. 1, pp. 125–150.
136. D. J. Daley, "Various concepts of orderliness for point processes," in *Stochastic Geometry* (E. F. Harding and D. G. Kendall, eds.), pp. 148–161. Wiley, New York, 1974.
137. H. E. Daniels, "The covering circle of a sample from a circular distribution," *Biometrika* **39** (1952), 137–143.
138. L. Danzer, B. Grünbaum, and V. Klee, "Helly's theorem and its relatives," in *Convexity* (*Proc. Symp. Pure Math.*, **7**, *Seattle, 1961*), pp. 101–180. Amer. Math. Soc., Providence, R.I., 1963.
139. D. A. Darling, "On a problem of Rényi," *Period. Math. Hungar.* **3** (1973), 5–7.
140. R. Davidson, "Construction of line processes: second-order properties," in *Stochastic Geometry* (E. F. Harding and D. G. Kenda,l, eds.), pp. 55–75. Wiley New York, 1974. (This book contains all of Davidson's published and a great part of his unpublished work on stochastic line processes.)
141. N. G. De Bruin, "Asymptotic distribution of lattice points in a rectangle," *SIAM Rev.* **7** (1965), 274–275.

ISBN 0-201-13500-0

141a. R. T. De Hoff, "The determination of the size distribution of ellipsoidal particles from measurements made in random plane sections," *Trans. Met. Soc. AIME* **224** (1962), 474.

141b. R. T. De Hoff, "The estimation of particle size distribution from simple counting measurements made on random plane sections," *Trans. Met. Soc. AIME* **233** (1965), 25–29.

141c. R. T. De Hoff, "The quantitative estimation of mean surface curvature," *Trans. Met. Soc. AIME* **239** (1967), 617–621.

141d. R. T. De Hoff, "The evolution of particulate structures," Spec. Suppl. (*Proc. Symp. Statist. Probabilistic Problems in Metallurgy, Seattle, 1971*), *Advances in Appl. Probability* (1972), 188–198.

141e. R. T. De Hoff and P. Bousquet, "Estimation of the size distribution of triaxial ellipsoidal particles from the distribution of linear intercepts," *J. Microscopy* **92** (1970), 119–135.

142. R. T. De Hoff and F. N. Rhines, "Determination of number of particles per unit volume," *Trans. Met. Soc. AIME* **221** (1961), 975–982.

143. R. Deltheil, "Sur la théorie des probabilités géométriques," *Ann. Fac. Sci. Univ. Toulouse* (3) **11** (1919), 1–65.

144. R. Deltheil, *Probabilités géométriques*. Gauthier-Villars, Paris, 1926.

145. G. De Rahm, "Sur un procédé de formation d'invariants intégraux," *Jber. Deutsch. Math. Verein* **49** (1939), 156–161.

145a. H. L. De Vries, "Über Koeffizientenprobleme bei Eilinien und über die Heinzsche Konstanz," *Math. Z.* **112** (1969), 101–106.

146. D. Z. Djorovic, "A problem in geometrical probability," *Publ. Fac. Electrotec. Univ. Belgrade, Ser. Mat.-Fis.* **175–179** (1967), 26–33.

147. C. T. J. Dodson, "Statistical analysis of patterns of deformation in flat bounded fibrous networks," *J. Phys.* **D3** (1970), 269–276.

148. C. T. J. Dodson, "Spatial variability and the theory of sampling in random fibrous networks," *J. Roy. Statist. Soc.* **B33** (1971), 88–94.

149. C. Domb, "The problem of random intervals on a line," *Proc. Cambridge Phil. Soc.* **43** (1947), 329–341.

150. C. Domb, "A note on the series expansion method for clustering problems," *Biometrika* **59** (1972), 209–211.

151. R. Dontot, "Sur les invariants intégraux et quelques points d'optique géométrique," *Bull. Soc. Math. France* **42** (1914), 53–91.

152. G. I. Drinfel'd, "Integral geometry," *Progr. Math.* **12** (*Algebra and Geometry*, R. V. Gamkreilidze, ed.; English transl.: N. H. Choksy), pp. 173–215. Plenum Press, New York and London, 1972.

153. G. I. Drinfel'd, "On measure in Lie groups," *Kharkov Gos. Univ. Ucen. Zap.* **29** (1949); *Zap. Mat. Otdel. Fiz.-Mat. Fak. Kharkov Mat. Obsc.* **21** (1949), 47–57.

154. R. J. Duffin, "Some problems of mathematics and science," *Bull. Amer. Math. Soc.* **80** (1974), 1053–1070.

154a. R. J. Duffin, R. A. Meussner, and F. N. Rhines, "Statistics of particle measurement and of particle growth," *Carnegie-Mellon Univ., Technical Report* **32**, April 1953.

155. S. W. Dufour, Intersections of Random Convex Regions, Ph.D. dissertation, Dept. of Statist., Stanford Univ., 1972.

156. A. Dvoretzky and H. Robbins, "On the parking problem," *Publ. Math. Inst. Hungar. Akad. Sci.* **A9** (1964), 209–226.

157. A. Dvoretzky and C. A. Rogers, "Absolute and conditional convergence in normed

ISBN 0-201-13500-0

linear spaces," *Proc. Natl. Acad. Sci. USA* **36** (1950), 192–197.
158. L. L. Eberhardt, "Some developments in distance sampling," *Biometrics* **23** (1967), 207–216.
159. J. Echarte Reula, "Medidas en espacios foliados y en espacios homogeneos," *Publ. Rev. Acad. Ci. Zaragoza Ser. 2* **21** (1966).
160. B. Efron, "The convex hull of a random set of points," *Biometrika* **52** (1965), 331–344.
161. H. G. Eggleston, "Note on a conjecture of L. A. Santalo," *Mathematika* **8** (1961), 63–65.
162. H. G. Eggleston, *Convexity*. Cambridge Univ. Press, Cambridge, England, 1958.
163. H. G. Eggleston, *Problems in Euclidean Space: Application of Convexity*. Pergamon Press, London, 1957.
164. P. Eggleston and W. O. Kermack, "A problem in the random distribution of particles," *Proc. Roy. Soc. Edinburgh* **62** (1944), 103–115.
165. L. P. Eisenhart, *Riemannian Geometry*. Princeton Univ. Press, Princeton, N.J., 1949.
166. H. Elias (ed.), *Stereology* (*Proc. 2nd Internat. Congr. Stereology*). Springer, Berlin, 1967.
167. D. Fairthone, "The distance between random points in two concentric circles," *Biometrika* **51** (1964), 275–277.
168. I. Fáry, "Sur la courbure totale d'une courbe gauche faisant un noeud," *Bull. Soc. Math. France* **77** (1949), 128–138.
169. I. Fáry, "Sur certaines inégalités géométriques," *Acta Sci. Math.* (*Szeged*) **12** (1950), 117–124.
170. I. Fáry, "Functionals related to mixed volumes," *Illinois J. Math.* **5** (1961), 425–430.
171. H. Fast and A. Gotz, "Sur l'intégrabilité riemannienne de la fonction de Crofton," *Ann. Soc. Polon. Math.* **25** (1952), 301–322.
172. J. Favard, "Définition de la longueur et de l'aire," *C. R. Acad. Sci. Paris* **194** (1932), 344–346.
173. H. Federer, "Coincidence functions and their integrals," *Trans. Amer. Math. Soc.* **59** (1946), 441–466.
174. H. Federer, "The (ϕ, k)-rectifiable subsets on n space," *Trans. Amer. Math. Soc.* **62** (1947), 114–192.
175. H. Federer, "Dimension and measure," *Trans. Amer. Math. Soc.* **62** (1947), 536–547.
176. H. Federer, "Some integral geometric theorems," *Trans. Amer. Math. Soc.* **77** (1954), 238–261.
177. H. Federer, "Curvature measures," *Trans. Amer. Math. Soc.* **93** (1959), 418–191.
178. H. Federer, *Geometric Measure Theory*. Springer, Berlin, 1969.
179. L. Fejes Toth, "Über einen geometrischen Satz," *Math. Z.* **46** (1940), 83–85.
180. L. Fejes Toth, *Lagerung in der Ebene, auf der Kugel und im Raum*. Springer, Berlin, 1953.
181. L. Fejes Toth, "Close packing of segments," *Ann. Univ. Sci. Budapest. Eötvös Sect. Math.* **10** (1967), 57–60.
182. L. Fejes Toth and H. Hadwiger, "Mittlere Trefferzahlen und geometrische Wahrscheinlichkeiten," *Experientia* **3** (1947), 366–369.
183. L. Fejes Toth and H. Hadwiger, "Über Mittelwerte in einem Eibereichsystem," *Bull. Inst. Polytech. Jassy* **1948**, 29–35.
184. W. Feller, "Some geometric inequalities," *Duke Math. J.* **9** (1942), 885–892.
185. W. Feller, *An Introduction to Probability Theory and Its Applications*. Wiley, New York, 1950.

ISBN 0-201-13500-0

186. W. Fenchel, "On the differential geometry of closed curves," *Bull. Amer. Math. Soc.* **57** (1951), 44–54.
187. W. Fenchel, "Über Krümmung und Windung geschlossener Raumkurven," *Math. Annal.* **101** (1929), 238–252.
188. T. Figiel, "Some remarks on Dvoretzky's theorem on almost spherical sections of convex bodies," *Colloq. Math.* **24** (1971–1972), 241–252.
189. J. P. Fillmore, "Barbier's theorem in the Lobatchewski plane," *Proc. Amer. Math. Soc.* **24** (1970), 705–709.
190. D. Filipescu, "On some integral formulae for convex bodies," *Stud. Cerc. Mat.* **22** (1970), 1013–1031.
191. D. Filipescu, "Integral formulae for systems of random convex figures in Euclidean space," *Bull. Inst. Politechn. Bucuresti* **34** (1972), 21–36.
191a. D. Filipescu, "Integral formulas relative to convex figures in the euclidean space E_2," *Stud. Cerc. Mat.* **23** (1971), 693–709.
191b. D. Filipescu, "On some integral formulas relative to convex curves in a riemannian space V_2 of constant curvature," *Stud. Cerc. Mat.* **23** (1971), 561–578.
192. W. J. Firey, "The mixed area of a convex body and its polar reciprocal," *Israel J. Math.* **1** (1963), 201–202.
193. W. J. Firey, "An integral-geometric meaning for lower order area functions of convex bodies," *Mathematika* **19** (1972), 205–212.
194. W. J. Firey, "Support flats to convex bodies," *Geometria Dedicata* **2** (1973) 225–248.
195. W. J. Firey, "Approximating convex bodies by algebraic ones," *Arch. Math.* **25** (1974), 424–425.
196. L. D. Fisher, "The convex hull of a sample," *Bull. Amer. Math. Soc.* **72** (1966), 555–558.
197. L. D. Fisher, "Limiting convex hulls of samples; theory and function space examples," *Z. Wahrscheinlichkeits.* **18** (1971), 281–297.
198. R. A. Fisher, "On the similarity of the distributions found for the test of significance in harmonic analysis and in Steven's problem in geometrical probability," *Ann. Eugenics* **10** (1940), 14–17.
199. R. A. Fisher and R. E. Miles, "The role of the spatial patterns in the competition between crop plants and weeds: A theoretical analysis," *Math. Biosci.* **18** (1973), 335–350.
200. F. J. Flaherty, "Curvature measures for piecewise linear manifolds," *Bull. Amer. Math. Soc.* **79** (1973), 100–102.
201. H. Flanders, *Differential Forms with Applications to the Physical Sciences.* Academic Press, New York, 1963.
202. H. Flanders, "Differential forms," in *Studies in Global Geometry and Analysis* (S. S. Chern, ed.) pp. 57–95. Prentice-Hall, Englewood Cliffs, N.J., 1967.
203. H. Flanders, "A proof of Minkowski's inequality for convex curves," *Amer. Math. Monthly* **75** (1968), 581–593.
204. W. H. Fleming, *Functions of Several Variables.* Addison-Wesley, Reading, Mass., 1965.
205. E. Fratila, "Une méthode statistique d'évaluer la longueur d'un arc de courbe spatial," *Rev. Roumaine Math. Pures Appl.* **16** (1971), 493–498.
206. M. Fujiwara, "Ein Satz über konvexe geschlossene Kurven," *Sci. Repts. Tôhoku Univ.* **9** (1920), 289–294.
207. K. Fukunaga, *Introduction to Statistical Pattern Recognition.* Academic Press, New York, 1972.

ISBN 0-201-13500-0

208. R. L. Fullman, "Measurement of particle sizes in opaque bodies," *J. Metals* **5** (1953), 447–452.

209. R. L. Fullman, "Measurement of approximately cylindrical particles in opaque samples," *J. Metals* **5** (1953), 1267–1268.

209a. H. Furstenberg and I. Tzkoni, "Spherical functions and integral geometry," *Israel J. Math.* **10** (1971), 327–338.

210. P. Funk, "Über eine geometrische Anwendung der Abelschen Integralgleichung," *Math. Ann.* **77** (1916), 129–135.

211. F. Gaeta, "Sobre la subordinacion de la geometria integral a la teoria de la representacion de grupos mediante transformaciones lineales," *Contrib. Ci. Univ. Buenos Aires, Fac. Ci. Ex. Nat. Ser. Mat.* **2** (1960), 31–87.

212. J. Gani, "A car parking problem," *Period. Math. Hungar.* **2** (1972), 61–72.

213. F. Garwood, "The variance of the overlap of geometrical figures with reference to a bombing problem," *Biometrika* **34** (1947), 1–17.

214. F. Garwood, "An application of the theory of probability to vehicular controlled traffic," *J. Roy. Statist. Soc.* Suppl. 7 (1960), 65–67.

215. F. Garwood, "The vectorial representation of the frequency of encounters of freely flowing vehicles," *J. Appl. Probability* **11** (1974), 797–808.

216. E. Gaspar, "Formulas integrales referentes a la interseccion de una figura plana con bandas paralelas variables," *Publ. Inst. Mat. Univ. Litoral, Rosario, Argentina* **2** (1940), 113–118.

217. E. P. Geciauskas, "Distribution of the distance between two points in an ovaloid," *Litovsk. Mat. Sb.* **7** (1967), 35–36.

218. E. P. Geciauskas, "The method of integral geometry for finding the distribution of distances in an ovaloid," *Litovsk. Mat. Sb.* **9** (1969), 481–482.

219. E. P. Geciauskas, "Search by an oval," *Theor. Probability Appl.* **9** (1965), 635–637 (translation).

220. J. Geffroy, "Contribution à la théorie des valeurs extrêmes," *Publ. Inst. Statist. Univ. Paris* **8** (1959), 123–185.

221. J. Geffroy, "Localisation asymptotique du lophy polyhedre d'appui d'un échantillon laplacien en *k* dimensions," *Publ. Inst. Statist. Univ. Paris* **10** (1961), 3, 212–228.

222. I. M. Gelfand and M. I. Graev, "The geometry of homogeneous spaces, group representations in homogeneous spaces and questions in integral geometry related to them, I," *Trans. Moscow. Math. Soc.* **8** (1959), 321–390.

223. I. M. Gelfand, M. I. Graev, and Z. Ya. Shapiro, "Differential forms and integral geometry," *Functional Anal. Appl.* **3** (1969), 101–114 (translation).

224. I. M. Gelfand, M. I. Graev, and N. Ya. Vilenkin, *Generalized Functions*, Vol. 5: *Integral Geometry and Representation Theory*. Academic Press, New York, 1966.

224a. J. Gerriets and F. Poole, "Convex regions which cover arcs of constant length," *Amer. Math. Monthly* **81** (1974), 36–41.

225. B. Ghosh, "Random distances within a rectangle and between two rectangles," *Bull. Calcutta Math. Soc.* **43** (1951), 17–24.

226. H. Giger, "Zufallsmoirés," *Optica Acta* **15** (1968), 511–519.

227. H. Giger, "Fundamental equations in stereology, I," *Metrika* **16** (1970), 43–57; II, **18** (1972), 84–93.

228. H. Giger and H. Hadwiger, "Über Treffzahlwahrscheinlichkeiten im Eikörperfeld," *Z. Wahrscheinlichkeits.* **10** (1968), 329–334.

229. H. Giger and H. Riedwyl, "Bestimmung der Größenverteilung von Kugeln aus Schnittkreisradien," *Biometrische Z.* **12** (1970), 156–162.

230. E. N. Gilbert, "Random subdivisions of space into crystals," *Ann. Math. Statist.*

ISBN 0-201-13500-0

33 (1962), 958–972.
231. E. N. Gilbert, "The probability of covering a sphere with n circular caps," *Biometrika* **52** (1965), 323–330.
232. J. R. Goldman, "Stochastic point processes: limit theorems," *Ann. Math. Statist.* **38** (1967), 323–330.
233. A. W. Goodman and R. E. Goodman, "A circle covering problem," *Amer. Math. Monthly* **52** (1945), 494–498.
234. S. Goudsmidt, "Random distributions of lines in a plane," *Rev. Modern Phys.* **17** (1945), 321–322.
235. J. W. Green, "On the determination of a function in the plane by its integrals over straight lines," *Proc. Amer. Math. Soc.* **9** (1958), 758–762.
236. L. W. Green, "Auf Wiedersehenflächen," *Ann. of Math.* (2) **78** (1963), 289–299.
237. U. Grenander, "Statistical geometry: a tool for pattern analysis," *Bull. Amer. Math. Soc.* **79** (1973), 829–856.
238. N. T. Gridgeman, "Geometric probability and the number π," *Scripta Math.* **25** (1960), 183–195.
239. P. A. Griffiths, "On the Bezout problem for entire analytic sets," *Ann. of Math.* (2) **100** (1974), 533–552.
240. P. A. Griffiths, "On Cartan's method of Lie groups and moving frames as applied to uniqueness and existence questions in differential geometry," *Duke Math. J.* **41** (1974), 775–814.
241. H. Groemer, "Über die mittlere Dichte von Massenverteilungen," *Monatsh. Math.* **70** (1966), 437–443.
242. H. Groemer, "On plane sections and projection of convex sets," *Canad. J. Math.* **21** (1969), 1331–1337.
243. H. Groemer, "Eulersche Charakteristik, Projektionen und Quermassintegrale," *Math. Annal.* **198** (1972), 23–56.
244. H. Groemer, "On some mean values associated with a randomly selected simplex in a convex set," *Pacific J. Math.* **45** (1973), 525–533.
245. H. Groemer, "On the mean value of the volume of a random polytope in a convex set," *Arch. Math. (Basel)* **25** (1974), 86–90.
246. B. Grünbaum, "Measures of symmetry for convex sets," in *Convexity* (*Proc. Symp. Pure Math. 7thSeattle, 1961*), pp. 233–270. Amer. Math. Soc., Providence, R.I., 1963.
247. B. Grünbaum, *Convex Polytopes*. Wiley, New York, 1967.
248. B. Grünbaum, "Polytopes, graphs and complexes", *Bull. Amer. Math. Soc.* **76** (1970), 1131–1201.
249. G. Grünwald and P. Turan, "Über den Blochschen Satz," *Acta Litt. Sci. R. Univ. Hungar.* **8** (1936), 238.
250. W. C. Guenther, "Circular probability problems," *Amer. Math. Monthly* **68** (1961), 541–544.
251. W. C. Guenther and P. J. Terragno, "A review of the literature on a class of coverage problems," *Ann. Math. Statist.* **35** (1964), 232–260.
252. H. Guggenheimer, *Differential Geometry*. McGraw-Hill, New York, 1963.
253. H. Guggenheimer, "Pseudo-Minkowski differential geometry," *Ann. Mat. Pura Appl.* (4) **70** (1965), 305–370.
254. H. Guggenheimer, *Plane Geometry and Its Groups*. Holden-Day, San Francisco, 1967.
255. H. Guggenheimer, "Integral geometry in the Laguerre group," *Atti Congr. Internaz. Geom. Diff.* (*Bologna*) **1967**, 1–3.
256. H. Guggenheimer, "Hill equations with coexisting periodic solutions," *J. Diff.*

ISBN 0-201-13500-0

Equations **5** (1969), 159–166.
257. H. Guggenheimer, "Bemerkung zur Aufgabe 493," *Elem. Math.* **24** (1969), No. 1.
258. H. Guggenheimer, "Polar reciprocal convex bodies," *Israel J. Math.* **14** (1973), 309–316.
259. H. Guggenheimer, "A formula of Furstenberg–Tzkonis type," *Israel J. Math.* **14** (1973), 281–282.
260. R. S. Guter, "On the probability of detecting a region by a linear search," *Theor. Probability Appl.* **9** (1964), 331–333.
261. H. Hadwiger, "Über Mittelwerte im Figurengitter," *Comment. Math. Helv.* **11** (1939), 221–233.
262. H. Hadwiger, "Über statistische Flächen- und Längenmessung," *Mitt. Naturforsch. Gesellsch. Bern* **53–58** (1939).
263. H. Hadwiger, "Überdeckung ebener Bereiche durch Kreise und Quadrate," *Comment. Math. Helv.* **13** (1941), 195–200.
264. H. Hadwiger, "Gegenseitige Bedeckbarkeit zweier Eibereiche und Isoperimetrie," *Vierteljschr. Naturforsch. Gesellsch. Zürich* **86** (1941), 152–156.
265. H. Hadwiger, "Die erweiterten Steinerschen Formeln für ebene und sphärische Bereiche," *Comment. Math. Helv.* **18** (1945), 59–72.
266. H. Hadwiger, "Nonseparable convex systems," *Amer. Math. Monthly* **54** (1947), 583–585.
266a. H. Hadwiger, "Un valor medio integral de la caracteristica de Euler para ovalos moviles," *Rev. Un. Mat. Argentina* **13** (1948), 66–72.
267. H. Hadwiger, "Neue Integralrelationen für Eikörperpaare," *Acta Sci. Math. (Szeged)* **13** (1950), 252–257.
268. H. Hadwiger, "Einige Anwendungen eines Funktionalsatzes für konvexe Körper in der räumlichen Integralgeometrie," *Monatsh. Math.* **54** (1950), 345–353.
269. H. Hadwiger, "Über zwei quadratische Distanzintegrale für Eikörper," *Arch. Math. (Basel)* **3** (1952), 142–144.
270. H. Hadwiger, *Altes und Neues über konvexe Körper*. Birkhauser, Basel and Stuttgart, 1955.
271. H. Hadwiger, "Über Gitter und Polyeder," *Monatsh. Math.* **57** (1953), 246–254.
272. H. Hadwiger, "Eulers Charakteristik und kombinatorische Geometrie," *J. Reine Angew. Math. (Crelle)* **194** (1955), 101–110.
273. H. Hadwiger, "Integralsätze im Konvexring," *Abh. Math. Sem. Univ. Hamburg* **20** (1956), 136–154.
274. H. Hadwiger, *Vorlesungen über Inhalt, Oberfläche und Isoperimetrie*. Springer, Berlin, 1957.
275. H. Hadwiger, "Zur Axiomatik der innermathematischen Wahrscheinlichkeitstheorie," *Mitt. Verein. Schweiz. Versich.-Math.* **58** (1958), 151–165.
276. H. Hadwiger, "Normale Körper im euklidischen Raum und ihre topologischen und metrischen Eigenschaften," *Math. Z.* **71** (1959), 124–140.
277. H. Hadwiger, "Volumen und Oberfläche eines Eikörpers der keine Gitterpunkte überdeckt," *Math. Z.* **116** (1970), 191–196.
278. H. Hadwiger, "Gitterperiodische Punktmengen und Isoperimetrie," *Monatsh. Math.* **76** (1972), 410–418.
279. H. Hadwiger and C. Meier, "Studien zur vektoriellen Integralgeometrie," *Math. Nachr.* **56** (1973), 261–268.
280. H. Hadwiger and R. Schneider, "Vektorielle Integralgeometrie," *Elem. Math.* **26** (1971), 49–52.
281. H. Hadwiger and F. Streit, "Über Wahrscheinlichkeiten räumlicher Bündelungs-

ISBN 0-201-13500-0

erscheinungen," *Monatsh. Math.* **74** (1970), 30–40.

282. H. Hadwiger, H. Debrunner, and V. Klee, *Combinatorial Geometry in the Plane.* Holt, New York, 1964.
283. M. Haimovici, "Géométrie intégrale sur les surfaces courbes," *C. R. Acad. Sci. Paris* **203** (1936), 230–232.
284. M. Haimovici, "Géométrie intégrale sur les surfaces courbes," *Ann. Sci. Univ. Jassy* **23** (1937), 57–74.
285. M. Haimovici, "Généralisation d'une formule de Crofton dans un espace de Riemann à *n* dimensions," *C. R. Acad. Sci. Roumaine* **1** (1936), 291–196.
286. M. Halperin, "Some asymptotic results for a coverage problem," *Ann. Math. Statist.* **31** (1960), 1063–1076.
287. J. Hammer, "On a general area–perimeter relation for two-dimensional lattices," *Amer. Math. Monthly* **71** (1964), 534–537.
288. J. M. Hammersley, "The distribution of distances in a hypersphere," *Ann. Math. Statist.* **21** (1950), 447–452.
289. J. M. Hammersley, "On counters with random head time, I," *Proc. Cambridge Phil. Soc.* **49** (1953), 623–637.
290. J. M. Hammersley and K. *W.* Morton, "A new Monte-Carlo tecnnique: antithetic variates," *Proc. Cambridge Phil. Soc.* **52** (1956), 449–475.
291. H. Hancock, *Development of the Minkowski Geometry of Numbers*, Vol. I. Dover, New York, 1939.
292. D. C. Handscomb, "On the random disorientation of two cubes," *Canad. J. Math.* **10** (1958), 85–88.
293. H. Happel, "Einige Probleme über geometrische Wahrscheinlichkeiten," *Z. Math. Phys.* **61** (1912), 43–56.
294. E. F. Harding and D. G. Kendall (eds.), *Stochastic Geometry.* Wiley, New York, 1974.
295. P. Hartman and A. Wintner, "On the needle problem of Laplace and its generalizations," *Bol. Mat.* **14** (1941), 260–263.
296. E. Heil, "Abschätzung für einige Affininvarianten konvexer Kurven," *Monatsh. Math.* **71** (1967), 405–423.

296a. E. Heil, Zur affinen Differentialgeometrie der Eilinien, Ph.D. dissertation, Darmstadt, 1965.

297. S. Helgason, "Differential operators on homogeneous spaces," *Acta Math.* **102** (1959), 239–299.
298. S. Helgason, *Differential Geometry and Symmetric Spaces.* Academic Press, New York, 1962.
299. S. Helgason, "Duality and Radon transform for symmetric spaces," *Amer. J. Math.* **85** (1963), 667–692.
300. S. Helgason, "A duality in integral geometry: some generalizations of the Radon transform," *Bull. Amer. Math. Soc.* **70** (1964), 435–446.
301. S. Helgason, "A duality in integral geometry on symmetric spaces," *U.S.–Japan Sem. Diff. Geometry* (*Kyoto, 1965*), pp. 37–56, Nippon Hyoronsha, Tokyo, 1966.
302. S. Helgason, "The Radon transform on Euclidean spaces, compact two-point homogeneous spaces and Grassmann manifolds," *Acta Math.* **113** (1965), 153–180.
303. S. Helgason, "Fundamental solutions of invariant differential operators on symmetric spaces," *Amer. J. Math.* **86** (1964), 565–601.
304. S. Helgason, "A duality for symmetric spaces with applications to group representations," *Advances in Math.* **5** (1970), 1–154.
305. S. Helgason, "The surjectivity of invariant differential operators on symmetric

ISBN 0-201-13500-0

spaces, I," *Ann. of Math.* (2) **98** (1973), 451–479.

305a. J. M. Hemmersley, "Stochastic models for the distribution of particles in space," *Advances in Appl. Probability* Spec. Suppl. (*Proc. Symp. Statist. Probabilistic Problems in Metallurgy, Seattle, 1971*) (1972), 47–68.

306. G. Herglotz, *Lectures on Geometrical Probability* (mimeographed notes), Göttingen, 1933.

307. G. Herglotz, "Über die Steinersche Formel für Parallelfläche," *Abh. Math. Sem. Univ. Hamburg* **15** (1943), 165–177.

308. R. Hermann, "Remarks on the foundations of integral geometry," *Rend. Circ. Mat. Palermo, II* **9** (1960), 91–96.

309. J. E. Hilliard, "Determination of structural anisotropy," in *Stereology* (*Proc. 2nd Internat. Congr. Stereology*), pp. 219–227. Springer, Berlin, 1967.

309a. J. E. Hilliard, "Stereology: an experimental viewpoint," *Advances in Appl. Probability* Spec. Suppl. (*Proc. Symp. Statist. Probabilistic Problems in Metallurgy, Seattle, 1971*) (1972), 92–111.

309b. J. E. Hilliard, "Direct determination of the moments of the size distribution of particles in an opaque sample," *Trans. Met. Soc. AIME* **242** (1968), 1373–1380.

310. D. F. Holcomb, M. Iwasawa, and F. D. K. Roberts, "Clustering of randomly placed spheres," *Biometrika* **59** (1972), 207–209.

311. P. Holgate, "The distance from a random point to the nearest point of a closely packed lattice," *Biometrika* **52** (1965), 261–263.

312. K. Horneffer, "Eine Crofton-Formel und der Satz von Stokes," *Izv. Akad. Nauk Armjan SSR Ser. Mat.* **5** (1970), 235–250.

313. K. Horneffer, "Über die Integration von Differentialformen mittels integralgeometrischer Masse," *J. Diff. Geometry* **5** (1971), 451–466.

314. H. Hornich, "Eine allgemeine Ungleichung für Kurven," *Monatsh. Math. Phys.* **47** (1939), 432–438.

315. M. Horowitz, "Probability of random paths across elementary geometrical shapes," *J. Appl. Probability* **2** (1965), 169–177.

316. B. Hostinsky, "Sur les probabilités géométriques," *Publ. Fac. Sci. Univ. Masaryk, Brno* **50** (1925).

317. B. Hostinsky, "Probabilités relatives à la position d'une sphere à centre fixe," *J. Math. Pures Appl.* (*Liouville*) (9) **8** (1929), 35–43.

318. B. Hostinsky, "Sur une nouvelle solution du problème de l'aiguille," *Bull. Sci. Math.* **44** (1920), 126–136.

319. A. Hurwitz, "Sur quelques applications géométriques des séries de Fourier," *Ann. École Normale Sup.* (3) **19** (1902), 357–408.

320. I. M. Jaglom and W. G. Boltjanski, *Konvexe Figuren.* Deutsch. Verlag Wiss., Berlin, 1956.

321. A. J. Jakeman and R. S. Anderssen, "A note on numerical methods for the thin section model," *Proc. 8th Internat. Congr. Electron Microscopy* (*Canberra, 1974*) Vol. 2, pp. 4–5.

322. J. Jakeman and R. S. Anderssen, "On computational stereology, *Proc. 6th Austral. Computer Conf.* (*Sydney, 1974*) Vol. 2, pp. 353–362.

323. A. T. James, "Normal multivariate analysis and the orthogonal group," *Ann. Math. Statist.* **25** (1954), 40–75.

324. F. John, "Bestimmung einer Funktion aus ihren Integralen über gewisse Mannigfaltigkeiten," *Math. Ann.* **100** (1934), 488–520.

324a. F. John, *Plane Waves and Spherical Means Applied to Partial Differential Equations.* Wiley (Interscience), New York, 1955.

ISBN 0-201-13500-0

325. W. A. Johnson and R. F. Mehl, "Reaction kinetics in processes of nucleation and growth," *Trans. Amer. Inst. Min. Met. Eng.* **135** (1939), 416–458.

326. D. Jusupov, "On the integral geometry method in the theory of nonregular curves," *Izv. Vyss. Ucebn. Zaved. Matematika* **119** (4) (1972), 120–122.

327. M. Kac, "Toeplitz matrices, transformation kernels and a related problem in probability theory," *Duke Math. J.* **21** (1954), 501–509.

328. M. Kac, E. R. van Kampen, and A. Wintner, "On Buffon's problem and its generalizations," *Amer. J. Math.* **61** (1939), 672–676.

329. B. C. Kahan, "A practical demonstration of a needle experiment to give a number of concurrent estimates of π," *J. Roy. Statist. Soc.* **A124** (1961), 227–239.

330. J. Karamata, "On the distribution of intersections of diagonals in regular polygons," Univ. of Wisconsin Research Center Tech. Rept. 281 (Dec. 1961).

331. E. Kelly, "Lower-dimensional horocycles and the Radon transforms on symmetric spaces," *Trans. Amer. Math. Soc.* (to appear).

332. D. G. Kendall, "On the number of lattice points inside a random oval," *Quart. J. Math.* (2) **19** (1948), 1–26.

333. D. G. Kendall, "Foundations of a theory of random sets," in *Stochastic Geometry* (E. F. Harding and D. G. Kendall, eds.), pp. 322–356. Wiley, New York, 1974.

334. D. G. Kendall and R. A. Rankin, "On the number of points of a given lattice in a random hypersphere," *Quart. J. Math.* (2) **4** (1953), 178–189.

335. M. G. Kendall and P. A. P. Moran, *Geometrical Probability*. Griffin, London, 1963.

336. J. F. C. Kingman, "Mean free paths in a convex reflecting region," *J. Appl. Probability* **2** (1965), 162–268.

337. J. F. C. Kingman, "Random secants of a convex body," *J. Appl. Probability* **6** (1969), 660–672.

338. V. Klee, "What is the expected volume of a simplex whose vertices are chosen at random from a given convex body?" *Amer. Math. Monthly* **76** (1969), 186–188.

339. E. Knothe, "Über Ungleichungen bei Sehnenpotenzintegralen," *Deutsche Math.* **2** (1937), 544–551.

340. E. Knothe, "Konjugierte Normalensysteme und Stacheln von Eiflächen," *Arch. Math. (Basel)* **19** (1968), 214–224.

341. E. Knothe, "Bemerkungen zur Isoperimetrie des Kreises auf der Kugel," *Arch. Math. (Basel)* **22** (1971), 325–327.

341a. S. Kobayashi and K. Nomizu, *Foundations of Differential Geometry*, Interscience Publ., New York, vol. I, 1963; vol. II, 1969.

342. A. Kolmogoroff, *Grundbegriffe der Wahrscheinlichkeitsrechnung* (*Ergeb. Math.* **2**, 1933). English transl. Chelsea, New York, 1950.

343. C. W. Kosten, "The mean free path in room acoustics," *Acustica* **10** (1960), 245–250.

344. G. Kowalewski, *Einführung in die Determinantentheorie*. Berlin, 1942.

345. K. Krickeberg, "Invariance properties of the correlation measure of line processes, *Izv. Akad. Nauk Armjan. SSR Ser. Mat.* **5** (1970), 251–262. Reproduced in *Stochastic Geometry* (E. F. Harding and D. G. Kendall, eds.), pp. 76–88. Wiley, New York, 1974.

346. K. Krickeberg, "Moments of point processes," in *Stochastic Geometry* (E. F. Harding and D. G. Kendall, eds.), pp. 89–113. Wiley, New York, 1974.

346a. K. Krickeberg, "The Cox process," *Symposia Matematica* (*Convegno di calcolo delle probabilità, INDAM, Roma, 1971*), pp. 151–167. Academic Press, London, 1972.

347. J. P. V. Krishma, "The theory of probability distributions of points on a lattice,"

ISBN 0-201-13500-0

Ann. Math. Statist. **21** (1950), 198–217.

348. T. Kubota, "Über konvex geschlossene Mannigfaltigkeiten im n-dimensionalen Raum," *Sci. Repts. Tôhoku Univ.* **14** (1925), 85–99.

349. T. Kubota and D. Hemmi, "Some problems of minima concerning the oval," *J. Math. Soc. Japan* **5** (1953), 372–389.

349a. R. D. Kulle, "Messung des spezifischen Flächeninhaltes," *Nachr. Akad. Wiss. Göttingen II, Math. Phys. Kl.* **1971**, No. 10, 209–215.

350. R. D. Kulle and A. Reich, Flächenmessung mit gleichverteilten Folgen, *Nachr. Akad. Wiss. Göttingen II, Math.-Phys. Kl.* **1973**, Heft 12.

351. M. Kurita, "An extension of Poincaré's formula in integral geometry," *Nagoya Math. J.* **2** (1951), 55–61.

352. M. Kurita, *Integral Geometry*. Nagoya Univ., 1956 (japanese).

353. M. Kurita, "On the volume in homogeneous spaces," *Nagoya Math. J.* **15** (1959), 201–217.

354. M. Kurita, "On the vector in homogeneous spaces," *Nagoya Math. J.* **5** (1953), 1–33.

355. E. Langford, "Probability that a random triangle is obtuse," *Biometrika* **56** (1969), 689–690.

356. E. Langford, "A problem in geometrical probability," *Math. Mag.* **43** (1970), 237–244.

356a. M. M. Lavrentev and A. G. Buhgeim, "A certain class of problems of integral geometry," *Dokl. Akad. Nauk SSSR* **211** (1973), 38–39; English transl.: *Soviet Math. Dokl.* **14** (1973), 957–959.

357. H. Lebesgue, "Exposition d'une mémoire de M. W. Crofton," *Nouv. Ann. Math.* (4) **12** (1912), 481–502.

358. K. Legrady, "Symplektische Integralgeometrie," *Ann. Mat. Pura Appl.* (4) **41** (1956), 139–159.

359. K. Legrady, "Sobre la determinacion de functionales en geometria integral," *Rev. Un. Mat. Argentina* **19** (1960), 175–178.

360. J. Lehner, *A Short Course in Automorphic Functions*. Holt, New York, 1966.

361. C. G. Lekkerkerker, *Geometry of Numbers*. North-Holland, Amsterdam, 1969.

362. J. Lengauer, "Geometrische Wahrscheinlichkeitsprobleme," *Progr. Gymn. (Würzburg)* **1899**.

363. H. Lenz, "Mengenalgebra und Eulersche Charakteristik," *Abh. Math. Sem. Univ. Hamburg* **34** (1969–1970), 135–147.

364. H. Levine, "A theorem on holomorphic mappings into complex projective space," *Ann. of Math.* (2) **71** (1960), 529–535.

365. P. Levy, "Sur la division d'un segment par des points choisis au hasard," *C. R. Acad. Sci. Paris* **208** (1939), 147.

366. G. van der Lijn, "Sur la mesure d'un ensemble autre qu'un ensemble de points et son application au problème du Buffon," *Bull. Soc. Roy. Sci. Liège* **2** (1933), 104–113.

367. D. V. Little, "A third note on recent research in geometrical probability," *Advances in Appl. Probability* **6** (1974), 103–130.

368. L. H. Loomis and S. Sternberg, *Advanced Calculus*. Addison-Wesley, Reading, Mass., 1968.

369. R. D. Lord, "The distribution of distances in a hypersphere," *Ann. Math. Statist.* **25** (1954), 794–798.

370. R. Luccioni, "Geometria integral en espacios proyectivos," *Rev. Mat. Fis. Téor. Univ. Tucumán* **15** (1964), 53–80.

ISBN 0-201-13500-0

371. R. Luccioni, "Sobre la existencia de medida para hipercuádricas singulares en espacios proyectivos," *Rev. Mat. Fis. Téor. Univ. Tucumán* **14** (1962), 269–276.
372. A. V. Lucenko, "Measurability of n-point sets with respect to projective groups in the plane," *Ukrain. Geometr. Sb.* **11** (1971), 53–63.
373. A. V. Lucenko, "The measure of sets of geometric elements and their subsets," *Ukrain. Geometr. Sb.* **1** (1965), 39–57.
374. A. V. Lucenko and L. M. Jurtova, "Measures of sets of pairs on the plane that are invariant under projective transformation groups," *Ukrain. Geometr. Sb.* **5–6** (1968), 99–102.
375. G. Lüko, "On the mean length of the chords of a closed curve," *Israel J. Math.* **4** (1966), 23–32.
375a. E. Lutwak, "A general Bieberbach inequality," *Proc. Cambridge Phil. Soc.* **78** (1975), 493–495.
376. W. Maak, "Integralgeometrie 18: Grundlagen der ebenen Integralgeometrie," *Abh. Math. Sem. Univ. Hamburg* **12** (1938), 83–110.
377. W. Maak, "Integralgeometrie 27: Über stetige Kurven," *Abh. Math. Sem. Univ. Hamburg* **12** (1938), 163–178.
378. W. Maak, "Schnittpunktzahl rektifizierbarer und nicht rektifizierbarer Kurven," *Math. Ann.* **118** (1942), 229–304.
379. W. Maak, "Oberflächenintegral und Stokes-Formel im gewöhnlichen Raum (Integralgeometrie 29)," *Math. Ann.* **116** (1939), 574–597.
380. W. Maak, "Über Längen- und Inhaltsmessung," *Nachr. Akad. Wiss. Göttingen II, Math. Phys. Kl.* **1970**, 57–66.
381. A. M. Macbeath and C. A. Rogers, "Siegel's mean value theorem in the geometry of numbers," *Proc. Cambridge Phil. Soc.* **54** (1958), 139–151; **51** (1955), 567–576; **54** (1958), 322–326.
382. C. Mack, "An exact formula for $Q_k(n)$, the probable number of k-aggregates in a random distribution of n points," *Phil. Mag.* (7) **39** (1948), 778–790.
383. C. Mack, "The expected number of aggregates in a random distribution of points," *Proc. Cambridge Phil. Soc.* **45** (1949), 285–292.
384. C. Mack, "The effect of overlapping in bacterial counts of incubated colonies," *Biometrika* **40** (1953), 220–222.
385. C. Mack, "The expected number of clumps when convex laminae are placed at random and with random orientation on a plane area," *Proc. Cambridge Phil. Soc.* **50** (1954), 581–585.
386. J. K. Mackenzie, "Sequential filling of a line by intervals placed at random and its application to linear adsorption," *J. Chem. Phys.* **37** (1962), 723.
387. K. Mahler, "Ein Übertragungsprinzip für konvexe Körper," *Casopis Pest. Mat. Fys.* **68** (1939), 93–102.
388. C. L. Mallows and J. M. Clark, "Linear intercept distributions do not characterize plane sets," *J. Appl. Probability* **7** (1970), 240–244.
389. D. Mannion, "Random space-filling in one dimension," *Publ. Math. Inst. Hungar. Acad. Sci.* **A9** (1964), 143–154.
390. L. Mantel, "An extension of the Buffon needle problem," *Ann. Math. Statist.* **22** (1951), 314–315; **24** (1953), 674–677.
391. F. H. Marriot, "Buffon's needle problem for non-random directions," *Biometrics* **27** (1971), 233–235.
392. G. Masotti-Biggiogero, "La geometria integrale," *Rend. Sem. Mat. Fis. Milano* **25** (1953–1954), 3–70.
393. G. Masotti-Biggiogero, "Su alcune formule di geometria integrale," *Rend. Mat.*

ISBN 0-201-13500-0

(5) **14** (1955), 280–288.

394. G. Masotti-Biggiogero, "Sulla geometria integrale: generalizzazione di formule di Crofton, Lebesgue e Santalo," *Rev. Un. Mat. Argentina* **17** (1955), 125–134.
395. G. Masotti-Biggiogero, "Sulla geometria integrale: nuove formule relative agli ovaloidi, in *Scritti Math. in Onore di Filippo Sibirani*, pp. 173–179. Cesari Zuffi, Bologna, 1957.
396. G. Masotti-Biggiogero, "Nuove formule di geometria integrale relative agli ovali," *Ann. Mat. Pura Appl.* (4) **58** (1962), 85–108.
397. G. Masotti-Biggiogero, "Nuove formule di geometria integrale relative agli ovaloide," *Rend. Sc. Ist. Lombardo* **A96** (1962), 666–685.
398. M. Masuyama, "On a fundamental formula in bulk sampling from the viewpoint of integral geometry," *Rept. Statist. Appl. Res. Un. Japan. Sci. Engrs.* **4** (1956), 85–89.
399. G. Matheron, "Ensembles fermés aléatoires, ensembles semi-markoviens et polyèdres poissoniens," *Advances in Appl. Probability* **4** (1972), 508–543.
400. G. Matheron, "Random sets theory and its applications to stereology," *J. Microscopy* **95** (1972), 15–25.
401. G. Matheron, "Hyperplanes poissoniens et compacts de Steiner," *Advances in Appl. Probability* **6** (1974), 563–579.

401a. G. Matheron, *Random Sets and Integral Geometry*. Wiley, New York, 1975.

402. M. Matschinski, "Considérations statistiques sur les polygones et les polyèdres," *Publ. Inst. Statist. Univ. Paris* **3** (1954), 179–201.
403. M. Matschinski, "Détermination expérimentale des probabilités du nombre des côtés des polygones couvrant chaotiquement un plan sans lacunes," *Publ. Inst. Statist. Univ. Paris* **18** (1969), 246–265.
404. G. A. McIntyre, "Estimation of plant density using line transects," *Ecology* **41** (1955), 90–96.
405. J. L. Meijering, "Interface area, edge length and number of vertices in crystal aggregates with random nucleation," *Philips Research Rept.* **8** (1953), 270–290.
406. Z. A. Melzak, "Self-intersections in continuous random walk," *Bull. Amer. Math. Soc.* **70** (1970), 1251–1252.
407. R. E. Miles, "Random polygons determined by random lines in a plane, I, II," *Proc. Natl. Acad. Sci. USA* **52** (1964), 901–907, 1157–1160.
408. R. E. Miles, "On random rotations in R^3," *Biometrika* **52** (1965), 636–639.
409. R. E. Miles, "Probability distribution of a network of triangles (a solution to problem 67–15)," *SIAM Rev.* **11** (1969), 399–402.
410. R. E. Miles, "Poisson flats in Euclidean space, Pt. I: A finite number of random uniform flats," *Advances in Appl. Probability* **1** (1969), 211–237.
411. R. E. Miles, "On the homogeneous planar Poisson point process," *Math. Biosci.* **6** (1970), 85–127.
412. R. E. Miles, "The asymptotic values of certain coverage probabilities," *Biometrika* **56** (1969), 661–680.
413. R. E. Miles, "Random points, sets and tessellations on the surface of a sphere," *Sankhyá* **A33** (1971), 145–174.
414. R. E. Miles, "Poisson flats in Euclidean spaces, Pt. II: Homogeneous Poisson flats and the complementary theorem, *Advances in Appl. Probability* **3** (1971), 1–43.
415. R. E. Miles, "A synopsis of Poisson flats in Euclidean spaces," *Izv. Akad. Nauk Arm. SSR Ser. Mat.* **5** (1970), 263–285. Reprinted in *Stochastic Geometry* (E. F. Harding and D. G. Kendall, eds.), pp. 202–227. Wiley, New York, 1974.
416. R. E. Miles, "Multidimensional perspectives on stereology," *J. Microscopy* **95** (1972), 181–195.

ISBN 0-201-13500-0

417. R. E. Miles, "On the elimination of edge effects in planar sampling," in *Stochastic Geometry* (E. F. Harding and D. G. Kendall, eds.), pp. 229–247. Wiley, New York, 1974.
418. R. E. Miles, "The various aggregates of random polygons determined by random lines in a plane," *Advances in Math.* **10** (1973), 256–290.
419. R. E. Miles, "The random division of space," *Advances in Appl. Probability* Suppl. (1972), 243–266.
420. R. E. Miles, "Isotropic random simplices," *Advances in Appl. Probability* **3** (1971), 353–382.
421. R. E. Miles, "A simple derivation of a formula of Furstenberg and Tzkoni," *Israel J. Math.* **14** (1973), 278–280.
421a. R. E. Miles, "The fundamental formula of Blaschke in integral geometry and geometrical probability and its iteration, for domains with fixed orientations," *Austral. J. Statist.* **16**(2) (1974), 111–118.
421b. R. E. Miles, "Direct derivation of certain surface integral formulae for the mean projection of a convex set," *Advances in Appl. Probability* **7** (1975), 818–829.
422. E. Mohr, "Bemerkung zu Mises' Behandlungen des Nadelproblems von Buffon," *Deutsche Math.* **6** (1941), 108–113.
423. P. G. Moore, "Spacing in plant populations," *Ecology* **35** (1954), 222–227.
424. P. A. P. Moran, "Measuring the surface area of a convex body," *Ann. of Math.* **45** (1944), 783–789.
425. P. A. P. Moran, "The random division of an interval, I," *J. Roy. Statist. Soc. Suppl.* **9** (1947), 92–98; II, **B13** (1951), 147–150; III, **B15** (1953), 77–80.
426. P. A. P. Moran, "Numerical integration by systematic sampling," *Proc. Cambridge Phil. Soc.* **46** (1950), 111–115.
427. P. A. P. Moran, "Measuring the length of a curve," *Biometrika* **53** (1966), 359–364.
428. P. A. P. Moran, "A note on recent research in geometrical probability," *J. Appl. Probability* **3** (1966), 453–463.
429. P. A. P. Moran, "A second note on recent research in geometrical probability," *Advances in Appl. Probability* **1** (1969), 73–89.
430. P. A. P. Moran, "The probabilistic basis of stereology," *Advances in Appl. Probability* Spec. Suppl. (*Proc. Symp. Probabilistic Problems in Metallurgy, Seattle, 1971*) (1972), 69–91.
431. P. A. P. Moran and S. Fazekas de St. Groth, "Random circles on a sphere," *Biometrika* **49** (1962), 289–396.
432. G. W. Morgenthaler, "Some circular coverage problems," *Biometrika* **48** (1961), 313–324.
433. M. Morimoto, "Sur les transformations horosphériques généralisées dans les espaces homogènes," *J. Fac. Sci. Univ. Tokyo Sect. IA* **13** (1966), 65–83.
434. R. R. A. Morton, "The expected number and angle of intersections between random curves in a plane," *J. Appl. Probability* **3** (1966), 559–562.
435. V. M. Morton, "The determination of angular distribution of planes in space," *Proc. Roy. Soc.* **A302** (1967), 51–68.
436. G. D. Mostow, "Homogeneous spaces with finite invariant measure," *Ann. of Math.* (2) **75** (1962), 17–37..
437. A. Müller, "Integralgeometrie 16: Dichten linearer Mannigfaltigkeiten im euklidischen und nichteuklidischen R_n," *Math. Z.* **42** (1936), 101–124.
438. H. R. Müller, "Über lineare und quadratische Momente konvexer Bereiche," *Nachr. Österr. Math. Ges.* **21–22** (1952), 47.
439. H. R. Müller, "Über Momente ersten und zweiten Grades in der Integralgeometrie,"

ISBN 0-201-13500-0

Rend. Circ. Mat. Palermo II **2** (1953), 119–140.

440. J. P. Mullooly, "A one-dimensional random space-filling problem," *J. Appl. Probability* **5** (1968), 427–435.

441. M. E. Munroe, *A Modern Multidimensional Calculus*. Addison-Wesley, Reading, Mass., 1963.

442. E. J. Myers, "Size distribution of cubic particles," in *Stereology* (*Proc. 2nd Internat. Congr. Stereology*), pp. 187–188. Springer, Berlin, 1967.

443. L. Nachbin, *The Haar Integral*. Van Nostrand, Princeton, N.J., 1965.

443a. G. Nagy, "State of the art in pattern recognition," *Proc. IEEE* **56** (1968), 836–862.

444. J. I. Naus, "Clustering of random points in two dimensions," *Biometrika* **52** (1965), 263–267.

445. F. Neumann, "Center affine invariants of plane curves in connection with the theory of the second-order linear differential equations," *Arch. Math. Brno* **4** (1968), 201–216.

446. F. Neuman, "A note on Santalo's isoperimetric problem," *Rev. Mat. Fis. Teor. Univ. Tucumán* **20** (1970), 204–206.

447. F. Neuman, "Closed plane curves and differential equations," *Rend. Mat.* 3 (1970), 423–433.

448. F. Neuman, "Linear differential equations of the second order and their applications," *Rend. Mat.* (6) **4** (1971), 559–617.

449. P. E. Ney, "A random space-filling problem," *Ann. Math. Statist.* **33** (1962), 702–718.

449a. W. L. Nicholson and K. R. Merckx, "Unfolding particle size distribution," *Technometrics* **11** (1970), 707–723.

450. W. L. Nicholson, "Estimation and linear properties of particle size distributions," *Biometrika* **57** (1970), 273–297.

451. A. Nijenhuis, "On Chern's kinematic formula in integral geometry," *J. Diff. Geometry* **9** (1974), 475–482.

452. G. Nöbeling, "Über die Flächenmaße im Euklidischen Raum," *Math. Ann.* **118** (1943), 687–701.

453. G. Nöbeling, "Über die Flächeninhalte dehnungsbeschränkter Flächen," *Math. Z.* **48** (1943), 747–771.

454. G. Nöbeling, "Über die Hauptformel der ebenen Kinematik von L. A. Santalo und W. Blaschke, I, II," *Math. Ann.* **120** (1949), 585–614, 615–633.

455. G. Nöbeling, "Die Formel von Poincaré für beliebige Kontinuen," *Abh. Math. Sem. Univ. Hamburg* **15** (1943), 120–126.

456. G. Nöbeling, "Verallgemeinerung eines Satzes von Herrn W. Maak," *Abh. Math. Sem. Univ. Hamburg* **17** (1951), 95–97.

457. A. B. J. Novikoff, "Integral geometry as a tool in pattern perception," in *Principles of Self-Organization* (H. von Foerster and G. W. Zopf, Jr., eds.), pp. 347–368. Pergamon Press, London, 1962.

458. N. Obrechkoff, "On hyperbolic integral geometry," *C. R. Acad. Bulgare Sci. Math. Nat.* **2** (1949), 1–4.

459. A. M. Odlyzko, "On lattice points inside convex bodies," *Amer. Math. Monthly* **80** (1973), 915–918.

460. D. Ohmann, "Ungleichungen zwischen den Quermassintegralen beschränkter Punktmengen, I," *Math. Ann.* **124** (1952), 265–276; II, **127** (1954), 1–7; III, **130** (1956), 386–393.

461. D. Ohmann, "Eine Verallgemeinerung der Steinerschen Formel," *Math. Ann.* **129** (1955), 209–212.

ISBN 0-201-13500-0

462. D. Ohmann, "Extremalprobleme für konvexe Bereiche der euklidischen Ebene," *Math. Z.* **55** (1952), 347–352.
463. O. G. Owens, "The integral geometric definition of arc length for two-dimensional Finsler spaces," *Trans. Amer. Math. Soc.* **73** (1952), 198–210.
464. S. Oshio, "On mean values and geometrical probabilities in E_3," *Sci. Rept. Kanazawa Univ.* **3** (1955), 35–43.
465. S. Oshio, "On mean values and geometrical probabilities in E_n," *Sci. Rept. Kanazawa Univ.* **3** (1955), 199–207.
466. R. S. Palais, "A global formulation of the Lie theory of transformation groups," *Amer. Math. Soc. Memoirs* No. 22 (1957).
467. I. Palasti, "On some random space-filling problems," *Publ. Math. Inst. Hung. Acad. Sci.* **5** (1960), 353–360.
468. J. E. Paloheimo, "On a theory of search," *Biometrika* **58** (1971), 61–75.
469. F. Papangelou, "On the Palm probabilities of processes of points and processes of lines," in *Stochastic Geometry* (E. F. Harding and D. K. Kendall, eds.), pp. 114–147. Wiley, New York, 1974.
470. B. Parzen, [1] *Modern Probability Theory and Its Applications.* Wiley, New York, 1960.
471. B. Petkantschin, "Zusammenhänge zwischen den Dichten der linearen Unterräume im n-dimensionalen Raume," *Abh. Math. Sem. Univ. Hamburg* **11** (1936), 249–310.
472. E. E. Petrov, "The Radon transform in matrix spaces and in Grassmann manifolds," *Soviet Math. Dokl.* **8** (1967), 1504–1507.
473. C. M. Petty, "Centroid surfaces," *Pacific J. Math.* **11** (1961), 1535–1547.
474. C. M. Petty, "Projection bodies," *Proc. Colloq. on Convexity, Copenhagen, 1965*, pp. 234–241. Copenhagen Univ. Math. Inst., 1967.
475. C. M. Petty, "Isoperimetric problems," *Proc. Conf. on Convexity and Combinatorial Geometry* pp. 26–41. Univ. of Oklahoma Press, Norman, Okla., 1972.
476. C. M. Petty, "Geominimal surface area," *Geometria Dedicata* **3** (1974), 77–97.
477. J. R. Philip, "Some integral equations in geometrical probability," *Biometrika* **53** (1966), 365–374.
478. E. M. Philofsky and J. E. Hilliard, "The measurement of the orientation distribution and arear arrays," *Quart. Appl. Math.* **27** (1969), 79–86.
479. A. Pleijel, "Zwei kurze Beweise der isoperimetrischen Ungleichung," *Arch. Math. (Basel)* **7** (1956), 317–319.
480. A. Pleijel, "On konvexa kurvor," *Nordisk Math. Tidskr.* **3** (1955), 57–64.
481. W. F. Pohl, "On some integral formulas for space curves and their generalization," *Amer. J. Math.* **90** (1968), 1321–1345.
482. W. F. Pohl, "Extrinsic complex projective geometry," *Proc. Conf. Complex Analysis, Minneapolis, 1964* pp. 18–29.
483. W. F. Pohl, "The self-linking number of a closed space curve," *J. Math. Mech.* **17** (1968), 975–986.
484. H. Poincaré, *Calcul des probabilités*, 2nd ed. Gauthier-Villars, Paris, 1912.
485. H. Poincaré, "Théorie des groupes fuchsiens," *Acta Math.* **1** (1882), 1–62.
486. G. Polya, "Zahlentheoretisches und Wahrscheinlichkeitstheoretisches über die Sichtweite im Walde," *Arch. Math. Phys.* **27** (1918), 135–142.
487. G. Polya, "Über geometrische Wahrscheinlichkeiten," *Sitz. Akad. Wiss. Wien* **124** (1917), 319–328.
488. G. Polya, "Über geometrische Wahrscheinlichkeiten in konvexen Körpern," *Ber. Verh. Sächs. Ges. Wiss. Leipzig* **69** (1917), 457–458.
489. G. Polya and G. Szego, *Aufgaben und Lehrsätze aus der Analysis*, Vols. 1 and 2.

ISBN 0-201-13500-0

Springer, Berlin, 1925.

490. G. Polya and G. Szego, *Isoperimetric Inequalities in Mathematical Physics* (Ann. of Math. Studies No. 27). Princeton Univ. Press, Princeton, N.J., 1951.
491. G. Poole and J. Gerriets, "Minimum covers for arcs of constant length," *Bull. Amer. Math. Soc.* **79** (1973), 462–463.
492. A. Prekopa, "On the number of vertices of random convex polyhedra," *Period. Math. Hungar.* **2** (1972), 259–282.
493. J. Radon, "Über die Bestimmung von Funktionen durch ihre Integralwerte längs gewisser Mannigfaltigkeiten," *Ber. Verh. Sächs. Akad. Wiss. Leipzig. Math. Natur. Kl.* **69** (1917), 262–277.
494. E. Raimondi, "Sobre un problema de probabilidades geométricas," *Rev. Un. Mat. Argentina* **7** (1940–1941), 106–109.
495. E. Raimondi, "Sobre los pares de secantes de un poligono," *Rev. Un. Mat. Argentina* **7** (1940–1941), 133–134.
496. H. Raynaud, "Sur l'enveloppe convexe des nuages des points aléatoires dans R_n, I," *J. Appl. Probability* **7** (1970), 35–48.
497. L. Redei and B. v. Sz. Nagy, "Eine Verallgemeinerung der Inhaltsformel von Heron," *Publ. Math. Debrecen* **1** (1949), 42–50.
498. W. J. Reed, "Random points in a simplex," *Pacific J. Math.* **54** (1974), 183–198.
499. W. P. Reid, "Distribution of sizes of spheres in a solid from a study of slices of the solid," *J. Math. Phys.* **35** (1955), 95–102.
500. H. Reiter, *Classical Harmonic Analysis and Locally Compact Groups* (Oxford Math. Monographs). Oxford Univ. Press, London and New York, 1968.
501. A. Rényi, "On a new axiomatic theory of probability," *Acta Math. Acad. Sci. Hungar.* **6** (1955), 285–335.
502. A. Rényi, "On a one-dimensional problem concerning random place filling," *Magyar Tud. Akad. Mat. Intézet Közl.* **3** (1958), 109–127 (in Hungarian).
503. A. Rényi, *Wahrscheinlichkeitsrechnung*. Deutsch. Verlag Wiss., Berlin, 1962.
504. A. Rényi and R. Sulanke, "Über die konvexe Hülle von n zufällig gewählten Punkten, I," *Z. Wahrscheinlichkeits.* **2** (1963), 75–84; II, **3** (1964), 138–147.
505. A. Rényi and R. Sulanke, "Zufällige konvexe Polygone in einem Ringgebiet," *Z. Wahrscheinlichkeits.* **9** (1968), 146–157.
506. Y. G. Reshetnyak, "Integral geometric methods in the theory of curves," *Proc. 3rd All-Union Math. Congr.* Vol. 1, p. 164. Akad. Nauk SSSR, Moscow, 1956 (in Russian).
507. J. Rey Pastor and L. A. Santalo, *Geometria Integral*. Espasa-Calpe, Buenos Aires, Argentina, 1951.
508. J. Rezek, "A contribution to embracing the basic concepts of the integral geometry within the scope of ideas of Lie's group theory," *Casopis Pest. Mat. Fys.* **75** (1950), 17–26.
509. H. Rhee, "A representation of the solutions of the Darboux equation in odd-dimensional spaces," *Trans. Amer. Math. Soc.* **150** (1970), 491–498.
510. P. I. Richards, "Averages for polygons formed by random lines," *Proc. Natl. Acad. Sci. USA* **52** (1964), 1160–1164.
511. S. A. Roach, *The Theory of Random Clumping*. Methuen, London, 1968.
512. H. E. Robbins, "On the measure of a random set, I," *Ann. Math. Statist.* **15** (1944), 70–74; II, **16** (1945), 342–347.
513. H. E. Robbins, "Acknowledgement of priority," *Ann. Math. Statist.* **18** (1947), 297.
514. F. D. K. Roberts, "Nearest neighbours in a Poisson ensemble," *Biometrika* **56** (1969), 401–406.

ISBN 0-201-13500-0

515. F. D. K. Roberts and S. H. Storey, "A three-dimensional cluster problem," *Biometrika* **55** (1968), 258–260.

516. H. Rohde, "Integralgeometrie 33: Unitäre Integralgeometrie," *Abh. Math. Sem. Univ. Hamburg* **13** (1940), 295–318.

517. V. G. Romanov, *Integral Geometry and Inverse Problems for Hyperbolic Equations* (Springer Tracts in Natural Philosophy, Vol. 26). Springer, Berlin, 1974.

518. H. Ruben, "An intrinsic formula for volume," *J. Reine Angew. Math.* (*Crelle*) **226** (1967), 116–119.

519. A. M. Russell and N. S. Josephson, "Measurement of area by counting," *J. Appl. Probability* **2** (1965), 339–351.

520. H. Sachs, "Ungleichungen für Umfang, Flächeninhalt und Trägheitsmoment konvexer Kurven," *Acta Math. Acad. Sci. Hungar.* **11** (1960), 103–115.

521. J. Sancho San Roman, "Sobre la existencia de medidas relativamente invariantes en un espacio de Klein," *Actas 2nd Coloq. Geometria Dif., Santiago de Compostela, 1967* pp. 39–44.

522. L. A. Santalo, "Algunas propiedades de las curvas esféricas y una caracteristica de la esfera," *Rev. Mat. Hispano-Amer.* **10** (1935), 1–4.

523. L. A. Santalo, "Geometria integral 4: Sobre la medida cinemática en el plano," *Abh. Math. Sem. Univ. Hamburg* **11** (1936), 222–236.

524. L. A. Santalo, "Integralgeometrie 5: Über das kinematische Maß im Raum," *Actualités Sci. Indust.* **357** (1936).

525. L. A. Santalo, "Geometria integral 7: Nuevas aplicaciones de la medida cinemática en el plano y en el espacio," *Rev. Acad. Ci. Madrid* **33** (1936), 451–477, 481–504.

526. L. A. Santalo, "Unos problemas referentes a probabilidades geométricas," *Rev. Mat. Hisp.-Amer.* (2) **11** (1936), 87–97.

527. L. A. Santalo, "Geometria integral 15: Formula fundamental de la medida cinemática para cilindros y planos paralelos moviles," *Abh. Math. Sem. Univ. Hamburg* **12** (1938), 38–41.

528. L. A. Santalo, "Geometria integral de figuras ilimitadas," *Publ. Inst. Mat. Univ. Litoral, Rosario, Argentina* **1**, No. 2 (1939).

529. L. A. Santalo, "Geometria integral 31: Sobre valores medios y probabilidades geométricas," *Abh. Math. Sem. Univ. Hamburg* **13** (1940), 284–294.

530. L. A. Santalo, "Géométrie intégrale 32: Quelques formules intégrales dans le plan et dans l'espace," *Abh. Math. Sem. Univ. Hamburg* **13** (1940), 344–356.

531. L. A. Santalo, "Una demonstracion de la propiedad isoperimétrica del circulo," *Publ. Inst. Mat. Univ. Litoral, Rosario, Argentina* **2**, No. 3 (1940).

532. L. A. Santalo, "Sur quelques problèmes de probabilités géométriques," *Tohoku Math. J.* **47** (1940), 159–171.

533. L. A. Santalo, "Beweise eines Satzes von Bottema über Eilinien," *Tôhoku Math. J.* **48** (1941), 221–224.

534. L. A. Santalo, "Un esquema de valores medios en la teoria de probabilidades geométricas," *Rev. Ci.* (*Lima*) **43** (1941), 147–154.

535. L. A. Santalo, "A theorem and an inequality referring to rectifiable curves," *Amer. J. Math.* **53** (1941), 635–644.

536. L. A. Santalo, "Generalizacion de un problema de probabilidades geométricas," *Rev. Un. Mat. Argentina* **7** (1941), 129–132.

537. L. A. Santalo, "Algunos valores medios y desigualdades referentes a curvas situadas sobre la superficie esférica," *Rev. Un. Mat. Argentina* **8** (1942), 113–125.

538. L. A. Santalo, "La desigualdad isoperimétrica sobre superficies de curvatura constante negativa," *Rev. Mat. Fis. Teor. Univ. Tucumán* **3** (1942), 243–259.

ISBN 0-201-13500-0

539. L. A. Santalo, "Integral formulas in Crofton's style on the sphere and some inequalities referring to spherical curves," *Duke Math. J.* **9** (1942), 707–722.
540. L. A. Santalo, "Una formula integral referente a figuras convexas," *Rev. Un. Mat. Argentina* **8** (1942), 165–169.
541. L. A. Santalo, "Integral geometry on surfaces of constant negative curvature," *Duke Math. J.* **10** (1943), 687–704.
542. "Sobre la distribucion probable de corpusculos en un cuerpo deducida de la distribucion en sus secciones," *Rev. Un. Mat. Argentina* **9** (1943), 145–164.
543. L. A. Santalo, "Acotaciones sobre la longitud de una curva o para el numero de puntos necesarios para cubrir aproximadamente un domino," *An. Acad. Brasil. Ci.* **16** (1944), 112–121.
544. L. A. Santalo, "Valor medio del numero de regiones en que un cuerpo del espacio es dividido por *n* planos arbitrarios," *Rev. Un. Mat. Argentina* **10** (1945), 111–108.
545. L. A. Santalo, "Note on convex curves on the sphere," *Bull. Amer. Math. Soc.* **50** (1944), 528–534.
546. L. A. Santalo, "Unas formulas integrales referentes a cuerpos convexos," *Rev. Un. Mat. Argentina* **12** (1946–1947), 78–87.
547. L. A. Santalo, "Sobre el circulo de radio máximo contenido en un recinto," *Rev. Un. Mat. Argentina* **10** (1945), 155–162.
548. L. A. Santalo, "Note on convex curves on the hyperbolic plane," *Bull. Amer. Math. Soc.* **51** (1945), 405–412.
549. L. A. Santalo, "Sobre la longitud de una curva del espacio como valor medio de las longitudes de sus proyecciones ortogonales," *Math. Notae* **6** (1946), 158–166.
550. L. A. Santalo, "Sobre figuras planas hiperconvexas," *Summa Brasil. Math.* **1** (1946), fasc. 11.
551. L. A. Santalo, "Sobre la medida de figuras convexas congruentes contenidas en el interior de un rectángulo o de un triángulo," *Actas Acad. Ci. Lima* **10** (1947), 103–116.
552. L. A. Santalo, "On the first two moments of the measure of a random set," *Ann. Math. Statist.* **18** (1947), 37–49.
553. L. A. Santalo, "Sobre la distribucion de planos en el espacio," *Rev. Un. Mat. Argentina* **13** (1948), 120–124.
554. L. A. Santalo, "Geometria integral en los espacios tridimensionales de curvatura constante," *Math. Notae* **9** (1949), 10–28.
555. L. A. Santalo, "Integral geometry on surfaces," *Duke Math. J.* **16** (1949), 361–375.
556. L. A. Santalo, "Un invariante afin para los cuerpos convexos del espacio de *n* dimensiones," *Portugal. Math.* **8** (1949), 155–161.
557. L. A. Santalo, "Integral geometry in projective and affine spaces," *Ann. of Math.* (2) **51** (1950), 739–755.
558. L. A. Santalo, "On parallel hypersurfaces in the elliptic and hyperbolic *n*-dimensional space," *Proc. Amer. Math. Soc.* **1** (1950), 325–330.
559. L. A. Santalo, "Unas formulas integrales y una definicion de área *q*-dimensional de un conjunto de puntos," *Rev. Mat. Fis. Teor. Univ. Tucumán* **7** (1950), 271–282.
560. L. A. Santalo, "Sobre unas formulas integrales y valores medios referentes a figuras convexas moviles en el plano," *Publ. Fac. Ci. Univ. Buenos Aires, Ser. Mat.* No. 2 (1950).
561. L. A. Santalo, "Integral geometry in general spaces," *Proc. Internat. Congr. Math., Cambridge, Mass., 1950* Vol. 1, pp. 483–489. The American Math. Soc., Providence, R.I., 1952.
562. L. A. Santalo, "Problemas de geometria integral," *UNESCO Symp. sobre Algunos*

ISBN 0-201-13500-0

problemas matem. que se están estudiando en América Latina, Montevideo, 1952 pp. 23–40.

563. L. A. Santalo, "Generalizacion de una desigualdad isoperimetrica de Feller," *Rev. Un. Mat. Argentina* **16** (1954), 78–81.
564. L. A. Santalo, "Integral geometry in Hermitian spaces," *Amer. J. Math.* **74** (1952), 423–434.
565. L. A. Santalo, "Geometria integral en espacios de curvatura constante," *Publ. Com. Nac. Energia Atomica, Ser. Mat.* **1** (1952), fasc. 1.
566. L. A. Santalo, "Algunos valores medios sobre la semiesfera," *Math. Notae* **12** (1952), 32–37.
567. L. A. Santalo, "Measure of sets of geodesics in a riemannian space and applications to integral formulas in elliptic and hyperbolic spaces," *Summa Brasil. Math.* **13** (1952), fasc. 1.
568. L. A. Santalo, Introduction to integral geometry, *Actualités Sci. Indust.* **1198** (1952).
569. L. A. Santalo, "On the kinematic formula in spaces of constant curvature," *Proc. Internat. Congr. Math., Amsterdam, 1954* Vol. 2, pp. 251–252. North-Holland Publ. Co. Amsterdam, 1957.
570. L. A. Santalo, "Cuestiones sobre geometria diferencial e integral en espacios de curvatura constante," *Rend. Sem. Mat. Torino* **14** (1955), 277–295.
571. L. A. Santalo, "On geometry of numbers," *Japan. J. Math.* **7** (1955), 208–213.
572. L. A. Santalo, "Sur la mesure des espaces linéaires qui coupent un corps convexe et problèmes qui s'y rattachent," *Colloq. sur les questions de réalité en géométrie Liège, 1955* pp. 177–190. Masson, Paris, 1956.
573. L. A. Santalo, "Sobre la distribucion de los tamanos de los corpusculos contenidos en un cuerpo a partir de la distribucion en sus secciones," *Trabajos Estadist.* **6** (1955), 181–196.
574. L. A. Santalo, "On the mean curvatures of a flattened convex body," *Rev. Fac. Sci. Univ. Istanbul* **21** (1956), 189–194.
575. L. A. Santalo, "Un nuevo invariante afin para cuerpos convexos del plano y del espacio," *Math. Notae* **16** (1958), 78–91.
576. L. A. Santalo, "Sobre los sistemas completos de desigualdades entre los elementos de una figura convexa del plano," *Math. Notae* **17** (1958), 82–104.
577. L. A. Santalo, "Two applications of the integral geometry in affine and projective spaces," *Publ. Math. Debrecen* **7** (1960), 226–237.
578. L. A. Santalo, "On the measure of sets of parallel linear subspaces in affine space," *Canad. J. Math.* **14** (1962), 313–319.
579. L. A. Santalo, "Sobre la formula fundamental cinemática de la geometria integral en espacios de curvatura constante," *Math. Notae* **18** (1963), 79–94.
580. L. A. Santalo, "Una relacion entre las curvaturas medias de cuerpos convexos paralelos en espacios de curvatura constante," *Rev. Un. Mat. Argentina* **21** (1963), 131–137.
581. L. A. Santalo, "Integral geometry on the projective groups of the plane depending on more than three parameters," *An. Sti. Univ. "Al. I. Cuza" Iasi, Sect. Ia Mat. (N.S.)* **11B** (1965), 307–335.
582. L. A. Santalo, "Valores medios para poligonos formados por rectas al azar en el plano hiperbolico," *Rev. Mat. Fis. Téor. Univ. Tucumán* **16** (1966), 29–44.
583. L. A. Santalo, "Horocycles and convex sets in the hyperbolic plane," *Arch. Math. (Basel)* **28** (1967), 529–533.
584. L. A. Santalo, "Horospheres and convex bodies in hyperbolic space," *Proc. Amer. Math. Soc.* **19** (1968), 390–395.

ISBN 0-201-13500-0

585. L. A. Santalo, "Grupos del plano respecto de los cuales, los conjuntos de puntos y de rectas admiten una medida invariante," *Rev. Un. Mat. Argentina* **23** (1967), 119–148.

586. L. A. Santalo, "Integral geometry," in *Studies in Global Geometry and Analysis* (S. S. Chern, ed.), pp. 147–195. Math. Assoc. Amer., Washington, D.C., 1967.

587. L. A. Santalo, "On some geometric inequalities in the style of Fary," *Amer. J. Math.* **91** (1969), 25–41.

588. L. A. Santalo, "Curvaturas absolutas y totales de variedades contenidas en un espacio euclidiano," *Acta Ci. Compostelana* **5** (1969), 149–158.

589. L. A. Santalo, "Convexidad en el plano hiperbolico," *Rev. Mat. Fis. Teor. Univ. Tucumán* **19** (1969), 174–183.

590. L. A. Santalo, "Mean values and curvatures," *Izv. Akad. Nauk Armjan. SSR Ser. Mat.* **5** (1970), 286–195. Reprinted in *Stochastic Geometry* (E. F. Harding and D. G. Kendall, eds.), pp. 165–175. Wiley, New York, 1974.

591. L. A. Santalo, "Total curvatures of compact manifolds immersed in euclidean space," *Symp. Mat. Ist. Naz. Alta Mat. Roma* **14** (1974), 364–390.

592. L. A. Santalo and I. Yanez, "Averages for polygons formed by random lines in euclidean and hyperbolic planes," *J. Appl. Probability* **9** (1972), 140–157.

593. D. B. Sawyer, "On the covering of lattice points by convex regions," *Quart. J. Math. Oxford Ser.* 2, **4** (1953), 284–292.

594. J. J. Schäffer, "Smallest lattice-point covering convex set," *Math. Ann.* **129** (1955), 265–273.

595. E. Scheil, "Die Berechnung der Anzahl und Größenverteilung kugelförmiger Körper mit Hilfe der durch ebene Schnitte erhaltenen Schnittkreise," *Z. Anorg. Allgem. Chem.* **201** (1931), 259–264.

596. B. Schiffman, "Applications of geometric measure theory to value distribution theory of meromorphic maps," in *Value Distribution Theory* (R. O. Kujala and A. L. Willer III, eds.), Pt. A, pp. 63–95. Dekker, New York, 1974.

597. L. Schläfli, *Gesammelte mathematische Abhandlungen*, Vol. 1. Birkhäuser, Basel, 1950.

598. W. M. Schmidt, "Some results in probabilistic geometry," *Z. Wahrscheinichkeits.* **9** (1968), 158–162.

598a. W. M. Schmidt, "Volume, surface area and the number of integer points covered by a convex set," *Arch. Math. (Basel)* **23** (1972), 537–543.

599. R. Schneider, "On Steiner points of convex bodies," *Israel J. Math.* **9** (1971), 241–249.

600. R. Schneider, "Krümmungsschwerpunkte konvexer Körper, I, II," *Abh. Math. Sem. Univ. Hamburg* **37** (1972), 112–132, 204–217.

601. E. F. Schuster, "Buffon's needle experiment," *Amer. Math. Monthly* **81** (1974), 26–29.

602. D. G. Scott, "Packing of spheres," *Nature* **188** (1961), 908–909; **194** (1962), 956–957.

603. P. R. Scott, "A lattice problem in the plane," *Mathematika* **20** (1973), 247–252.

604. S. Sherman, "A comparison of linear measures in the plane," *Duke Math. J.* **9** (1942), 1–9.

605. J. A. Shohat and J. D. Tamarkin, *The Problem of Moments*. Amer. Math. Soc., Providence, R.I., 1943.

606. M. Sholander, "On certain minimum problems in the theory of convex curves," *Trans. Amer. Math. Soc.* **23** (1952), 139–173.

607. L. P. Sibasov, "Integral geometry in Grassmann manifolds," *Moskov. Oblast. Ped. Inst. Ucen. Zap.* **262** (1969), 256–270.

ISBN 0-201-13500-0

608. F. Sibirani, "Alcune probabilità geometriche," *Mem. Accad. Sci. Ist. Bologna, Cl. Sci. Fis.* (10), **1**, (1944), 113–123.
609. Z. A. Sidak, "On the mean number and size of opaque particles in transparent bodies," in *Studies in Mathematical Statistics: Theory and Applications* (K. Sarkadi and I. Vincze, eds.), Acad. Kiadr., Budapest, 1968.
610. C. L. Siegel, "Symplectic geometry," *Amer. J. Math.* **65** (1943), 1–86.
610a. C. L. Siegel, "A mean value theorem in geometry of numbers," *Ann. of Math.* (2) **46** (1945), 340–347.
611. L. Silberstein, "Aggregates in random distribution of points," *Phil. Mag.* (7) **36** (1945), 319–336.
612. M. Silver, "On extremal figures admissible relative to rectangular lattices," *Pacific J. Math.* **40** (1972), 451–457.
613. J. G. Skellam, "The mathematical foundations underlying the use of line transects in animal ecology," *Biometrics* **14** (1958), 385–400.
614. V. V. Slavskii, "A certain integral geometry relation in surface theory," *Sibirsk. Mat. Z.* **13** (1972), 645–658.
615. C. S. Smith and L. Gutman, "Measurement of internal boundaries in three-dimensional structure of random sectioning," *J. Metals* **5** (1953), 81–87.
616. H. Solomon, "A coverage distribution," *Ann. Math. Statist.* **21** (1950), 139–140.
617. H. Solomon, "Distribution of the measure of a random two-dimensional set," *Ann. Math. Statist.* **24** (1953), 650–656.
618. H. Solomon, "Random packing density," *Proc. 5th Berkeley Symp. Math. Statist. and Probability* pp. 119–154. Univ. of Calif. Press, Berkeley, 1966.
619. H. Solomon and P. C. C. Wang, "Non-homogeneous Poisson fields of random lines with applications to traffic flow," Stanford Univ. Statist. Dept. Tech. Rept. No. 22 (1971).
620. F. Spiter and H. Widom, "The circumference of a convex polygon," *Proc. Amer. Math. Soc.* **12** (1961), 506–509.
621. G. Stanilow, "Zur Integralgeometrie im euklidischen Raum E_3," *Math. Nachr.* **43** (1970), 181–183.
622. G. Stanilow, "The integral geometry of generalized biaxial spaces," *Bülgar. Akad. Nauk. Otdel. Mat. Fiz. Nauk. Izv. Mat. Inst.* **11** (1970), 39–53.
623. G. Stanilow and R. Sulanke, "Crofton type integral formulae in the theory of line congruences in the euclidean space E_n," *Bülgar. Akad. Nauk. Otdel. Mat. Fiz. Nauk. Izv. Mat. Inst.* **11** (1970), 27–37.
624. S. K. Stein, "Averages over a pair of convex surfaces," *Michigan Math. J.* **15** (1968), 377–380.
625. H. Steinhaus, "Sur la portée pratique et théorique de quelques théorèmes sur la mesure des ensembles de droites," *C. R. 1er Congr. Mathématiciens des Pays Slaves, Warszawa, 1929* pp. 348–354.
626. H. Steinhaus, "Length, shape and area," *Colloq. Math.* **3** (1954), 1–13.
627. J. A. Steppe and E. Wong, "Invariant recognition of geometric shapes," *Internat. Conf. Method. Pattern Recognition, Hawaii, 1968.*
628. S. Sternberg, *Lectures on Differential Geometry*. Prentice-Hall, Englewood Cliffs, N.J., 1964.
629. S. Sternberg, *Advanced Calculus* (with L. H. Loomis). Addison-Wesley, Reading, Mass., 1968.
630. W. L. Stevens, "Solution to a geometrical problem in probability," *Ann. Eugenics* **9** (1939), 315–320.
631. M. I. Stoka, "Masura unei multimi da varietati dintr-un spati R_n," *Bull. Stint.*

ISBN 0-201-13500-0

Acad. R. P. Romine **7** (1955), 903–937.

632. M. I. Stoka, "Asupra masurii multimii cerculiror din plan," *Gaz. Mat. Fiz.* **A10** (1955), 556–559.
633. M. I. Stoka, "Invariantii integrali ai unui grup Lie de transformari," *An. Univ. Bucaresti Mat.-Mec.* **20** (1958), 33–35.
634. M. I. Stoka, "Geometria integrale in uno spazio euclideo E_n," *Boll. Un. Mat. Ital.* **13** (1958), 470–485.
635. M. I. Stoka, "Masura familitor de varietati dintr-un spatiu E_n," *Stud. Cerc. Mat.* (2) **9** (1958), 547–558.
636. M. I. Stoka, "Asupra grupurilor G_r masurabili dintr-un spatiu E_n," *Comm. Accad. R. P. Romina* **9**, 1 (1959), 5–10.
637. M. I. Stoka, "Géométrie integrale dans un espace E_n," *Rev. Roumaine Math. Pures Appl.* **4** (1959), 123–156.
638. M. I. Stoka, "Congruence de variétés mesurables dans un espace E_n," *Rev. Roumaine Math. Pures Appl.* **4** (1959), 431–449.
639. M. I. Stoka, "Famiglie de varietà misurabili in un spazio E_n," *Rend. Circ. Mat. Palermo* **8** (1959), 192–205.
640. M. I. Stoka, "Sulla misura cinematica in un spazio euclideo E_n," *Boll. Un. Mat. Ital.* **14** (1959), 467–476.
641. M. I. Stoka, "Asupra grupurilor de miscare ale spatilor riemanniene V_2 cu curbura constanta," *Stud. Cerc. Mat.* **11** (1960), 207–228.
642. M. I. Stoka, "Geometrie integrala intr-un spatiu riemannian V_n," *Stud. Cerc. Mat.* **11** (1960), 381–395.
643. M. I. Stoka, "Das Maß der Untersysteme von Mannigfaltigkeiten in einem Raum X_n," *Rev. Roumaine Math. Pures Appl.* (2) **5** (1960), 275–286.
644. M. I. Stoka, "Géométrie intégrale dans un espace E_3," *Rev. Mat. Fis. Teor. Univ. Tucumán* **14** (1962), 25–59.
645. M. I. Stoka, "Familii de varietati mesurabili intr-un spatiu riemannian V_3 cu curbura constanta," *Stud. Cerc. Mat.* **12** (1963), 365–376.
646. M. I. Stoka, *Géométrie Intégrale* (Mémorial des Sciences Mathématiques, Fasc. 165). Gauthier-Villars, Paris, 1968.
647. M. I. Stoka, *Geometria Integrala.* Ed. Acad. R. S. Romania, 1967.
648. M. I. Stoka, "Alcune formule integrali concernenti i corpi convessi dello spazio euclideo E_3," *Rend. Sem. Mat. Torino* **28** (1969), 95–105.
649. M. I. Stoka, "Une extension du problème de l'aiguille de Buffon dans l'espace euclidien R_n," *Boll. Un. Mat. Ital.* (4) **10** (1974), 386–389.
650. M. I. Stoka, "Alcune formule integrali concernenti le curve convesse di uno spazio riemanniano V_2 dalla curvatura constanta," *Rend. Sem. Mat. Messina* **12** (1968–1970), 63–78.
651. M. I. Stoka, "La variance d'une variable aléatoire associé à une famille des ovales du plan euclidien," *Acad. Roy. Belg. Bull. Cl. Sci.* (5) **59** (1973), 178–184.
652. M. I. Stoka, "Géométrie integrale dans le plan projectif," *Rev. Roumaine Math. Pures Appl.* **19** (1974), 79–96.
653. M. I. Stoka, "Sur quelques probabilités géométriques liées aux systèmes de corps convexes dans un espace euclidien E_n," *Boll. Un. Mat. Ital.* (4) **11** (1975), 22–29.
654. M. I. Stoka and R. Theodorescu, *Probabilitate si geometrie.* Bucaresti, 1966.
655. W. Stoll, "A Bezout estimate for complete intersections," *Ann. of Math.* (2) **96** (1972), 361–401.
656. W. Stoll, "Deficit and Bezout estimates," in *Value Distribution Theory* (R. O. Kujala and A. L. Willer III, eds.), Pt. B, pp. 1–272, Dekker, New York, 1973.

ISBN 0-201-13500-0

657. F. Streit, "On multiple integral geometric integrals and their applications to probability theory," *Canad. J. Math.* **22** (1970), 151–163.
658. F. Streit, "Mean value formulae for a class of random sets," *J. Roy. Statist. Soc.* **35** (1973), 437–444.
659. R. Sulanke, "Die Verteilung der Sehnenlängen an ebenen und räumlichen Figuren," *Math. Nachr.* **23** (1961), 51–74.
660. R. Sulanke, "Schnittpunkte zufälliger Geraden," *Arch. Math.* **16** (1965), 320–324.
661. R. Sulanke, "Croftonsche Formeln in Kleinschen Räumen," *Math. Nachr.* **32** (1966), 218–224.
662. R. Sulanke, "Integralgeometrie ebener Kurvennetze," *Acta Math. Acad. Sci. Hungar.* **17** (1966), 233–261.
663. R. Sulanke, "Croftonsche Formeln für Strahlensysteme des euklidischen Raumes," *Math. Nachr.* **38** (1968), 299–307.
664. R. Sulanke and P. Wintgen, *Differentialgeometrie und Faserbunde*. Birkhäuser, Basel, 1972.
665. R. Sulanke and P. Wintgen, "Zufälliger konvexer Polyeder im N-dimensionalen euklidischen Raum," *Period. Math. Hungar.* **2** (1972), 215–221.
665a. J. J. Sylvester, "On a funicular solution of Buffon's needle problem," *Acta Math.* **14** (1890), 185-205.
666. L. Takacs, "On the probability distribution of the measure of the union of random sets placed in a Euclidean space," *Ann. Univ. Sci. Budapest, Eötvos Sect. Mat.* **1** (1958), 89–95.
667. G. M. Tallis, "Estimating the distribution of spherical and elliptical bodies in conglomerates from plane sections," *Biometrics* **26** (1970), 87–103.
668. C. Tate, "Aggregates in a random distribution of points in a cylinder," *SIAM J. Appl. Math.* **17** (1969), 1177–1189.
669. H. R. Thomson, "Distribution of distance to nth neighbour in a population of randomly distributed individuals," *Ecology* **37** (1956), 391–394.
670. H. Tietze, "Würfelspiel und Integralgeometrie," *S.-B. Mat. Kl. Bayer. Akad. München* **1945/46** (1947), 131–158.
671. R. Trandafir, "Familles de courbes mesurables du plan euclidien, I," *Rev. Roumaine Math. Pures Appl.* (2) **11** (1966), 1009–1013; II, **12** (1967).
672. R. Trandafir, "Familii de varietati cu unu, doi, si trei parametri masurabili din spatiul euclidean E_3," *Stud. Cerc. Mat.* **19** (1967), 619–634.
673. R. Trandafir, "Probleme de acoperiri in spatiul euclidean E_3," *Stud. Cerc. Mat.* **19** (1967), 1105–1111.
674. R. Trandafir, "Probleme de acoperiri in spatii riemannien V_2 cu curbure constante," *Stud. Cerc. Mat.* **19** (1967).
675. R. Trandafir, "Problems of integral geometry of lattices in an euclidean space E_3," *Boll. Un. Mat. Ital.* (3) **22** (1967), 228–235.
676. R. Trandafir, "Problems of integral geometry of lattices in a riemannian space V_2 with constant curvature," *Boll. Un. Mat. Ital.* (3) **23** (1968), 244–248.
677. R. Trandafir, "Problemi di geometria della reti di un spazio omogeneo," *Rend. Sem. Mat. Messina* **12** (1971), 3–25.
677a. R. Trandafir, "Probleme de geometrie integrala de retelelor dintr-un spatiu omogen," *Stud. Cerc. Mat.* **23** (1971), 425–466.
678, F. Tricomi, "Densita di un continuo di punti o di rette e densita di una correspondenza," *Rend. Acad. Naz. Lincei* (6) **23** (1936), 313–316.
679. M. Tsuji, "Theorems in the geometry of numbers for Fuchsian groups," *J. Math. Soc. Japan* **4** (1952), 189–193.

ISBN 0-201-13500-0

680. M. Tsuji, "Analogue of Blichfeldt's theorem for Fuchsian groups," *Comment. Math. Univ. St. Paul* **5** (1956), 17–24.
681. S. Ueno, "On the densities in a two-dimensional generalized space," *Mem. Fac. Sci. Kyusyu Univ.* **A9** (1955), 65–77.
682. S. Ueno, H. Hombu, and J. Naito, "Some integral geometric inequalities," *Mem. Fac. Sci. Kyusyu Univ.* **A6** (1951), 97–106.
682a. E. E. Underwood, *Quantitative Stereology.* Addison-Wesley, Reading, Mass., 1970.
683. F. Valentine, *Convex Sets.* McGraw-Hill, New York, 1964.
684. K. Valsala, "Zur Theorie der geometrischen Wahrscheinlichkeiten," *Ann. Acad. Sci. Fenn.* **A51** (1938), No. 6, 1–22.
685. O. Varga, "Integralgeometrie 3: Croftons Formeln für den Raum," *Math. Z.* **40** (1935), 387–405.
686. O. Varga, "Integralgeometrie 8: Über die Masse von linearen Mannigfaltigkeiten im projektiven Raum P_n," *Rev. Mat. Hisp. Amer.* (2) **10** (1935), 241–264; **11** (1936), 1–14.
687. O. Varga, "Integralgeometrie 19: Über Mittelwerte an dem Durchschnitt bewegter Flächen, *Math. Z.* **41** (1936), 768–784.
688. O. Varga, "Über die Integralinvarianten, die zu einer Kurve in der Hermiteschen Geometrie gehören," *Acta Litt. Sci. Szeged* **9** (1939), 88–102.
689. G. P. Ventikos, "On the mean value of a straight segment in a plane convex region," *Bull. Soc. Math. Grèce* **22** (1946), 195–197.
690. E. Vidal Abascal, "A generalization of Steiner's formula," *Bull. Amer. Math. Soc.* **53** (1947), 841–844.
691. E. Vidal Abascal, "Geometria integral sobre superficies curvas," *Publ. Observ. Astron. Santiago de Compostela* **7** (1950).
692. E. Vidal Abascal, "Sobre los fundamentos de la geometria integral," *Rev. Mat. Hisp.-Amer.* (4) **12** (1952), 290–310.
693. E. Vidal Abascal, "Sobre los fundamentos de la geometria integral," *Mem. Acad. Ci. Madrid* **4** (1953), 1–29.
694. E. Vidal Abascal, "A generalization of integral invariants," *Proc. Amer. Math. Soc.* **10** (1959), 721–727.
695. E. Vidal Abascal, "Generalizacion de los invariantes integrales y aplicacion a la geometriá integral en los espacios de Klein y de Riemann," *Collect. Math.* **12** (1960), 71–102.
696. E. Vidal Abascal, "Sobre algunos problemas en relacion con la medida en espacios foliados," *Acta 1er Coloq. Geometria Dif., Santiago de Compostela, 1963* pp. 63–82.
697. E. Vidal Abascal, "Mesures définies sur les espaces des feuilles d'un feuilletage," *Rend. Circ. Mat. Palermo* (2) **15** (1966), 247–256.
698. E. Vidal Abascal, "Sur les feuilletages réguliers et les problèmes qui s'y reportent," *Acta Ci. Compostelana* **3** (1966), 3–12.
699. E. Vidal Abascal, "On regular foliations," *Ann. Inst. Fourier (Grenoble)* **17** (1967), 129–133.
700. E. Vidal Abascal, "Cuestiones en relacion con las medidas en espacios foliados," *Acta 2nd Coloq. Geometria Dif., Santiago de Compostela, 1967* pp. 59–63.
701. U. Viet, "Umkehrung eines Satzes von H. Brunn über Mittelpunktbereiche," *Math.-Phys. Semesterber.* **5** (1956), 141–142.
702. P. Vincensini, *Corps convexes séries linéaires, domaines vectoriels* (Mémorial des Sciences Mathématiques, Fasc. 94). Gauthier-Villars, Paris, 1938.
703. O. P. Vinogradov and I. P. Zaregradski, "Geometrical probabilities in the theory of optimal search," *Izv. Akad. Nauk Armjan. SSR Ser. Mat.* **5** (1970), 207–218.

ISBN 0-201-13500-0

704. G. Voronoi, "Nouvelles applications des paramètres continus à la théorie des formes quadratiques, *J. Reine Angew. Math.* **134** (1908), 198–287.
705. D. F. Votaw, "The probability distribution of the measure of a random linear set," *Ann. Math. Statist.* **17** (1946), 240–244.
706. G. Vranceanu, "The measurability of Lie groups," *Ann. Polon. Math.* (2) **15** (1964), 179–188.
707. A. G. Walters, "The distribution of projected areas of fragments," *Proc. Cambridge Phil. Soc.* **43** (1947), 342–347.
708. G. S. Watson, "Estimating functionals of particle size distributions," *Biometrika* **58** (1971), 483–490.
709. G. S. Watson, "Orientation statistics in the earth sciences," *Bull. Geol. Inst. Univ. Uppsala* (2) **9** (1970), 73–89.
710. A. Weil, "*L'intégration dans les groupes topologiques et ses applications*," *Actualités Sci. Indust.* **869** 1938.
711. A. Weil, "Review of Chern's article [105], *Math. Reviews* **3** (1942), 253.
712. A. Weil, "Sur quelques résultats de Siegel," *Summa Brasil. Math.* **1** (1946), 21–39.
713. J. L. Weiner, "A generalization of the isoperimetric inequality on the 2-sphere," *Indiana Univ. Math. J.* **24** (1974), 243–248.
714. J. G. Wendel, "A problem in geometric probability," *Math. Scand.* **11** (1962), 109–111.
715. J. E. Wetzel, "Covering balls for curves of constant length," *Enseignement Math.* (2) **17** (1971), 275–277.
716. H. Weyl, "On the volume of tubes," *Amer. J. Math.* **61** (1939), 461–472.
717. H. Weyl, *Meromorphic Functions and Analytic Curves* (Ann. of Math. Studies No. 12). Princeton Univ. Press, Princeton, N.J., 1943.
718. H. Weyl and J. Weyl, "Meromorphic curves," *Ann. of Math.* **39** (1938), 516–538.
719. D. J. White, "On the largest cap in a convex set of a sphere," *Proc. London Math. Soc.* (3) **17** (1967), 157–162.
720. J. H. White, "Self-linking and the Gauss integral in higher dimensions," *Amer. J. Math.* **91** (1969), 693–728.
721. J. H. White, "Geometric inequalities for surfaces in euclidean three-space," *Rev. Roumaine Math. Pures Appl.* **17** (1972), 1487–1494.
722. J. H. White, "Some differential invariants of submanifolds of euclidean space," *J. Diff. Geometry* **5** (1971), 357–369.
723. H. Whitney, *Geometric Integration Theory*. Princeton Univ. Press, Princeton, N.J., 1957.
724. S. D. Wicksell, "The corpuscle problem, Pt. I: A mathematical study of a biometric problem," *Biometrika* **17** (1925), 84–99; Pt. II, 18 (1926), 151–172.
725. T. J. Willmore, *An Introduction to Differential Geometry*. Oxford Univ. Press, London and New York, 1959.
726. J. M. Wills, "Ein Satz über konvexe Mengen und Gitterpunkte," *Monatsh. Math.* **72** (1968), 451–463.
727. J. M. Wills, "Gitterpunkte und Volumen-Oberfläche-Verhältnis konvexer Mengen," *Arch. Math.* (*Basel*) **22** (1971), 445–448.
728. W. Wirtinger, "Eine Determinantenidentität und ihre Anwendungen auf analytische Gebilde in Euklidischer und Hermitescher Maßbestimmung," *Monatsh. Math. Phys.* **44** (1936), 343–365.
729. J. Wolfowitz, "The distribution of plane angles of contact," *Quart. J. Appl. Math.* **7** (1949), 117–120.
730. H. H. Wu, *The equidistribution theory of holomorphic curves* (Ann. of Math.

ISBN 0-201-13500-0

Studies, No. 64). Princeton Univ. Press, Princeton, N.J., 1970.

731. T. J. Wu, "Integralgeometrie 26: Über die kinematische Hauptformel," *Math. Z.* **43** (1937), 212–227.

732. T. J. Wu, "Integralgeometrie 28: Über elliptische Geometrie," *Math. Z.* **43** (1938), 495–521.

732a. T. J. Wu, "Der Dual der Grundformel in der Integralgeometrie," *J. Chinese Math. Soc.* **2** (1940), 199–204.

733. I. M. Yaglom, "Integral geometry in the set of line elements," Appendix to the Russian translation of L. A. Santalo's *Introduction to Integral Geometry*, pp. 153–169, Moscow, 1956.

734. L. M. Yurtova and A. V. Lutsenko, "Measures of sets of pairs," *Ukrain. Geometr. Sb.* 1968.

735. O. Zoll, "Über Flächen mit Scharen geschlossener geodetischer Linien," *Math. Ann.* **57** (1903), 108–133.

736. H. Ziezold, "Über die Eckenzahl zufälliger konvexer Polygone," *Izv. Akad. Nauk Armjan. SSR. Ser. Mat.* **5** (1970), 296–312.

ISBN 0-201-13500-0

Author Index

Author Index

Subject Index

Subject Index

ISBN 0-201-13500-0

ISBN 0-201-13500-0

ISBN 0-201-13500-0

ISBN 0-201-13500-0

ISBN 0-201-13500-0